Wombs with a View

Lawrence D. Longo • Lawrence P. Reynolds

Wombs with a View

Illustrations of the Gravid Uterus from the Renaissance through the Nineteenth Century

With a Foreword by Kent L. Thornburg

and

An Afterword by Catherine Y. Spong

 Springer

Lawrence D. Longo (Deceased)
Center for Perinatal Biology
Loma Linda University School of Medicine
Loma Linda, CA, USA

Lawrence P. Reynolds
Center for Nutrition and Pregnancy, and Animal
  Sciences
North Dakota State University
Fargo, ND, USA

ISBN 978-3-319-23566-0   ISBN 978-3-319-23567-7   (eBook)
DOI 10.1007/978-3-319-23567-7

Library of Congress Control Number: 2015956790

Printed on acid-free paper

This Springer imprint is published by Springer Nature
The registered company is Springer International Publishing AG Switzerland

Dedication

"… to focus on the pursuit of truth wherever it may lead, to promote scholarship, to contribute to betterment of the human condition, to keep in mind the considerable social relevance of our work, this then is our inspiration, our responsibility, and our legacy for the future." Lawrence D. Longo In: Advances in Fetal and Neonatal Physiology, Zhang L. and Ducsay C.A. (eds.), NY: Springer, 2014.

For Lawrence P. Reynolds, I dedicate this volume to my co-author, Dr. Lawrence D. Longo (1926–2016), who was Distinguished Professor of Obstetrics and Gynecology, Physiology and Pharmacology, and Biochemistry, and Founding Director of the Center for Perinatal Biology, School of Medicine, Loma Linda University, Loma Linda, California.

I was deeply saddened to learn on the morning of the 6th of January 2016 that Dr. Longo had passed the previous evening. My first thought was that the world had lost someone and something very valuable! Dr. Longo was certainly one of the bright lights of our profession, and besides that a profoundly wonderful human being and a mentor beyond compare. I was very fortunate to have gotten to know him and to work with him over the last 10 years—one of the highlights of my career! This volume is what we jokingly called The Great Book, and what Dr. Longo termed his Magnum Opus (Great Work). I will always remember his unyielding curiosity, his unfathomable store of knowledge of Obstetric and Gynecological investigation, and his incredible generosity. The latter is exemplified by the Reproductive Scientist Development Program for which Dr. Longo prepared the original grant application, which was first funded in 1988 by the National Institutes of Health, and for which he served as Director and Co-Director until 2013.

Although working as a basic scientist in developmental physiology, with a focus on cerebral blood flow, Dr. Longo also published several volumes on fetal physiology, most recently *Stress and Developmental Programming of Health and Disease: Beyond Phenomenology* (Public Health in the 21st Century) (2015) with co-author Lubo Zhang, and *The Rise of Fetal and Neonatal Physiology: Basic Science to Clinical Care* (Perspectives in Physiology) (2013). In addition to a number of articles on nineteenth-century obstetrics and gynecology, Dr. Longo published *A Short History of the Society for Gynecologic Investigation, 1953–1983* (1983); edited *A Treatise on the Management of Pregnant and Lying-in Women* by Charles White (1987); and with T.K.A.B. Eskes edited *Classics in Obstetrics and Gynecology, Innovative Papers That Have Contributed to Current Clinical Practice* (1993). With Philip M. Teigen, he edited *"Dearest G …, Yours W.O.": William Osler's Letters from Egypt to Grace Revere Osler* (2002). The latest historical volume that he edited is *Our Lords the Sick* (McGovern Lectures in the History of Medicine and Medical Humanism) (2004). Before his death, he was completing a volume, *Robert Patterson Harris of Philadelphia: The "Statistical Method" and the Evolution of Cesarean Section.*

For Lawrence D. Longo and Lawrence P. Reynolds, we also dedicate this volume to the lives and memories of George W. Corner, George L. Streeter, Howard B. Adelmann, Samuel R.M. Reynolds, Elizabeth M. Ramsey, Harland W. Mossman, and others of the great anatomists/embryologists by whom we were inspired.

Foreword

Kent L. Thornburg[*]

This foreword is a personal invitation for you, the reader, to join with two of America's finest scientists in the field of pregnancy and reproduction on a journey backward in time when heroes of that day enlightened their contemporaries with the beauty of the anatomical features of pregnancy and childbirth.

Scientists of every era struggle to escape the walls of knowledge within which they find themselves constrained. Forward movement in medical science often requires a person of creative brilliance to discover a new pathway that turns a once ignored lightless field of inquiry into one which illuminates fundamental mechanisms of biology and brings a new arsenal of weapons efficacious in the war against devastating disease. Longo and Reynolds offer within these pages vignettes of pioneers of the past who were the brilliant pacesetters of their day. Some like da Vinci are well sung, but others previously unknown to me, like Rueff, are unsung heroes, who nevertheless escaped the walls of dogma and social restraint to enlighten contemporary intelligentsia regarding the intricacies of human reproduction.

This book is not only a compendium of important historical images related to human reproduction; it is a history of technology and artistry. The volume carries with it the development of tools applied to imagery as much as the evolution of comprehension of anatomy and reproductive function. Drawings, primitive at first and then more sophisticated, and paintings, flat and distorted and later well-proportioned, represent the evolution of thought and skill that preceded the primitive silver halide photograph which eventually became the digital images of the twenty-first century with their high-resolution sharpness, perfect lighting, and color balance. Even in the earliest days of this technological progression, centuries ago, humans sought to record in some fashion what they saw in real life.

But because of rapid advances in high-resolution digital technology, modern biology has entered a new phase from which it can never return. No longer will a boring textbook drawing capture the mind of a student. When I was training in biology, I was made to draw the organs of many animals in the phylogenetic scale to appreciate their relatedness. Most modern students learn better by absorbing the detail of a full color image than by contemplating a colorless concept. The impact of color undoubtedly reflects the image-rich specialization of the human brain. And so, scientists now want to see every single aspect of biological action with their own eyes, preferably in real time. Every chemical reaction, every movement of every organelle in every cell, or every movement of every nutrient that leaves the mother's blood and enters the fetus is now meant to be followed with the human eye. The test of reality in science is becoming that which can be visualized. But in the distant past, few people had access to images that accurately depicted the growth and development of the unborn baby. But to those

[*]Dr. Thornburg is M. Lowell Edward Chair, Professor of Medicine in the Knight Cardiovascular Institute, Director of the Center for Developmental Health, and Director of the Bog and Charlee Moore Institute for Nutrition and Wellness, Oregon Health and Sciences University, Portland, OR, USA.

few, hand-drawn images were likely as enlightening then as a multicolored image of a cell decorated with its rainbow of colored organelles is today.

Access to images and video clips is now so readily available on the Internet that children of the twenty-first century might forget that there was a time when human reproduction was little understood. Watching the entire birth process now only requires a few clicks on a portable tablet. But that was not always so. There was a time when the intricate biological processes driving sexual activity, fertilization, embryo development, placentation, fetal organ growth and maturity, and the final powerful expulsion of the fetus were virtually unknown. The work by Longo and Reynolds enlightens the arduous journey taken by human scientific trailblazers in their quest to illustrate the processes of pregnancy and the reproduction of our own species.

In contemporary society, we may be unprepared to face the emotional impact wrought by the renderings of these ancient artists. Thus, one must not confuse the stark boldness of the now primitive images offered in this work with the degrading depictions of a woman's body drawn by the misogynistic graffitist on some dark and lonely wall. Rather, the amazing beauty of human structure which motivated the early artists to draw also afforded them the opportunity to share their excitement of biological discovery. These early artists used their best efforts to convey the relationships between a mother, the developing child, her/his placenta, and the womb in which it was nurtured.

The state of affairs during pregnancy has taken on a new layer of importance in the late twentieth century. Recent discoveries have shown that the flow of nutrients from mother to baby through the placenta endures beyond life in the womb and provides the pillars of health for at least two future generations. Our understanding of the functional importance of the placenta has its roots in the scholarly reports of old, herein highlighted by Longo and Reynolds, that motivate modern day women and men to join the quest of discovery that will bring health and prosperity to the children of tomorrow.

This book celebrates the joy, excitement, and spiritual fulfillment that accompanies the impartation of life from mother and father to a whole new human being. Visualizing the long journey of enlightenment through which we as a race discovered how new human life emerges from parental contributions also offers us a glimpse of that fulfillment shared by every proud parent and grandparent over the ages.

On my coffee table, or on the bookshelf nearby, are several bound volumes depicting various collections of art highly valued by aficionados across the globe. The images they contain serve as a reminder of the heights to which the artistic soul can reach. As much as I admire the beauty of a high-resolution image on my vibrant computer screen, I am old fashioned enough to want a bound copy of this book among those highly prized on my shelf. Should the Louvre Museum in Paris, the National Gallery of Art in Washington, D.C., or the National Gallery, London, wish to assemble a collection of original images depicting the historical quest of humans to understand their own reproduction, the works revealed in this volume by Longo and Reynolds would be a grand beginning.

Preface

"Education is not the filling of a pail but the lighting of a fire."

(attributed to William Butler Yeats [1865–1939])

We would like to think that it was fortuitous "meeting of the minds." One of us (L.D.L.) had been planning a volume on the history of depictions of the *fetus in utero* and certain concepts of fetal physiology. The other (L.P.R.) had been thinking about a similar work on the evolution of ideas concerning the placenta. In a chance encounter, we discovered one another's interests and decided to explore a collaboration. Our goal in the preparation of this volume was to provide a permanent archive of some of the most beautiful drawings ever made of the gravid uterus with fetus and placenta, which could serve future generations of investigators, educators, and students of reproduction.

For each author whose work is depicted in this volume, in almost every instance we have used the first edition or first illustrated edition. In our commentary, we have endeavored to place each volume and illustration in historical perspective, noting the significance of that image, but also giving some background on the life and work of the author. Insofar as possible, we also have included accurate citation to relevant bibliographic references. For most of the works cited, we have included additional references of relevance for the reader who may wish to explore these in greater depth.

As academic, basic science investigators, we suffer no illusions about the significance of our personal contributions to the study of the biology of gestation. Rather, we have simply extended the work of our predecessors, standing on the shoulders of the true "giants" of this field. These include many cited in this volume, and in addition, Howard Bernhart Adelmann (1898–1988), Emmanuel Ciprian Amoroso (1901–1982), Francis Maitland Balfour (1851–1882), George Washington Corner (1889–1981), Sir John Hammond (1889–1964), Franklin Paine Mall (1862–1917), Samuel Robert Means Reynolds (1903–1982), Elizabeth Mapelsden Ramsey (1906–1993), George Linius Streeter (1873–1948), and numerous others.

Several individuals were of great assistance in preparing this work. Especially important was Dominic Budicin of Riverside, CA, who worked with us to photograph with precision the plates from the works depicted. James Ponder, Loma Linda University Office of Public Relations, also contributed by photographing for us the illustrations of Jenty (1757) and Snip (1793).

For Lawrence P. Reynolds, I owe a debt of gratitude to many. My high school biology teacher, Mr. David Conatser, who helped light the fire, and my mentors, Drs. Charles Weems, Stephen Ford, and Calvin Ferrell, who fanned the flames. Enduring gratitude to Dr. Robert Melampy (deceased) who began my journey down this road by exhorting me to read the "older" literature, including Joseph Needham's *Chemical Embryology* (1931). Thanks to Dr. Joseph Gall—his beautiful compendium *Views of the Cell: A Pictorial History* (Bethesda, MD: The American Society for Cell Biology, 1996) inspired the format for this volume. My parents and my father- and mother-in-law (both deceased), whose encouragement was unflagging and whose support played no small part in my education. Undying love and gratitude to my wonderful wife, Kay, who supported all these efforts with understanding and patience.

For Lawrence D. Longo, I also am in debt to many dedicated teachers and mentors. In particular, I have learned much of value from my lovely remarkable wife Betty Jeanne and from each of my four wonderful children: April Celeste, Lawrence Anthony, Elisabeth Lynn, and Camilla Giselle. For both L.P.R. and myself, we thank Jimin Suh, administrative assistant to L.D.L., who was invaluable in the preparation of the manuscript.

Lastly, each of us believes that we owe a special debt of gratitude to our colleague and coauthor, each of whom served as mentor for the other in exploring this fascinating facet of biology and science.

Loma Linda, CA, USA   Lawrence D. Longo
Fargo, ND, USA   Lawrence P. Reynolds

Abbreviations for References Cited

Austin	Austin, R.B. *Early American Medical Imprints*: *A Guide to Works Printed in the United States, 1668–1820*. Washington, DC, U.S. Department of Health, Education, and Welfare, National Library of Medicine, 1961
Bakken	Overmier, J.A. & Senior, J.E. *Books and Manuscripts of the Bakken*. Metuchen, NJ, Scarecrow Press, 1992
Blake	Blake, J.B. *A Short Title Catalogue of Eighteenth Century Printed Books in the National Library of Medicine*. U.S. Department of Health, Education, and Welfare, Public Health Service, National Institutes of Health, National Library of Medicine, Bethesda MD, DHEW Publication No. (NIH) 79-104, 1979
Castiglioni	Castiglioni, A. *A History of Medicine*. New York, Alfred A. Knopf, 1941
Choulant and Choulant-Frank	Choulant, L. *Geschichte und bibliographie der anatomischen abbildung nach ihrer beziehung auf anatomische wissenschaft und bildende kunst*. Leipzig, R. Weigel, 1852. Choulant, L. *History and bibliography of anatomic illustration in its relation to anatomic science and the graphic arts*. Translated by M. Frank. Chicago, University Press, 1920
Cole	Cole, F.J. *The History of Anatomical Injections*. In: Studies in the History and Method of Science. Vol. 2, Oxford, Oxford University Press, 1921, pp. 285–343
Cushing	*The Harvey Cushing Collection of Books and Manuscripts* New York, Schuman's, 1943
Cutter & Viets	Cutter, I.S. & Viets, H.R. *A short history of midwifery*. Philadelphia, PA, WB Saunders & Co. 1964
D.N.B.	*Dictionary of National Biography*. Oxford, Oxford University Press, 2004
D.S.B.	Gillispie, C.C., Holmes, F.L., & American Council of Learned Societies. *Dictionary of Scientific Biography*. New York, Scribner, 1990
Durling	Durling, R.J. *A Catalogue of Sixteenth Century Printed Books in the National Library of Medicine*. Bethesda, MD [For sale by the Supt. of Docs., U.S. Govt. Print Off., Washington], 1967
Garrison	Garrison, F.H. *An Introduction to the History of Medicine*. Philadelphia, W.B. Saunders, 1921
Garrison Morton (GM)	Morton, L.T. & Norman, J.M. *Morton's Medical Biography: an annotated check-list of texts illustrating the history of medicine* (Garrison and Morton). 5th ed. Aldershot, Hants, England, Scolar Press, 1991

Grolier, *Medicine*	Norman, H.F. & Mayo, H. *One Hundred Books Famous in Medicine*: *Based on an exhibition held at the Grolier Club 20 September–23 November 1994*. New York, The Grolier Club, 1995
Heirs of Hippocrates	Eimas, R. (ed). *Heirs of Hippocrates*: *Development of Medicine in a Catalogue of Historic Books in the Hardin Library for the Health Sciences, the University of Iowa*. 3rd ed. Iowa City, University of Iowa Press, 1990
Hellman	Hellman, A.M. *A collection of early obstetrical books*. An historical essay with bibliographical descriptions of 37 items, including 25 editions of Roesslin's Rosengarten. Published by the author, 1952
Hirsch	Biographisches Lexikon der hervorragenden Ärzte aller Zeiten und Völker, Vols. I–VI, eds. Hirsch, August, Vernich, A.L.A. and Gurlt, E.J. Wien und Liepzig, Urban & Schwarzenberg, 1884–1885
Krivatsy	Krivatsy, P. *A Catalogue of Seventeenth Century Printed Books in the National Library of Medicine*. Bethesda, MD, U.S. Dept. of Health and Human Services, Public Health Service, National Institutes of Health, National Library of Medicine, 1989
Le Fanu	Le Fanu W.R. *The writings of William Hunter F.R.S. – bibliography*. Bibliotheck 1:3–14, 1958
Lilly	LeFanu, W.R. (ed). *Notable Medical Books from the Lilly Library Indiana University*. Indianapolis, the Lilly Research Laboratories, 1976
Lindeboom	Lindeboom, G.A. *Dutch Medical Biography, A Biographical Dictionary of Dutch Physicians and Surgeons, 1475–1975*. Amsterdam, Rodopi, 1984
Munk	Munk, W. *The Roll of the Royal College of Physicians of London*. London, Longman, Green, Longman & Roberts, 1861
Norman	Hook, D.H., Norman, J.M. & Norman, H.F. *The Haskell F. Norman Library of Science and Medicine*. San Francisco, J. Norman, 1991
Osler	Osler, W. *Bibliotheca Osleriana*: *A Catalogue of Books Illustrating the History of Medicine and Science*. Oxford, at the Clarendon Press, 1929
PMM	Carter, J. & Muir, P.H. *Printing and the Mind of Man*: *A Descriptive Catalogue Illustrating the Impact of Print on the Evolution of Western Civilization during Five Centuries*. London, Cassell, 1967
Ricci	Ricci, J.V. *The development of gynecological surgery and instruments....* Philadelphia, Blakiston, 1949. (GM 6310)
Russell	Russell, K.F. *British Anatomy 1525-1800. A bibliography*. Melbourne, Melbourne University Press, 1963. Russell, K.F. *British Anatomy 1525-1800. A bibliography of works published in Britain, America and on the Continent*. 2nd Ed. Winchester, St. Paul's bibliographies, 1987
Speert	Speert, H. *Iconographia Gyniatrica; A Pictorial History of Gynecology and Obstetrics*. Philadelphia, F.A. Davis, 1973
Spencer	Spencer HR. *History of British Midwifery, From 1650 to 1800. The Fitz-Patrick Lectures for 1927 delivered before the Royal College of Physicians of London*. London, J. Bale & Sons & Danielsson, Ltd., 1927
Waller	Waller, E. *Bibliotheca Walleriana: the books illustrating the history of medicine and science collected by Erik Waller, and bequeathed to the Library of the Royal University of Uppsala; a catalogue compiled by Hans Sallander*. Stockholm, Almqvist & Wiksell, 1955
Wellcome	Wellcome Historical Medical Library. *A Catalogue of Printed Books in the Wellcome Historical Medical Library*. 4 vols. London, Wellcome Historical Medical Library, 1962–95

Contents

Introduction

"As thou knowest not what is the way of the spirit, nor how the bones do grow in the womb of her that is with child: even so thou knowest not the works of God who maketh all." Ecclesiastes 11:5

"I will praise thee; for I am fearfully and wonderfully made: marvelous are thy works…" Psalms 139:14

These verses from Ecclesiastes and the Psalms articulate clearly the profound mystery of embryonic and fetal development, which has fascinated humankind since earliest times. Indeed, as astronomy and mathematics are to the physical sciences, the study of embryology and reproduction is one of the earliest of the biological sciences. One of the most mysterious and beautiful phenomenon is the development of a single cell, the fertilized egg, into a living, breathing, sentient human being with several trillion cells of about 200 distinct types. In frogs, fish, and other creatures, this transpires in full view of the observer. Within humans and other mammals however, this occurs hidden away within the dark, moist folds of the uterine cavity. In both instances, the events are similar, commencing with the totipotent fertilized ovum, followed by the first few cell divisions that give rise to the 16 cell morula, each cell of which remains totipotent. A few more cell divisions result in the structured blastocyst of 50 to 150 cells in the form of a shell surrounding a hollow space. From its inception, the blastocyst contains an outer covering of the placental-forming trophectoderm, or trophoblast, an inner cell mass of 20 to 30 pluripotent cells, the embryonic stem cells, and a fluid-filled cavity, the blastocoele. The embryonic stem cells have the ability to multiply indefinitely and to differentiate into any of about 200 cell types in the various tissues and organs. During the embryonic stages succeeding the blastocyst, the plasticity of developmental potency gradually diminishes until, near the end, most of the cells in mature tissues and organs are committed to their biological roles, and are unable to multiply or develop further. A fundamental "shift" in our understanding of cell commitment came with somatic cell nuclear transfer, the first successful application of which resulted in "Dolly" the sheep (Campbell 2002) and subsequently with the advent of "induced pluripotent stem [adult] stem cells" (Okita and Yamanaka 2011).

Certainly, the history of obstetrics and gynecology with illustrations of the fetus *in utero* traces back to the very beginnings of mankind. As Harvey Graham [pseudonym for Isaac Harvey Flack (1912–1966)] observed, "Everyone [of us] is a miracle in miniature with a personal history linked with the history of men and women who lived, loved, and gave birth to others since ever there were men and women" (Graham 1950, p. 1). Graham's thesis was especially profound because we now understand that one's mitochondrial genome can, in fact, be used to trace our maternal ancestry (mitochondrial genetic material comes from the oocyte, or egg, and not from the sperm; Stoneking and Soodyall 1966).

Relatively early in the beginning of a new century and a new millennium, it may be of value to pause and reflect on our origins and our progress. During recent decades in the reproductive sciences, as with medicine as a whole, the advances have been breathtaking. Previously unimagined achievements have been made in genetic analysis and engineering, diagnostic abilities, reproductive biology and endocrinology, stem cell biology, and assisted reproductive technology, resulting in science fiction-like marvels of contributions to reproduction and clinical care. These advances have not occurred in a vacuum, however, but rather have depended upon concurrent advances in cellular and molecular biology, engineering, computer sciences, and communications technologies, as well as advances in illustration, both in print and electronic format. As recorded by Johann Ludwig Choulant (1791–1861), one must consider anatomic illustration from both the viewpoint of "the aid rendered to anatomic science by the graphic arts [and] the aid rendered to the graphic arts by anatomic science" (Choulant 1920, p. 24).

As citizens of the twenty-first century, we are accustomed to viewing photographs from the most fundamental aspects of physics, chemistry, cellular and molecular biology to the outermost reaches of the universe. We live in a world of images, an iconographic jungle. For those, however, who lived prior to the twentieth century, pictorial images depicting contemporary understanding of science, including that of

© Springer International Publishing Switzerland 2016
L.D. Longo, L.P. Reynolds, *Wombs with a View*, DOI 10.1007/978-3-319-23567-7_1

reproduction, were uncommon and seldom seen. Of course, critical to such illustration was the anatomizing of female bodies ("the secrets of women," as opposed to the more common practice in males), a topic with a rich history (Park 2006). For the most part, these anatomic images were inaccessible to all but a few scholars, being confined to university libraries or the libraries of certain palaces or cathedrals. Although in some cases whimsical, naïve, or simplistic, the images have a charming beauty that resonates with our modern day mind. In addition, for those who share an ontological desire to understand the epistemology, the origins of knowledge, we believe viewing such images and considering their context can be a salutary and edifying experience. As in other works of the premodern period, such as those of the botanical sciences and herbals, the subjects may appear to have been plucked from nature and rendered in splendid, unnatural isolation. Although "decontextualized," one cannot be immune from the elegance and beauty of the forms, and the mystery with which one is filled in contemplating their particulars.

Today, representations and understandings of pregnancy and the developing fetus have been greatly enlarged by visual information available through contemporary medical technologies such as diagnostic ultrasound imaging, magnetic resonance imaging, and computed tomography. Some have argued that this has diminished the pregnant mother's role to that of a "carrier" or "receptacle." In contradiction, we would argue that the images of the developing infant stress the wonder of reproduction, and the unimaginable [opportune] and resilient capabilities of the maternal organism.

In terms of the history of the illustrations of human development, one may ask how to make sense of it all (Herrlinger 1970). We must remember that the history of science is no more a part of science than the history of mathematics or philosophy are part of those disciplines. Thus, we recognize that although the history of illustrations of embryology and fetal development may be of great interest, it is not an understanding of these phenomena in themselves, but rather it is merely part of the venture of discovery. In addition, we must make clear that our goal is not to be encyclopedic, but rather to emphasize those works that were most original, most intelligent, and most significant. That is not to say, that in a few instances we have not included rather obscure volumes that are essentially unknown, in part, to bring them to light. In speaking of books that become immortal, Thomas Woodrow Wilson (1856–1924), who was to become the twenty-eighth President of the United States (1913–21), wrote,

When once a book has become immortal, we think that we can see why it became so. It contained, we perceive, a casting of thought which could not but arrest and retain men's attention; it said some things once and for all because it gave them their best saying. Or else it spoke with a grace or with a fire of imagina-

tion, with a sweet cadence of phrase and a full harmony of tone, which have made it equally dear to all generations of those who love the free play of fanciful thought or the incomparable music of perfected human speech. Or perhaps it uttered with full candor and simplicity some universal sentiment; perchance pictured something in the tragedy or the comedy of man's life as it was never pictured before, and must on that account be read and read again as not to be superseded. There must be something special, we judge, either in its form or in its substance, to account for its unwonted fame and fortune. (Wilson 1891)

This present volume begs the question, "Why a 'picture book' such as this?" First, we each have a long-standing interest in the history of the study of the gravid uterus. As this interest developed, we came to realize that much of the early work in this area, as depicted in illustrative form, is 'lost;' that is, the early illustrations are not readily available because of the rarity of the texts in which they are contained. Even those illustrations that are widely available and viewed, like Leonardo's illustration of the human fetus in the opened uterus (sometimes referred to as "The Babe in the Womb" – pg. 35), are often poor quality reproductions. We believe these illustrations not only are important historically, but instructive, as they provide insights into the evolution of ideas. They also represent some of the most beautiful and detailed depictions of the gravid uterus ever made. In addition, we believe there is value in making these illustrations more widely available and placing them into their historical context. Beyond a mere narrative of dates, events, and the great people, history can also help us learn many lessons of life. As the English physician-rhetorician Sir Thomas Browne (1605–1682) observed, "…the past is made contemporary through our reading and thinking; the past is a bookshelf open to all" (Browne 1642). In a sense, it is from great works of the past that we confirm the continuity of human experience through the ages. Along this line, although contemporary digital technology allows us to have a 'virtual' experience and exploration, the physical durability and touch of old volumes can help us to appreciate the miracle and joy of shared understanding. Hopefully, the reader can be transported across the late fifteenth through nineteenth centuries as an armchair traveler through reproductive medicine.

Second, despite photography being well established by the mid-nineteenth century (Newhal 1982), and photomicrography by the later part of that century (Gall 1996), because of its ability to convey an idea, in a clear and powerful manner, the tradition of drawn illustration, nowadays usually referred to as a "schematic" (a modern, industrial term), has continued to the present. Nonetheless, as cautioned by well respected researchers, even these modern illustrations increasingly are becoming lost, not because they are truly unavailable but, rather, because they are inaccessible electronically. Thus, we believe there is a need to preserve and present more recent, but still historically significant, illustrations as well.

The miracle of reproduction, with the delivery of intelligent, sentient, productive human beings, has been recorded since the beginning of writing and illustration. Thus, the history of the reproductive sciences comprises one of the richest records of any aspect of medicine. Too often, history is recited from a national perspective; however the great ideas and achievements, almost without exception, have entanglements that cross borders, as well as involving multiple disciplines. Thus, in the present volume our survey encompasses most of Europe, and is contingent with the development of Western Civilization. Since the late fifteenth century, printed books have included illustrations of the gravid uterus and its contents, the fetus and placenta. Initially depicted by woodblock prints, in the seventeenth century these were replaced by copper plate engravings with much finer detail, and in the eighteenth century these were transformed into steel plate engravings with added advances in lithography and typography. Such improvements in illustration have continued to the present day. A particularly striking feature of this evolution of fine printing has been the power of the images depicted. In concert with the advance in illustrative technologies, especially during the last half century, has been the ever increasing knowledge and understanding of the fine structure and function of these organs and tissues, and their interrelations (Gall 1996).

Following Johann Gutenberg's (ca. 1395–1468) invention of movable type in the mid-fifteenth century, it was not until the early sixteenth century that schematic anatomic illustrations appeared more than rarely, and print technology caught up with the manuscript culture of previous centuries. Of course writing, whether Egyptian hieroglyphs or Chinese and Japanese characters, were stylized drawings before they were reduced to their present forms, and thus illustration plays a vital role in communication and in the history of thought. In the complex schema of life, medical illustration thus unites art and medicine (Herrlinger 1970).

With the Renaissance and its humanistic revival of classical art, literature, and learning in the fifteenth and sixteenth centuries, increased attention was given to the human form. Johannes de Ketham (ca. 1470) in his *Fasciculus medicinae* displayed the earliest anatomic illustrations in a printed work (Ketham 1491), albeit schematic and stylistic in form. With the 1543 publication of *De humani corporis fabrica libri septum* [The structure of the human body in seven books] by Andreas Vesalius (1514–1564), an appreciation and understanding of human anatomy was placed on a solid foundation (Vesalius 1543). Although containing numerous errors, Vesalius' work was a monumental improvement over the depictions of Galen (130–200 CE) and later workers, many of which were based on the dissection of dogs, primates, or other mammals.

A topic of importance to a volume such as this is the relation of the history of art to the history of medical illustration. Although consideration of this topic is beyond the limits of this volume, in his 1936 essay "*The historical aspect of art and medicine*," the medical historian Henry Ernest Sigerist (1891–1957) examined the evolution of medicine and art as depicted throughout the ages (Sigerist 1936). Both are important aspects of culture, society, and civilization. Art, the creation of an imaginative mind, is a product of its time, yet timeless. It forever strikes cords of emotion, and speaks to us in a language we understand. In contrast, medicine, a compilation and synthesis of bioscientific thought, evolves so that the theories and thinking of one age are barely understood by the next, and most often are seen as being in error and even irrelevant. Sigerist examined the artistic depiction of medical themes during Romanesque (Ninth to Twelfth Century), Gothic (Twelfth to Fifteenth Century), Renaissance (Fourteenth to Sixteenth Century), Baroque (Sixteenth to Seventeenth Century), and the age of Romanticism (late Eighteenth to Nineteenth Century). He observed that if artistic style reveals a whole *Weltauschauung* [conception of the world, world outlook or ideology], that this must be evident in medicine as well. Sigerist noted,

> A history of anatomical illustrations seen from the point of view not of the history of medicine but of the history of art would be an extremely worthwhile undertaking that would place the whole problem in an entirely new light. (Sigerist 1936, p. 291)

Later, this theme was expanded upon by George Alfred Leon Sarton (1884–1956) (Sarton 1941). As noted, we make no pretense of attempting a history of art or illustrations in books. Rather our goal is to point to seminal contributions, important in their time, that have contributed to an understanding of embryonic and fetal development as well as to obstetrics, and as such have helped to enlarge our appreciation of the miracle of reproduction. In considering the history of thought and of our cultural development, art has played a critical role in the communication of ideas. Anatomical and medical illustrations have been and continue to be vital to this development, and here we draw attention to works that have combined science and art in a reasonably well defined field. As an aside, we believe that preservation of works such as those depicted in this volume serves to emphasize the crucial nature and aspects of books, as a physical object, as contrasted to a digital image on a computer screen, for contemplation, learning, and inspiration.

During the Renaissance, the mother of modern civilization, a period beginning in Italy in the fourteenth century, science emerged as an authority in its own right, and discovery and invention were cause for delight. A key element in the Renaissance was the separation in jurisdiction between divine and natural philosophy. This latter discipline had its roots in nature, and that which could be verified empirically became the province of reason and science. In concert with these developments was the revolution in art, which resulted from scientific studies of the mathematics of composition, the geometry of perspective, the illusions of chiaroscuro (the sixteenth century technique of printing shades of light and dark using several wood blocks), and a deeper understanding of anatomy. In part preceding, or at least contemporary with, the art of anatomical studies was that of botanical illustration, associated with the publication of herbals.

Overall, in reviewing these images we considered several questions that relate to illustrations of the developing fetus from the late fifteenth through the nineteenth centuries:

To what extent do these art forms reflect the society and culture of the times?

In what ways did midwives contribute to the advancement of knowledge?

What can we learn from images of the placenta and of twins, from ideas concerning fertilization and development of the embryo, and from the depictions of monsters (or prodigies as they were called)?

In terms of pregnancy and illustrations of the *gravid* (Latin, heavy, weight) uterus, a particularly fruitful period occurred during the mid-eighteenth century in Great Britain. Although the Enlightenment constituted a philosophical movement, its concern with a critical examination of previously accepted doctrines from the point of view of rationalism became a dominant force in medicine. In part related to the rise of the male midwife, a number of workers, both those in academic positions and independent private practitioners, sought to gain a deeper understanding of pregnancy, its practice, and its complications. This intellectual search resulted in an outpouring of midwifery treatises with lectures, catalogues of obstetrical forceps, cephalotribes and other instruments of fetal destruction, and textbooks. Of particular note, were the great obstetrical folio atlases published during the second half of the century by William Smellie (1697–1763; 1754), Charles Nicholas Jenty (ca. 1720–1770; 1757), and William Hunter (1718–1783; 1774). With their near-life-sized engravings, drawn for each volume by Jonnes van Riemdijk (also Jan van Rymsdyk; fl. 1750–1788), and accompanying texts, these volumes presented the gravid uterus and its contents in a new light.

The gravid uterus and its contents must be viewed as much more than the fetus and its placenta. Although self evident, the goal of pregnancy is to produce an infant that as the Roman satirist Juvenal (circa 60–140) stated, *mens sana in corpore sano* [a sound mind in a healthy body], with prospects for a long and fruitful life. As demonstrated by research in the reproductive sciences during the past half-century (Longo 2013), successful pregnancy, with birth of a healthy, viable infant, involves complicated interactions among elements of the "maternal-placental-fetal unit."

Although beyond the scope of the present volume, some salient aspects of these interrelations and interactions are presented in the introductions to each of the major sections. Thus, as suggested by the title of this volume, in addition to that of the fetus, one must consider illustrations of the placenta, as that organ is so vital to development of the conceptus. We have given some details of the history of ideas concerning the placenta elsewhere (Longo and Reynolds 2010). From the standpoint of depictions of "the gravid uterus…," an important aspect is the development of ideas in embryology, i.e., that development prior to eight weeks gestation when the major fetal organs and the placenta are formed. The science of embryology has been of interest for centuries, the earliest work involving the study of bird embryos. In the present volume, we examine early depictions of embryonic development. In regard to teratology and malformations, since earliest times birth defects and monstrosities in humans and animals also have attracted attention, and to a great extent they were regarded as of supernatural origin.

As noted, many of the early illustrations (and even those more recent) have been "lost" because they exist only in rare volumes (or only as individual prints), and even those images which usually are presented are of relatively poor quality. A related purpose, therefore, in assembling this compendium, has been to "resurrect" some of the most important and beautiful of these illustrations, and to attempt to place them into their historical context. The intellectual justification for this work, however, is not dependent upon the abundance of arcane and fascinating aspects of intrauterine development. Rather, it is a function of the similarity of approach to their subject by the majority of writers. Importantly, each individual contribution must be viewed within the context of the broad sweep of European, and to a less extent American, culture, society, and political history.

Of course, there are several ways in which such a series might be organized. We appreciate that life is a continuum; however, for the purposes of this work, we have chosen to group much of the material by centuries, with later sections on midwifery, embryology, and teratology. Although one would argue that history cannot be sliced salami-like into epochs, many professional historians hold that historical periods such as the Renaissance, the Enlightenment, and periods thereafter are "sets of coherences." In a sense, this view has helped us trace the evolution of ideas and depictions in a reasonably logical manner. Overall, the present volume makes no claims of being an "ode to optimism" or "progressivism," a study in "technological determinism," nor a Whig depiction. Rather, we would hope that it presents in a reasonably lucid and logical format a synthesis of some of the notable achievements in the evolution of thought regarding embryonic and fetal development as expressed in images presented in these volumes intended for professionals.

In closing, perhaps we should say a word about a work such as this in terms of historiography; that is, the history and analysis of the writing of history. By its nature, history is selective, biased, and an incomplete reconstruction of the past. The historian must make assumptions about former ages and the accuracy of information, its synthesis, its interpretation, and appreciate the complexity of the task of writing "history." Much of medical history to mid-twentieth century considered the "great men" and their accomplishments, and such writing tended to be hagiographic, emphasizing the progressive nature of the field of knowledge under study, with advancement for the better. Beginning in the mid-nineteenth century with Leopold von Ranke, historicism, that is, history written without preconception, based on archival material, with consideration of the past in its own terms and eschewing a more modern interpretation, came into being. Much of this history concerned nations, politics, society, and diplomacy. The view was that the study of history can lead to the discovery of general laws of development. With the work of Henry Ernest Sigerist and his school, in the early- to mid-twentieth century, social history, that concentrating on social groups, their roles in economic and cultural structures and developments, and their interrelations, became prominent. Finally, beginning in mid-twentieth century, with historians such as Michael Foucault (1926–1984), intellectual history, the historical study of ideas in a comprehensive sense, expanded the horizons of historical thought and analysis. Intellectual historians have moved beyond emphasizing the great men or deeds, the scientific analysis of archives, the social forces, or social institutions, to the evolution of ideas and the philosophy of an era. Although to a great extent the present work centers on the great men and women, and their specific accomplishments and contributions, we have attempted to include elements of historicism, social history, and intellectual history. We leave it to the reader to judge the success of our efforts.

References

Browne, T. *Religio Medici*. London, Crooke, 1642.

Campbell, K.H.S. A background to nuclear transfer and its applications in agriculture and human therapeutic medicine. *J Anat* 200:267–275, 2002.

Choulant, L. *History and bibliography of anatomic illustration in its relation to anatomic science and the graphic arts*. Translated and edited with notes and a biography by Mortimer Frank; a biographical sketch of the translator and two additional sections by Fielding H. Garrison, M.D. and Edward C. Streeter, M.D. Chicago, IL, University of Chicago Press, 1920.

Gall, J. *Views of the Cell: A Pictorial History*, Bethesda, MD: The American Society for Cell Biology, 1996.

Graham, H. *Eternal Eve. The History of Gynaecology & Obstetrics*. London, W. Heinemann, 1948–1950.

Herrlinger, R. *History of Medical Illustration. From antiquity to 1600*. New York, Editions Medicina Rara, 1970.

Hunter, W. *Anatomia uteri humani gravidi tabulis Illustrata … The Anatomy of the Human Gravid Uterus Explained by Figures*. Birmingham, John Baskerville, 1774.

Ketham, J. *Fasciculus de medicinae*. Venetiis, per Johannem & Gregorius fraters de Forlivio, 1491.

Longo, LD. *The Rise of Fetal and Neonatal Physiology: Basic Science to Clinical Care*. New York, Springer, 2013.

Longo, L.D. and L.P. Reynolds. Some historical aspects of understanding placental development, structure and function. *Int J Develop Biol* 54:237–255, 2010.

Newhal, B. *The History of Photography*. New York: The Museum of Modern Art, 1982.

Okita, K. and Yamanaka, S. Induced pluripotent stem cells: opportunities and challenges. *Phil Trans R Soc B* 366:2198–2207, 2011.

Park, K. *Secrets of women: gender, generation, and the origins of human dissection*. New York, Zone Books, 2006.

Sarton, G. The Fielding Garrison Lecture. The history of medicine versus the history of art. *Bull Hist Med* 10:123–135, 1941.

Sigerist, H.E. The historical aspect of art and medicine. *Bull Hist Med* 4:271–297, 1936. Later reprinted in *History of medicine. Vol. 1. Primitive and archaic medicine*. New York: Oxford Univ. Press, 1951.

Smellie, W. *A Sett of Anatomical Tables, with Explanations, and an Abridgment of the Practice of Midwifery, with a view to illustrate a Treatise on that Subject and Collection of Cases*. London, 1754.

Stoneking, M., Soodyall, H. Human evolution and the mitochondrial genome. *Curr Opin Genet Dev* 6: 731–736, 1966.

Vesalius, A. *De humani corporis fabrica libri septum*. Basel, Ioannis Oporini [Johannes Oporinus], 1543.

Wilson, W. How books have become immortal. *The Atlantic* 68:406–413, 1891.

The Fetus in Utero Including Twins and the Placenta

The Fetus

By definition, the fetus is a developing mammal or other vivparous vertebrate. In humans, this is the unborn from the eighth week of pregnancy following fertilization (or 10 weeks after the onset of the last menstrual period), when in the embryo the major organs have been formed and continue their development until the moment of birth. At this time, the fetus is about 5 cm (2 in.) in length, weighs ~8 g, and the developing head constitutes about one-half the total mass. The word *fetus* derives from Latin, bearing, bringing forth, or hatching of young. From weeks 11–17 of pregnancy the brain, heart, and other organs continue to develop, and subtleties appear in the various structures such as development of the genitalia and centers of ossification in bones. At about 16 weeks, a woman who has been pregnant previously will feel fetal movements, "quickening;" this may not occur, however, until about 20 weeks in those women who are nulliparous (having not delivered before). From weeks 18–27, development continues with the appearance of many structures such as the eyelashes and eyebrows, finger and toe nails, and a fine lanugo hair covering the body. From weeks 28–40 (term), nerve growth with myelinization continues, the amount of body fat increases, the cerebral electrocorticogram takes on cyclic activity, e.g., low voltage high frequency, during which metabolism of the brain increases, (and which is associated with rapid eye movements and other muscular activity), versus a high voltage low frequency sleep state. At this time, the fetus is capable of life independent of the mother's womb; survival for the most immature has been enhanced greatly by advances in neonatal intensive care. At the end of the 36th weeks of gestation almost all infants born can survive independently of neonatal intensive care.

At term, the gravid, i.e., weighty uterus, with fetus, placenta, and amniotic fluid weighs 5–7 kg. The fetus/newborn weighs about half of this total, depending upon ethnicity, size of the parents, parity of the mother, and maternal nutritional state, with males weighing about 100 g more than females. As noted earlier, rather than a passive passenger in the drama of reproduction, the fetus plays a dynamic role in assuring its own growth, development, and, at the appropriate time, its delivery into the world. The so called "maternal-placental-fetal unit," constitutes a vibrant, vital system of communication with exchange of nutrients from the mother, production of essential hormones by both placenta and fetus (particularly the pituitary gland and adrenal cortex), and other interactions that effect each of the major tissue compartments. Only recently is the extreme complexity of these hormonal interactions being unraveled. Within the uterus, the fetus experiences a position of protection from the external world, never to be experienced again in life.

Rather than being just a "small adult," in addition to its dependence for life upon the placenta and its unique circulation, the fetus is distinct in many respects. Critical to its development are unique aspects of the heart and circulation. As may be expected, gestation is associated with considerable variation in rates of fetal growth and maturation, these being a function of genome, ethnicity, mother's weight, body mass index, exposure to poor diet and/or toxins (such as tobacco products, alcohol, recreational drugs), altitude of residence, and other factors. Dysregulation of placental structure and/or function, utero-placental blood flow, and other issues may compromise nutrient delivery from mother to fetus. Any of these factors may result in intrauterine growth restriction (IUGR), the fetus being small for gestational age (SGA).

In terms of pregnancy and illustrations of the gravid uterus, as noted the first anatomical illustration to appear in a printed book was that of Johannes de Ketham (died ca. 1470) of 1491. For the most part, the drawings of the fetus in books of the early sixteenth century [Eucharius Rösslin (ca. 1470–1520; 1513), Jacob Rueff (1500–1558; 1554), and others] were "cartoons," many copied from medieval manuscripts. With the seventeenth century, illustrations for the volumes of Adriaan van de Spieghel (1578–1638; 1626), Hendrik van Roonhuyse (1625–1672; 1663), François Mauriceau (1637–1709; 1668), and others, became more realistic and detailed. A particularly

© Springer International Publishing Switzerland 2016
L.D. Longo, L.P. Reynolds, *Wombs with a View*, DOI 10.1007/978-3-319-23567-7_2

fruitful period for anatomical illustration occurred during the mid-eighteenth century in Great Britain. Although the Enlightenment constituted a philosophical movement, its concern with a critical examination of previously accepted doctrines from the point of view of rationalism became a dominant force in medicine. In part, related to the rise of the male midwife, a number of workers, both those in academic positions and independent practitioners, sought to gain a deeper understanding of pregnancy, its practice, and its complications. This resulted in an outpouring of midwifery treatises with lectures, catalogues of obstetrical forceps, cephalotribes and other instruments of fetal destruction, and textbooks. As noted in the introduction, of particular note were the great obstetrical folio atlases published during the second half of the century by William Smellie (1754), Charles Nicholas Jenty (1757), and William Hunter (1774), which, with their near-life-sized engravings and accompanying texts, presented the gravid uterus and its contents in a new light.

Twins

From ancient times, and in the folklore of many cultures and ethnic groups, from Africa to Asia, throughout Europe and the Americas, twins (two infants resulting from one pregnancy) have been regarded as a gift of the gods, mystically endowed, healing magicians. From the fifth dynasty (2494–2345 BCE) in Egyptian art these were depicted the twin goddesses *Isis*, goddess of fertility and patron of nature and magic and *Nephthys* [Nebt-het], a "useful" and "excellent" goddess, (although she possessed attributes of an ominous nature – *Isis* married her brother *Osiris*, while *Nephthys* married her brother *Set*; *Osiris*, considered a wise and generous king, later was murdered by *Set*). These twins were the daughters of *Geb*, god of the earth, and *Nut*, goddess of the overarching sky. Worship of the goddess *Isis* as the ideal woman became widespread throughout the Greek and Roman Mediterranean world. In the Hebrew Scriptures, Genesis 25, Rebekah, wife of Isaac suffered tortuous pain as a result of her twin sons Esau and Jacob commencing their life-long disputes within her uterus (these twins also may have been the first recorded examples of twin-twin transfusion syndrome). Later, Genesis Chapter 38 records the rather complicated story of Tamar giving birth to Perez and Zarah following impregnation by her father-in-law Judah.

In Greek and Roman mythology, *Castor* and *Polydeuces* (Pollux in Latin), the twin sons of *Leda* and *Zeus* (Jupiter, in Latin), were divinely endowed, and named *Dioscuri* [Greek, sons of Zeus]. An alternate legend holds that while these twins had the same mother, *Pollux* was fathered by *Zeus*, and therefore immortal, while *Castor's* father was *Tyndareus*, and thus mortal. When *Castor* died, *Pollux* implored *Zeus* to keep them together and share immortality, and *Zeus* transformed the brothers into the constellation *Gemini* [Latin, twins]. As such, they were believed to be the patrons of sailors, who invoked them to send favorable winds. Roman mythology presents the twin offspring, *Remus* and *Romulus* of the Vestal Virgin priestess *Ilia* (also Rhea Silvia) and *Mars*, the god of war. *Romulus* is believed to have killed his brother, and in 753 B.C.E. founded Rome. He also was believed to have become the god *Quirinus*, whom the Romans worshipped as divine. In the twin myths of many native peoples, one twin may be good and the other evil. During the middle ages and the early modern period, views varied on the causes of twins.

Twins can be either dizygotic (fraternal) or monozygotic (identical). The incidence of the former varies widely throughout the world from 12 and 6 per 1000 live births in the U.S.A. and Japan to about 40+ per 1000 live births among the Yoruba of West Africa. The incidence of monozygotic twins, in contrast, is relatively constant at 3 per 1000 live births throughout the world. While dizygotic twins develop from two eggs fertilized independently by different sperm, monozygotic twins originate from a single egg that upon fertilization forms a zygote that then divides into two (or more) embryos. Despite having essentially identical genetic DNA, because of environmental and other epigenetic factors, monozygotic twins may differ phenotypically to a considerable degree. Conjoined, or "Siamese" twins are monozygotic twins whose bodies are joined together. This union can occur at the side, as in the case of Chang and Eng Bunker (1811–1874), born in Siam (later Thailand) in 1811, and who gained fame in the U.S.A. Alternatively, when the zygote starts to divide relatively late in development (after day 13) and division is incomplete, union of such twins can occur at any part of the body – head, thorax, or abdomen.

Because of the limited space within the mother's uterus, twins, or other "multiples," commonly deliver 3 or more weeks early. As noted elsewhere, extreme premature birth can result in serious health consequences. Twins have a much higher death rate than do singleton infants, both prior to birth and as infants, accounting for about 15 % of neonatal deaths. In part, this higher mortality rate is related to the twin-to-twin transfusion syndrome seen in monozygotic twins with a monochorial placenta, which results from vascular (artery to artery or artery to vein) communications in the infants' placenta. Of importance, the mothers of twins also are subject to a higher mortality rate, as a consequence of preeclampsia and placental abruption (premature separation of the placenta).

Several woodcuts of twins were depicted in the first printed work devoted to obstetrics, *Der schwangern frauwen…* (1513) by Eucharius Rosslin (ca. 1470–1526), and in most of the illustrated volumes that followed. Conjoined twins were illustrated in the works of Conrad Wolffhart [Lycosthenes] (1518–1561; 1557), Fortunio Liceti (1577–1657; 1634), Ulisse Aldrovandi (1522–1605; 1642), and many others.

The Placenta

As the title of this volume suggests, in tracing the evolution of ideas and illustrations of the gravid uterus and its contents, one must consider the placenta as well as the fetus. Ideas concerning development of the placenta and its function are inextricably interwoven into the history of embryology and fetal development and these are almost inseparable. Several major reviews have presented a detailed overview of concepts of that organ, its morphology and fine structure, and many aspects of the relation of the fetus to its "lung-in-utero" and "bundle of life," and its comparison among different species (Boyd and Hamilton 1970, pp 1–19; Ramsey 1982). Further details of this history have been given by the authors of the present volume (Longo and Reynolds 2008). The placenta, a fetomaternal organ characteristic of mammalian (eutherian) pregnancy, differs from other organs in being formed by the apposition and interaction of both fetal and maternal tissues for physiologic exchange, shows extreme diversity in its structure among the species, and is of limited lifespan. The term placenta was introduced in its present connotation by Matteo Realdo Columbo [Columbus] (ca. 1510–1559) in his *De re anatomica* [Concerning anatomy] (1559), and derived from the Greek *plakous* or *plakóenta*, meaning a flat surface or cake. From the placenta, the umbilical cord courses to the fetal mid abdomen, the *umbilicus* [Latin, navel] or *omphalos* [Greek, navel]. As an aside, the ancient Greeks used the term *omphalos* to refer to a sacred, rounded stone in the Temple of Apollo at Delphi, which they believed marked the center of the earth.

Since earliest times, the placenta has been recognized as being of the greatest importance, and at the same time quite mysterious—even somewhat mystical. For example, in many cultures the placenta has been held as an *alter ego* [my second self], a symbol for the preservation of health and good fortune, and as a talisman in case of danger. In some societies, a sympathetic animism exists between the placenta and the future adult (Ploss and Bartels 1935). In early Egypt, the placenta was believed to be the seat of the "External Soul." The Hebrew scriptures include several references to the placenta, sometimes referred to as the "Bundle of Life" and "External Soul" (For instance, I Samuel 25:29) (Stirrat 1998). In the folklore of many peoples of the Pacific Islands, Australasia and Africa, the placenta is regarded variously as a sibling of the infant, a companion, or soul or otherwise possessing supernatural properties (Longo 1963). For instance, in Africa today, one tribe with which one of us is familiar, the Yoruba people of

Western Nigeria, buries the placenta near the entrance to the home, so that the child "will always look back to its father" (Longo 1964).

The Greeks recognized the importance of the *placenta* [flat cake] in fetal nutrition, and named the outermost embryonic membranes *chorion* [membrane] and the innermost membrane encompassing the fetus *amnion* [bowl]. The Greek philosopher-biologist Aristotle (384–322 BCE) may have been the first to use the term *chorion*, and he also recognized the yolk sac of lower vertebrates (see Aristotle 1831–1870). Because Aristotle based many of his ideas on findings in ruminants and other animals, considerable confusion about many topics was perpetuated. Nonetheless, he did much to establish the science of the study of fetal membranes. In his great embryological treatise *De generatione animalium* [On the generation of animals] (circa 340 BCE), Aristotle stated that "The [umbilical] vessels join on the uterus like the roots of plants and through them the embryo receives its nourishment" (Aristotle 1836–1870). The Greek physician-anatomist Claudius Galenus [Galen] (ca. 130–201), of Pergamon, considered embryological development in his treatises *De formatu foetus* in utero [intrauterine development of the fetus], *De uteri dissectione* [dissection of the uterus], and *De usu partium* [on births] (Galen 1914). In these works, Galen maintained that the uterine vessels open their mouths and unite with the fetal vessels in the chorionic membranes, thus establishing direct communication between the mother and the fetus. This was concordant with his view that the arteries supplied "vital spirits" or "spiritual blood" to maintain the innate heat of the tissues, whereas the veins supplied the "alimentary blood" to provide pabulum or foodstuffs. Until the discovery of the circulation of the blood in the early seventeenth century, Galen's views on this subject were held as dogma (Galen 1914).

If there was a single dispute that captured the core of developmental biology from the early sixteenth through the eighteenth centuries, it was the debate over the degree to which the maternal and fetal placental blood vessels actually were interconnected. In his monumental work on the "*Fabric of the human body…*," one of the most seminal volumes in the development of modern medicine and science, Andreas Vesalius described the human uterus consisting of a single chamber, rather than having two horns as in many mammals, and its rich vasculature. Vesalius upheld the Galenic doctrine of anastomosis between maternal and fetal vessels. Mistakingly, he illustrated the placenta as being zonary or girdle-like, as in the dog (Vesalius 1543), but corrected this in the second edition (Vesalius 1555). In his compendium on anatomy, Matteo Renaldo Colombo, a pupil of Vesalius and his successor as professor at the University of Padua, coined the term *placenta*, and described many of its features including the importance of the umbilical vessels (Colombo 1559). Gabriele Falloppio (1523–1562) in his *Observationes anatomica*, [anatomical observations] 1561, noted that the human fetus has a single umbilical vein, in contrast to two in ruminants. Giolio Cesare Aranzi [Arantius] (1530–1589) in his *De humano foetu* (1564) was the first to maintain the separation of maternal and fetal circulations in the placenta, and that that organ acts to purify fetal blood.

In essence, it was William Harvey's (1578–1657) discovery of the circulation of the blood by the pumping action of the heart (Harvey 1628) that eventually forced investigators to the realization that the flow of blood in the maternal uterine and fetal umbilical vessels must constitute separate circulations. A century and a half later, this was demonstrated by the injection studies of William and John Hunter (1728–1793; 1786; 1794). As noted, the products of conception must be viewed, not as an individual parts list, but rather as a whole, as a complex unit involving mother, placenta, and fetus. The placental arm of this communication network involves maternal blood in the intervillous space which bathes the fetal villous trophoblast, fetal blood within villous capillaries, hormonal conversion and synthesis by trophoblastic tissue that interacts with decidua lining the gravid uterus, and other maternal tissues. This fetal villous trophoblast is quite efficient in extracting and transferring to the fetal blood, nutrients from blood of the mother. Also to be considered are the fetal membranes, the amnion and chorion laeve, that are active in this complex system of communication. Aspects related to the decidua for consideration are ideas regarding implantation and the development of the amniotic and chorionic membranes, the concept of the intervillous space through which the maternal placental circulation flows, placental villi with their vasculature and "trophoblast" cellular coverings, and classification of placental types. Each of these topics requires independent consideration.

We have chosen to divide this segment of the book on The Fetus in Utero including Twins and the Placenta into centuries, realizing that a 100-year period, as with a human being or other entity, is much more than the sum of its parts. In dividing time into such epochs, we understand that it is for convenience, to have a segment of time that we can explore and from which we can learn.

References

Benirschke, K. The placenta: how to examine it and what you can learn. *Contemp Ob Gyn* 17: 117–119, 1981.

Boyd, J.D. and W.J. Hamilton. Development and structure of the human placenta from the end of the third month of gestation. *J Obstet Gynaecol Br Commonw* 74:161–226, 1967.

Boyd, J.D. and W.J. Hamilton. *The human placenta*. Cambridge, Heffer, 1970.

Longo, L.D. Medicine and medical education in Nigeria. *New Eng. J. Med.* 268–1044-1055, 1963.

Longo, L.D. Socio-cultural practices relating to obstetrics and gynecology in a community of West Africa. *Am J Obstet Gynecol* 89:470–475, 1964.

Longo, L.D. and L.P. Reynolds. Some historical aspects of understanding placental development, structure, and function. *Int J Dev Biol* 54:237–255, 2008.

Ploss, H.H., M. Bartels and P. Bartels. *Woman: an historical, gynaecological and anthropological compendium. Eric John Dingwall (Ed)*. London, W. Heinemann, 1935.

Ramsey, E.M. *The placenta: human and animal*. New York, Praeger, 1982.

Stirrat, G.M. The bundle of life – the placenta in ancient history and modern science. In: *The Yearbook of Obstetrics and Gynaecology. P.M. Shaughn O'Brien (Ed), Vol. 6.*. London, Royal College of Obstetricians and Gynaecologists, 1998, pp. 1–14.

The Renaissance

As noted in the introduction, the present work makes no pretense of being a history of anatomy or anatomical illustration. It may be of value, however, to consider briefly some seminal contributions to the development of anatomy and dissection. The first detailed records are Greek in origin, with Aristotle's (384–322 BCE) rather detailed descriptions of invertebrate and vertebrate creatures. To a limited extent, the Greeks practiced human dissection. Following the death of Alexander the Great (356–323 BCE), King of Macedonia and conqueror of Greece, the Persian Empire, and Egypt, the center for anatomical study shifted from Athens to Alexandria. Although not encouraged by the Romans, human dissection continued at Alexandria until the second century C.E. Because of the Roman concerns, Galen (c. 130–200) of Pergamon, who served as surgeon to the gladiators, based most of his studies on mammals including the dog, sheep, and Barbary ape. Galen was held in such high esteem that his anatomical errors dominated medical thinking for almost a millennium and a half. During these centuries many of the Greek and Roman anatomical and other texts were preserved by Arabic scholars, and during the eleventh and twelfth centuries these began to be re-translated into Latin. In science, earth, air, fire, and water were considered the four fundamental elements or constituents of the universe. In medieval medicine and physiology, four fluids, or humors, were believed to regulate bodily function: blood, phlegm, choler (yellow bile), and black bile. The relative dominance of any one of these was believed to determine a person's character and general health. Thus, ones disposition might be sanguine, phlegmatic, choleric, or melancholic, respectively. Disease was believed to arise from an imbalance of these humors, hence contemporary therapy (e.g., bloodletting, emetics, and laxatives) were directed towards reestablishing humoral equilibrium.

The first modern anatomical work was that of Mondino de'Luzzi [Mundinus] (c. 1275–1326) whose *Anothomia* appeared in 1316, and was published as a book *Anothomia papiae* in 1478. The earliest contemporary illustration of a dissection survives in a manuscript of Henri de Mondeville (1260–1320) of Montpellier, who is known to have taught with the aid of illustrations (these first appeared in a printed book in 1889). Although anatomical knowledge advanced during the fourteenth and fifteenth centuries, this was not reflected in diagrams and illustrations, which persisted in depicting traditional views, rather than reality as it was observed.

The Renaissance [rebirth], the humanistic revival of classical art, literature, and learning, with rediscovery of the dignity of the individual, originated in Italy during the fourteenth century, and then over the next century spread throughout Europe. Dominating this period was a belief in the need to return to the classic culture of the Greeks and Romans, and classical antiquity became a rediscovered source of cultural and artistic inspiration. With the introduction of Aristotelianism into the Latin west, natural philosophy became central to the intellectual enterprise, and science emerged as an authority in its own right, as opposed to the authoritarianism of the Church. Socially and culturally, several major plagues (e.g., the Black Death, Plague of 1348–1351) and epidemic disease decimated much of the population of Europe. With the introduction of gunpowder, military activity enlarged, such as the Hundred Years' War between England and France (1337–1453), and artillery played an increasingly important role, making wars exceedingly expensive and deadly. From mid-fifteenth century onward, Western nation states began to explore the globe with increasing vigor. The Spanish and Portuguese Empires began to spread, Christopher Columbus (1451–1502) discovered the new world, and the Byzantine Empire came to an end. Thus, the European empires came to dominate a vast portion of the earth. With the Renaissance emerged stronger states, the spread of princely controls, and the rapid expansion of trade. With the beginning of this age of discovery, from Middle English arose the modern English language.

Some world leaders of note included Henry V (1387–1422) of England, who in 1405 won the battle of Agincourt; Constantine XI (1404–1453), the last Byzantine Emperor; the French heroine Joan of Arc (1412–1431); Isabella I of Castile (1451–1501) co-ruler with Ferdinand V (1452–1516) of Aragon, responsible for the unification of Spain, aiding Christopher Columbus (1451–1506), and promoting the inquisition; Henry VII (1457–1509), King of England and founder of the Tudor dynasty; and Vasco da Gama (ca. 1460s–1524) the Portuguese navigator who reached India by sea.

The fourteenth century had seen a flourishing in architecture, with the rise of many of the great cathedrals and palaces in Europe and Britain. During the early part of the fifteenth century, the technology of printmaking flourished. This, combined with the emergence of paper mills in Germany and Italy, made paper and prints widely available. Highly skilled *Formschneiders* [block cutters], carved woodblocks (and in some cases metal), combined with the mid-century introduction of movable type by Johann Gutenberg (c. 1395–1468), allowed wide availablitty of printed books and unprecedented spread of knowledge and ideas. After Gutenberg, a print culture replaced the oral and handwritten manuscript tradition of previous centuries. International commerce and economics developed both on land and via sea routes. Universities flourished as centers of knowledge and intellectual ferment. The Florentine architect Filippo Brunelleschi (c. 1377–1446) invented one-point perspective, (as well as engineering construction of the cupola of the cathedral *Santa Maria del*

Fiore [Holy Mary of the Flowers] in Florence) which, in turn, lead to innovation in illustration in architecture, engineering, and medical science. Nonetheless, for the most part, illustrations were without background, and with little or no shading or other embellishment.

In Europe, the Renaissance continued with advances in engineering, science, literature, and art. Francis I (1494–1547; King of France 1515–1547), although a relatively weak monarch was an enthusiastic patron of art and literature. In Germany, the Protestant Reformation, which shattered the unity of Western Christendom and the hegemony of the Catholic Church, commenced with Martin Luther (1483–1546) posting his 95 Theses on the church door at Wittenberg, Saxony (1517). Unlike Luther, the French reformer John Calvin (1509–1564) emphasized the importance of an institutional church and system of theology. As a key player in the Counter-Reformation, Ignatius Loyola (1491–1556), Spanish soldier-ecclesiastic in Paris, founded the Society of Jesus, the Jesuits (1534). The Ottoman Empire expanded to gain control of Egypt, Arabia, and the Levant (countries bordering on the eastern Mediterranean). Hernando Cortés (1485–1547) led the Spanish conquest of the Aztec Empire to found New Spain, while the conquistador Francisco Pizarro (ca. 1470s–1541) annihilated the Inca Empire. The Italian Renaissance was considered ended with the 1527 sack of Rome. Under Henry VIII (1491–1547; Monarch 1509–1547), in 1534 the Church of England broke from the Roman Catholic Church. That same year, the explorer Jacques Cartier (1491–1557) claimed the St. Lawrence River Valley for France. After 1558 under Elizabeth I (1533–1603), the Elizabethan era was considered the height of the English Renaissance. During the last four decades of the century, the French Wars of Religion raged between Catholics and Huguenots. These conflicts were abated in 1598 when, with the Edict of Nantes, Henry IV ("Henry of Navarre"; 1553–1610) of France granted religious and civil liberties to the Huguenots (almost a century later (1685) this was revoked by Louis XIV (1638–1715)). The English navigator Francis Drake (ca. 1540s–1596) circumnavigated the globe, and under Drake's command in 1588 the English repulsed the Spanish Armada. Under Sultan Suleiman I (Suleimar the Magnificent) (1520–1566) legal reforms were instituted, and the Ottoman Empire reached its zenith. Charles V (1500–1566) King of Spain and Holy Roman Emperor greatly expanded the Spanish Empire.

The Italian Andrea Palladio (1518–1580) inaugurated a new conception of architectural design that was to have incalculable influence. Michael Eyquem Montaigne (1533–1592) helped to define the essay and contributed to philosophy, as did Francis Bacon (1561–1626). The Holy Roman Emperor, Rudolf II (1552–1612) of the Hapsburg Dynasty, assembled an outstanding art collection and *Wunderkammern* [cabinet of curiosities]. The court of Rudolf II became one of the most avant-garde outposts of the Hapsburg Empire, with the palace serving as a center of Renaissance intellectual activity. Despite the increasing influence of art and science, a touch of medieval mysticism persisted which, among the inteligencia and elite, became not only a way of representing personal identities, but contributed to the development of natural history as well as fostering increased interest in magic, alchemy, and astrology. The *Meistersinger* [mastersinger] Hans Sachs (1494–1576) authored a multitude of poems and dramatic works. The English dramatist and poet William Shakespeare (1564–1616) redefined the theater.

The scientific revolution of sixteenth century Europe sparked essentially uninterrupted development in science and engineering, with one discovery leading to others. This revolution comprised several major aspects. In 1543 in his *De revolunibus...*, the Polish astronomer Nicolaus Copernicus (1473–1543) enunciated the principle of the solar system being heliocentric (sun-centered) rather than geocentric (earth-centered), thus laying the foundation for modern astronomy. With the acceptance of Copernicanism, a new science emerged including mechanical philosophy, the application of mathematics to nature, and the continuing development of natural history. The Danish astronomer Tycho Brahe (1546–1601) precisely observed and recorded stellar and planetary positions. Learned physicians of this era established a tradition important to medicine, extolling the "great men," Hippocrates and Galen, recalling Hippocratic-style aphorisms and Aristotelian-style argumentation, affirming practices such as bloodletting based on the humoral doctrine, and honing their anatomical curiosity. Philippus Aureolus Paracelsus (1493–1541), a Swiss physician-alchemist, contributed innovations in chemical therapeutics. While Gerolamo Cardano [Cardanus] (1501–1576) published several encyclopedias of medicine, natural science, and alchemy, Leonhart Fuchs (1501–1556) with Pietro Andrea Mattioli (1500–1577) contributed to medical botany. Ambrose Paré (1510–1590) revolutionized the field of surgery, and in England barbers and surgeons united as the "Commonality of the Barbers and Surgeons."

In art, the naturalism of the Renaissance was replaced by more objective illustration. In art and architecture, Baroque ornamentation with elaborate and ornate scrolls, curves, and flamboyant devices, named after the originator of the style Federigo Barocci (1528–1612), became widely popular. Albrecht Dürer (1471–1528), Titian (Tiziano Vecelli 1477–1576), Peter Paul Rubens (1577–1640), and others, with their *formschneider* (master block cutters) used this style to great advantage in their remarkable woodcuts. With advances in print making and improvement of woodcut techniques, prints were produced widely. Anatomists collaborated with artists to execute drawings from dissections, and with their accurate illustrations, revolutionized anatomy. The collaboration of anatomist, artist, and printer is epitomized by the publication in 1543 of the most prominent work of Renaissance anatomy, *De Humani Corporis Fabrica...*by the Flemish anatomist-surgeon Andreas Vesalius (1514–1564). This work created modern anatomy. Working early in the century, the incomparable Florentine artist and inventor Leonardo da Vinci (1452–1519) conducted his breathtaking studies in anatomy and, with Michelangelo Buonarrti (1475–1564), set a new standard in their portrayals of the human figure. Unfortunately, Leonardo's drawings did not come to wide attention until two and a half centuries after his death. Also in art, Dürer invented etching, the method of preparing metal plates from which one can print designs and illustrations. During the renaissance, artists became anatomists by necessity as they worked to develop a more lifelike, sculptural portrayal of the human figure. To their great credit, the Italian artists helped to formulate a consistent vocabulary of anatomical terms with which their discoveries could be recorded.

By the early sixteenth century, prints replaced medieval manuscripts as a seemingly inexhaustible source of motifs, and as they became widely distributed, had a significant impact on the history of art. The Renaissance revival of classical antiquity was nurtured by prints that spread knowledge of ancient Roman buildings and sculpture. Woodcuts, engravings (prints made from an engraved plate), and etchings (images made from acid treated metal plates), the later two techniques being by the intaglio process, helped to spread knowledge of new styles, inventions, and discoveries.

Johannes de Ketham [Kirchheim]

Fasciculus de medicinae. Venetiis, per Johannem & Gregorius fraters de Forlivio, 1491.

Ketham, Kirchheim, von Kircheim, or other variant spellings (d ca 1470), is believed to have been a German physician who practiced in Venice; however, that has never been established. In the mid-fifteenth century, Ketham collated a series of medieval anatomical and physiological treatises that included 12 illustrated medical tracts. Originally appearing in manuscript form, these treatises contain advice to physicians on the conduct of medical practice, uroscopy (or urinoscopy, the diagnosis and/or prognosis from observation of a patient's urine), venesection (blood letting), surgery, wound treatment, and midwifery. Each section of the volume includes a traditional diagram: urine glasses arranged in a circle, relating to the four humors (blood, yellow bile, black bile, and phlegm); a male figure marked with sites for bloodletting; parts of the body subject to wounds made by various weapons; and body parts governed by constellations of the Zodiac. Ketham's "medical letters" or "medical miscellany" was published in 1491 in Venice, after his death, and was a translation into Italian by Sebastiano Manillo of Ketham's *Fasciculus medicinae.* This translation contained illustrations fom new woodcuts that were superior to to earlier, Latin, manuscripts (see Plate, left). These illustrations represent the first anatomic illustrations in a printed book, and is based on traditional representation in medieval manuscripts, rather than on original dissection and observation.

The figures from earlier woodcuts represent the medieval prototype of a crouching pregnant woman with arms raised and legs akimbo depicts the opened thoracic and abdominal cavities, the latter of which contains the opened thick-walled uterus with a fetus at about 4 or 5 month's gestation with its hands in front of its face. The title above the figure reads, "Third Table on woman." On the various parts of the figure and plate are named the diseases to which the parts are subject, and in some cases the name of that part. For instance,

above the breasts are the notations "Cancer of the breasts" and "Tumor of the breasts." On the breasts reads "Lack of Milk." In the box below the figure is noted, "Certain signs of conception." "The first sign is… after coition [the woman] feels a cold and pains in the kidneys…. Also if the color of the face changes beyond the usual manner…. Also if she longs for certain foods such as earth or coals…. Furthermore, if you wish to know whether a male or female has been conceived, if the color of the face is slightly red, if the abdomen becomes swollen and round, and if thick, well done and digested milk comes from the breasts, and when placed on some clean and smooth surface, it does not separate but stays together, then the concept is male." The margins contains additional explanations.

The text includes a series of statements such as, "For the expulsion of the newly conceived embryo from a woman. Take castor oil and boil it in wine or in broth, drink in the morning and evening, and if it is a male embryo it will come out, but if it is a female you will in no way expel it". Another aphorism reads, "About the conception of the embryo: in the first month comes the coagulation of the blood, in the second the shaping of the body, in the third the connection of the body and soul, in the fourth it receives nails, in the fifth it receives the resemblance of the father or mother, in the sixth begins the contraction of the nerves, in the seventh the marrow is consolidated, in the eighth the bones and nerves are strengthened, in the ninth nature moves it, the infant is filled with the benefit of everything that comes from darkness to light". Others address exciting libido, the treatment of both the lack of menstruation and its excessive flow, rules for easy parturition, the treatment of sterility, tests of virginity, and growth of the fetus in the womb.

Several years later in 1493/4, an expanded text was translated into Italian by Sebastiano Manilio. This included four additional and more realistic woodcuts, one in color. These are scenes of medical practice in a setting of fifteenth-century Venice. One is of a physician at his desk, with a man, an old woman, and a boy waiting to consult him.

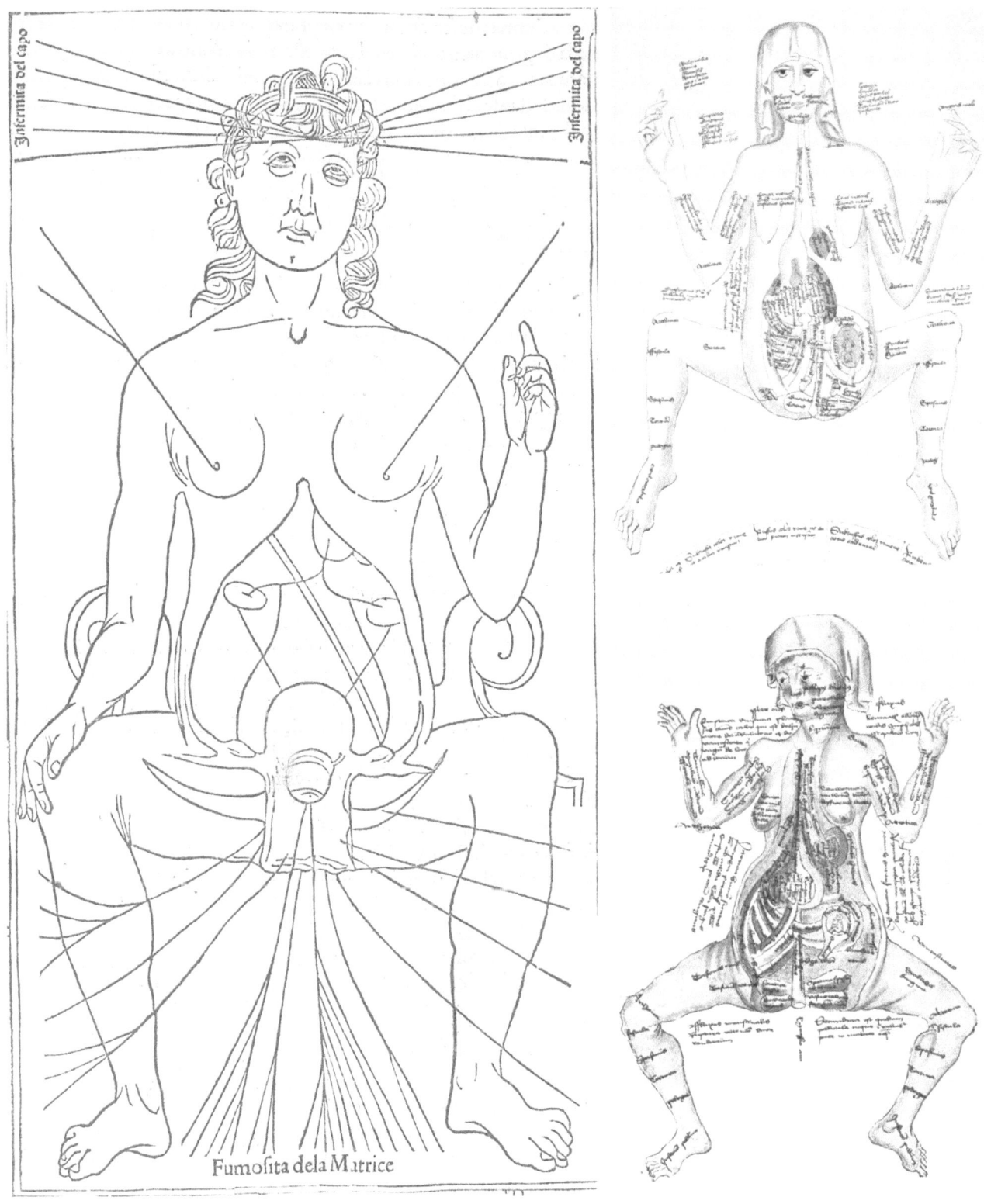

PLATES FROM KETHAM, 1491, AND EARLIER MANUSCRIPTS.
The left-hand figure is from the Italian translation in 1491. The other two figures are from earlier manuscripts,
as follows: *lower right,* figure of a pregnant woman, ca 1450, from Copenhagen Royal Library (ms Ny K1.Sm1.846);
upper right, ca 1470, the same from Munich City Library (Cod. germanicus 597 fo. 259)

Another shows a consultation of five long-robed doctors attended by two youths carrying urine glasses. To illustrate the treatise on the plague, there is a sickroom where a physician holds a sponge over his nose and mouth as he feels a plague patient's pulse. The final woodcut depicts an anatomy lesson. The professor lectures while the demonstrator points out to a small audience the organs exposed by the bare-armed dissector.

The text and formalized pictures provide some understanding of the haughty and rigid attitudes that must have prevailed during the fifteenth century among those who practiced medicine and surgery. Flasks containing urine, for instance, were not touched by physicians but only by their assistants. The possession and use of books, these pictures seem to show, set medical practitioners apart from common men, who patiently awaited attention while the physician consulted his texts. In latter editions, the illustrations are superior in quality to those of the original, and several are attributed to the School of Gentile Bellini (ca. 1429–1507). A number of other editions followed.

References

Garrison Morton 363; Choulant, 1962, pp 115–119; PMM 36.

Ketham, J. de. *Fasciculo di medicina*. Venetiis. *Johannes & Gregorius de Gregoriis de Forlivio*, 1493/4. (GM 363.1).

Ketham J. de. *The Fasiculus medicinae of Johannes de Ketham Alemanus. Facsimile of the First (Venetian) Edition of 1491 with Introduction by Karl Sudhoff. Translated and adapted by Charles [J.] Singer with XIII Plates*. Milian, R. Lier & Co. 1924.

Ketham, J. de. *The Fasciculo di Medicina. With an introduction by Charles Singer*. Florence, R. Lier & Co., 1925.

Ketham, J. de. *The Fasciculus medicinae of Johannes de Ketham Alemanus. Facsimilie of the First (Venetian) Edition of 1491, with English Translation by Luke Demaitre. Commentary by Karl Sudhoff. Translated and adapted by Charles Singer*. Birmingham, AL, The Classics of Medicine Library, 1988.

Ketham, Johannes de. *Fasiculo de medicina*. (Venice: Zuane & Gregorio di Gregorii, 1494). Historical Anatomies on the Web, U.S. National Library of Medicine, Bethesda, MD (http://www.nlm.nih.gov/exhibition/historicalanatomies/ketham_bio.html). Last accessed 04 July 2013.

Singer, C.J. *The evolution of anatomy. A short history of anatomical and physiological discovery to Harvey*. London, Kegan Paul, 1925 (GM 454).

Caius Suetonius Tranquillus

(Vita Caesarum) cum *Philippi Beroaldi et Marci Antonii Sabellici commentariis.* Cum *figures nuper additis.* Venetiis, per Ioannem Rubeum, 1506.

Probably the earliest illustration of cesarean section to appear in a printed book was not in a medical work, but rather in the *Lives of the Twelve Caesars*, a collective biography of the Roman Empires' first leaders by the scholar-historian Suetonius (ca.70–ca.150 CE). For many years it was popularly believed that Julius Caesar (100–44 BCE) was delivered by this means. In his *Natural History*, which contained all that was known at this time in biology, zoology, medicine, anthropology, and other natural sciences, Gaius Plinius Secundus (23–79), Pliny "the Elder", in speaking about the "signs" which accompany a child's birth wrote that the first of those to bear the surname of Caesar was so named because his mother's belly had been cut [Latin, *caesus*]. Pliny here intended to explain the origin of the family surname. According to him, it had been introduced by an unknown ancestor who had come into life through the cut belly of his mother. However, since antiquity the more convenient legend that Julius Caesar was born by section became widespread.

This is almost certain not to be true, as his mother, Aurelia Cotta (120–54 BCE), was alive when Caesar undertook the invasion of Britain, and no recorded cases of a woman surviving this operation appeared until the sixteenth century. Suetonius perpetuated the myth. The term Cesarean, however, may have derived from the royal law started during the reign of Numa Pompilius (715–672 BCE), a legendary king of Rome, that the child be removed from the womb of a woman who died late in pregnancy; a law that continued under the rule of the Caesars. Jacques Guillemeau (1550–1613), a pupil of Ambroise Paré (1510–1590) apparently was the first to use the term "section" in connection with the operation, which he held was unwarranted because of its excessive mortality (Guillemeau 1609).

The other Emperors about whom Suetonius wrote were Augustus (Gaius Octavius Augustus, 63 BCE-14 CE; Emperor, 44 BCE–14 CE) adopted son of Julius Caesar, and, following the Republic which had been destroyed under the dictatorship of Caesar, founder of the imperial Roman government; Tiberius (Claudius Nero Caesar, 42 BCE-37 CE; Emperor 14–37) adopted son of Augustus; Caligula (Gaius Caesar Germanicus, 12–41; Emperor, 37–41) who made Britain a province; Claudius (Tiberius Claudius Drusus Nero Germanicus, 10 BCE-54 CE; Emperor, 41–54); Nero (Claudius Caesar Drusus Germanicus, 37–68; Emperor, 54–68); Galba (Servius Sulpicius, 3 BCE 69 CE; Emperor, 68–69); Otho (Marcus Salvius, 32–69; Emperor, 69); Vitellius (Aulus Vitellius Germanicus, 15–69; Emperor, 69); Vespasian (Titus Flavius Sabinus Vespasianus, 9–79; Emperor, 69–79) who was responsible for reforms and stability of the Empire; Titus (Titus Flavius Sabinus Vespasianus, ca. 40–81; Emperor, 79–81) was the conqueror of Jerusalem in 70 CE and under whose rule the Empire prospered. Suetonius called him "the darling of the human race"; and Domitian (Titus Flavius Domitianus Augustus, 51–96; Emperor, 81–96). Rather than history per se, Suetonius' *Lives* abounds in anecdotes of which he portrays the characters of the Caesars, and presents insights into the customs and manner of the times. As noted by one writer, "The banquet hall and the bedchamber figure more largely in his narrative than do the forum and the camps. And, as the real man lurks inevitably beneath the purple which conceals him, this method gives us a series of indelible portraits... drawn... with the objective art of the skilled cartoonist who sharply differentiates individuals by stressing their most prominent characteristics" (Suetonius 1930, p. viii). A subsidiary theme of Suetonius' treatise is the issue of how one in authority deals with immense power.

During the reign of Vespasian, Suetonius was born in *Hippo Regius* in north Africa (now part of Algeria), the son of Suetonius Laetus, a military tribune and member of the Roman elite. Sent to Rome for his education, he was mentored by the senator-historian Pliny the Younger (ca. 61–ca.112). Under Emperor Trajan (Marcus Ulpius Traianus, 53–117; Emperor, 98-117), Suetonius was appointed *a bybliothecis* [head of the libraries] and *a studius* [a documentalist]. Under Emperor Hadrian (76–138), he served as *ab epistulis* [minister of letters], the Emperor's private secretary. In this position he had access to documents in the Imperial archives, which, with his social contacts, may have provided ample material for his work. The Roman historian and orator Publius Cornelius Tacitus (ca. 55-ca. 117) was a contemporary of Suetonius. An English edition of *The History of the Twelve Caesars* was published in 1677.

References

Guillemeau, J. *De l'heureux accouchement des femmes.* Paris, Nicolas Buon, 1609. (GM 6145.1).

Plinius Secundus. *Historia naturalis. Libri XXXVII.* Venice, Johann von Speier, 1469. (GM 89).

Pliny the Elder. *Natural history.* Translated by H. Rackham, W.H.S. Jones, and D.E. Eichholz in *The Loeb Classical Library.* 10 vols. Cambridge, Mass., Harvard University Press; London, W. Heinemann, 1947–1963.

Suetonius, T.C. *The history of the twelve Caesars, emperours of Rome. A very early translation by Andrew Marvell into English.* London, Printed by J.M. for John Starkey, 1677.

[Suetonius, T.C.] *Suetonius' Lives of the Twelve Caesars. Newly translated with an introduction by H.M. Bird. Illustrated by Frank C. Papé.* New York, Argus Books, 1930.

Suetonius-Tranquillus, C. *The historie of Twelve Caesars, Emperors of Rome... Newly translated... by Philemon Holland from the edition of 1606.* London, Frederick Etchells & Hugh Macdonald, 1931.

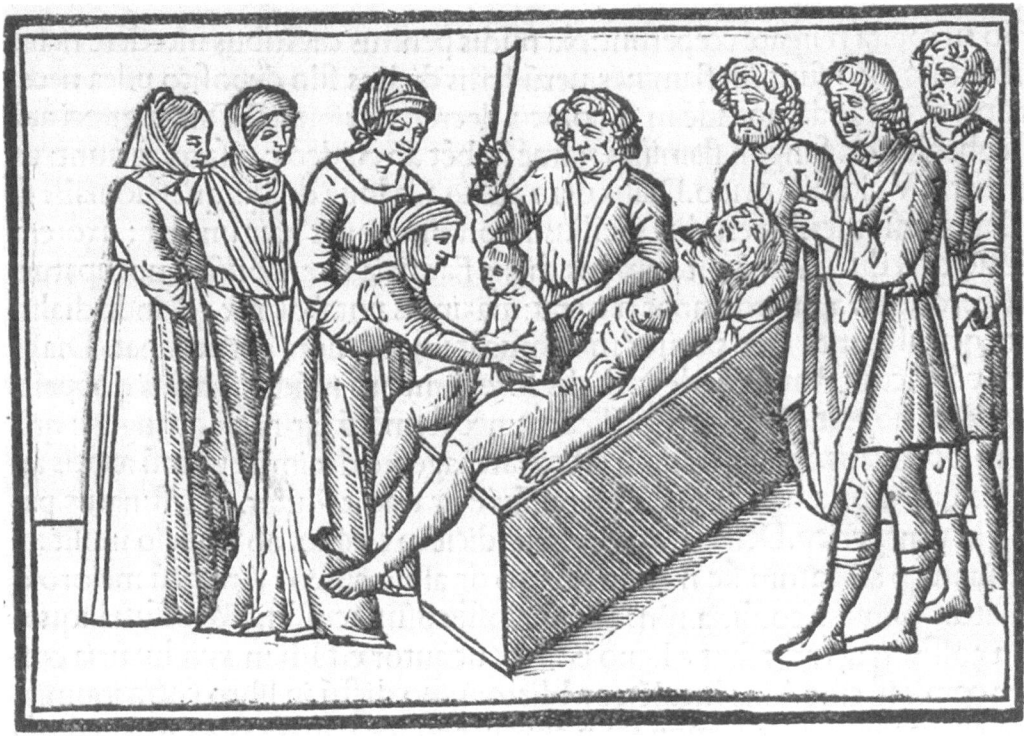

FIGURE FROM TRANQUILIUS, 1506

Leonardo da Vinci

Quaderni D' anatomia. Vol. III. Fogli della Royal Library di Windsor, Organi della Generazione, Embrione. C.L. Vangensten, A.M. Fonahn and H. Hopstock, editors. Christiania, J. Dybwad, 1913.

With the Renaissance, the revival of naturalism in the visual arts in Italy initiated a resurgence of the study of anatomy. Many artists, including Andrea del Verrochio (1435–1488), Michaelangelo Buonarroti (1475–1564), Raphael Raffaello [Sanzio] (1483–1520), and Albrecht Dürer (1471–1528) conducted dissections to increase their knowledge of human anatomy. However, da Vinci (1452–1519) alone is considered one of the greatest of biologists for his careful and quantitative approach to investigation, and reliance on his own dissection and observation. Following his painting Lisa Gherardini, the wife of Francesco del Giocondo, the *Mona Lisa* (1503–1506), Leonardo became an assistant to the anatomist Marcantonio Della Torre (1481–1511) professor of anatomy at the University of Pavia. With Della Torre, he worked on a large series of plates of human anatomy which they planned to publish. Included in these plates, is the first accurate illustration of the fetus in utero and other aspects of the gravid uterus and reproduction. He combined science with aesthetics, and for many of his anatomical studies the illustrations are depicted from several perspectives. In this regard, da Vinci was more than 300 years ahead of his time (Needham 1934).

In this drawing of the fetus in the breech position, Leonardo depicted the uterus with its blood vessels, as well as the fetal membranes and umbilical cord. Although he also depicts the single placental disc (upon which the fetus is sitting; the coiling umbilical cord can be seen just under the placental disc), he mistakenly depicted the placental attachment as cotyledonary (he clearly depicts the fetal cotyledons interdigitating with the maternal crypts in the upper right portion of drawing). In fact, this cotyledonary or "placento-mal" arrangement exists only in ruminants (antelope, cattle, deer, giraffes, goats, and sheep) (Ramsey 1982), which he also had studied (see Figure, next page). Leonardo observed, "Just as the fingers of the hand are interwoven, one in the interval of the other... so the fleshy villi of these little sponges [cotyledons] are interwoven like burrs, one half with the other". In regards to the fetus, da Vinci noted, in his char-

acteristic left-handed, mirror writing, "See how the great vessels of the mother pass into the uterus...." Incorrectly maintaining that the fetal heart did not beat, he also stated that the fetus did not breathe, in view of the fact that, being immersed in water, it would drown if it did so, and further that breathing was unnecessary as the fetus was, "...vivified and nourished by the life and food of the mother.

This food nourishes this creature not otherwise than it does with other members of the mother, that is, the hands, feet and other members." Presciently, Leonardo observed, "the vessels of the infant do not ramify in the substance of the uterus of its mother, but in the secundines [placenta], which takes the place of a shirt in the interior of the uterus, which it coats, and which it is connected (but not united) by means of the cotyledons" (O'Malley and Saunders 1952, pp. 472–476). As noted by Joseph Needham (1900–1995), in one sentence da Vinci states the error that the human placenta is cotyledonous, while correctly affirming that the fetal circulation is not continuous with that of the mother (Needham 1934, p. 78).

The first to make quantitative measurements of the infant, da Vinci also recorded that the child "grows daily far more within the body of its mother than when it is outside of the body... [and] in the first 9 months, it does not double the size of the 9 months when it found itself within the mother's body. Nor in 18 months has it doubled the size it was at 9 months... and thus in every 9 months diminishing the quantity of such increase till it has come to its greatest height" (Needham 1934, pp. 79–80). Leonardo's illustrations are believed to have been drawn from about 1510–1512. However, because Della Torre died of the plague, they remained unpublished and had no impact on contemporary medical thought. Not until the late 1770s, were da Vinci's manuscripts and drawings discovered and exhibited at Windsor Castle.

In addition to being one of the world's greatest artists, as a polymath Leonardo was a botanist, architect, city planner and inventor of mechanical devices and engineering marvels. One writer observed, "Leonardo arrives at the conclusion that there is but one natural law which governs the world, Necessity. Necessity is nature's master and guardian, it is Necessity that makes the eternal laws" (Hopstock 1921). Among the wisdom attributed to Leonardo is his saying, "I have been impressed by the urgency of doing. Knowing is not enough; we must apply. Being willing is not enough; we must do."

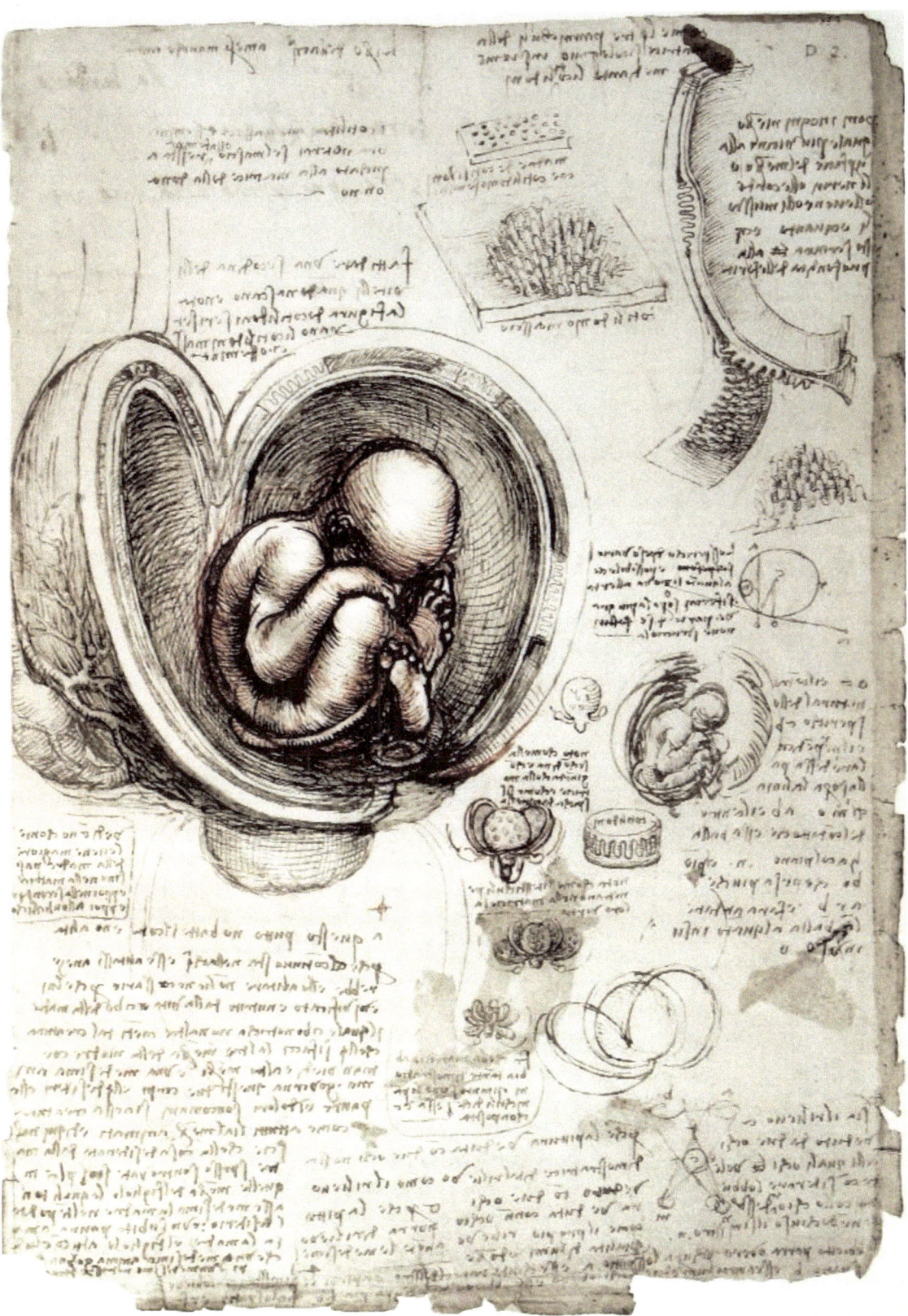

Figure from da Vinci, 1510–1512

References

Choulant-Frank, pp 99–105; Garrison Morton 365; Osler 517.

Buhrer, E.M. (Ed). *The unknown Leonardo*. London, Hutchison and Co., 1974.

Hopstock, H. Leonardo as an anatomist. In: *Studies in the history of medicine*. C. Singer (Ed). Oxford, Clarendon Press, 1921, pp. 153–191.

Keele, K.D. et al. Leonardo da Vinci. In: *Dictionary of Scientific Biography*. Vol VIII. C.C. Gillispie (Ed). New York, Charles Scribner's Sons, 1973, pp. 192–245.

MacCurdy, E. *The notebooks of Leonardo da Vinci*. Vols 1–3. London, Reprint Society, 1954.

Needham, J. *A history of embryology*. Cambridge, Cambridge University Press, 1934, pp. 77–81.

O'Malley, C.D. and J.B. de C.M. Saunders. *Leonardo da Vinci on the human body, the anatomical, physiological, and embryological drawings of Leonardo da Vinci; with translation, emendations, and a biographical introduction*. New York, Henry Schuman, 1952.

Ramsey, E.M. *The placenta: human and animal*. New York, Praeger, 1982.

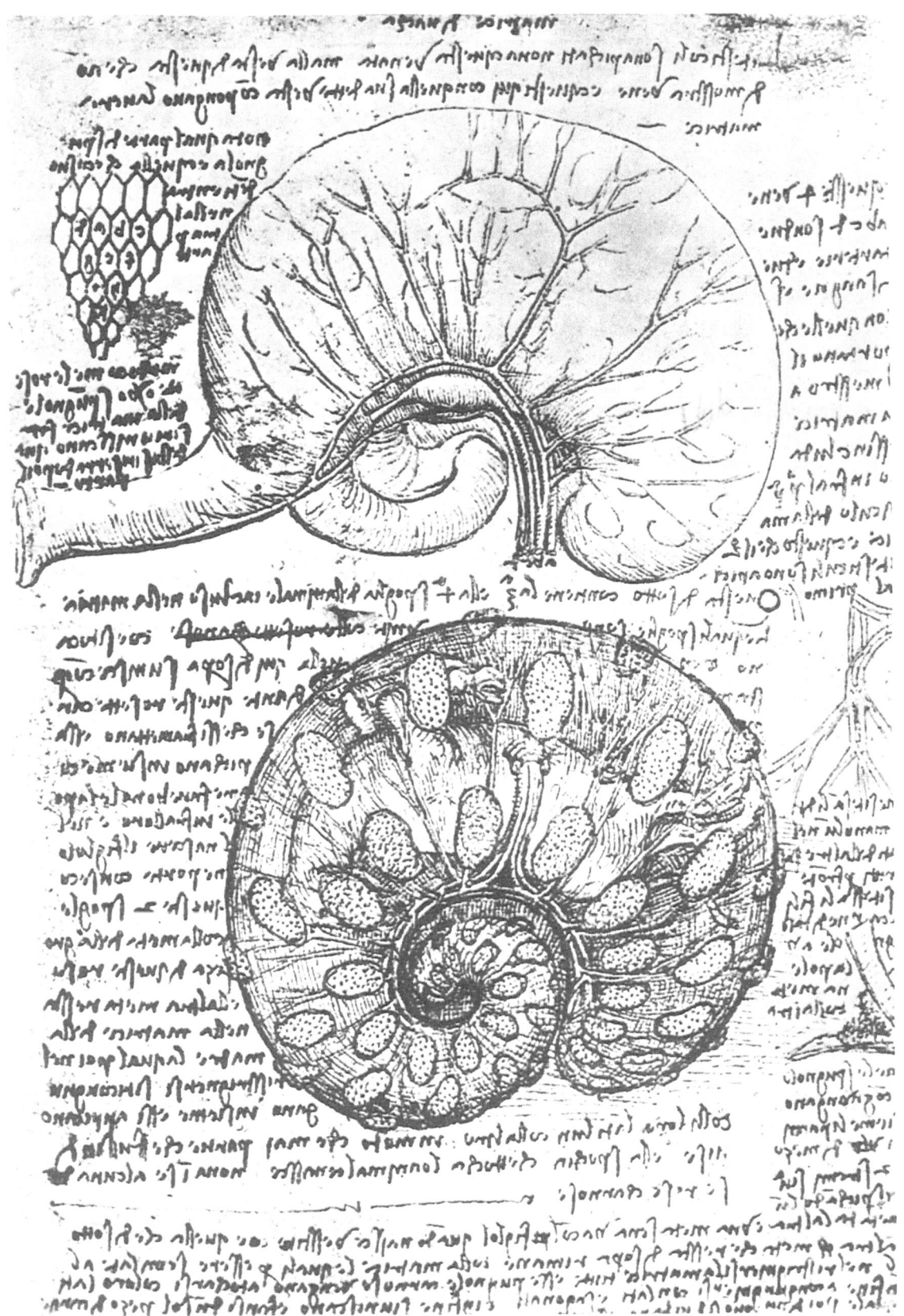

FIGURE FROM DAVINCI, 1510–1512

Eucharius Rösslin

Der schwangern Frauwen und Hebammen Roszgarten. Strassburg, Martinus Flach, Junior, 1513.

Rösslin (Röslin, Roesslin] (ca. 1470–1526), the *Stadtartz,* town physician and supervisor of midwives in Worms, was the first to publish an illustrated obstetrical manual for the use of midwives. He wrote in the vernacular rather than in Latin, to aid in the instruction of midwives and practitioners. With the exception of the *Fraunbuchlein* of Ortolff von Bayerlant (or Bayerland; ca. 1400) a volume of only 13 pages published about 1495, this is the first printed book dealing exclusively with obstetrics apart from medicine and surgery. In this "A Garden of Roses for Pregnant Women and Midwives", a lovely dedicatory woodcut (signed with the initials M C, attributed to the Frankfurt artist Martin Kaldenbach) shows the author presenting the book to Countess Katherine (ca 1465–1526), Duchess of Brunswick and Lüneburg, Saxony, who is believed to have encouraged Rösslin to produce this work and to whom the volume was dedicated.

This work of 114 pages contains little that is original. Rather, Rösslin referred at length to the medical authors of antiquity including Hippocrates (460–375 B.C.E.), Soranus of Ephesus (98–138), Galen (130–200), Rhazes (ca. 854–930), Avicenna (980–1037), Albertus Magnus (1193–1280), and others. In his preface in verse, Rösslin inveighed against contemporary obstetrical practice, and the ignorance, carelessness, and superstition of the midwives, who were responsible for the unnecessary deaths of many newborns. Infant mortality Rösslin labeled murder, for which the guilty ones deserved to be buried alive, or "broken on the wheel," rather than receiving an honorarium for their services. In its 12 chapters Rösslin's instructions are simple and direct. He discussed both natural and unnatural births, considered difficult delivery including podalic version, and described maternal positions for delivery and methods of assisting at delivery. In addition, he included instructions on the care of the infant from birth until weaning, and outlined treatment of diseases of infancy.

Twenty woodcut illustrations, for the most part derived from those found in manuscripts of works on gynecology by Soranus of Ephesus and Moschion [Muscio, Mustio] (fl circa 500). The small illustrations of the *fetus* in utero apparently were copies from the *Codex Palatina,* a thirteenth century manuscript of Moschion. In Rösslin's work the 16 illustrations were increased to 17 by adding figures of conjoined twins. It also showed for the first time, printed figures of the birth chair

and the lying-in chamber. The legend on the plate shown of a parturient on the birth stool with two attendants reads

> "Come to my aide in the time of a difficult agonizing birth which is marked by great anxieties, worries, and distress. As this has been reported in chapter after chapter, so it must be acknowledged so here it is written."

In rather stylized cartoons, Rösslin illustrated many fetal positions including those of normal and Siamese twins. For the Siamese twin figures shown, Rösslin states in discussing various causes of difficult deliveries, "it is because there is more than one infant/or …[it] has more limbs than is natural/especially two heads as for example in this .xii. year in the County of Werdenberg/a baby was born with two heads/of which a figure is here drawn/…." Regarding the other two figures, he wrote, "…if the infant appears with both hands/then the midwife should take hold of both shoulders…/and lift the child back again/And…extend the babies hands up next to its sides/and then take hold of the head/and help it out. Furthermore if the infant presents its behind/Then the midwife should put in her hand and lift the infant up/and lead it out by the feet. However where possible if she can shove the baby/so it comes with its head down/this would be much better…." Some of the plates are of high artistic merit and were copied by subsequent writers including Ambroise Paré (1549) and Jacob Rueff (1554).

This first edition of the *Roszgarten* was followed by two other undated editions, one printed by Henricus Gran of Hagenauf; and one by Arnst von Aich of Cologne, which have been dated 1515 and about 1518, respectively. The first Latin edition "*De partu hominis* …" was published in 1532 by Rösslin's son (also named Eucharius) who succeeded his father as town physician of Frankfurt am Main. In accordance with the contemporary custom, he used the Greek form of his name, and styled himself "Rhodion" [Roesschen or Little Rose]. The first English edition "*The Byrth of mankynde, otherwise named the womans booke*" translated by Richard Jonas was published in London in 1540. Rösslin's work became enormously popular, and went through over 100 editions including translations. For nearly two centuries it served as the authoritative treatise on obstetrics throughout Europe. Probably no medical book in history has been so widely translated and distributed.

Eucharius Rösslin became apothecary of Freiburg in 1493. In 1506, he was elected physician to the City of Frankfurt am Main, and in 1508 entered the service at the court of Katherine. In 1517 he returned to Frankfurt as the town physician and supervisor of midwives, serving in that post until his death in 1526.

Der Frauwen

Jll mã zů hilff kõmē in schwårer myßlicher har//
w   ter geburt/die mit grossen sorgen /angstē vñ nõtē
     beschicht/wie dañ da von gemeldet ist/ in .yviij.
stuckē nacheinander/So můß man merckē solichs so hie
nach geschriben stadt.

Figure of "Birth Chair" from Rösslin, 1513

References

Garrison Morton 6138 incorrectly cites the Hagenau edition, which was dated by Benzing as 1515; Cutter & Viets, p 6, 178; Durling, 3893; Hellman 1; Norman, 33.

Aveling, J.H. Account of the Earliest English work on Midwifery and the Diseases of Women. *Obstet J Gr Brit & Ireland* 2:73–83, 1874.

Ballantyne, J.W. The "Byrth of Mankynde" (its author and editions). *J. Obstet Gynaecol Brit Emp.* 10:297–325, 1906.

Ballantyne, J.W. The "Byrth of Mankynde" (its contents) *J Obstet Gynaecol Brit Emp* 12:175–194, and 255–274, 1907.

Ballantyne, J.W. Further copies of Jonas' and Raynalde's "Birth of Mankynde" (with illustrations). *J. Obstet Gynaecol Brit Emp.* 17:329–332, 1910.

Crummer, L. The copper plates in Raynalde and Germinus. *Proc Roy Soc Med* 20:53–56, 1927.

Ingerslev, E. Rösslin's "Rosengarten": Its relation to the past (the Muscio manuscripts and Soranos), particularly with regard to podalic version. *J Obstet Gynaec Brit Emp* 15:1–12, 73–92, 1909.

Ortolff von Bayerland. *Buchlein der schwangern Frawen.* [Augsburg, anonymous printer, circa 1495].

Pare'. A. *Briefve collection de láadministration anatomique: avec la mainere de conjoindres les os…* Paris, G. Cavellat, 1549 (GM 6140).

Power, Sir D'Arcy. *The Birth of Mankind or the Woman's Book. A Bibliographical Study. The Library 4th Ser* 8:1–37, London, The Bibliographical Society, 1927.

Power, Sir D'Arcy. *The Foundations of Medical History.* Baltimore, Williams & Wilkins, 1931, pp 123–146.

Rösslin, E. *Der swangern Frawen und Hebammen Roszgarten.* [Hagenan], Heinrich Gran [1513 or 1515]. (Hellman 2, 3).

Rösslin, E. *Der swangern Frawen und Hebammen Roszgarten.* [Colone, Arnt von Aich, circa 1518].

Rösslin, E. *De partu hominis, et quae circa ipsum accident.* Medici, Fran. Chri, Egen, Francofurti. Xix., Octobris, 1532.

Rösslin, E. The byrth of mankynde, translated out of the Latin into Englysshe…London, T.R., 1540. (Hellman 19).

Rösslin, E. *The birth of mankynde, otherwyse named the womens booke. Newly set fourth, corrected and augmented…by Thomas Raynalde Phisitan.* London, R. Jugge, 1565. (Hellman 21).

Rösslin, E. When midwifery became the male physician's province. The sixteenth century handbook *The Rose Garden for Pregnant Women and Midwives*, Newly Englished. Translated from the German and with an introduction by Wendy Arons. Jefferson, NC, McFarland & Co., 1994.

Rueff, J. *Ein schön lustig Trostbüchle von den Empfengkhussen und Geburten der Mehschen…*Tiguri, Apud Frosch [overum], 1554. (GM 6141).

Rosegarten

¶ Wo aber dz kind erscheynt vñ kompt mit vnnatürlicher geburt/mit bedē füessen/ vnd seint die hend vnnd arm/neben den beinen hinab gestreckt (als dise figur anzeuge ist)so soll die hebam die arm vñ hēd des kindes/schicklichē wysen/ fügē vñ schybē/mitt salbē vñ andern dingē die glat machē/ Also dz die hend vñ arm des kindes/gestreckt bleibē neben des kinds seitē vndersich hinab an die dicke der bein/ Vñd darnach sol sie im von stadt helffen. Wo aber es möglich wer/dz die hebam die füeß des kindes senfftiklichē vñ subtri-lichē vbersich wyse/also dz iñ wēdig in mūter leib/die solen des kindes füeßlin/geschyben wurdē/gegē d mūter nabel/ vñ sein heuptlin/gegē seiner mūter ruckē/vñdsich gegē dē vßgang gestürtzt vñ gewen-der/wer vyl bösser.

¶ Wo aber dz kind erscheint mit beyden füessen/ vnd hatt die hend nit neben im/vndersich hinab gestreckt/ als oben stadt/sonder vbersich/

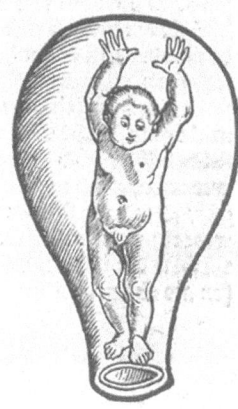

Der frawen

¶ Wo aber die zwy-ling kōmen/mitt den füessen/ Sol sie aber-mals thūn fleiß anke-ren/eins nach dē an-dern vßfüren/in mas-sen als obstadt.

¶ So aber der zwyling ei-ner kompt mit dē haupt/ der ander mitt den füesse Sol abermals die heb-am fleiß ankerē/dem nech sten zū erstē helffen/vnnd dz ander nit vlassen/Vñ das soll also geschehen/on quetzung ir beyder.

FIGURES OF THE "FETUS IN UTERO" FROM RÖSSLIN, 1513

Andreas Vesalius

De humani corporis fabrica libri septum. Basel, Ioannis Oporini [Johannes Oporinus], 1543.

By basing his epic work *De humani corporis fabrica...* [The structure of the human body in seven books] on humans, rather than animals, Vesalius (1514–1564) revolutionized the understanding of anatomy. This encyclopedic opus was a milestone in that its "books", one for each major system (bones, muscles, arteries, spinal cord and nerves, abdomen, thorax, head and brain), were based on his own dissections and observations, Vesalius gave a more complete and accurate depiction of human anatomy than any of his predecessors.

The woodblock prints are the most famous anatomical illustrations of all time. Vesalius explained the dissections, and insisted that examination of the body must be performed by the physician himself, rather than by a diener or assistant. By undermining the religious-like reverence for authority in science, Vesalius prepared the way for independent observations in anatomy and clinical medicine. Thus, the *Fabrica* is one of the most seminal volumes in the development of modern medicine and science.

In Book 1, Vesalius accurately described the human pelvis for the first time. He also demonstrated that it was impossible for the pelvic bones to separate during labor, as was commonly believed at that time. In Book V, he first carefully described the uterus, and confirmed Giacomo Berengario da Carpi's (1460–1530) observation that it contained a single chamber rather than two uterine horns as in most mammals. Vesalius also described the insertion of the tubes, the ligaments, and the vasculature. In this depiction of the human female genital tract, he clearly recognized the rich vascular supply of the uterus. Vesalius upheld the Galenic doctrine of anastomosis between maternal and fetal vessels. Also mistakenly in this original edition, he illustrated the placenta as being zonary or girdle-like, as in the dog (see figure on middle left). He corrected this, however, in the second edition (Vesalius 1555) (see figure on lower left). In his compendium on anatomy, Matteo Renaldo Colombo [Columbus] (ca. 1510–1559), a pupil of Vesalius and his successor as professor of the University of Padua, first used the term *placenta* (Colombo 1559).

The illustrated female organs of reproduction show the uterus as bifid and split longitudinally, from a parous woman who had been hanged. The legends may be translated: "A, A,

B, B—Sinuses of the fundus uteri. C, D—A line, somewhat like a suture, projecting slightly into the fundus uteri. E, E—The thickness of the inner and proper tunic of the fundus uteri. F, F—A portion of the inner fundus uteri; projecting downward, from its surface. G, G—Orifice of the fundus uteri. H, H—Second and external covering of the fundus uteri reflected from the peritoneum. I, I et cetera—By this we indicate the membranes on both sides which are reflected from the peritoneum and contain the uterus. K—The substance of the cervix uteri. L—A part of the neck of the bladder." (The urethra is incorrectly shown opening into the vagina).

A native of Brussels, Vesalius was educated at Louvain, and studied medicine at Montpellier and Paris. He then returned to Louvain to teach anatomy, after which he served briefly as army surgeon to Charles V (1500–1558). Vesalius then became public prosector of anatomy at the University of Padua. When he published the *Fabrica,* he was 29 years old. The illustrations have been attributed to Titian's pupils Jan Stephan van Calcar [Kalkar] (ca. 1499–ca. 1550) and Domenico Campagnola (1500–1564), although this is a subject of some debate. A second and more complete edition was published in 1555. As a consequence of this work, many of the erroneous doctrines of Galen [Claudius Galenus (130–200)] and medieval writers were overturned. This shattering of the ostensibly infallible truths of Galenic doctrine provoked bitter controversy.

In part as a consequence of publishing this volume, Vesalius abruptly left his post at Padua and became court physician to the Holy Roman Emperor Charles V (1500–1558, Emperor 1519–1556), and later to his son, Phillip II (1527–1598; Monarch 1556–1598). Several years later, in 1564 Vesalius left for a trip to Jerusalem in the Holy Land. On the return voyage from this pilgrimage he died on the Greek island of Zákynthos. Lost for several centuries, the original woodblocks were discovered in the early 1930s in the basement of the medical library of the University of Munich. In collaboration with the New York Academy of Medicine, a special edition of these woodblock illustrations was published in 1934. A decade later, these woodblocks were destroyed during the fire bombing of Munich during the latter part of World War II (1939–1945).

As William Osler (1849–1919) observed, "The *Fabrica...* is the first modern work of a creator....to have completed before his 29th year of a task of this magnitude, at such a period when dissections were difficult to make and authority difficult to resist, is a feat to which the literary history of the profession offers no parallel...[The] *Fabrica* is one of the great books of the world" (Osler, 1929, p. 58).

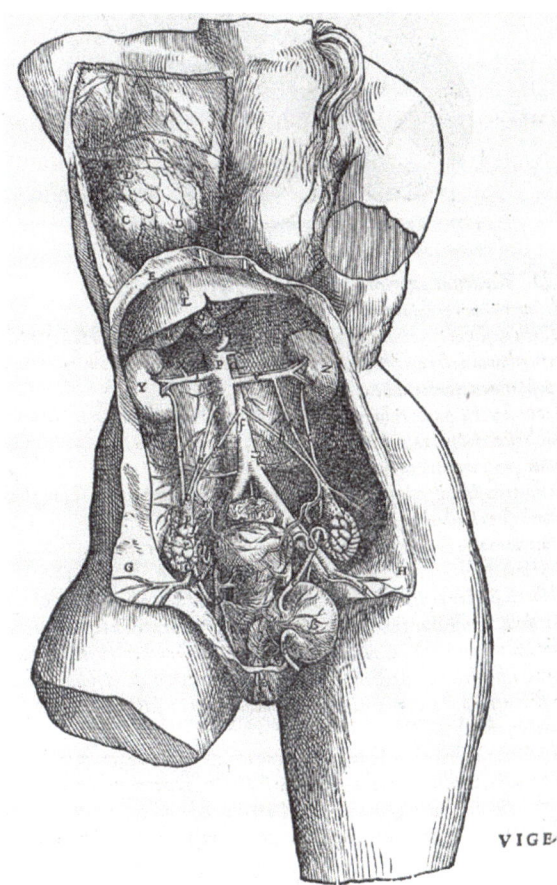

VIGE·

TRIGESIMA QVINTI LIBRI FIGVRA,
QVATVOR PECVLIARIBVS COMPLEXA TABVLIS.

PRIMA.   SECVNDA.   TERTIA.   QVARTA.

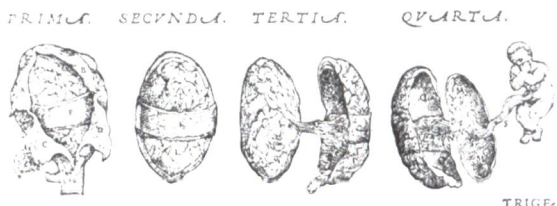

TRIGE·

TRIGESIMA QVINTI LIBRI FIGV-
RA, QVATVOR PECVLIARIBVS COMPLEXA TABVLIS.

PRIMA TRI-
GESIMÆ FI-
gura tabella.   SECUNDA
                          TABELLA.

TERTIA
TABELLA.   QUARTA
                          TABELLA.

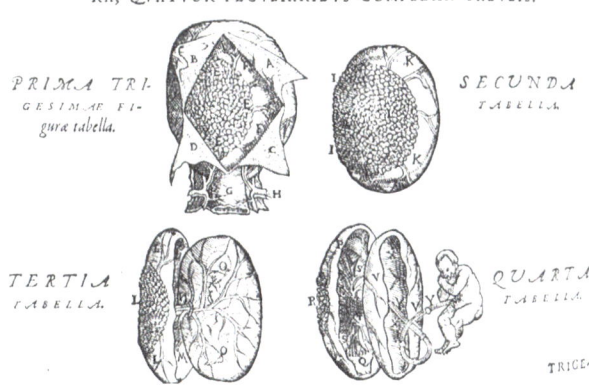

TRICE·

FIGURES FROM VESALIUS, 1543.
The *top plate* shows the female reproductive and urinary systems (Courtesy, U.S. National Library of Medicine). The *middle plate*, from the original edition shows the placenta as zonary (as in carnivores), but the *bottom plate*, from the second, 1555, edition, correctly shows the placenta as discoid

References

Choulant-Frank, pp 168–199; Durling 4577; Garrison Morton 375; Norman 18; Osler 567; Russell, p 830 ff.

Colombo, M.R. *De re anatomica libri XV.* Venetiis, Ex typographia Nicolai Bevilacquae, 1559. (GM 378.1)

Cushing, H. *A bio-bibliography of Andreas Vesalius.* Hamden, Conn., Archon Books, 1962.

Mossman, H.W. Comparative anatomy. In: *Biology of the Uterus.* 2nd Ed. R.M. Wynn and W.P. Jollie (Eds). New York, Plenum, 1989.

O'Malley, C.D. *Andreas Vesalius of Brussels, 1514–1564.* Berkeley, University of California Press, 1964.

O'Malley, C.D. Andreas Vesalius. In: *Dictionary of Scientific Biography.* Vol XIV. Charles Coulston Gillispie (Ed). New York, Charles Scribner's Sons, 1976, pp. 3–12.

Osler, W. *Bibliotheca Osleriana: A Catalogue of Books Illustrating the Histry of Medicine and Sciences.* Oxford: Clarendon Press, 1929.

Vesalius A. *De humani corporis fabrica libri septem.* Basel, Ioannis Oporini, 1555. (GM 377).

Vesalius, A. *Icones Anatomicae.* [New York, McFarlane], Library of the New York Academy of Medicine, 1934.

Charles Estienne [Étienne; Carlos Stephanus]

De dissectione partium corporis humani libri tres.…Parisiis, Apud S. Colinaeum, 1545.

Estienne (ca. 1504–1564), a graduate of medicine of the University of Paris, and later lecturer in anatomy in the Faculty of Medicine, was son of Henri Estienne (ca. 1460–1520), founder of a prominent family of printers. He studied anatomy under Jacques Dubois-Sylvius (1478–1555), as did Andreas Vesalius (1514–1564). In his "Concerning dissection of parts of the human body in three books", many of the approximately 60 full-page woodcuts, superb examples of Mannerist art, are signed by the artist Jollat—Jean "Mercure" Jollat (fl. 1530–1545) (or with his sign of Mercury), and the surgeon artist Estienne de la Rivière (d. 1569). The figures of women are believed to have been derived from those of Giovanni Jacopo Caraglio (1505–1565) and prepared from drawings by Perino del Vaga (ca. 1500–1547) and by Giovanni Battista di Jacopo Rosso (known as Rosso Fiorentino or "Il Rosso"; ca. 1495–1540) (see Kellett 1955, 1957).

Although Estienne commenced this work in about 1530 with de la Rivière, because of a dispute over priority and authorship, and with a lawsuit brought by the latter against Estienne, its publication was delayed for a decade and a half. Thus, rather than having been published in 1539 as planned, it was eclipsed in 1543 by the *Fabrica…* of Vesalius, and was finally published 2 years later. Following the father's death, his mother married Simon de Colines (ca. 1480–1546) who then ran the family business until 1526, when Charles' brother Robert (ca. 1503–1559) assumed leadership. Colines then established his own printing house, and published this volume. The woodcuts are examples of the expressive style called early or Florentine Mannerism. As Choulant noted about the plates, which generally represent the whole body with a great many elaborations, "the rendering of the actual anatomic position is small and indistinct".

Importantly, Estienne challenged and corrected many of the teachings of Galen (130–200). In the introduction he states that, "one should not believe in books on anatomy but far more in one's own eyes". Estienne also observed, "that of all things in the universe from which the power and the workmanship of God could be contemplated, the human body was the best so far…." Under Robert's leadership, the family firm became printer for the King, Francis I (1494–

1547). Then, because of his controversial religious publications, being accused of "Protestantizing", and attacks by opponents of the King, in 1550 Robert sought asylum in Geneva. Following Robert's departure, Charles then headed the family press in Paris. Nonetheless, despite publishing seminal works in medicine, agriculture and other subjects (Estienne 1554, 1556), the business failed. Because of bankruptcy and heavy indebtedness, in 1561 Charles was incarcerated in the debtor's prison, *le Châtelet* [small castle], where he died. Estienne's son Henri (1531–1598) published a specialized lexicographical Greek-Latin diction-ary in which he defined a number of anatomical terms on which he had been working with his father before the latter's death (Estienne 1564). This work had considerable influence on contemporary anatomical terminology.

References

Choulant-Frank, pp 152–155; Durling 1391; Garrison Morton 378; Norman 728; Osler 2541.

Estienne, C. *De Latinis et Graecis nominibus arborum, fruticum, herbarum, piscium, & avium liber.*… Lutetiae, Ex officina Roberti Stephani,1 544. (Durling 1393).

Estienne, C. *Dictionarium historicum ac poeticum.* Lutetiae, cura ac diligentia Caroli Stephani, 1553.

Estienne, C. *L'agriculture et maison rustique.*… Paris, ____1554.

Estienne, C. *Thesaurus M. Tulii Ciceronis.* Paris, Apud C. Stephanum, 1557.

Estienne, H. *Dictionarium medicum.*… [Genevae], Henricus Stephanus, 1564. (Durling 1402; GM 6791).

Herrlinger, R. History of Medical Illustration. From antiquity to 1600. New York, Editions Medicina Rara, 1970.

Kellett, C.E. Perino del Vaga et les illustrations pour l'anatomie d'Estienne. *Aesculape* 37: 74–89, 1955.

Kellett, C.E. A note on Rosso and the illustrations to Charles Estienne's *De Dissectione. J Hist Med Allied Sci* 12:325–336, 1957.

Kellet, C.E. Two anatomies. *Med Hist* 8:342–353, 1964.

Newsom, B. Estienne's *De dissectione. J S C Med Assoc* 91:318, 1995.

Ruth, G. Charles Estienne: Contemporary of Vesalius. *Med Hist* 8:354–359, 1964.

Tubbs, R.S. and E.G. Salter. Charles Estienne (Carlos Stephanus) (ca. 1504–1564): Physician and anatomist. *Clin Anat* 19:4–7, 2006.

FIGURE FROM ESTIENNE, 1545.
Shown in the figure are the female internal genitalia

Jacob Rueff

Ein schön lustig Trostbüchle von dem Empfengknussen und Geburten der Menschen. Tiguri, Apud Frosch [overum], 1554.

Jacob Rueff [Rüff, Ruoff] (1500–1558), the city physician of Zurich, was responsible for instruction and examination of midwives of the Canton. He improved upon Rösslin's manual of 1513, in part by stressing the importance of knowledge of the anatomy of the female pelvis and genatalia and reproductive organs. His book for midwives and pregnant women in the German vernacular, "… *a comforting booklet of encouragement concerning the conception and birth of man, and its frequent accidents and hindrances, et cetera,"* appeared the same year as the work in Latin, *De conceptu et generatione…* (Rueff 1554). The illustrations, many derived from Rösslin (1513) and Vesalius (1543), were the first in an obstetric book to be based on anatomic reality, rather than showing diagrammatic figures in a bottle or balloon.

In the Preface, Rueff dedicated the work "To all grave and modest matrons, especially to such as have to doe with women in that great danger of childe-birth, as also, to all young practitioners in Physick and Chirurgery, whom these matters may concerne…my labours I bequeath…. And whose helpe upon occasion of extreame necessity may be useful and good, both for mother, child, and midwife". (Translation from *The Expert Midwife* (Rueff 1637)). The text comprised six sections, or "books". The first deals with the physiology of impregnation and conception, and the development and nutrition of the fetus. The second describes and illustrates the uterus and the condition of the fetus within it, and includes a chapter of precepts for pregnant women. Book three discusses parturition, with rules and medicaments for alleviating delay and difficult birth, and for the care of the mother and infant. Rueff portrayed the birth stool with drapery, rather than wooden boards, on the lower portion "So that the child will not be injured and so that… women assisting the midwife can insert their hands." It includes a brief chapter on obstetric instruments, such as the speculum and both smooth and toothed forceps for extraction of a dead fetus. Book four contains 15 short chapters, each of which describe the management of various forms of unusual presentations, and of twins. These are illustrated by traditional birth figures in which the fetuses are shown as grown children, although the artist has added more anatomic detail than that of earlier drawings. Rueff discussed cephalic version by combined external and internal version, and described manual delivery of the placenta. Book five discusses false conception, tumors of the uterus, abortion and its treatment, and the signs of conception. Book six outlines some of the causes of sterility and describes the principal diseases of the uterus, again offering prescriptions and appropriate remedies.

Rueff illustrated the concepts of Aristotle (384–322 BCE) regarding development. Shown in this volume are a) the intrauterine coagulum of semen and menstrual blood; b) this developing mass surrounded by the fetal membranes; c) the early brain, heart, and liver, shown by "three tiny white points, not unlike coagulated milk"; d) four major blood vessels branching from the heart; e) beginning outline of the cranium; f) major blood vessels outlining the fetal body; g) a near-term fetus. Rueff is stated to have besought the Burgomaster to present copies to all midwives and nurses in the Canton, and they were obliged to have an appropriate section read aloud "by a well-read woman," if possible, during the first stage of labour at any confinement they attended.

Although Rueff believed strongly in astrologic influences on pregnancy, particular in the development of monsters, his book, with Rösslin's, had a great influence on improving obstetric care. Some of the illustrations depict contemporary ideas about embryology in mammals. Not only an outstanding obstetrician and surgeon, Rueff wrote on astronomy, history, drama, and poetry. Following Rueff's death, the Frankfurt publisher Sigmund Feyerabend engaged the celebrated Swiss artist Jost Amman (1539–1591) to illustrate a second edition, published in 1580. The remarkable and dramatic Amman woodcuts, depict many different aspects of childbirth, from engaging the midwife, to actual delivery of the child, to husband and wife celebrating their good fortune. The section on monstrosities is illustrated by over 30 woodcuts. Two scenes show a pregnant noblewoman and a woman giving birth with the midwife in attendance. Shown here (next page) is the parturient on a birth tool. Two female attendants are helping to support her upper body, while a midwife performs the delivery. In the background to the right an astrologer, using the positions of the stars and planets as his guide, foretells the child's future. The right foreground illustrates the pitcher of warm water and tub for the infant's bath. There also are 26 woodcuts of the uterus and developing foetus, five of obstetrical instruments, and three of abdominal organs. There are also 31 of human deformities and monsters. With Amman's fine woodcuts, this is ranked as one of the most famous illustrated medical works of the sixteenth century. The work was widely translated. In the first English edition *The expert midwife…,* the title page announces that the book had been "translated into English for the general good and benefit of this Nation" (Rueff 1637).

LIBER

Fœtus. tus & dici & esse incipit, quod ueteres quidam his etiã
carminibus complexi sunt.
Sex in lacte dies, ter sunt in
sanguine trini.
Bisseni carnẽ, terseni mem-
bra figurant.
Et aliter.
Iniectum semen, sex primis
certe diebus
Est quasi lac; reliquisᶜᵇ no-
uem sit sanguis; at inde
Cõsolidat duodena dies; bis
nona deinceps
Effigiat; tempusᶜᵇ sequens
producit ad ortum.
Talis enim prædicto tempore figura constt.

FIGURE FROM RUEFF, 1554

References

Garrison Morton 463, 6141; Cutter & Viets, p 44, 188–190; Durling 3977.

Dunn, P.M. Jacob Rueff (1500–1558) of Zurich and *The expert midwife. Arch Dis Child Fetal Neonatal Ed* 85: F222–F224, 2001.

Rösslin, E. *Der schwangern Frauwen und Hebammen Roszgarten.* Strassburg, Martin Flach, Junior, 1513.

Rueff J. *De conceptu et generatione hominis et iis quae circa haec potissimum consyderantur.* Zurich, Christopher Froschoverus, 1554. (Hellman, 29).

Rueff J. *De conceptu et generatione hominis*: de matrice et eius partibus.... Frankfurt: [Sigismund Feyerabend] 1580. (Hellman 30).

Rueff, J. *Hebammen Buch, Darus man alle Heimligkeit das Weiblichen Geschlects erlehren, welcherley gestalt der Meusch in Mutter Leib empfangen...* Frankfurt am Main, (Sigmund Feyerabend), 1580. (Osler 3848 is 1600 ed).

Rueff J. *The expert midwife, or an excellent and most necessary treatise of the generation and birth of man.* London, printed by E. Griffin for S Burton..., 1637. (Osler 3849).

Vesaluis, A. *De humani corporis fabrica libri septum.* Basel, Ioannis Oporni [Johannes Oporinus], 1543.

LIBER

LIBER QVARTVS.

DE VARIETATIBVS NON NA-
turalis partus, & earundem curis.

Quan-

FIGURE FROM RUEFF, 1554

Juan Valverde de Hamusco [Amusco]

Historia de la composicion del cuerpo humano...Rome, Antonio Salamanca, y Antonio Lafrerij, 1556.

Valverde's (ca. 1525–ca. 1587) work, the first great medical book in Spanish (although published in Rome), appropriated all but four of his 42 plates from the "*Fabrica*..." (1543) of Andreas Vesalius (1514–1564). Nonetheless, Valverde made many revisions to Vesalius' text, and included several original descriptions.

Figure XXX is of a pregnant woman with her anterior abdominal wall opened to reveal her pregnant uterus (L), and was not copied from Vesalius. It also depicts the surface of the uterine surface of the placenta (XXXI and XXXII) as well as the isolated placenta and vasculature (XXXIII and XXXIIII). The Spanish artist Gaspar Becerra (1520–1570) and pupil of Michelangelo Buonarroti (1475–1564) is believed to have drawn the illustrations, and the French engraver Nicolas Beatrizet [Beautrizet] (1515–1565) is believed to have prepared the copper plates. The work is dedicated to Cardinal Archbishop Juan Álvarez de Toledo (1488–1557) who may have been its patron. Translations were made into Latin, Italian, and Dutch. Valverde, from the Kingdom of Leon in Palencia, Spain, studied anatomy in Padua under Matteo Realdo Colombo (1516–1559), and in Rome with Bartolomeo Eustachi (ca. 1510–1574).

References

Choulant-Frank pp 205–208; Durling 4530; Garrison Morton 378.02.

Vesalius, A. *De humani corporis fabrica libri septum.* Basileae, ex off. Ioannis Oporini, 1543.

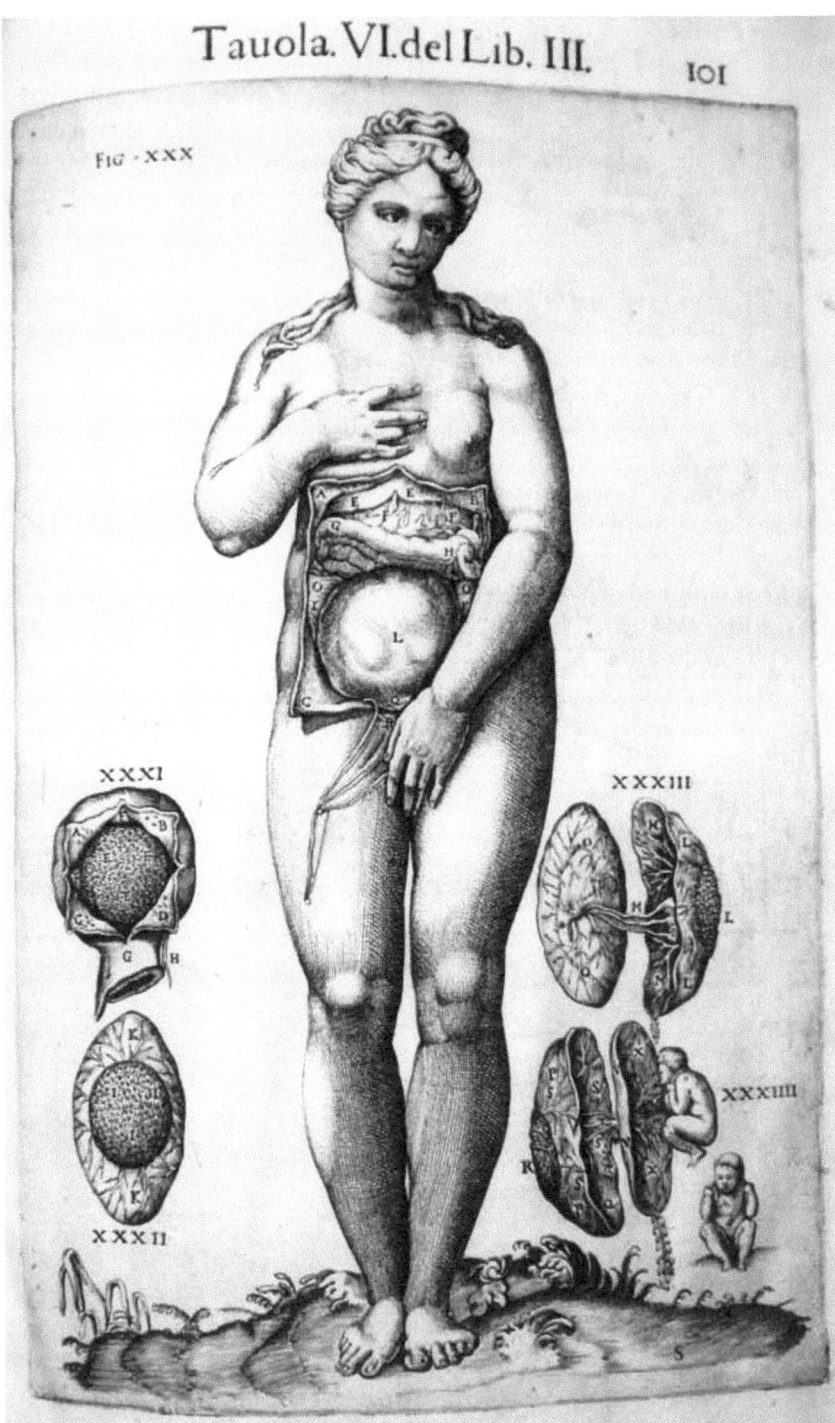

FIGURE FROM HAMUSCO, 1556

Geronimo Scipione Mercurio

La commare o riccoglitrice… Venettia, G.B. Ciotti, 1596.

With his "Instructions for the midwife," Mercurio (ca. 1550–1615) published the first Italian manual for midwives and book on obstetrics. In part, its importance lies in Mercurio's advocacy for Cesarean section in those cases of pelvic contraction with obstructed labor when the woman was still alive. This at a time when, because of the almost certain mortality, the operation was held in general disrepute. Here you see a description of "The reproductive tract of a Cesaeran, which is in a delicate position."

The work is divided into three parts; the first dealing with natural labor and care of the mother and newborn; the second with abnormal presentations; and the third with diseases that complicate pregnancy and effect the newborn. For difficult births, Mercurio also illustrated the delivery of a woman in bed (right-hand figure), in what now is referred to as the [Gustav Adolf] Walcher (1856–1935) or "hanging legs" position (Walcher 1899). He also advocated that the midwife remove the placenta manually by inserting her hand between it and the uterus, with her finger tips manipulating, "much as does the butcher remove the skin from a dead animal."

Born in Rome, Mercurio studied medicine at the Universities of Bologna and Padua prior to entering the Dominican monastery in Milano, where he became a monk and assumed the name Hieronymus. Finding little satisfaction in monastic life, Mercurio then left to become personal physician to a military officer, Girolamo da Lodrone (1526–1579). Following travels in France and Spain in 1571–1572, he returned to Italy to practice in Padua and Milano. During the last 15 years of his life Mercurio settled in Venice, living within a monastery and engaging in the practice of medicine and obstetrics.

References

Garrison Morton 6144; Speert.

Mercurio, G.S. *La commare o riccoglitrice…Diusa in tre libri. Ristampata corrota et accresciuta dall' isteso autore.* Venettia, Appreso Bio. Bat, Ciotti, 1601.

Walcher, G.A. Die Conjugata eines engen Beckens ist Keine Konstante Grosse, sondern lässt sich durch die Körperhaltung der Tragen verändern. *Zbl Gynäk* 13:892–893, 1889. (GM 6200).

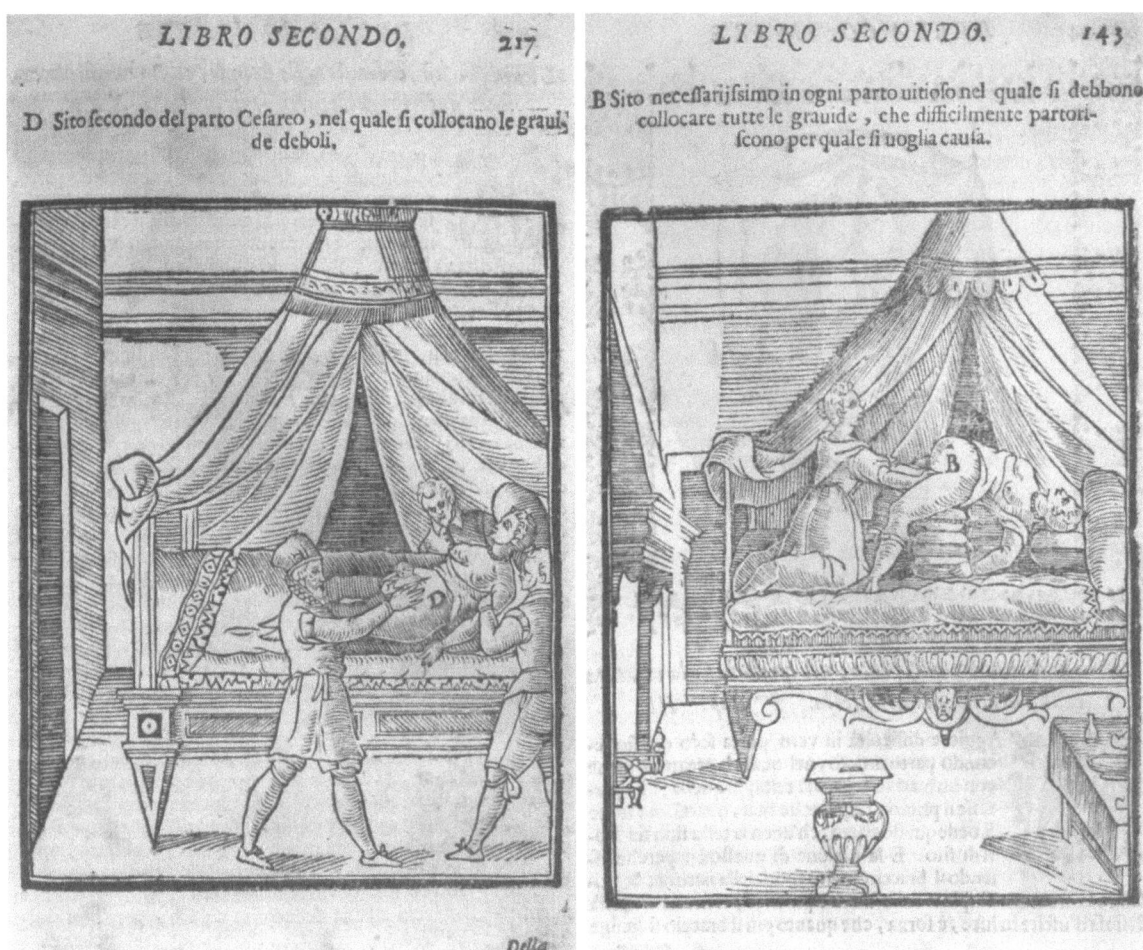

FIGURES FROM MERCURIO, 1596

Séverin Pineau

Opusculum physiologum & anatomicum in duos libellos distinctum. In quibus primum. De integritatis et corruptionis virginum notis... Parisiis, Steph. Prevosteau, 1597.

In his treatise on the anatomical signs of virginity (and manner in which to lose it), Pineau (ca. 1550–1619) wrote on embryology and obstetrics. The illustration on the left shows 12 day old embryos with a placenta, the latter greatly exaggerated in size. The figure on the right shows fetuses later in gestation, but again with a greatly exaggerated placenta. Additional illustrations show fetuses at several stages of gestation and a fetal skeleton. Because of Pineau's frank discussion of virginity, the plates were removed from many copies of this work, or the entire volume was confiscated. In his *History of Embryology*, Joseph Needham (1900–1995) castigated Pineau's illustrations as being "almost ludicrous" (Needham 1934, p. 95).

In this pre-Harveian era of understanding the true function of the heart and circulation of blood, the common concept of the heart, blood vessels, and their contents was that of Galen (Claudius Galenus, 130–200). Pineau also became involved in the controversy regarding the function of the heart and its septum separating the two ventricles. Pineau's work also is noteworthy, for in an edition published several decades later was included the "fine observation" of Pierre Gassendi (1592–1655), a theologian, philosopher, and mathematician on the vestigial *foramen ovale* (an oval opening between the right and left atria of the fetal heart which normally closes shortly after birth) in the adult heart. Rather than being a discovery of Gassendi himself, his report described an anatomic dissection he had witnessed 8 years previously in the anatomical amphitheater at Aix, where he was a professor of philosophy. In this exhibition, the demonstrator apparently forced a probe through the septum from the right to the left side of the heart, without actually puncturing it (Gassendi 1639). At this time it was believed that blood passed from the right to left ventricle through pores in the septum. Thus, this erroneous observation disputed the definitive studies of William Harvey (1578–1657) at least a decade earlier that had established definitively the circulation of blood, the true role of the heart as a pump, and the lack of communication between the two sides of the heart (Harvey 1628).

Pineau, a native of Chartres, became a prominent surgeon and *accoucheur* in Paris, and was appointed surgeon to King Louis XIII (1601–1643; King 1610–1643). He also was an accomplished lithotomist, and wrote on the removal of calculi from the urinary bladder (Pineau 1610).

References

Durling 3654; Garrison Morton 802; Heirs of Hippocrates 239; Krivatsy 9008.

Gassendi, P. *Elegans de septo cordis pervio observatio.* Published with *De integritatis et corruptionis virginum notis...* by Severinus Pineaus. Lugduni-Batavorum, Apud Franciscos Hegerum & Hackman, 1639.

Harvey, W. *Exercitatio anatomica de motu cordis et sanguinis in animalibus.* Francofurti, Sumptibus Guilielmi Fitzeri, 1628. (GM 759).

Pineau, S. *Discours touchant l'invention et vraye instruction pour l'opération et extraction du calcul de la vessie, à toutes sortes de personnes.* Parigi, 1610.

Pineau, S. *De integritatis et coruptionis virginum notis. De graviditate & partu natural; mulierum....* Lugduni Batavorum, Apud Franciscos Hegerum & Hackium, 1639. (Krivatsy 9008).

Pineau, S. *De integritatis et coruptionis virginum notis. De graviditate & partu natural; mulierum....* Lugduni Batavorum, Apud Franciscum Moyaert, 1650. (Krivatsy 9011).

Sampalmieri, A. Uno scrito poco noto di Severino Pineau, litotomista del XVI secolo. *Riv Stor Med* 11: 235–246, 1967.

Tallmadge, G.K. Pierre Gassendi and the *Elegans de septo cordis pervio observatio. Bull Hist Med* 7: 429–457, 1939.

Willius, F.A. and T.E. Keys. Pierre Gassendi (1592–1655). In: *Cardiac classics. A collection of classic works on the heart and circulation with comprehensive biographic accounts of the authors....* St. Louis, C.V. Mosby, 1941, pp. 83–85.

Prima figura fœtus duodecim dierum. pag.III.II2 Folio 220. 221.

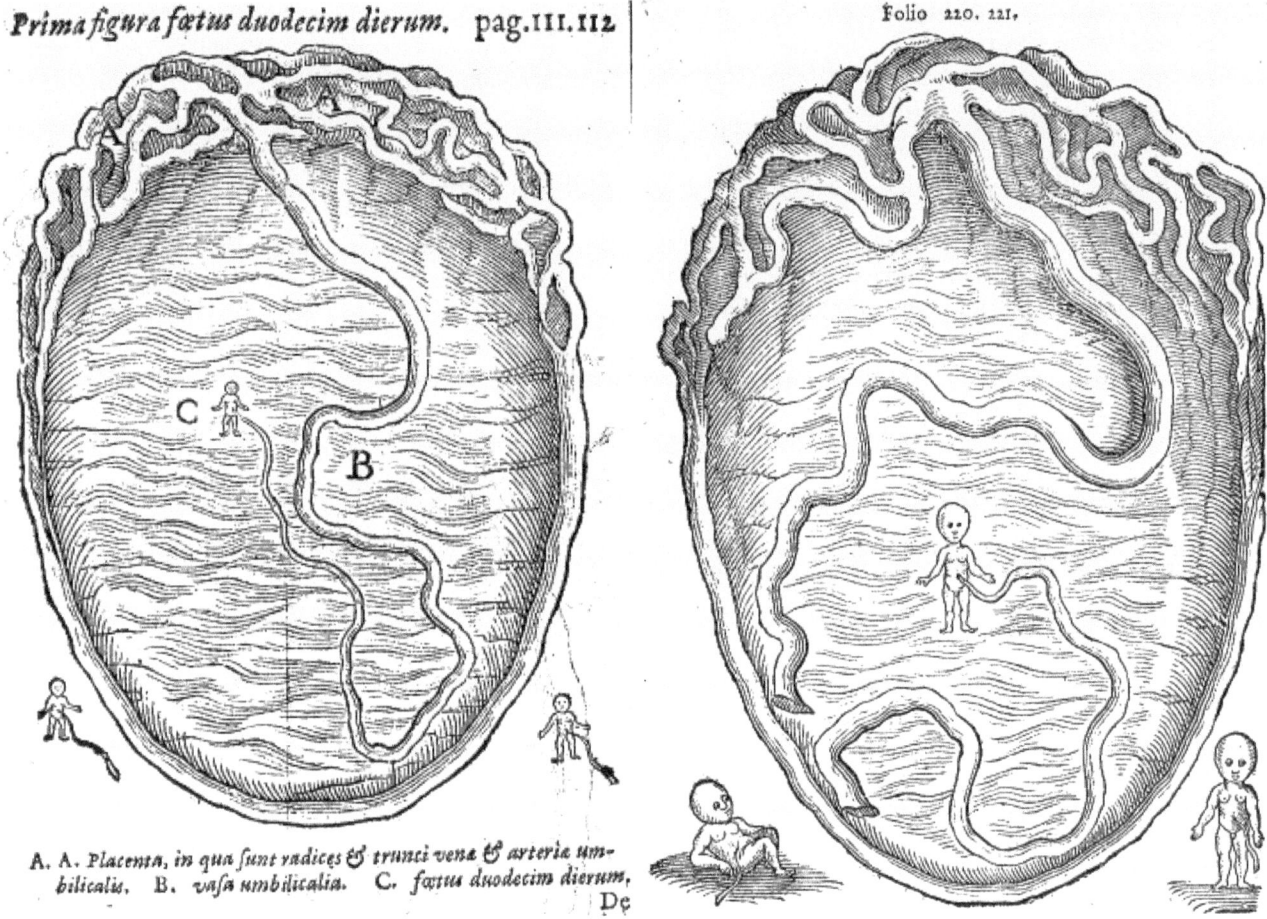

A. A. *Placenta, in qua sunt radices & trunci venæ & arteriæ um-*
bilicalis. B. *vasa umbilicalia.* C. *fœtus duodecim dierum.*
Dę

FIGURES FROM PINEAU, 1597

Additional Author(s) of Significance in the Sixteenth Century

Giulio Cesare Aranzi [Aranzio; Arantius]

De humano foetu libellus… Bononiae, J. Rubrii, 1564.

Aranzi (1530–1589) first wrote on the independence of the maternal and fetal circulations in the placenta, the experimental verification of which was made by John (1728–1793) and William Hunter (1718–1783) (see Hunter 1774). He also described the vascular shunts in the fetal circulation, e.g., the ductus arteriosus (between the left pulmonary artery and the descending aorta), and the ductus venosus or canal of Arantii (connecting the umbilical vein and the inferior vena cava and right atrium of the heart). He also discovered the *corpora* [bodies] *Arantii*, the nodules on the aortic valves of the heart. In addition, he maintained that during the progress of labor, the thickness of the upper portion (fundus) of the uterine wall increases, rather than decreases.

A graduate of the University of Bologna (1556), Aranzi became professor of both anatomy and surgery at that institution. In the 1587 third edition of *De humano foetu…* he gave the first description of a deformed pelvis. He also described the elevator muscle of the upper eyelid. In the brain, he first described the *pedes hippocampus* [sea horse; in fact he named that organ], the fourth ventricle, and the cerebellar cistern. Importantly, Aranzi pioneered the field of rhinoplastic surgery. Because he did not publish however, this is credited to his colleague at Bologna, Gaspar Tagliacozzi (1545–1599), who detailed many aspects of the procedure, including theoretical considerations, instrumentation, progressive steps, and postoperative care (Tagliacozzi 1597).

References

Garrison Morton 464; Durling 236.

Aranzi, G.C. *In Hippocratis librum De vulneribus capitis commentarius brevis….* Lugduni, Apud Ludovicum Cloquemin, 1580. (Durling, 237).

Aranzi, G.C. *De humano foetu liber tertio editus, ac recognitus….* Venetiis, Apud Jacobum Brechtanum, 1587. (Durling 234).

Aranzi, G.C. *De tumoribus secundum locus affectum.* Bologna, 1571.

Aranzi, G.C. *Observationes anatomicae.* Basel, 1579.

Dall'Osso, E. Giulio Cesare Aranzio. In: *Dictionary of Scientific Biography.* Vol I. Charles Coulston Gillispie (Ed.). New York, Charles Scribner's Sons, 1970, p 204.

Tagliacozzi, G. *De curtorum chirurgia per insitionem, libri duo. In quibus ea omnia….* Venetiis, Apud Gasparem Bindonum, juniorem, 1597. (GM 5734; Durling 4310).

François Rousset

Traitte nouveau de l'hysterotomotokie, ou enfantement caesarien. Paris, Denys du Val, 1581.

Rousset [Roussetus; Rossetus] (ca. 1530–ca. 1603), a surgeon at the University of Paris, in his "New treatise on opening the uterus and cesarean childbirth" recorded 15 cesarean sections with survival of the mother, performed by different individuals, during the preceding 80 years. Thus in this work, he first used the term "cesarean", and argued against the idea that this constituted the most dangerous operation in surgery. Rousset recommended the operation in those instances in which the fetus was dead and might present a focus of infection within the uterus. He also advocated this procedure in cases of uterine or tubal rupture when the fetus, within the peritoneal cavity, might give rise to abscess formation. He also recorded that pregnancy could follow the cesarean operation, as illustrated in the case of a woman upon whom the procedure had been performed six times. It is questioned whether Rousset himself ever performed the operation, and many of the case reports would appear to be anecdotal.

References

Garrison Morton 6236, see also 6013; Osler 2494.

Rousset, F. *Foetus vivi ex matre viva sine alterutrius vit'periculo c'sura a F. Rousseto Gallice conscripta. C. Bauhino Latine reddita variis historiis aucta. Adjecta est I. Albosii… Foetus per annos XXIIX in utero contenti et lapide facti historia elegantiss.* Basel, C. Waldkirch, 1591.

Rousset, F. *Partus Caesarei definit or Maronian Autoris au Libennum in Icomem Lithopaedii. In Isreal Spachii, Gynaeciorum sive de mulierum tum communibus, tum gravidarum, parientium, et puerperarum affectibus et morbis libri Graecorum, Arabum Latinorum veterum et recentium quotquot extant, partium hunc primum editi.* Argentinae, sumpt Lazori Zetzerni, 1597.

Rousset, F. *Exercitatio medica assertionis novae veri usus anastomoseon cardiacarum foetus ex utero materno trans ipsas trahentis aërem internum in suos pulmones motus respiratorii (contra communem opinionem) tu[n]c non expertes, & illum cordi eum appetenti, suique etiam tunc micantis motus compoti praeparaturos.* Paris, excudebat Dionysius Duvallius, sub pegaso, in vico Bellovaco, 1603.

The Seventeenth Century

This century saw major advances in philosophy and science, including mathematics, physics, chemistry, and biology, and the introduction of the experimental method. Perhaps no other century can equal its strides in the realm of knowledge, and it has been characterized as giving rise to the scientific revolution. In part, this resulted from centralized political authority and the influence of the royal courts, the Stuart Court in Britain, and the baroque court societies of Prussia, Denmark, and Italy – this through bonds forged by dynastic marriages, diplomacy, aristocratic tourism, and intense competition for cultural prestige. Also contributing to the flourishing of science were the skeptical attitudes towards authority. The resulting networks facilitated the movement of natural philosophers from one court to another. This resulting spread of ideas, information, and cultural fashions included natural philosophy. Perhaps surprisingly, for the most part the most prominent scientists and seminal thinkers were without university affiliation. Rather, through their work in scientific affiliations such as the Royal Society of London, the *Academia dei Lincei* of Rome, the *Collegium Naturae Curiosarum* ... in Germany, and other national societies, experimental science flourished, and laboratory based experiments became the valid means of scientific study. The simultaneous growth of overseas empires led to important contacts with Asia and the Americas, allowing some courts to gather observations and natural specimens on a global scale. The century also served as a canvas depicting war, conflagration, intolerance, and misery. With revolution and reform, the "state" and "nation," emerged as changing concepts.

Culturally and socially, political authority became more centralized and armies grew in size and force. The Dutch East India Company was founded (1602); James VI of Scotland united the crowns of Scotland and England (1603); the London Company established the Jamestown Settlement in North America (1607) which lead to the British colonization of the Americas. James I (1566–1625) the first Stuart King of England (1603–1625), as James VI King of Scotland (1567–1625) supported publication of the scholarly translation, King James Bible (1611). A series of religious conflicts that involved Protestant and Catholic rivalries and German constitutional issues, the savage "Thirty Years' War" (1618–1648), devastated Central Europe and left men weary, cautious, and skeptical. In 1609 Henry Hudson (ca. 1565–1611) anchored the "Half Moon" in New York Bay, and on Massachusetts Bay, in 1620 the Pilgrims established the Plymouth Colony (and towards the end of the century in 1691 united with other New England colonies to form the Massachusetts Bay Colony). Shortly thereafter, New Amsterdam was founded by the Dutch West India Company (1625); Portugal regained its independence from Spain (1640), thus ending the Iberian Union. Following the English Revolution of 1660, with the overthrow of Oliver Cromwell (1599–1658) and the Puritan government, the Commonwealth of England was restored with Charles II (1630–1685), the monarch, with the Restoration lasting until the end of the century. In 1664 British troops captured New Amsterdam, naming it New York. Two years later (1666), a year following the Great Plague (1665), the Great Fire almost annihilated London. The French explorer Robert Cavalier [Sieur de La Salle] (1643–1687) sailed the length of the Mississippi River and claimed Louisiana for France (1682); and the Salem witch trials (1692) racked Massachusetts. The century also witnessed the writings of Miguel de Cervantes Saavedra (1547–1616), John Donne (1572–1631), Sir Thomas Browne (1605–1682), John Milton (1608–1674), Molière [Jean Baptiste Poquelin] (1622–1673), and John Bunyan (1628–1688).

In science, the English philosopher and politician Francis Bacon (1561–1626) founded the modern scientific tradition with his formulation of the idea of planned and controlled experiment, what soon thereafter was called the "scientific method." A leader in this scientific revolution, the French philosopher and mathematician René Descartes (1596–1650) invented algebraic notation and Cartesian coordinates that enable one to give algebraic equations for geometrical figures. Several philosophies competed for attention, including the traditional Aristotelianism of the Scholastics, the still influential Neoplatonism of the Renaissance, and the "new" Cartesian mechanistic system. By a series of hypothesis-driven experiments, in his *De motu cordis...* (1628) the English physician-physiologist William Harvey (1578–1657) correctly described the function of the heart, with circulation of the blood. Marcello Malpighi (1628–1694) demonstrated capillaries, the presence of which made possible a more valid understanding of the manner in which blood circulates. Following his improvements in design of the refracting telescope, Galileo Galilei (1564–1642) performed his fundamental observations in astronomy (discovering four of the moons of Jupiter as well as observing the transit of Venus across the sun) that gave validity to a Copernican heliocentric solar system. In addition, he performed fundamental experiments and mathematical analysis in physics. The Dutch physicist and astronomer Christian Huygens (1629–1695) influenced greatly the development of science both in Europe and in England, with establishment of Anglo-Dutch intellectual networks. The English polymath Isaac Newton (1642–1727) described the theory of universal gravitation, terrestrial mechanics, a theory of color and light, invented the reflecting telescope, and, independently, along with the mathematician Gottfried Wilhelm Leibnitz (1646–1716), invented the calculus. In addition, Leibnitz with the

Portuguese-Jewish philosopher Benedict (Baruch) de Spinoza (1632–1677), whose parents had moved to the Netherlands to escape the Inquisition, contributed to philosophical religious thought. In 1660, the Royal Society of London was founded for "the Improvement of Natural Knowledge."

The later part of the century saw the rise of neoclassicism, the revival of classical aesthetics in art and literature. In art, Rembrandt Harmenszoon van Rijn (1606–1669), Giovanni Lorenzo Bernini (1598–1680), and many others were dominant forces of influence. It was during the seventeenth century that etching became the preferred medium of innovative printmakers, and engraving came to be used for book illustration. As the science of anatomy advanced with the accumulation of material and its interpretation, the center of biological and medical research shifted from Italy to the north—Holland, France, Germany, and England. This period saw the invention of the microscope, and its refinement by Antonj van Leeuwenhoek (1632–1723) and others, which opened new vistas for anatomical study at the cellular level. Sophisticated injection techniques developed by Frederik Ruysch (1638–1731) and others, demonstrated the presence of blood vessels in essentially all tissues and organs of the body, that allowed detailed study of vascular relationships, and other structures. With improved methods of preservation, the establishment of private *Wunderkammern* [cabinet of curiosities] with necroscopy collections, as well as public anatomical museums, flourished. Such collections contributed much to the spread of knowledge. Beyond serving as theaters of the rare, bizarre, and marvelous, natural philosophers believed that by classifying and comprehending nature they might glimpse the mind of the creator, and his ingenuity and goodness.

Girolamo Fabrizio [Hieronymi Fabrici; Fabricius ab Aquapendente]

De formato foetu. Venetiis. per Franciscum Bolzettam 1600 (Colophon, Laurentius Pasquatus, 1604).

In his magnificently illustrated embryological atlas, *De formato foetu* [The formed fetus], Fabricius (ca. 1533–1619) presented more accurately than anyone before, and for long thereafter, the relation of the fetus to the umbilical vessels, fetal membranes, placenta, and the uterus (see Tabulae VII, pg.). Fabricius also described several structures of the fetal circulatory system, including the right and left atria, foramen ovale, ductus arteriosus, vena cava, and pulmonary vein. This work contains 34 plates which illustrate, in some instances for the first time, various aspects of the anatomy of the uterus and fetus in human and in other species, including: dog, sheep, cow, horse, pig, rat, mouse, and shark (see Tabulae XV, pg.). The last plate depicts the development of serpents. Fabricius was the first to study and to illustrate the decidua of the human uterus, and the uterine crypts in animals, the latter which he interpreted as the open ends of uterine vessels. Because Fabricius mistook the cotyledon crypts of the sheep placenta as the opening of blood vessels, which match equivalent openings in the uterine caruncles, he concluded that there must be continuity between the maternal and fetal circulations, and that somehow the fetal vessels were "plugged into" those of the mother. He attacked Aranzi for questioning the Galenic doctrine regarding the confluent fetal and maternal vascular channels.

In the accompanying illustration, Tabulae III, Fig. VI, is shown the near-term fetus attached by umbilical cord (B) to the placenta (A), swimming in amniotic fluid (C). Also depicted are the fourfolds of uterine wall (D), neck of the uterus/cervix (E), and opened vagina (F). Figure VII in the lower right corner shows, within the opened uterine wall (C), the posterior aspect of a near-term fetus in cephalic (A) presentation awaiting delivery. The artist is unknown.

A student of philosophy, a primary purpose of Fabricius' studies of fetal development were written in conjunction with his anatomical works that were published later. This was in an attempt to understand more completely the function of various organs and other structures. A teleologist, he interpreted natural design in terms of utility. Fabricius' first embryological treatise, *De instrumentis seminis* [the implements of insemination], was never published. Later, he published his *De formatione ovi et pulli...* [the formation of the egg and chick], addressed the nature and properties of the seed and the generation and formation of the fetus (see Fabrizio 1621).

Because of their narrative nature, these works are believed to have been prepared from his classroom lectures. Joseph Needham (1900–1995) observed that although Fabricius compounded many errors in regards to development, his illustrations were "beautiful and accurate" (Needham 1934, p. 90).

A graduate of the University of Padua (circa 1559) and student of Gabriele Falloppio (1523–1562), following Falloppio's death Fabricius was appointed *supraordinario* [Professor] of Anatomy and later of Surgery, at that institution. In one of his other works, *De venarum ostiolis* [On the doorkeepers of the veins] (1603), he first accurately described the valves of the veins. Because of his Galenic concepts of function, Fabricius mistakenly believed these valves retarded the flow of blood from the right side of the heart to the tissues to facilitate nutrient absorption. Appreciation of the true function of valves of the veins played a key role in his student William Harvey's (1578–1657) development of the idea of the circulation of the blood. Fabricius also published a number of other works (see Zanobio 1971). He had planned to compile these into a monumental *Totius animalis fabricae theatrum* [exhibition of the entire structure of animals], which he did not live to complete. A distinguished consultant, he cared for many notables. For his services, he was made a Knight of St. Marks by the Republic of Venice.

References

Garrison Morton 465; Krivatsy 3827; see Norman 27 B.

Adelmann, H.B. *The embryological treatises of Hieronymus Fabricius of Aquapendente. The formation of the egg and of the chick [De Formatione Ovi et Pulli]. The formed fetus [de Formato Foetu]. A facsimile edition.* Ithaca, N.Y., Cornell Univ Press, 1942.

Fabrizio, G. *Pentateuchos cheirurgicum....* Francofurti ad Moenum, [Palthenius, Lechler und Fischer], 1592. (Krivatsy 3805).

Fabrizio, G. *De venarum ostiolis.* Patavii, ex typ. L. Pasquatin, 1603. (GM 757; Krivatsy 3831).

Fabricio, G. *Operationes chirurgicae....* Venetiis, Apud Paulum Megliettum, 1619. (Krivatsy 3805)

Fabrizio, G. *De formatione ovi et pulli....* Patavii, Ex officina Aloysii Bencii, 1621. (GM 466).

Falloppio, G. *Observationes anatomicae.* Venetiis, Apud M.A. Ulmum, 1561. (GM 1208, GM 1537).

Needham, J. *A history of embryology.* Cambridge, at the University Press, 1934.

Zanobio, B. Girolamo Fabrici (or Fabricius ab Aquapendente, Geronimo Fabrizio).... In: *Dictionary of Scientific Biography. Vol IV. Charles Coulston Gillispie (Ed).* New York, Charles Scribner's Sons, 1971, pp. 507–512.

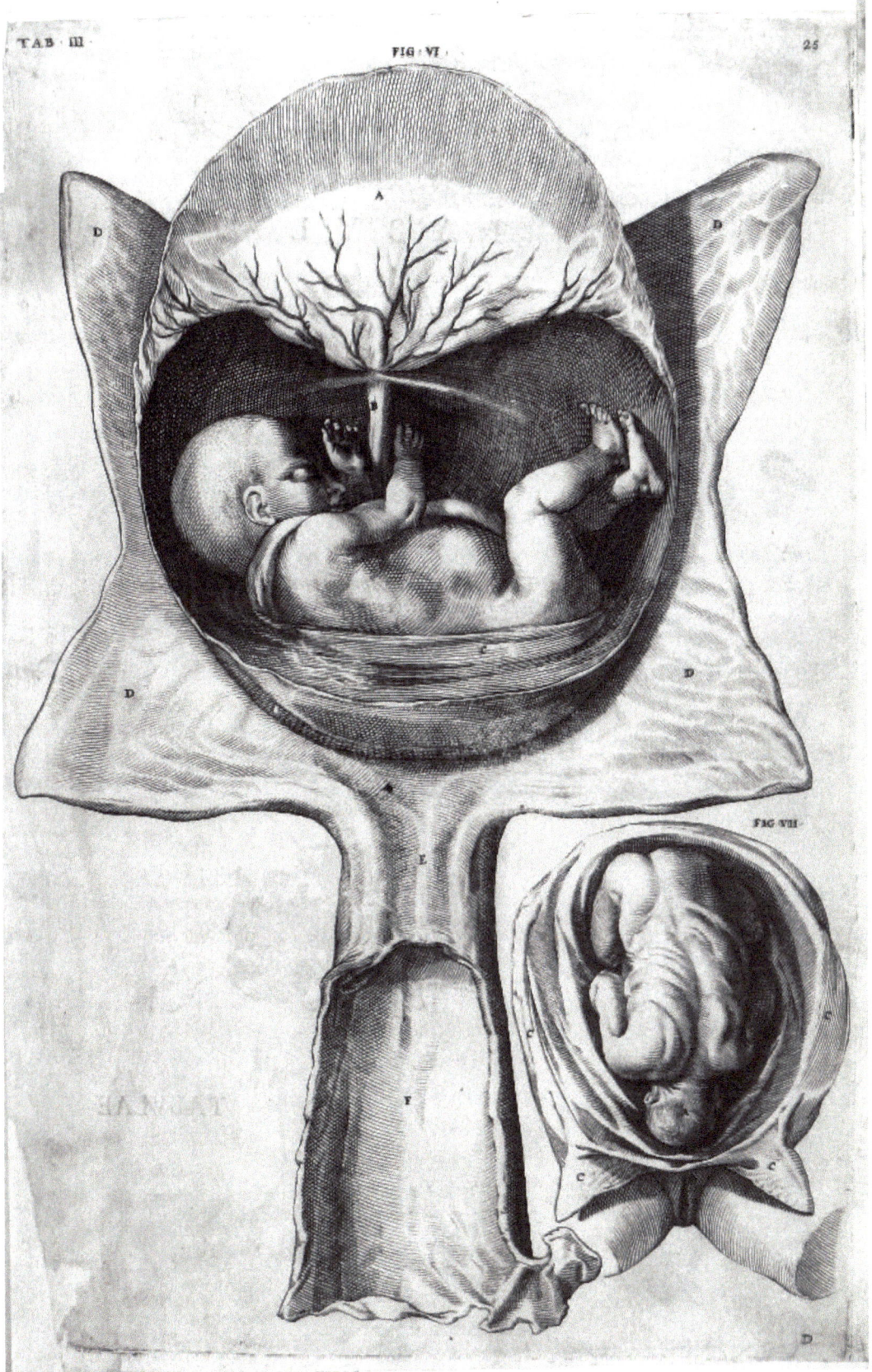

TAB III FROM FABRIZIO, 1604

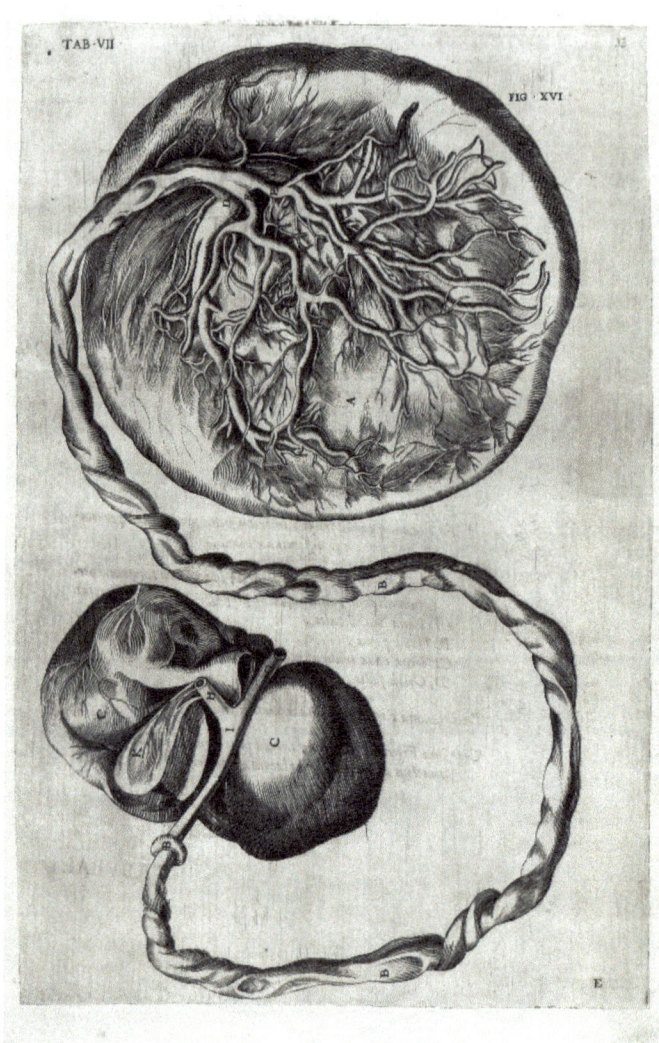

**Tabulae VII, Figure XVI shows the fetal surface of the human placental with the umbilical cord (B),
and fetal liver (C), umbilical vein (I) and portal vein (H)**

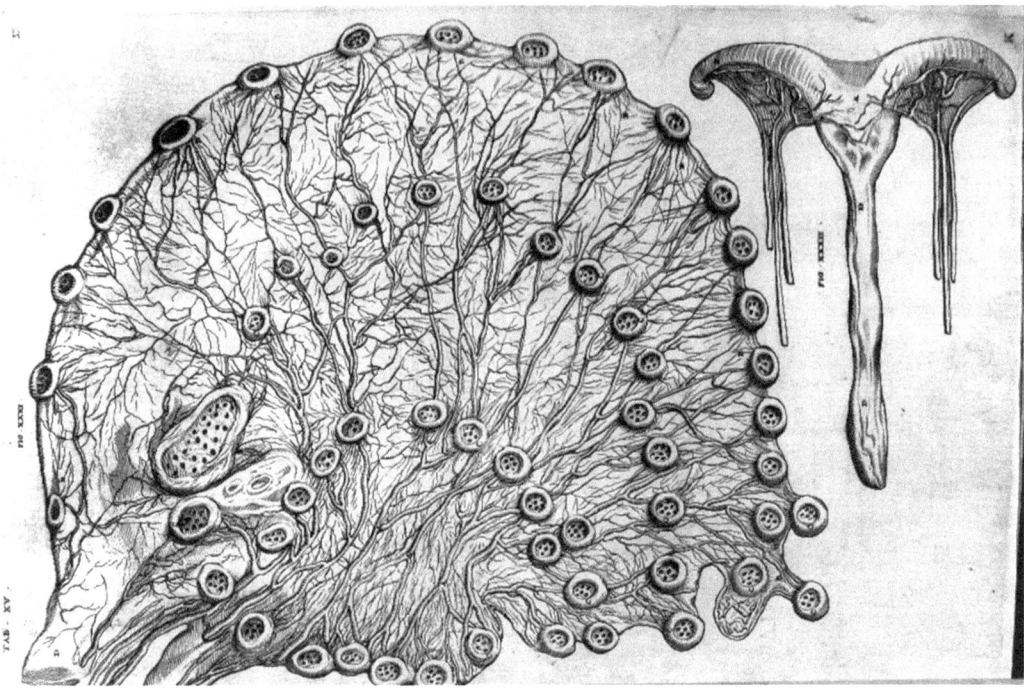

Tabulae XV shows the entire sheep reproductive tract (Fig. XXXII) as well as the placenta with its cotyledons, each of which is well supplied with blood vessels (Fig. XXXI)

Albertus Magnus [Albert von Bollstädt]

Daraus man alle Heimligkeit dess Weiblichen Geschlechts, erkennen kan, dessgleichen von ihrer Geburt/sampt mancherley Artzney der Krauter/auf von tugent der edlen Gestein vnd der Thier/mit sampt einem bewehrten Regiment fur das bose ding. Jtzundt [sic] abert auffs neuw gebessert.... Franckfurt am Mayn [durch Matthis Becker/in verlegung Gottfried Tampacks], 1608.

Albert von Bollstädt, "Saint Albert the Great" (ca. 1193–1280), included in the present work essays on the diseases of women, obstetrics and pediatrics; plants and herbs; animals, mineralogy, semi-precious stones, and a regimen of health. The title is printed in red and black with a large vignette of Adam and Eve, a vignette on the colophon, and numerous woodcuts in the text. These include an obstetrical chair, a delivery, the fetus including twins, plants, birds, and so forth. In this plate, the seated woman displays her abdominal cavity, including organs of reproduction with mulberry-like ovaries and a fetus *in utero*. The figure modeled after that of the *Fabrica...* of Andreas Vesalius (1514–1564) (Vesalius 1543), is essentially a copy of that seen in Jacob Rueff (1500–1558) (Rueff 1554), except that the fetus is slightly larger and the lounge chair is carved in a more elaborate manner. Albertus Magnus was one of the first after Aristotle to study embryology by opening incubated chicken eggs at various intervals of time, following embryonic development from the pulsitile red speck that was to develop into the heart, until the time of hatching. He also studied the development of insects, fish, and mammals.

Believed to be one of the most learned scholars of the thirteenth century, Albertus Magnus was of noble birth being eldest son of the Count of Bollstädt. He studied at the University of Padua, following which he became a Dominican friar. Albertus taught in Germany and Paris, becoming prior Provincial (1254–1257) of the German Dominicans of his order. He was an ardent naturalist and student of all the sciences. An encyclopedic writer, Albertus wrote on many aspects of learning, including theology, philosophy, ethics, logic, and science (including anthropology, psychology, minerals, plants, and animals). His work on mineralogy includes descriptions of numerous chemical substances, details of 95 precious stones or minerals, and many other particulars. The "Universal Doctor", he played an important role in rediscovering the works of Aristotle (384–322 BCE) and Greek science. In this regard, he combined elements of Aristotelianism, Neo-Platonism, and Christian, Muslim, and Jewish theology, into a great philosophical system. In this work, which occupied the last decade of his life, his chief aim was to reconcile Aristotelian thinking with Christian theology (1476, 1478), thus forging a "unified field theory" of Christian thought. For the history of modern science, his importance stems from his role in rediscovering the writings of Aristotle and introducing Greek and Arabic science into the universities.

Among the works attributed to Albertus, many were undoubtedly spurious. For instance, among these is *De secretis mulierum* [on the secrets of women]. Not only is this not likely to be a genuine work of his, because of the suggestiveness of the title and with its contents on gynecology and generation it was placed on the *Index Expurgatorius*, the catalogue of books from which passages regarded as against faith or morals must be removed before they could be read by those of the Roman Catholic faith. Nonetheless, the work was quite popular with nearly 50 editions being published. Several treatises on magic and the occult also were ascribed to him.

In his *Summa de eucharistide* [Above the Eucharist] (1474) Albertus reaffirmed the importance of Holy Communion as a sacrament commemorating the actions of Jesus Christ at the Last Supper. Considered a saint, he was beatified, one of the Blessed, in 1622. In 1931, he was canonized and proclaimed a Doctor of the Church, and in 1941 declared to be patron saint of the natural sciences. Among his pupils was the Dominican theologian and philosopher Thomas Aquinas (1225–1274). Two major compilations of Albertus' works were printed in France, one in 1651, the other in 1890–1899. Publication of another major compendium commenced in Germany in 1951.

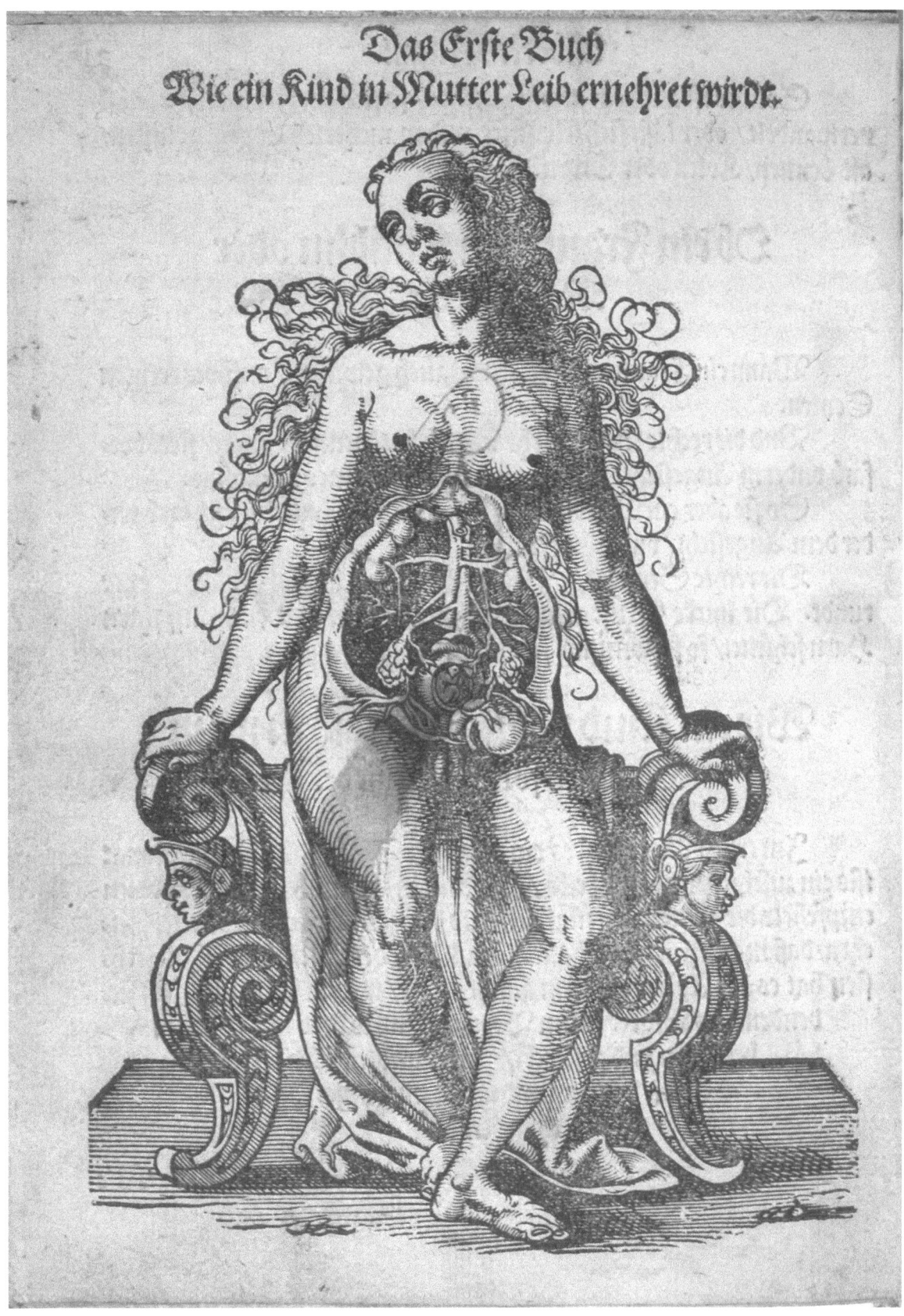

PLATE FROM MAGNUS, 1608

References

Castiglioni, 349–350, PMM, 17.

Albert, S.M. *Albert the Great*. Oxford, Blackfriars Publications, 1948.

Albertus Magnus. *Summa de eucharistiae sacramento*. Ulm, Johann Zainer, 1474.

Albertus Magnus. *Alberti Magni De mineralibus, liber primus incipit*. [Padua: Petrus Maufer for Antonius de Albricis], 1476.

Albertus Magnus. *De secretis mulierum et vivorum....* Cologne, 1476.

Albertus Magnus. *De animalibus*. Rome, Simon Nicolai Chardella de Lucca, 1478. (GM 276).

Albertus Magnus. *Secreta mulierum et vivorum (cum expositione Henrici de Saxonia)*. [Venice, Adam de Rottweil], 1478. (Schullian & Sommer, 20).

Albertus Magnus. *De secretis mulierum et virorum*. [Leipzig, Konrad Kachelofen, circa 1487–1495]. (Schullian & Sommer, 22A).

Albertus Magnus. *Philosophia pauperum, sive Philosophia naturalis*. Brescia, Baptista Farfengus, 1490. (Schullian & Sommer, 17).

Albertus Magnus. *De generatione et corruptione*. Venice, Joannes and Gregorius de Gregoriis, de Forlivio, 1495. (Schullian & Sommer, 9).

Albertus Magnus. *De virtutibus herbarum. De virtutibus lapidum. De virtutibus animalium et mirabilibus mundi. Item parvum regimen sanitatis valde utile*. Anvers, Govaert Bac, 1502. (Similar to Durling 83).

Albertus Magnus. *Opera omnia*. Petrus Jammy, Ed. 21 folio volumes. Lugduni, 1651.

Albertus Magnus. *De vegetabilibus libri VII, historiae naturalis pars XVIII. Ediitionem criticam ab Ernesto Meyero coeptam absolvit Carolus Jessen*. Berolini, Typis et impensis Georgii Reimeri, 1867.(GM 1792).

Albertus Magnus. *Opera omnia ex editione lugdunensi religiose castigate.....* Auguste Borgnet, Ed. 38 quarto volumes. Parisiis, Apud Ludovicum Vivès, 1890–1899.

Albertus Magnus. *Opera omnia ad fidem codicum manuscriptorum edenda*. Bernhard Geyer, Ed. Monasterii Westfalorum, In aedibus Aschendorff, 1951.

Kennedy, D.J. Albertus Magnus. In: *The Catholic Encyclopedia*. Vol I. New York, Robert Appleton Company, 1907, pp. 264–267.

Morris, W. On the artistic qualities of the woodcut books of Ulm and Augsburg in the fifteenth century. *Bibliographica* 1: 437–455, 1895.

Rueff, J. *Ein schön lustig Trostbüchle von dem Empfengknussen und Geburten der Menschen....*

Tiguri, Apud Frosch [overum], 1554. (GM 6141).

Vesalius, A. *De humani corporis fabrica libri septum*. Basel, Ioannis Oporni [Johannes Oporinus], 1543. (GM 375).

Wallace, W.A. Albertus Magnus, Saint. In: *Dictionary of Scientific Biography*. Vol I. Charles Coulston Gillispie (Ed.). New York, Charles Scribner's Sons, 1970, pp. 99–103.

Weisheipl, J.A. Albert the Great (Albertus Magnus), St. In: *New Catholic Encyclopedia*. New York, McGraw-Hill, 1967.

Johann Remmelin

Catoptrum microcosmicum, suis aere incisis visionibus splendens, cum historia, & pinace, de novo prodit... Augustae Vindelicorum, Davidis Francki, 1619.

In the sixteenth and seventeenth centuries, several anatomical works were published in which human body structure was illustrated with the use of superimposed flap overlays. This idea did not originate with anatomists, but had been developed for much earlier treatises on astrology, astronomy, and cosmography. An elegant volume using this method was the "microscopic mirror" of Remmelin (1583–1632) which presents the classical idea of man as a microcosm of the universe. An abbreviated version of Remmelin's work (the three plates with minimal text) first appeared 6 years previously, but without attribution to the author. Subsequently, a number of other editions were published including translations into English, Dutch, and German.

Eight separate *visions* [plates], which after being cut, were pasted together to form three large plates. The first of these show the bodies of a man and woman, together with the trunk of a pregnant woman. The second plate is that of a male body, while the third is of a female. Each plate contined several 'flaps' that when lifted illustrated the next layer below. This figure (illustrated) shows the three layers (two flaps) depicting the female reproductive system (images downloaded from the Hardin Library, University of Iowa; http://sdrc.lib.uiowa.edu/exhibits/imaging/remmelin/sub_plate3.htm).

Remmelin is believed to have drawn the figures, which were engraved by Lucas Kilian (1579–1637) of Augsburg. Johann Ludwig Choulant (1791–1861) states that the value of the illustrations "in very slight and even as a whole, they represent the clumsiest study of anatomy" (Choulant 1920, p 232). In contrast, Leroy Crummer (1872–1934) credits Remmelin with producing "the most carefully planned and executed" work of this type. He also provided a survey of the various editions (Crummer 1932, pp 136–139). Walton Brooks McDaniel II (1897–1975) has provided a useful analysis of some questions regarding publication of the earliest editions (McDaniel 1938). In addition to the figures, each plate includes explanatory material and allegorical figures. Apparently the volume was intended for lay persons to teach moral precepts, in addition illustrating some anatomical detail. Remmelin served as town physician at his birthplace, Ulm on the Danube, and later in Augsburg where he also helped to fight the Black Death (bubonic plague).

References

Choulant, pp 232–234; Krivatsy 9551.

Crummer, L. A check list of anatomical books illustrated with cuts with superimposed flaps. *Bull Med Libr Assoc* 20: 131–139, 1932.

Hagelin, O. *Rare and important medical books in the Library of the Swedish Society of Medicine: a descriptive and annotated catalogue.* Stockholm, Svenska Läkaresallskapet, 1989.

McDaniel, W.B. The affair of the "1613" printing of Johannes Rümelin's *Catoptron. Trans Stud Coll Physic Phila* 6: 60–72, 1938.

[Remmelin, J] *[Catoptrum microcosmicum] Visio prina [-tertia].* Augsburg, Stephen Michael Spelcher, 1613. (Krivatsy, 9548).

Remmelin, J. *A survey of the microcosme; or, The anatomie of the bodies of man and woman.... English edition by John Ireton....* London, Joseph Moxon, 1675. (Krivatsy, 9555).

Russell, K.F. Johann Remmelin. *Aust N Z J Surg* 23: 145–147, 1953.

Russell, K.F. *A bibliography of Johann Remmelin the anatomist.* East St Kilda, VIC, Australia, J.F. Russell, 1991.

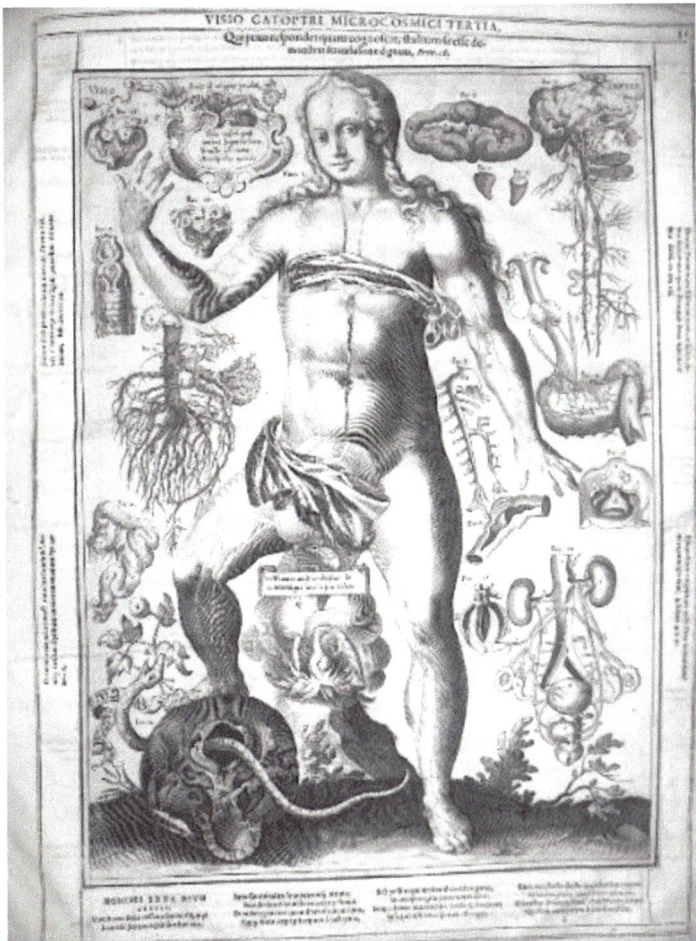

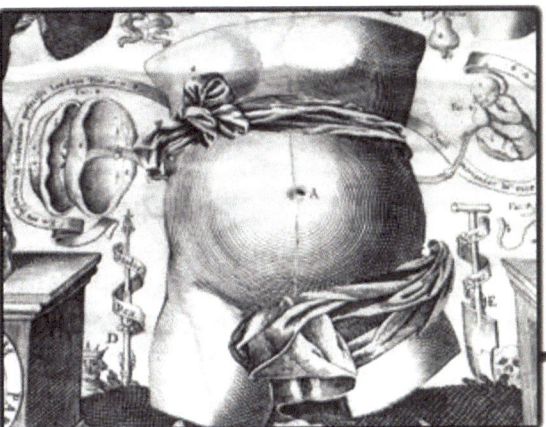

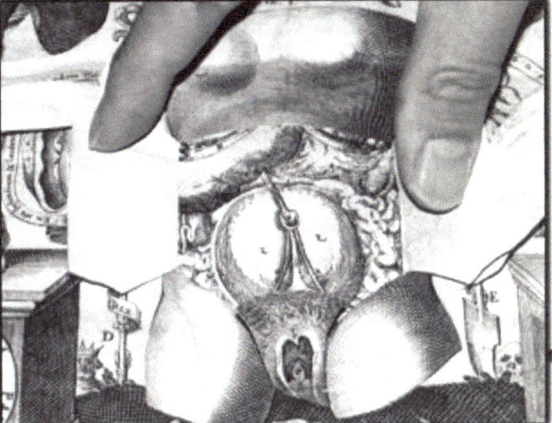

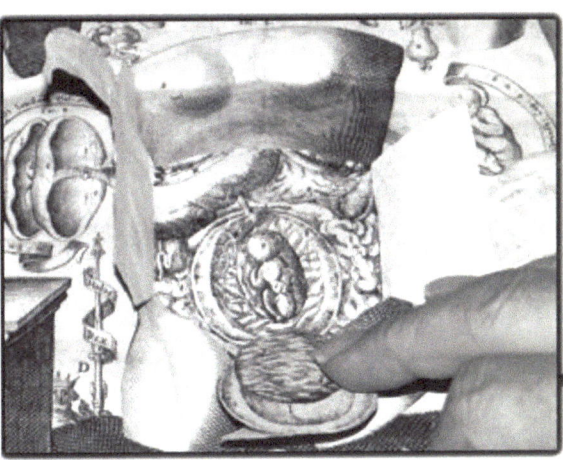

PLATE WITH FLAPS FROM REMMILIN, 1619. COURTESY HARDIN LIBRARY, UNIVERSITY OF IOWA

Adriaan van de Spieghel [Spigelius]

De formato foetu liber singularis, anaeis figuris exornatus.
Epistolae duae anatomicae.... Studio Liberalis Cremae.
Patavii, Apud lo Bap de Martinis & Livium Pasquatu, 1626.

During the seventeenth century, anatomic dissection reached a high state of competence. Concomitantly, a new technology, the use of copper plate engravings replaced wood blocks, allowing illustration to mark a new epoch of aesthetic sensibility. One of the most exquisite of the anatomic atlases published in that time, and the most striking illustrations in the history of anatomy, was that attributed to van der Spieghel [Spigelius] (1578–1625). In 1616, the Venetian Senate appointed Spieglius to succeed the recently deceased Giulio Casserio [Casserius] (1561–1616) as professor of Anatomy and Surgery at the University of Padua. Casserius had supervised the execution of superb Carreggio-like plates for a work he planned to publish, the *Theatrum anatomicum*. The artist is believed to have been Odoardo Fialetti (1573–1638), a pupil of Titian [Tiziano Vecelli (1477–1576)] and the engraver Francesco Valesio (ca. 1560–1648). Spigelius had written a treatise on the pregnant uterus, placenta, and infant. After the deaths of both Casserius and Spigelius, both of whom had studied under Girolamo Fabrizio [Fabricius ab Aquapendente] (ca. 1533–1619) in Padua, the Polish pupil or German physician Daniel Rindfleisch [Bucreitus] (ca 1570–1631) carried out his promise to Spigelius to publish his unillustrated treatise on anatomy, enlisting the help of Spieglius' son-in-law Liberalis Crema (fl. 1626) who had acquired a number of Casserius' plates from the latter's grandson. Thus, at his own expense Crema published with Bucreitus the *De formato foetu...*, which consisted of the text by Spiegelius and plates by Casserius.

The nine plates deal with the pregnant uterus, placenta, and the child, and are among Casserius' most beautiful engravings. Four depict full-length female figures against a landscape background, showing the gravid uterus and fetus in utero. Plate IIII illustrates a near-term pregnant nude, showing off her perfectly formed female infant (G) with placenta and umbilical vessels (D, E, F), chorion (A) and amnion (B) membranes, and opened uterine wall (C). Two plates show the uterine interior with placenta (Tab. V, next page), one shows the placenta itself (Tab. VI, next page) and three plates illustrate infants and their organs (e.g., Tab. VII). "The illustrations for this ambitious project marked a significant departure from the Vesalian prototype that had dominated illustrated anatomical texts throughout Europe during the last half of the sixteenth century. Its illustrations were innovative and inventive, but they had little influence on subsequent publications" (Cazort et al. 1996, p. 167).

At the time of donating his collection of books to the Boston Medical Library, Oliver Wendell Holmes (1809-1894) Professor of Anatomy at the Harvard Medical School observed regarding these 'eviscerated beauties' "The figures... will always attract attention, for the grace and beauty of the females who display their viscera as if they were jewels and laces. These are not likely to be overlooked by the lovers of undisguised nature and naked truth." (Holmes 1889). Several decades earlier, Holmes had penned a poem that was published only long after his death; it read,

> So the stout fetus, kicking and alive,
> Leaps from the fundus for his final dive.
> Tired of the prison where his legs were curled,
> He pants, like Rasselas, for a wider world.
> No more to him their wonted joys afford
> The fringed placenta and the knotted Cord.
> (Holmes, 1838? see note in References)

In his Foreword to a volume on *The Physiology of the Newborn Infant* (Smith 1945), Frederick Carpenter Irving (1883–1957), Harvard professor of obstetrics, included this poem in his essay, noting, "... never in the later life of man do such climatic changes occur in so short a time. The onset of puberty is gradual, the period of senescence consumes many years, even death itself for some may be a lingering event; a few seconds, however, suffice for the first breath, the adequate expansion of the lungs, and the adjustment of the circulation to pulmonary respiration.... Above all [obstetricians and pediatricians] should recognize that the newborn infant presents certain problems of its own and that it is not merely a very young baby" (Smith 1945, p vii).

Spigelius was the first to commence his work by describing the external female genitals (previously all such works had begun with the description of the uterus). He first observed the occurrence of milk in female breasts at birth, denied the presence of nerves in the umbilical cord, and argued against the idea that meconium in its intestines implied that the fetus ate *in utero*. Spigelius believed that maternal blood flows directly from the uterine vessels into the ends of the umbilical vein, thus rejecting Giulio Cesare Aranzio's (1530–1589) view. However, he correctly stated that the umbilical arteries terminate in the placenta. He also held that an important function of the placenta was to prevent more than minimal blood loss at the time of delivery, as would be the case if the two circulatins were joined by large vessels.

Spigelius, a native of Brussels, studied medicine at the universities of Louvain, Leiden, and Padua. After a short period of traveling and working in Belgium, Germany, and Moravia, he settled in Padua. He improved upon anatomical terminology, and his name is eponymized in the Spieglian lobe (caudate lobe) of the liver and van der Spieghel's line (*the linea semilunaris*). He also published the first extensive account of malaria (1624).

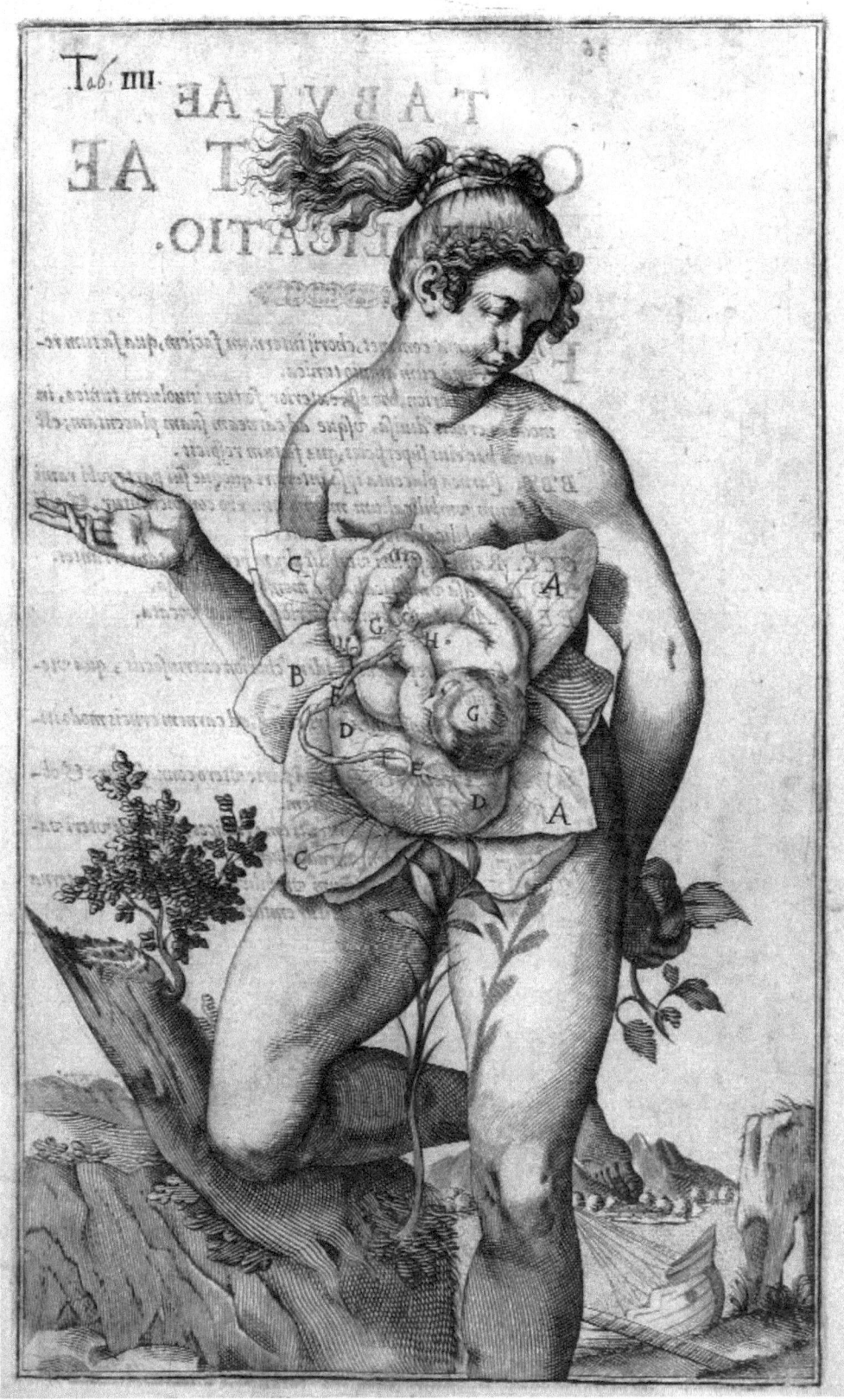

Tab III from Spigelius, 1626

References

See Garrison Morton 61.2; Choulaut-Frank, p. 226; Heirs of Hipprocates 282 or 415; Krivatsy 11295; Osler 4005, 4006.

Casseri G. *Tabulae anatomicae Ixxiix*, Venatiis, apud E. Devehinum, 1627.

Cazort, M., M. Kornell and K.B. Roberts. *The ingenious machine of nature: four centuries of art and anatomy*. Ottawa, Ottawa National Gallery of Canada, 1996.

Holmes, O.W. Address before the Boston Medical Library Association. *Boston Med Surg J* 120:129–130, 1889.

Holmes, O.W. *An unpublished poem, for the Boston Society of Medical Improvement*. February 7, 1838? (*Uncertainty as to date appears on page 321 of The Poetical Works of Oliver Wendell Holmes, Cambridge Edition, revised and with a new introduction by Eleanor M. Tilton*). Boston, Houghton Mifflin and Company, 1975. (see Smith and Nelson, 1976).

Lindeboom, G.A. Adriaan van den Spiegel. In: *Dictionary of Scientific Biography. Vol XII Charles Coulston Gillispie (Ed)*. New York, Charles Scribner's Sons, 1975, pp. 577–578.

Smith, C.A. *The physiology of the newborn infant*. Springfield, Ill., C.C. Thomas, 1945, p. vii.

Spieghel, A. van de. *Isagoges in rem herbarium libri duo....* Patavii, Apud Paulum Meiettum... 1606.

Speighel, A. van de. *De semitertiana libri quatuor*. Francofurti, Apud haeredes J.T. de Bry, 1624. (GM 5229).

Spieghel, A. van de. *De humani corporis fabrica libri decem, tabulis XCIIX aeri incisis elegantissmus... Daniel Bucretius....* Venetiis, Apud Euangelistam Deuchinum, 1627.

Spieghel, A. van de. *Opera quae extant, omnia. Ex recensione Joh. Antonidae vander Linden....* Amsterdami, Apud Johannem Blaeu, 1645. (GM 61.2).

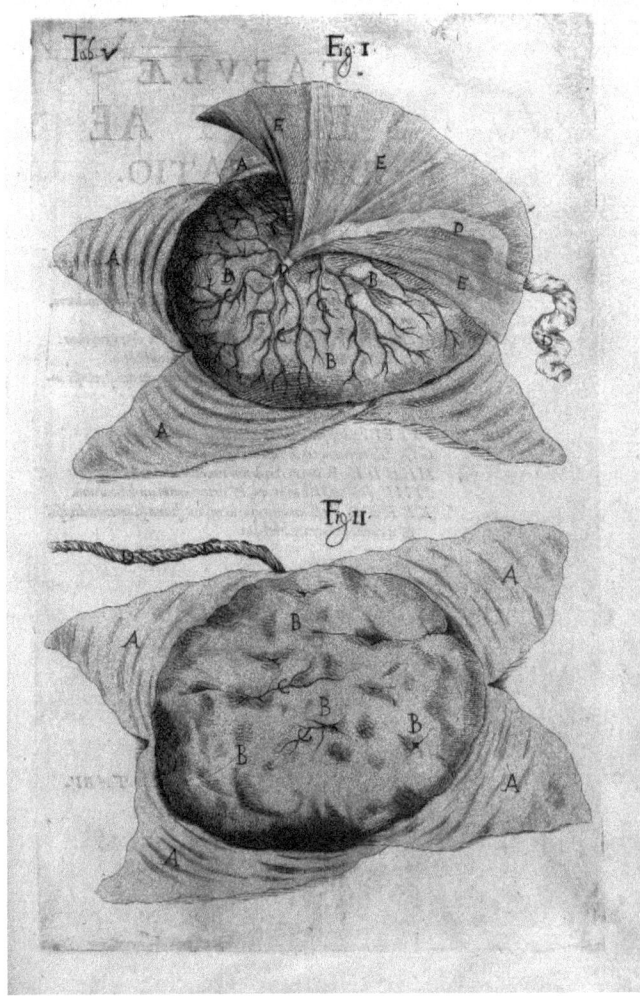

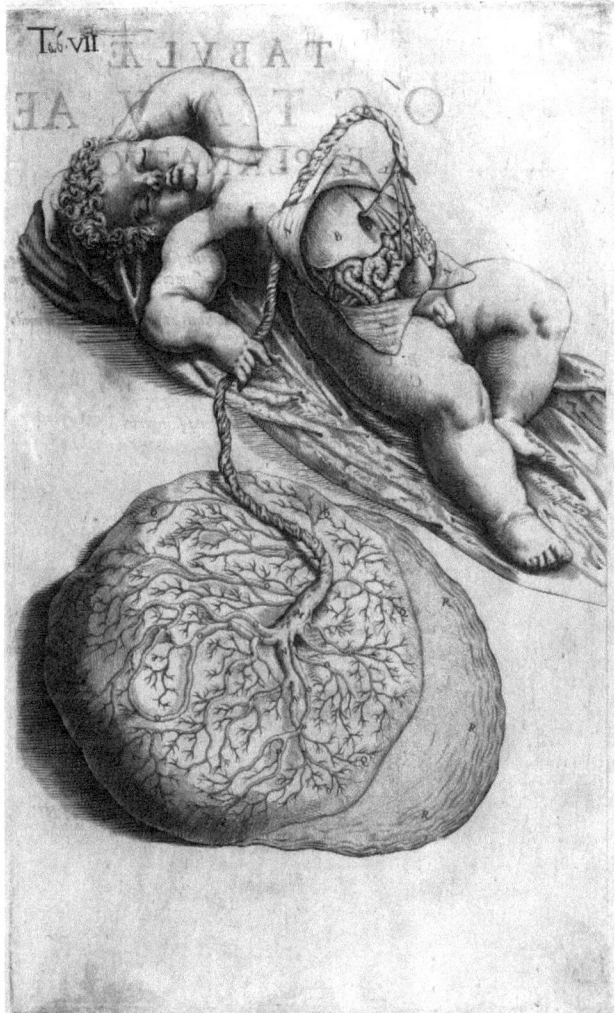

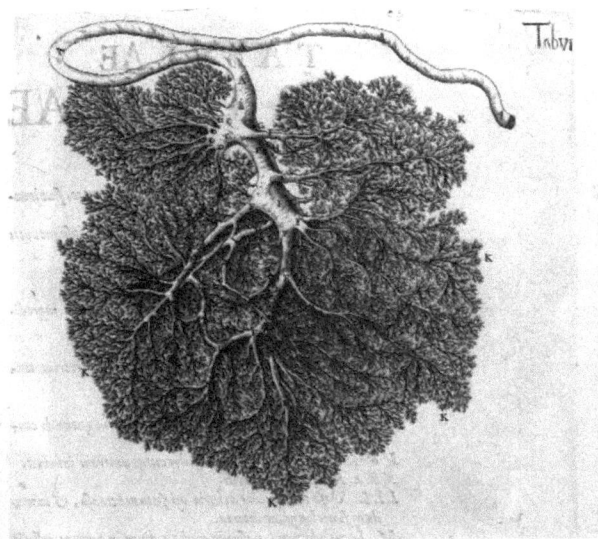

Tab V, VI, and VII from Spigelius, 1626

Jean Claude De La Courveé

De nutritione foetüs in ütero paradoxa. Dantisci [Danzig], Sumpt Georgii Försten, 1655.

In his studies, De La Courveé (1615–1664), one of the first adherents to William Harvey's *De Generatione...*(1651), gave considerable insight into the role of the placenta in fetal nutrition. Although supporting William Harvey's views on embryogenesis he held that the fetus breathes *in utero* and in addition that the placenta serves as a lung for the fetus. Incorrectly, he also taught that the fetus is nourished by the amniotic fluid. He also believed the placenta to be only loosely connected to the uterine wall, rather than being tightly united with it. Contemporary seventeenth century views on the source of fetal nutrition included the menstrual blood, amniotic fluid, which the fetus either drank or passed through pores in its skin, drinking uterine milk, or via the blood of the umbilical cord.

The allegorical title page of this work depicts two cherubims announcing the arrival of an infant with two suns shining, on a fanciful cart with the insignia *ubicunque* [wheresoever, anywhere] that is pulled by two doves. In the lower left hand corner is depicted a seated fetus draped in a uterus and fetal membranes, with the words *qui quasi flos egreditur* [he who comes forth as it were blooming in freshness]. An opened volume in the lower right hand gives De La Courvee's name as author, with the privilege of *Regina Polonia et Sueciae medico* [The Queen (or Princess) of Poland and her physician].

De La Courveé also believed that the fetus plays an active role in effecting its own delivery during the process of parturition. Although first suggested by Severin Pineau (1592), De La Courveé reported a case of symphysiotomy on a woman who had died during labor (p. 245).

De La Courveé studied medicine in Paris, and then practiced in Agenteuil in northern France. After serving as a physician to Queen Christina (1626–1689) of Sweden, he settled in Poland, where he published this volume on fetal nutrition.

References

Garrison Morton 6146.1; Krivatsy 6541.

Harvey, W. *Exercitationes de generatione animalium. Quibus accedunt quaedam de partu: de membranis ac humoribus uteri: & de conceptione.* Londoni, Octavian Pulleyn, 1651.

Pineau, S. *Oposculum physiologum, anatomicum...libris duobus distinctum.* Paris, 1592.

TITLE PAGE FROM DE LA COURVEÉ, 1604

Hendrik van Roonhuyse [Roonhuyze, Roon-Huyse]

Heel-konstige aanmerkkingen betreffende de gebreeken der vrouwen. Amsterdam, weduwe van Theunis Jacobsz, 1663.

In his "historical and surgical observations", van Roonhuyze (1625–1672), municipal surgeon of Amsterdam, is considered to have written the first textbook of operative gynecology. The plate illustrates a near-term pregnant woman displaying her opened abdominal wall (A, A) with uterus (F), exteriorized fetus, placenta (D), umbilical cord (E, E), intestines (B, B), and urinary bladder (G). Also shown are plates of van Roonhuyze's vertical curvilinear incision for Caesarean section through the left rectus muscle (p. 48, next page), and the sutured abdominal wall (p. 62, next page). (Note the incision sites on opposite sides of the midline). While admitting that the procedure was dangerous (it was outlawed in Paris at that time) he reported several successful cases, and stated that occasions arise when "there was no other means… to save the fruit and the mother".

van Roonhuyze, a graduate of the University of Amsterdam, apparently was the first to propose a rational operation for vesico-vaginal fistula, which he visualized with the aid of a *speculum vaginae*. He described the technique of carefully dissecting the urinary bladder free from the vaginal wall, and fastening the fistula edges together with quills which he tied together with silk threads. He did not record any successful cases, however. van Roonhuyze also wrote on vaginal prolapse, imperforate hymen, vaginal stenosis, and illustrated other operations and instruments that he employed. He also included a chapter on the surgical repair of harelip in young children (1673), and a work on the gout (1676). It is believed that in about 1663 while in The Netherlands, Hugh Chamberlen, the Elder (1630–ca. 1720) sold a portion of the one blade "lever" of the, until then, secret midwifery forceps, to Rogier van Roonhuyse (ca. 1650–1709) son of Hendrick (see Herbiniaux, 1782; Radcliffe, 1947).

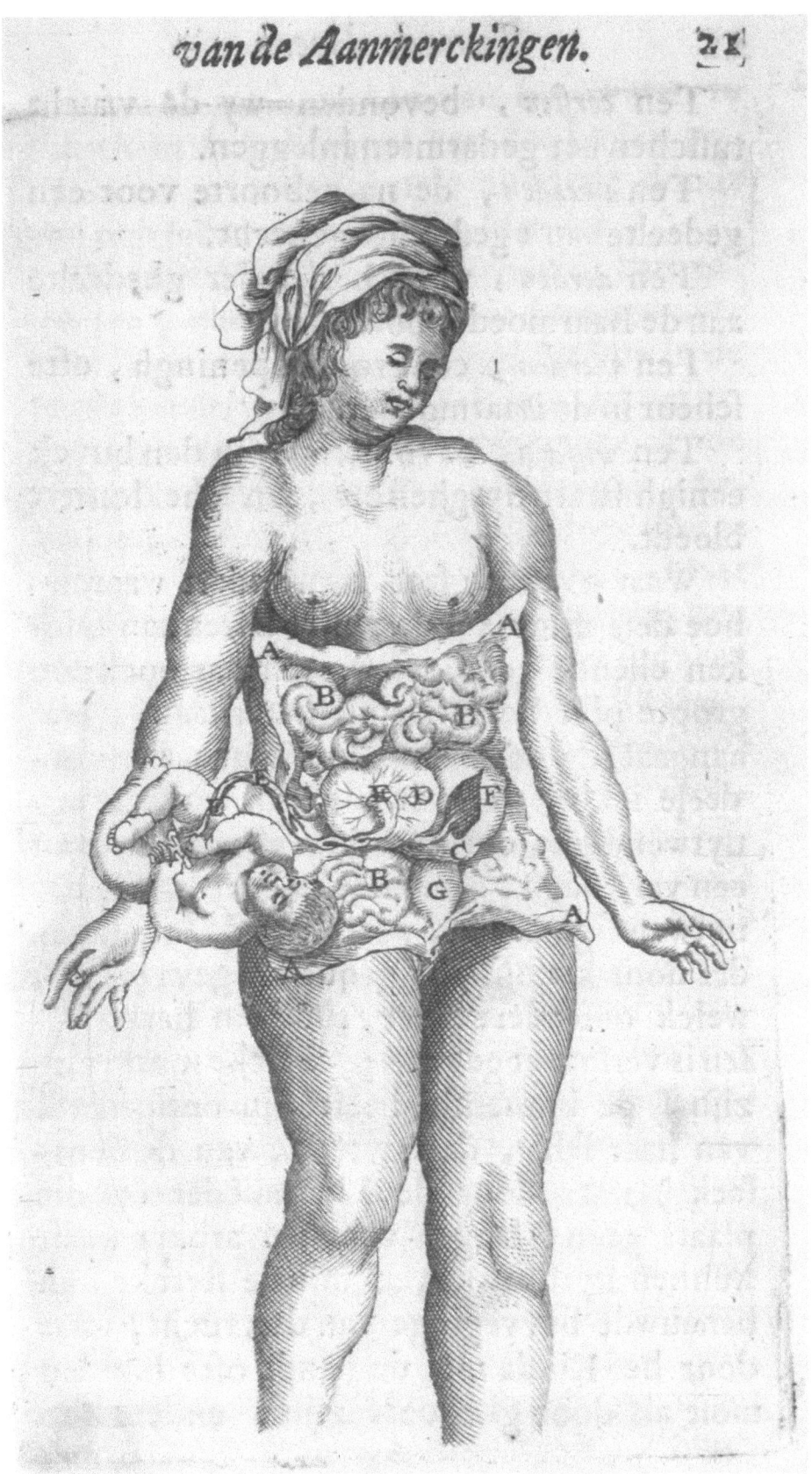

PLATES FROM ROONHUYZE, 1663

References

Cutter and Viets, p. 52; 191; Garrison Morton 6015; Krivatsy 9930

Busschof, H. and H. van Roonhuyze. *Two treatises, the one medical, of the gout, and its nature more narrowly search'd into than hitherto; together with a new way of discharging the same*. London, Printed by H.C. and sold by Moses Pitt, 1676.

Graham, H. *Eternal Eve*. London, Heinemann, 1950, p. 259.

Haeseker, B. Historical notes on 50 years of plastic surgery in the Netherlands. *Eur J Plast Surg* 23:163–167, 2000.

Herbiniaux, G. *Traité sur divers accouchemens laborieux, et sur les polypes de la matrice....* Bruxelles, J.L. de Boubers, 1782. (Blake, p. 208).

Leonardo, R.A. *History of gynecology*. New York, Froben Press, 1944.

Radcliffe, W. *The secret instrument (the birth of the midwifery forceps)*. London, William Heinemann, 1947, p. 30 ff. (GM 6311).

Ricci, J.V. *The genealogy of gynaecology; history of the development of gynaecology throughout the ages, 2000 B.C.–1800 A.D.* Philadelphia, Blakiston, 1943.

Roonhuyse, H.v. *Genees en heel-konstige aanmerkingen....* Amsteldam, De weduwe van Theunis Jacobsz Lootsman, 1672. (Krivatsy 9929).

Roonhuyse, H.v. *Historischer Heil-Curen in zwey Theile vergassete Anmerckungen....* Nürnberg, In Verlegung Michael und Johann Friederich Endtern, 1674. (Krivatsy 9932).

Roonhuyse, R. *Klaare bewyzen dat het berugte geheim in de vroedkunde*. Amsterdam, 1747. (Blake, p. 387).

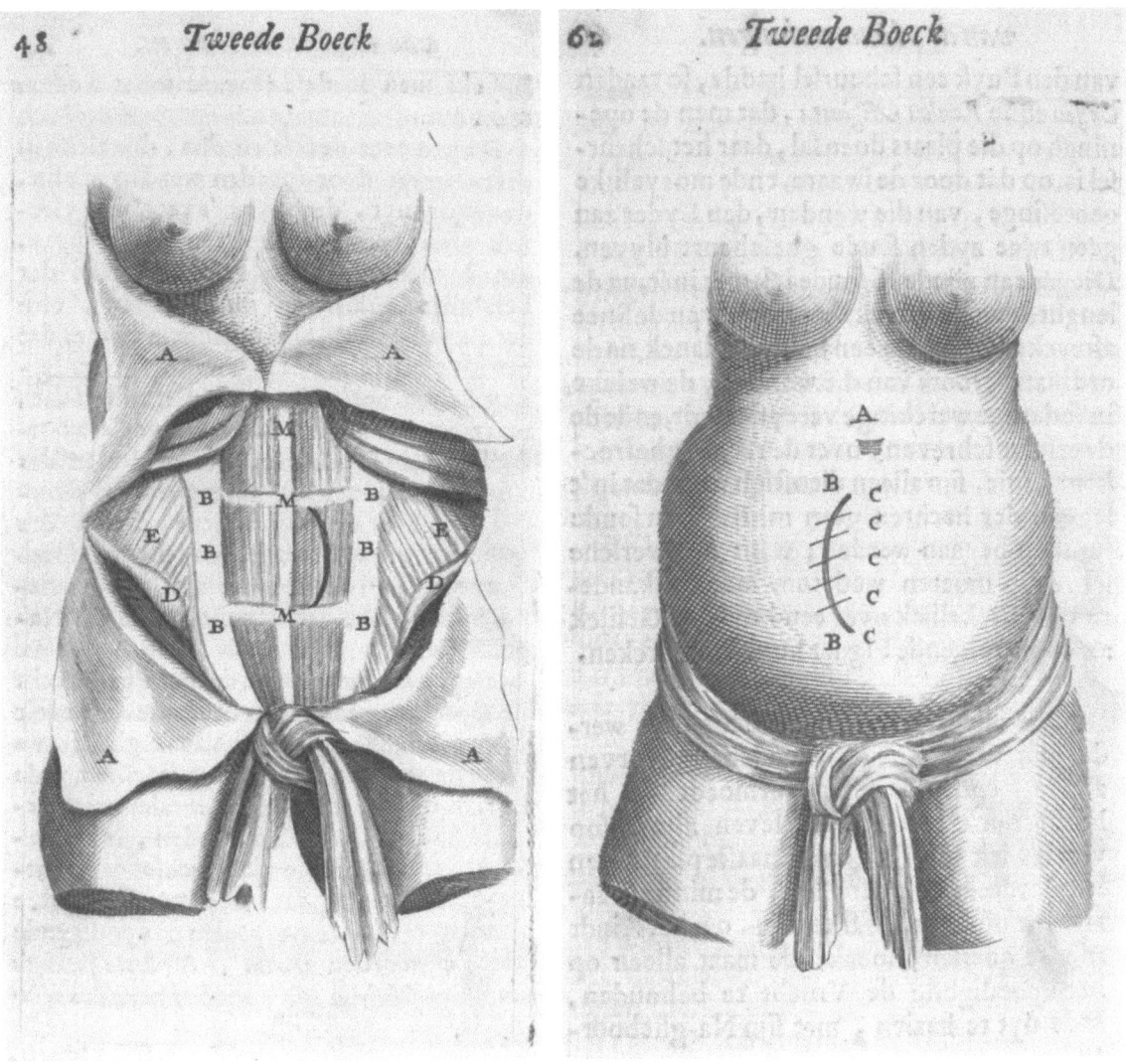

PLATE FROM ROONHUYZE, 1663

François Mauriceau

Des maladies des femmes grosses et accouchées. Avec la bonne et veritable méthode de les bien aider en leurs accouchemens naturels, ... Paris, Chez Jean Henault et al...., Imprimeries de Charles Coignara, 1668.

Alleged to be the most outstanding textbook of the mid-seventeenth century, was that of the *chirurgien accoucheur*, Mauriceau (1637–1709), *Des maladies des femmes ...* [The diseases of pregnant women and in child bed]. This work helped to establish obstetrics as a separate specialty and science, and through its various translations, exercised a dominant influence in latter seventeenth and early eighteenth century obstetrical practice throughout Europe. The text includes 30 copper plate engravings of birth figures and obstetric instruments.

Perhaps the first obstetric text in the modern sense, it treats the subject with logical order, clarity, and erudition. Its lengthy subtitle recommends it as useful for surgeons and necessary for midwives. Much of the work was a synthesis of prior teachings. In the preface, Mauriceau advised the readers, "The doctrine of books, which is one of the most wholesome effectual remedies we have to chase away ignorance, is wholly useless to men's wits, when not disposed to receive it." Although Mauriceau recommended the reading of other "learned" authors, he cautioned, "...the most part of them, having never practiced the art they under undertake to teach, resemble ... those geographers, who give us the description of many countries which they never have seen."

Following the introduction which describes anatomic landmarks, the work is divided into three sections which deal with diseases and abnormalities from conception to the end of pregnancy, and normal childbirth. In addition are chapters on the care of the mother and newborn infant, including the choice of a suitable wet nurse. Mauriceau devoted an entire chapter to the hygiene of pregnancy, beginning: "The pregnant women is like a ship upon a stormy sea full of white-caps, and the good pilot who is in charge must guide her with prudence if he is to avoid a shipwreck (p. 105)". He advocated, fresh air (p. 106), avoidance of extreme heat or cold (p. 106), freedom from smoke and foul odors (p. 106), and well-cooked wholesome food in small amounts at intervals rather than at one large meal (p. 108). He also advised low-heeled shoes (p. 110) to prevent the women from tripping (p. 110), and cautioned against whalebone corsets (p. 115) worn by women of the upper classes who wished to conceal their pregnancy. The expectant mother was admonished to help shorten the duration of labor by walking about the lying-in chamber, so that the weight of the child would help dilate the cervix, and thus that her pains would be stronger and more frequent. If a midwife perceived that the child was not presenting properly, Mauriceau cautioned her to "...send speedily for an expert and dextrous surgeon in the practice, and not delay, as too

many of them very often do, till it be reduced to extremity." He also admonished midwives to reassure their patients. In a critique of Mauriceau, one writer has stressed that sections of the treatise discussing relations with midwives, other surgeons, and physicians is an example of "blame narratives", written in cases of unfavorable outcome, to shift responsibility for a patient's death from the *accoucheur* to other attendants (McTavish 2006).

As may be evident, abnormal presentation of the near-term fetus may present difficulties for the *accoucheur* in terms of a safe delivery of mother and infant. In book II which describes childbirth that is both natural and that contrary to nature, the images of three fetuses in opened uteri illustrate transverse lie, shoulder presentation, and breech presentation.

Among important new ideas contributed by Mauriceau, were the abandonment of the birth stool with delivery of women in bed, treatment of various gestational periods, and discussion of many difficult cases. He also presented an analysis of the mechanism of labor, and maintained, contrary to the opinion of many at that time, that during labor the uterus is the active agent, while the fetus plays a passive role. He gave the earliest account of the prevention of congenital syphilis by antisyphilitic treatment during pregnancy. He was also the first to refer to tubal pregnancy, complications of prolapse of the umbilical cord, and epidemic puerperal fever. Mauriceau practiced podalic version, but condemned both cephalic version and Caesarean section. He gave rules for the management of placenta previa, and again, in contrast to his contemporaries, advanced the concept of primary repair of perineal lacerations. Mauriceau discredited the belief of Ambroise Paré (1510–1590) and others that the pubic bones separated during childbirth. He also denied that the uterus contains two cavities, and that the amniotic fluid is an accumulation of menstrual blood or milk. He argued against the common misconceptions that a child born at 7 months gestation had a greater chance of survival than one born at 8 months, and that because of the presence of two breasts, a woman can give birth only two children at a time. In the third edition of 1681, he first described management of the aftercoming head in breech delivery with the aid of an index finger in the infant's mouth, now referred to as the "Mauriceau-Smellie-Viet" maneuver. In the fourth edition of 1694, Mauriceau reported birthweight, noting that a singleton fetus at 9 months ordinarily weighs about 12 pounds, and he had even seen some weigh as much as 14 pounds (Mauriceau 1694). Mauriceau's somewhat excessive estimate of the weight of a normal term infant was topped by several other contemporary authors. It was not until mid-eighteenth century that the German obstetrician Johann Georg Roederer (1727–1763) recorded the correct weight and length of a newborn infant (Roederer 1753, see Cone

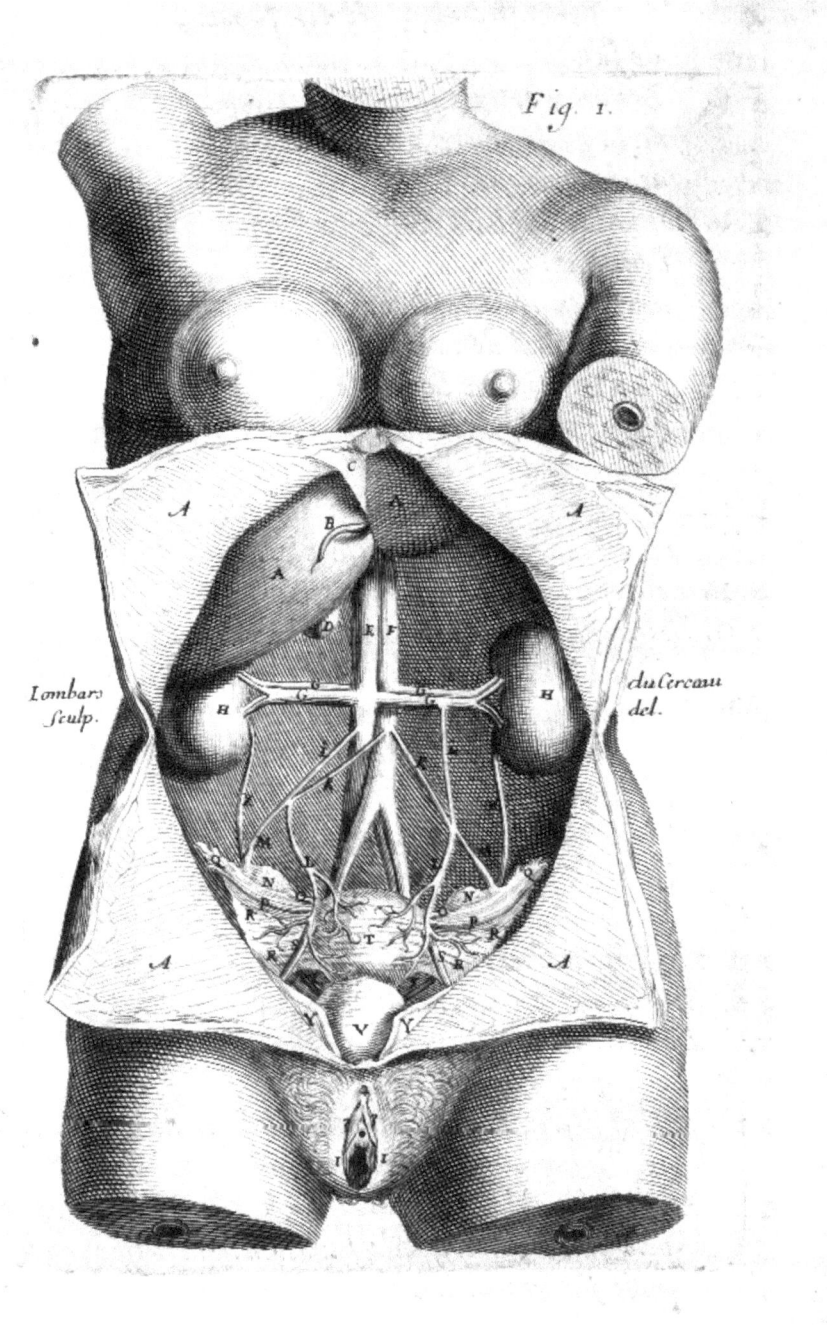

PLATE FROM MARICEAU, 1668

1961). With this fourth edition, Mauriceau appended a collection of 283 aphorisms concerning pregnancy, delivery, and the diseases of women.

Mauriceau was born in Paris, and practiced in the maternity wards of the *Hôtel Dieu*, where he gained considerable experience and established a brilliant reputation for himself. Although not a graduate in medicine, he was a sworn *chirurgien* of Saint-Come and was appointed *Maître accoucheur*. The volume also records the author's adventure with the celebrated Hugh Chamberlen (1630–1720) of the Huguenot family, who succeeded in keeping their invention of the obstetrical forceps a family secret for almost 200 years. Chamberlen translated the present work into English in 1673. From the third edition, Mauriceau published a Latin edition. The work was highly regarded for several generations.

References

Castiglioni, pp 555–6; Cutter & Viets, pp 51, 77–81; Garrison Morton 6147; Krivatsy 7588; Lilly, p 85; Norman, 33; Osler 5568.

Cone, T.E., Jr. De Pondere Infantum Recens Natorum [The history of weighing the newborn infant]. *Pediatrics* 28:490–498, 1961.

Mauriceau, F. *Traite des maladies des femmes grosses, et de celles qui sont nouvellement accouchées… Seconde edition*. Paris, Chez l'Auteur, 1675.

Mauriceau, F. *Traite des maladies des femmes grosses, et de celles qui sont accouchées… Troisíeme Edition*. Paris, chez l' Auteur, 1681.

Mauriceau, F. *Traite des maladies des femmes grosses, et de celles qui sont accouchées… Quatríeme Edition*. Paris, Houry, 1694.

Mauriceau, F. *The diseases of women with child, and in child bed…Translated by Hugh Chamberlen*. London, B. Billingsley, 1673.

Mauriceau, F. *The diseases of women with child, and in child bed…Translated by Hugh Chamberlen*. 2nd Edition. London, Printed by John Darby, 1683.

McTavish, L. Blame and vindication in the early modern birthing chamber. *Med Hist* 50: 447–464, 2006.

Robb, H. The writings of Mauriceau. *Bull Johns Hopkins Hosp* 6:51–57, 1895.

Roederer, J.G. *De pondere et longitudine infantum recens natorum. Comment Roy Soc Göttingen*, 1753, p. 140.

Speert, H. François Mauriceau and the Mauriceau maneuver. In: *Essays in eponymy, obstetric and gynecologic milestones*. New York, Macmillian Co., 1958, pp. 558–566.

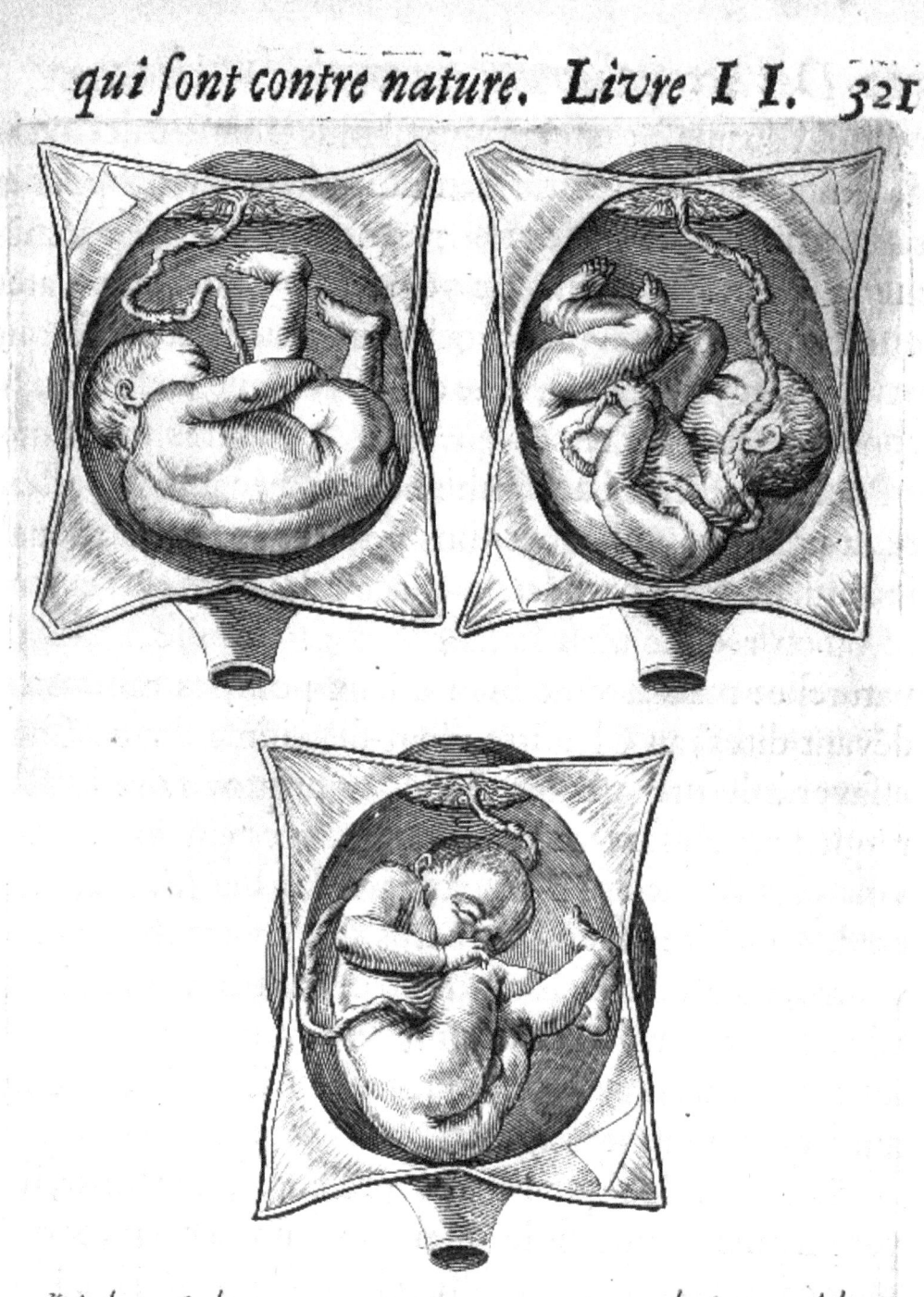

qui font contre nature. Livre II. 321

K.Audran ſcalp.   du Cerceau del.

PLATE FROM MARICEAU, 1668

Nicholaas Hoboken

Anatomia secundinae humanae, quindecim figuris ad vivum propriâ autoris manu delineates, illustrata... Trajecti ad Rhenum [Utrecht], apud Johannem Ribbium, 1669.

Hoboken (1632–1678), Professor of Medicine and Mathematics at the Universities of Utrecht and Harderwijk, first illustrated the vessels of the umbilical cord. Influenced by Fabricius ab Aquapendente's (1533–1619) demonstration of valves of the veins (1603), and William Harvey's (1578–1657) discovery of the circulation of the blood (1628), anatomists soon appreciated that the Galenic teaching that blood flow in the umbilical arteries and veins was in the same direction was untenable, and that the structure of these vessels differed from that of systemic vessels. Hoboken recognized intraluminal folds within the umbilical arteries, known as the "valves" or "nodes" of Hoboken. In Article VI, titled "The internal structure of the cord described in detail..." he wrote concerning these vessels,

1. And the interior of the cord being examined then, its substance is found to be all fibrous material, membrane, and fluid, and the smooth vessels contained within are arranged in artistic fashion. 2. Of the last, the two that proceed as arteries are thin and whitish; the third, more reddish to the view and thicker, returns alongside as a vein... 3. With respect to these structures, I intensified my efforts in examining and studying them, and was rewarded. Immediately I noticed (in addition to what I have already said concerning the differences in color, size, and thickness) a difference between the vein and the arteries: as I clearly discerned the vein almost in the middle of the section, maintaining a constant size. 4. The arteries in truth had the appearance of flexible tubes, greatly twisted into uneven shapes... 7. Next I was able to find a tiny nodule inside the vein: a tiny valve is also visible inside the same vessel. 8. But beyond the extent of this vessel, around the ramifications of the vein, and at their site of divergence in the placenta, one may be permitted to conjecture that there are various little valves. First, because the blood is held back by external pressure with the finger, and I was even able to move the blood contained within the cord back and forth. Even in the branches of this vein in the placenta this difference was observed: that I could move the blood toward the cord easily indeed, but less so in a reverse direction toward the placenta... 10. And I observed... that the various nodules, having the appearance of little buds, are joined together in the arteries and also in the cord. 11. While seeking these out, I discovered that these previously mentioned nodes are the result of collections of blood. See Figs. V, X, XIII. 12. And opening these, I found... circular valves, formed by the plication of a rather loose tunic, inhibiting the back flow of the arterial blood from the placenta toward the umbilicus of the fetus; the tunic of the little bud being distended in the middle by a copious collection of blood. See Figs. VI and VII" (Hoboken, 1669, pp. 28–34).

Considerable controversy has surrounded these "valves" and their function (Haller 1776). For instance, some have suggested they are past partum artifacts present only in the contracted vessels (Rouhault 1714; Verheyen 1710). In his treatise on the umbilical cord, Joszef Hyrtl (1810–1894) demonstrated circular valves in the umbilical artery, the semilunar variety in umbilical veins, and the dilitations seen on the outer surface of the arteries the *noduli Hobokenii* (Hyrtl 1870). More contemporary study has disclosed the markedly thickened media of the umbilical arteries, with the muscle fibers running in a spiral course (Stravinski 1876, Spivack 1936), and many believe these play an important role in the constriction of these vessels following birth to prevent excessive blood loss (Shordania 1929, Spivack 1936).

In his *Anatomia...*, Hoboken described the human placenta, umbilical cord, and fetal membranes. In nine plates with 15 figures which he drew by himself (eight of which related to the vascular folds), he described clearly the chorion as completely surrounding the amniotic membrane, and incorrectly emphasized that its most superficial layer is avascular not being penetrated by maternal vessels. These observations helped to establish the idea that vascular connections between the mother and fetus did not exist. He held that rather than being nourished by the mother's blood, the fetus was sustained by an "alimentary juice" secreted by this blood, and transferred to the fetus by "filtration and translocation". Hoboken stated that in the cow, during early months of gestation, the embryo was nourished by gelatinous juices secreted by glands lining the uterus, and these were absorbed by pores in the chorionic membrane. Later, the umbilical vessels from the fetus divide into 60–80 smaller vessels that radiate into the substance of the chorion, where they form protuberances called *placentulae* [little placentae] that approximate uterine wall structures called *glandulae* or *cotyledons* [caruncles of present day terminology]. Hoboken delineated the chorion as completely surrounding the outer membranes, not being penetrated by maternal vessels (Hoboken 1669, 1675).

A graduate of the University of Utrecht (1658, M.D. 1662), he was appointed chair of medicine and mathematics at Steinfurt, Westphalia where he became physician to Count Bentheim. In 1669, he became professor of medicine and of mathematics at the University of Harderwijk, which was not held in high esteem, but included among its graduates the

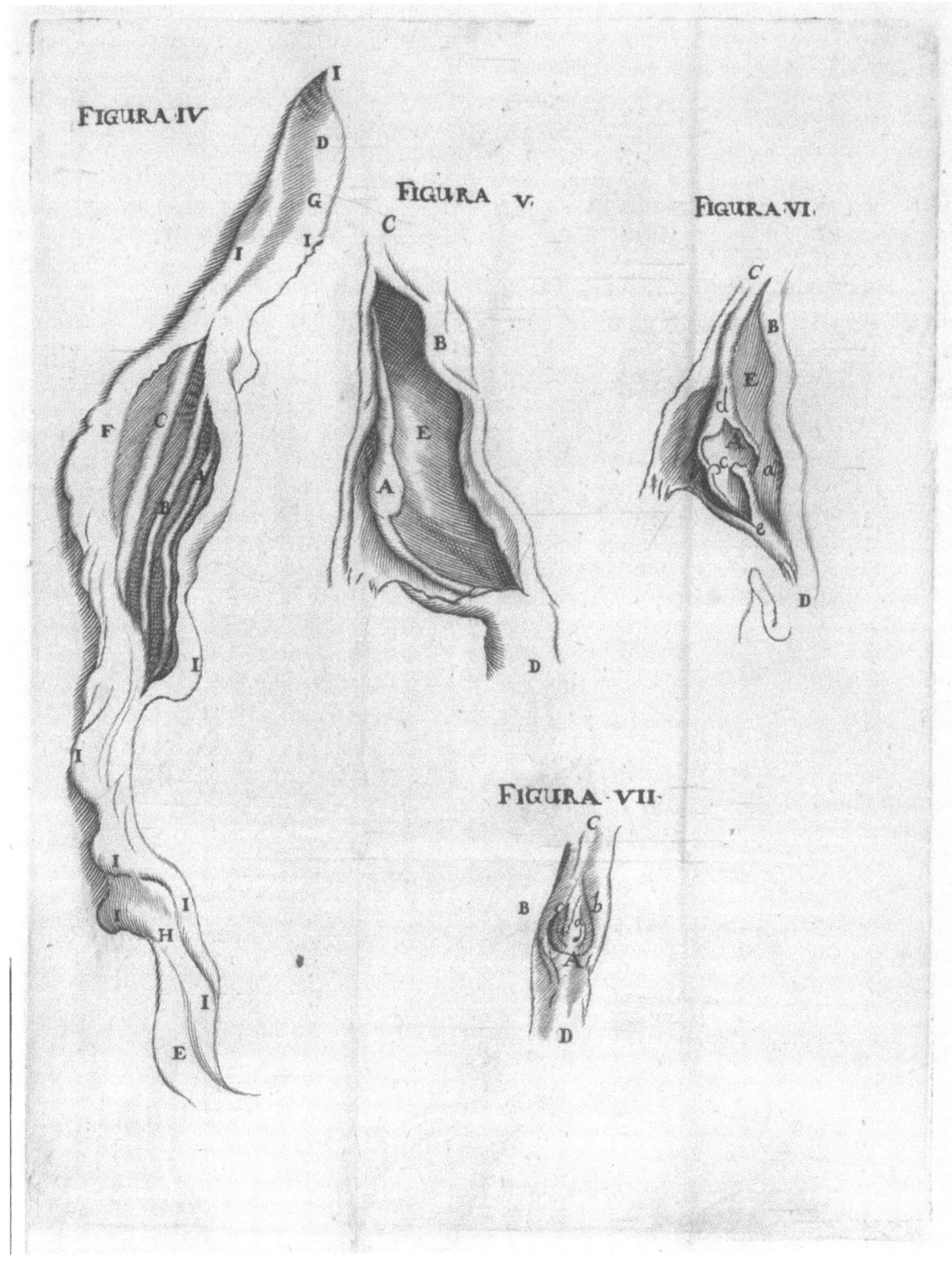

PLATE FROM HOBOKEN, 1669

physician Herman Boerhaave (1668–1738), and the Swedish zoologist Carl von Linné [Linnaeus] (1707–1778). Hoboken wrote on many topics including mathematics, philosophy, and politics. It is said that Hoboken was influenced greatly by Nicolaas Tulp (1593–1674), to whom he dedicated this volume, and who was notable as the demonstrator in Rembrandt Harmenszoon van Rijn's [Van Ryn] (1606–1669), "The Anatomy Lesson of Dr. Nicolaes Tulp" (1632), now in the *Mauritshuis* museum, Den Hague.

References

Cutter & Viets 222; Krivatsy 5745.

Fabrizio, G. *De venarum ostiolis*. Patavii, ex typ. L. Pasquati, 1603. (GM 757).

Haller, A.v. *Elementa physiologiae corporis humani. 8 vols.* Lausanne, 1757–1776. (GM 588).

Haller, A.v. *Bibliotheca medicinae practicae. 4 vols.* Basle, J. Schweighauser, Berne, E. Haller, 1776–1788. (GM 6747).

Harvey, W. *Exercitatio anatomica de mortu cordis et sanguinis in animalibus.* Francofurti, sumpt Guilielmi Fitzeri, 1628. (GM 759).

Hoboken, N. *Anatomia secundinae humanae repetia, aucta, roborata....* Ultrajecti [Utrècht], apud Johannem Ribbium, 1672.

Hoboken, N. *Anatomia secundinae vitulinae, triginta octo figuris... illustrata.* Ultrajecti, Apud Johannem Ribbium, 1675.

Hyrtl, J. *Die Blutgefässe der menschlichen Nachgeburt in normalen und abnormen Verhältnissen.* Wien, Braumüller, 1870, pp. 25–29.

Needham, J. *A history of embryology.* Cambridge, At the University Press, 1934, p. 175. (GM 533).

Rouhault, M. Du cordon umbilical. *Histoire de l'academie royale des sciences,* 1714, p. 312.

Shordania, J. Über das Gefässystem der Nabelschnur. *Ztschr für Anat u Entwicklgsch* 89:696–726, 1929.

Spivack, M. On the anatomy of the so-called "valves" of umbilical vessels, with especial reference to the "valvulae Hobokenii". *Anat Rec* 66:127–148, 1936.

Stravinski, [On the structure of the vessels of the umbilical cord and their occlusion after delivery. Dissertation]. St. Petersburg, 1876.

Verheyen, P. *Corporis humani anatomiae liber primus in quo tam veterum, quam recentiorum anatomicorum inventa.* Bruxellis, Apud Fratres t'Serstevens, 1710.

Verheyen, P. *Supplementum anatomicum, sive anatomiae corporis humani liber secundus; accedit descriptio anatomica partium foetui et recenter nato propriarum. Item, controversia de foramine ovali inter authorem et D. Mery.* (Lb. II, tract V). Bruxelles, t'Serstevens, 1710, p. 342.

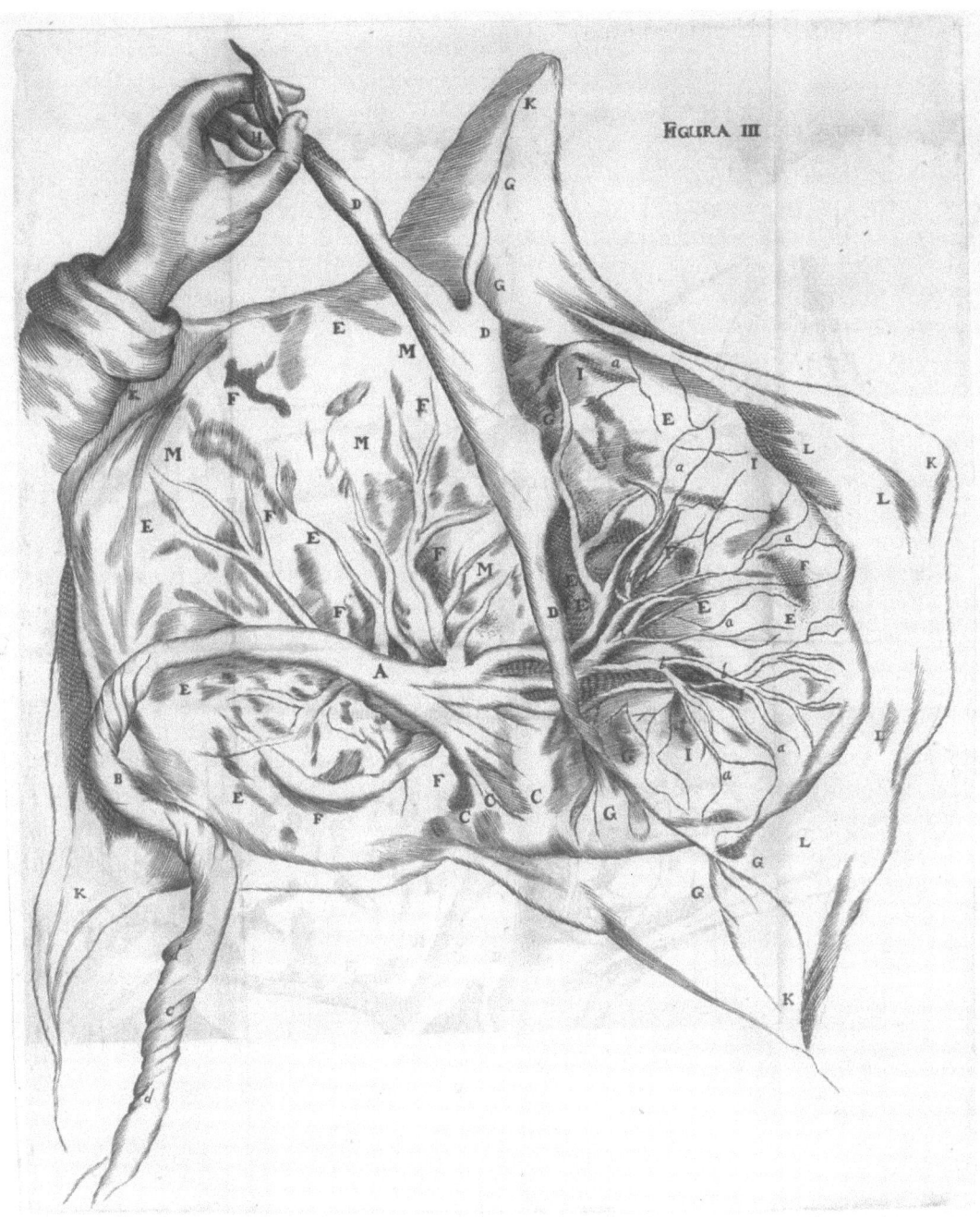

PLATE FROM HOBOKEN, 1669

William Sermon

The Ladies Companion, or, the English Midwife wherein is demonstrated the manner and order of how women ought to govern themselves during the whole time of their difficult labour ... also the various forms of the child's [sic] proceeding forth of the womb, in 17 copper cuts, with a discourse on the parts principally serving for generation/digested into a small volume... London, E. Thomas, 1671.

The seventeenth century man-midwife Sermon (ca. 1629–1679) recorded his motive for writing this volume, "The serious consideration of the intolerable misery that many women are daily incident to, occasioned chiefly by breeding and bringing forth children; and the want of help in such deplorable conditions, by reason of the unskillfulness of some which pretend the art of midwifery, & c, yet not in the least acquainted with the various diseases which frequently afflict the female sex in such times, hath been one principal motive to me at this time to undertake the publication of this treatise." In his first chapter, "the antiquity of midwives and what manner of women they ought to be," he deplored the practice of "cunning women" who "...desired to excel men, or at least would seem to go beyond them." (Sermon 1671). He observed, "Some ... (not wanting in ignorance), being over-hasty to busie themselves in matters they know not, destroy poor women by tearing the [fetal] membrane with their nails," and, in the process, sometimes hurting not only the mother, but damaging the infant, resulting in "... the death of many women and children too." (Sermon 1671).

A decade earlier, one of the first Englishmen to devote his practice entirely to obstetrics, Percivall Willughby (1596–1685) of Derby and London, completed a manuscript [*Observations in midwifery*] based on the patients that he had attended. Written in Latin, the manuscript which contains considerable wisdom was not published in English until 1863 (Willughby 1863).

References

Cutter & Viets pp 6–7, 178–179; Garrison Morton 6282.

Aveling, J.H. *English midwives, their history and prospects.* London, J. & A. Churchill, 1872, p. 40 ff. (GM 6282).

Dawson, J.B. A seventeenth century obstetrician Percival Willughby, Gentleman. *N Z Med J* 51:313–320, 1952.

Dunn, P.M. Dr Percivall Willughby, MD (1596–1685): pioneer "man" midwife of Derby. *Arch Dis Child* 76:F212-F213, 1997.

Phillips, M.H. Percival Willughby's observations in midwifery. *J Obstet Gynaecol Br Emp* 59:753–762, 1952.

Sermon, W. *A friend to the sick; or, The honest English mans preservation. Shewing the causes, symptoms, and cures of ... diseases ....* London, Printed by W. Downing for Edward Thomas, 1673. (Krivatsy 11024).

Thornton, J.L. *English midwives, their history and prospects by James Hobson Aveling... and biographical sketch of the author....* London, Hugh K. Elliott, 1967, p. 40ff.

Willughby, P. *Observations in midwifery, as also The country midwifes opusculum or vade mecum.* Henry Blenkinsop (Ed). Warwick, H.T. Cooke and Son, 1863.

Willughby, P. *Observations in midwifery by Percival Willughby (1596–1685). Edited from the original manuscript by Henry Blenkinsop 1863, with a new introduction by John L. Thornton, PLA.* East Ardsley, Wakefield, UK, S.R. Publishers Ltd., 1972.

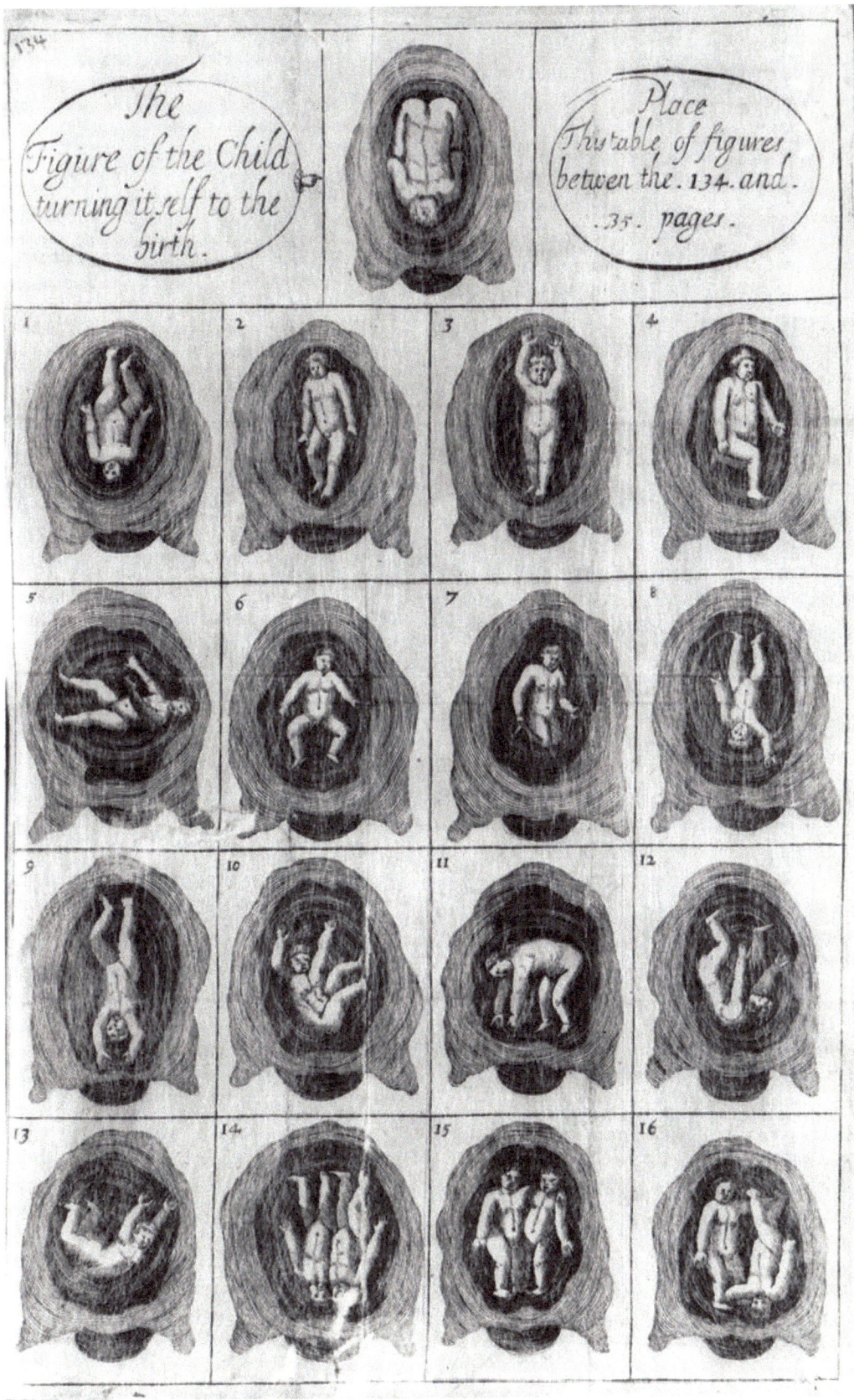

Figure from Sermon, 1671 (Courtesy The British Library)

Cosme Viardel

Observations sur la pratique des accouchemens naturels, contre nature, & monstrueux, avec une methode tres-facile pour secourir les femmes en toutes sortes d'accouchements, sans se servir de crochets, ni d'aucun instrument, que de la seule main... Paris, Edme Couterot; Nicolas Bessin; François Mauger, 1671.

In three "books" with 65 chapters on "Observations of the practice of obstetrics..." the Parisian physician Viardel (ca. 1640–1694) recorded a number of pregnancy complications, illustrating them with cases from his practice and various afflictions to which women were subject. Along with an engraved frontispiece, he included 17 plates by an unknown artist. In the preface, Viardel noted that it was because of the persuasion of some of his colleagues that he published these cases. He eschewed the use of instruments, emphasizing instead how the facile operator could use his hands to effect delivery of the most complicated cases. These included transverse lie with shoulder and arm presentations, breech presentations, successful conversion of a face to a cephalic presentation, as well as the management of placenta previa. For many complicated deliveries he advocated version and extraction, and was the first to recommend intrauterine flexion of the fetal arms at the elbow to facilitate breech extraction. Each case demonstrated a resourceful obstetrician, familiar with the complications of the art and skilled in the manipulations to effect delivery. In his rejection of instrumental delivery, Viardel avoided much of the brutality and malpractice in midwifery, at a time when mutilating and destructive instruments too frequently were rashly and unwisely used. He also described replacing an inverted uterus following manual removal of an adherent placenta. In addition, by use of continuous suture he successfully repaired a lacerated perineum 3 days following delivery.

In advance of his time, Viardel also detailed the high standards and qualifications for the *chirurgien accoucheur.*

> He should be well made, of middle age, both to have attained experience and to be able to support the labor and fatigue which he must undergo in his operations. He must be ambidextrous to operate equally with both hands, his hands must be long and slender, and the nails well cut, so as not to injure the womb in his operations. He must have good habits, but always modestly dressed, and not a boasting, bragging manner, so that he may have nothing to hinder him. Most of all, he should be virtuous, prudent, wise, well informed, a man of good sense, to invent methods on the spot, and to change the presentation when it is against nature. He should be soft in his words and agreeable in his conversation, so as to cheer the sick and encourage her in her pains or suffering, treating her kindly, making her understand that she shall soon be through her pains, and that he has come only to help and comfort her. But, above all, he must be prudent and discreet. Prudent to deliver his prognosis, he must be discreet and not reveal the secrets confided to him. We have to add that he should have perfect knowledge of anatomy, so as not to make mistakes in his operations .... In one word, he ought to be careful not to lose his humanity, and above all, to be charitable to the poor, and not to do his work for lucre and his own gain. In contrast, as the apostle said, one's work should be for the honor and glory of God.

(Viardel, 1671, pp. 279–282)

Little is known of Viardel's background or life. Despite his accomplishments and being an ingenious practitioner, he was maligned by François Mauriceau (1637–1709) as being dishonorable and unscrup-ulous. Viardel's single book was justly popular, setting forth succinctly his views on the practice of mid-wifery, and illustrating them with a collection of appropriate clinical observations.

References

Cutter & Viets p. 223; Durling, 12384; Krivatsy 12383; Waller 9946 (all without plates or with fewer than this copy).

Elliot, G.T. Jr. Historical and bibliographical notice of Cosmo Viardel. *New York Med J* 3:161–173, 1866.

Mauriceau, F. *Des maladies des femmes grosses et accouchées. Avec la bonne et veritable méthode de les bien aider en leurs accouchemens naturels, ...* Paris, Chez Jean Henault et al...., Imprimeries de Charles Coignara, 1668.

McTavish, L. Blame and vindication in the early modern birthing chamber. *Med Hist* 50:447–464, 2006.

Placet, E. *Etude historique sur les traites d'accouhcement de Viardel, Portal & Mauquest de la Motte.* Paris, Jouve, 1891.

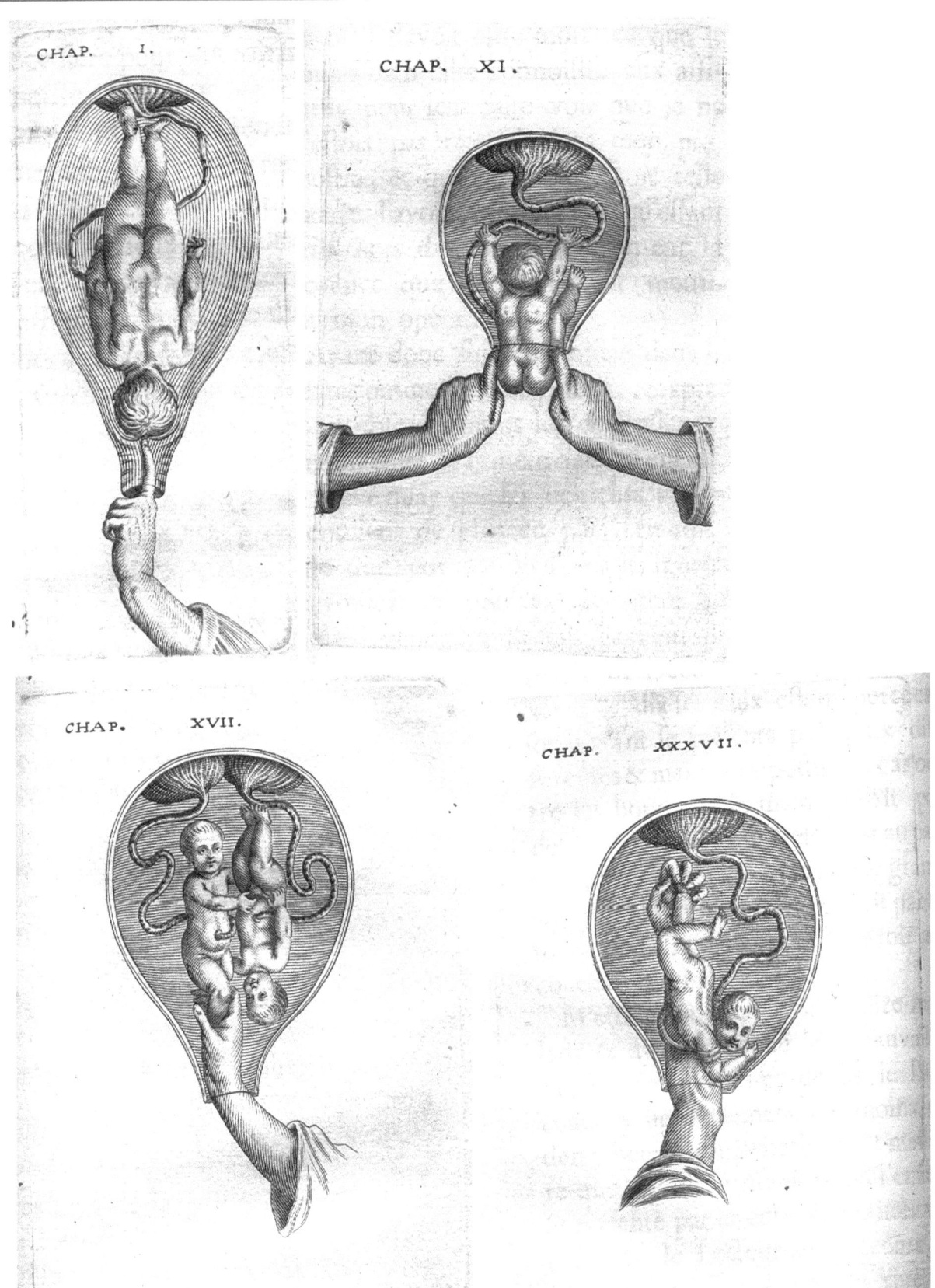

PLATES FROM VIARDEL, 1671

Johann Georg Sommer

Nohtwendiger Hebammen-Unterricht, wie eine Hebamme gegen schwangere, gebehrende und entbundene Weiber und deren Kinderlein, so wohl bey naturlichen als unnaturlichen Geburten sich zuerweisen; wie auch, wie sie bey solcher Weiber und Kinder Zufallen im Nohtfall einige diensame Mittel anzuwenden habe... Welchem noch beygefugt ist ein Weiber und Kinder Pfleg-Buchlein, wie auch eine Anleitung zur christlichen Kinderzucht... Arnstadt, Druckts Heinrich Meures, zufinden bey Matth. Bircknern in Jehna, 1676.

This pocket manual, "much-needed Instruction for Midwives, concerning how a midwife is to deal with pregnant women, women in childbirth, and women who have given birth, as well as their infants …," was first published by Sommer (1634–1705) in 1676. Its objective was to aid in the education of midwives, and help women of all social classes in the principality so that maternal and infant mortality would be reduced. It contains little that was original. The engraved frontispiece of this second edition shows a seventeenth-century "bionic" superwoman, managing her spinning-wheel with one hand and spoon-feeding an infant with the other, while breast-feeding a baby; two older children are praying. Another woodcut includes a full page birth-scene, with the reclined parturient with her hips raised, and a midwife standing by, presumably preparing to assist in the delivery. On the bedside table is an open devotional book, with the words *Gott unser Hülffe und Heil* [God our helper and healer]. A number of other woodcuts show birth figures of the fetuses in various positions. Sommer included six prayers for the women during pregnancy and childbirth. These included asking for divine assistance and a fruitful outcome in pregnancy, a prayer to recite during labor, one for a patient who is dying during childbirth, another to thank God for a happy outcome, one for the woman's older children to pray during their mothers labor, and a final prayer for the midwife in spiritual preparation for her role in assisting the delivery.

In his Preface, Sommer argued for the education of midwives, "because unfortunately it is clear that there are midwives who conduct their vital work not only with great ignorance, but also with little diligence". He continued, stressing the need of learning the art of craft from experienced teachers, rather than on an *ad hoc* basis by "practical experience." In the initial chapter of his handbook, Sommer considered the responsibility of local governing bodies to ensure that parturient women had access to well-trained midwives of piety, character, skill, and considerable experience,

who themself had borne children. The following chapters present explicit instruction on prenatal care, and the management of both normal and abnormal childbirth.

Sommer was court physician to Count Ludwig Günther (1621–1681) in Arnstadt, capital of the Barony of the principality Schwarzburg-Sondershausen. In Thuringia, the twin principality was Schwarzburg-Rudolstadt. These, with other German principalities, had suffered greatly from economic and population losses during the devastating Thirty Years War (1618–1648), and struggled to recover for decades. Ordinances were passed to professionalize midwifery, with the view to help restore and augment the population. In this era, a number of *Gebet-büchlein* [small prayer book or breviary] of devotional songs and prayers were compiled for the use of women during pregnancy and childbirth. Their object was to help the woman express her thanks for conceiving, her hopes and fears regarding her pregnancy, and to receive assurance from the Lord as to her and her infant's safety in the face of suffering. The term "*Habermännle*", which has been used in a generic sense for such volumes, was named for Johann Habermann (pseudo. Avenarius; 1516–1590) whose prayer book is a notable example of such works (Habermann 1567; 1672). Another example is that of a prolific hymn writer Ämilie Juliane Countess of Schwarzburg-Rudolstadt (1637–1706) (1683). (For review of such works and German midwifery of this era see Aiken 2003, 2004).

References

Krivatsy, 11205; Wellcome, p. 149.

Aiken, J.P. Gendered theologies of childbirth in early modern Germany and the devotional handbook for pregnant women by Aemilie Juliane, Countess of Schwarzburg-Rudolstadt (1683). *J Women's History* 15: 40–67, 2003.

Aikin, J.P. The welfare of pregnant and birthing women as a concern for male and female rulers: A case study. *Sixteenth Century J* 35: 9–41, 2004.

Gleixner, U. Die "Gute" und die "Böse". Hebammen als Amts-frauen auf dem Land (Altmark/Brandenburg, 18 Jahrhundert). In: *Weiber, Menscher, Frauenzimmer: Frauen in der ländlichen Gesellschaft, 1500–1800. Ed. Heide Wunder and Christina Vanja.* Göttingen, Vandenhoeck & Ruprecht, 1996, pp. 96–122.

Juliane, E. [Gräfin von Schwarzburg und Hohenstein] *Geistliches Weiber-Aqua-Vit; das ist Christliche Lieder und Gebete....* Rudolstadt, Christoph Fleischer, 1683.

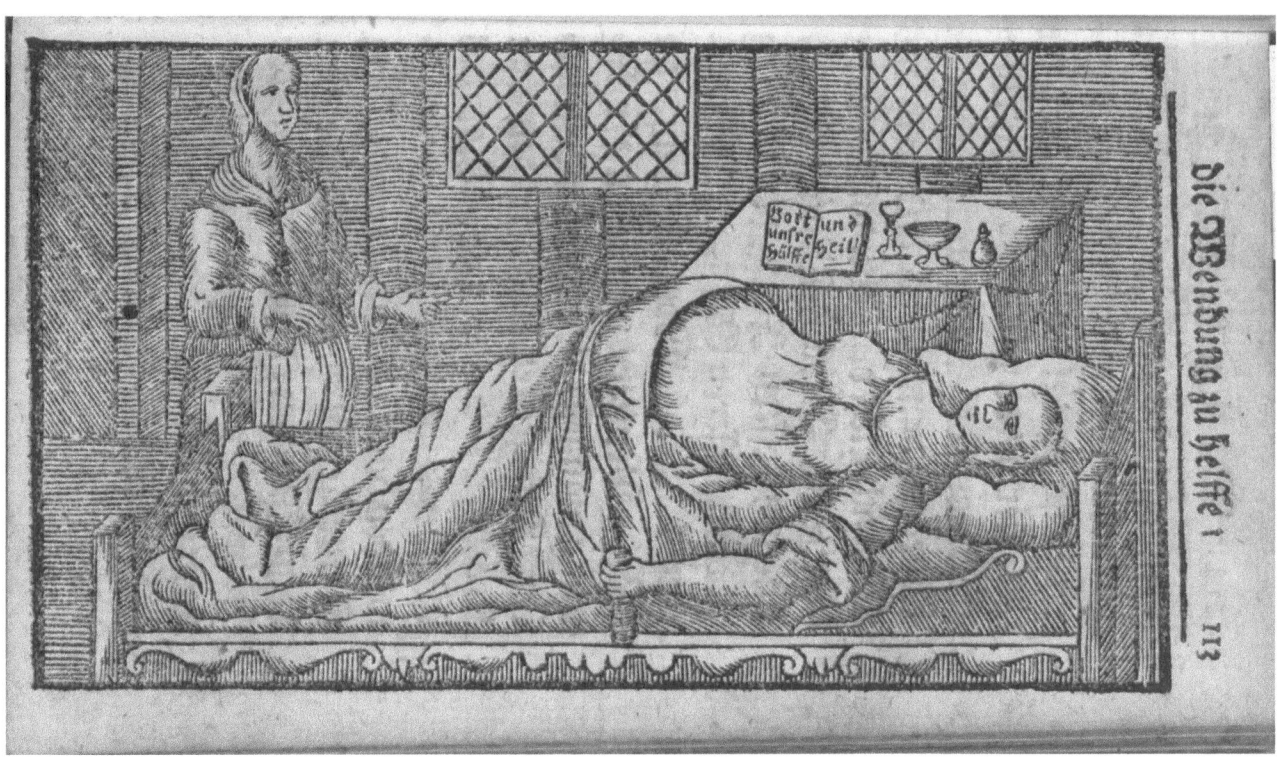

PLATE FROM SOMMER, 1676

Habermann, J. *Christliche Gebete für allerlei Noth und Stände der ganzen Christenheit.* Wittenberg, 1567.

Habermann, J. *Christliches Gebet-Buch/oder Morgen-und Abendsegen; auch andere Gebet/für alle Noht und Stände der Christenheit/auff alle Tag in der Wochen aussgetheilt....* Tübingen, Reiss, 1672.

Labouvie, E. *Andere Umstände: eine Kulturgeschichte der Geburt.* Köln, Böhlau, 1998.

Sommer, J.G. *Kurtzes und nützliches Weiber- und Kinder-Pflege-Büchlein, wie noch unverheyrathete Weibs-Personen....* Rudolstadt, In Verlegung des Autoris, 1691. (Krivatsy, 11207).

Sommer, J.G. *Nohtwendiger Hebammen-Schul; oder Gründlicher Unterricht, wie eine Hebamme gegen schwangere....* Arnstadt, sold by August Boethius in Gotha, 1693.

[part 2:] *Kurtzes und nutzliches Weiber-und Kinder-Pflege-Buchlein.* Rudolstadt, the author, 1691–1692; [part 3:] *Wohlgemeinte Anleitung zur Chrislichen Kinder Zucht* [Arnstadt], printed by Heinrich Meurer. Arnstadt, and Rudolstadt, 1693. (Krivatsy 11206).

Wunder, H. and C. Vanja. *Weiber, Menscher, Frauenzimmer: Frauen in der ländlichen Gesellschaft 1500–1800.* Göttingen, Vandenhoeck & Ruprecht, 1996.

Govert [Godfrid] Bidloo

Anatomia humani corporis, centum et quinque tabulis...ad vivum delineates. Amstelodami, Joannes a Sumpt Someren, etc., 1685.

One of the outstanding anatomical works of the seventeenth century was that of Bidloo (1649–1713), Professor of Anatomy at The Hague and at Leiden. In his *Anatomia...*, he presented with 105 illustrations one of the finest anatomical atlases of the Baroque period. The drawings were by Gérard de Lairesse, and the engravings are believed to have been executed by Pieter Van Gunst (1659–1724).

The plates of the reproductive system were exceptionally well done. For instance, Bidloo illustrated the female reproductive tract, including the clitoris, labia, vagina, the opened cervix and endometrial cavity, the Fallopian tubes, ovaries, and broad ligaments. Bidloo also included several plates of the fetus *in utero* and its anatomy. For instance, Plate 56 illustrates a near-term fetus within the uterine cavity. The anatomical features identified include: A, the placenta; D, uterine wall; G, amnion; H, umbilical cord; L, knee; M, hand; N, chest; P, foot. Several years later, this figure was reproduced by Justine Siegemundin (1636–1705) in her volume on midwifery (Siegemundin 1690).

In one of the most notorious cases of plagiarism in scientific publication, in 1698 the English surgeon—anatomist William Cowper (1666–1709) published Bidloo's work under the title "*The Anatomy of Humane Bodies...*" as his own, only adding nine perfunctory plates and translating the text into English. In a letter to the great microscopist-microbiologist Antonj van Leeuwenhoek (1632–1723), Bidloo made microscopic observations on the *Fasciola hepatica* and its ova found in the bile passages of the liver (Bidloo 1698). A parasitic flatworm, the liver fluke not uncommonly infects the hepatic ducts of sheep and cattle. In humans, the condition is known as fascioliosis. When William of Orange (1650–1702), *Stadholder* of Holland, visited England in 1688, prior to becoming William III, King of England and joint sovereign with Mary II (1662–1694), Bidloo accompanied him as his personal physician.

References

Choulant-Frank 250 ff; Garrison Morton. 385; Krivatsy 1238; see Russell 211 note.

[Bidloo, G.] *Letter from G. Bidloo to Antony van Leeuwenhoek. About the animals which are sometimes found in the liver of sheep and other beasts.* (Facsimile of the first Dutch edition, Delft, 1698). Ed. By J. Jansen, Nieuwkoop, B. De Graaf, 1972.

Cowper, W. The anatomy of humane bodies, with figures drawn after the life by some of the best masters of Europe, Oxford Sam[uel] Smith, 1698.

Siegemundin, J. *Die Chur-Brandenburgische Hoff-Wehe-Mutter, das ist: Ein höchst-nöthiger Unterricht, von schweren und unrecht-stehenden Geburten....* Cölln an der Spree, ... Ulrich Liebperten, 1690. (GM 6149; Krivatsy 11085).

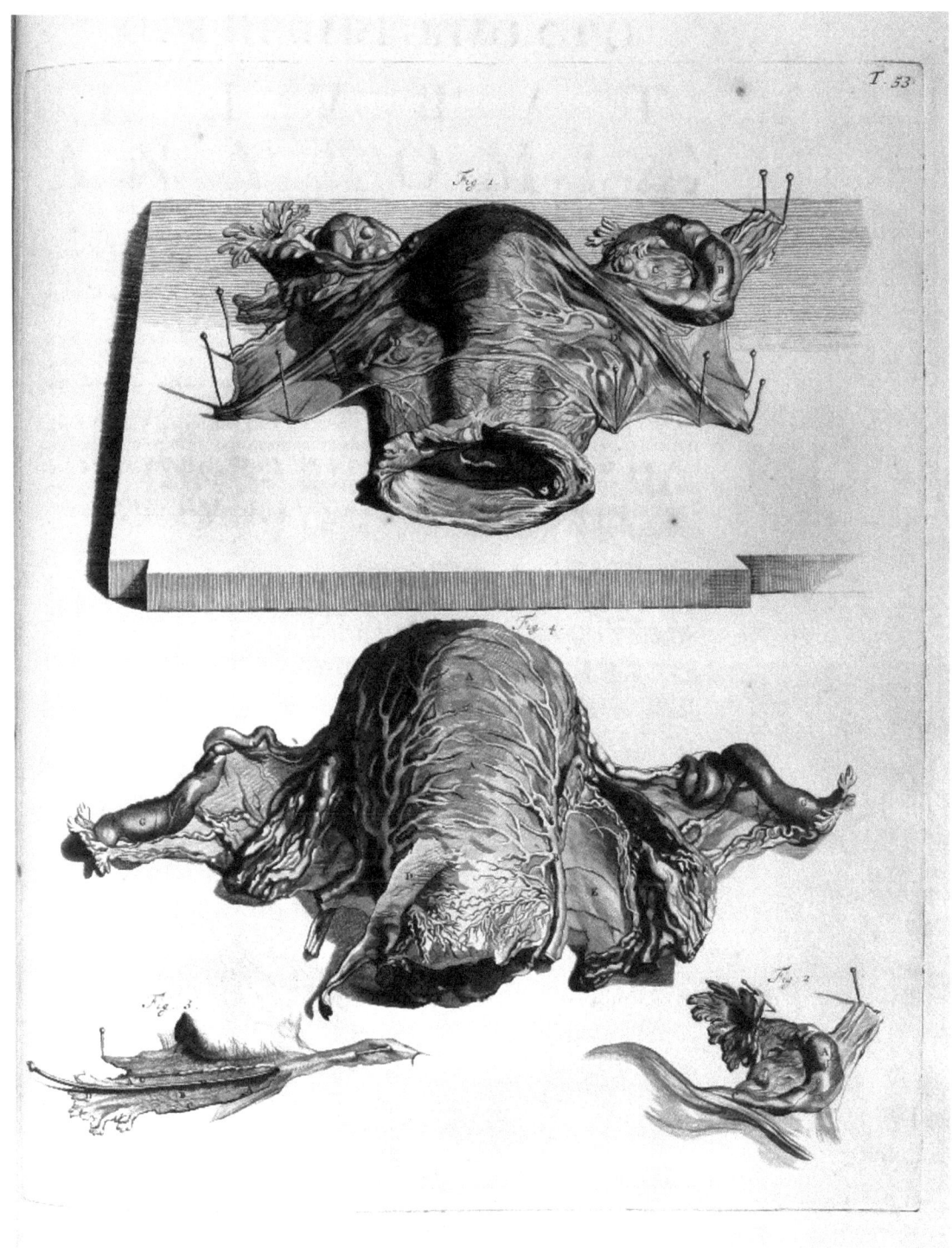

PLATE FROM BIDLOO, 1685

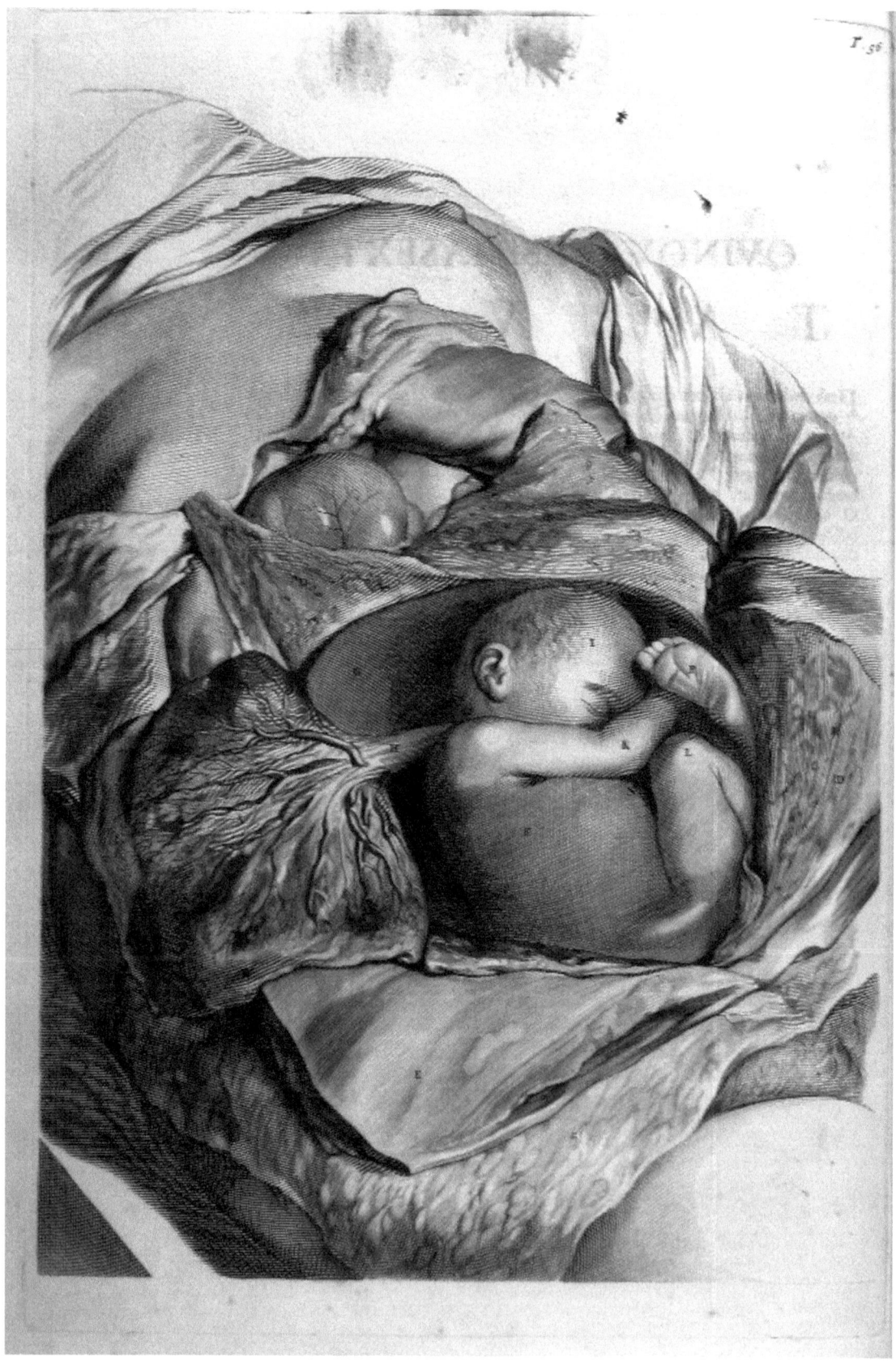

PLATE FROM BIDLOO, 1685

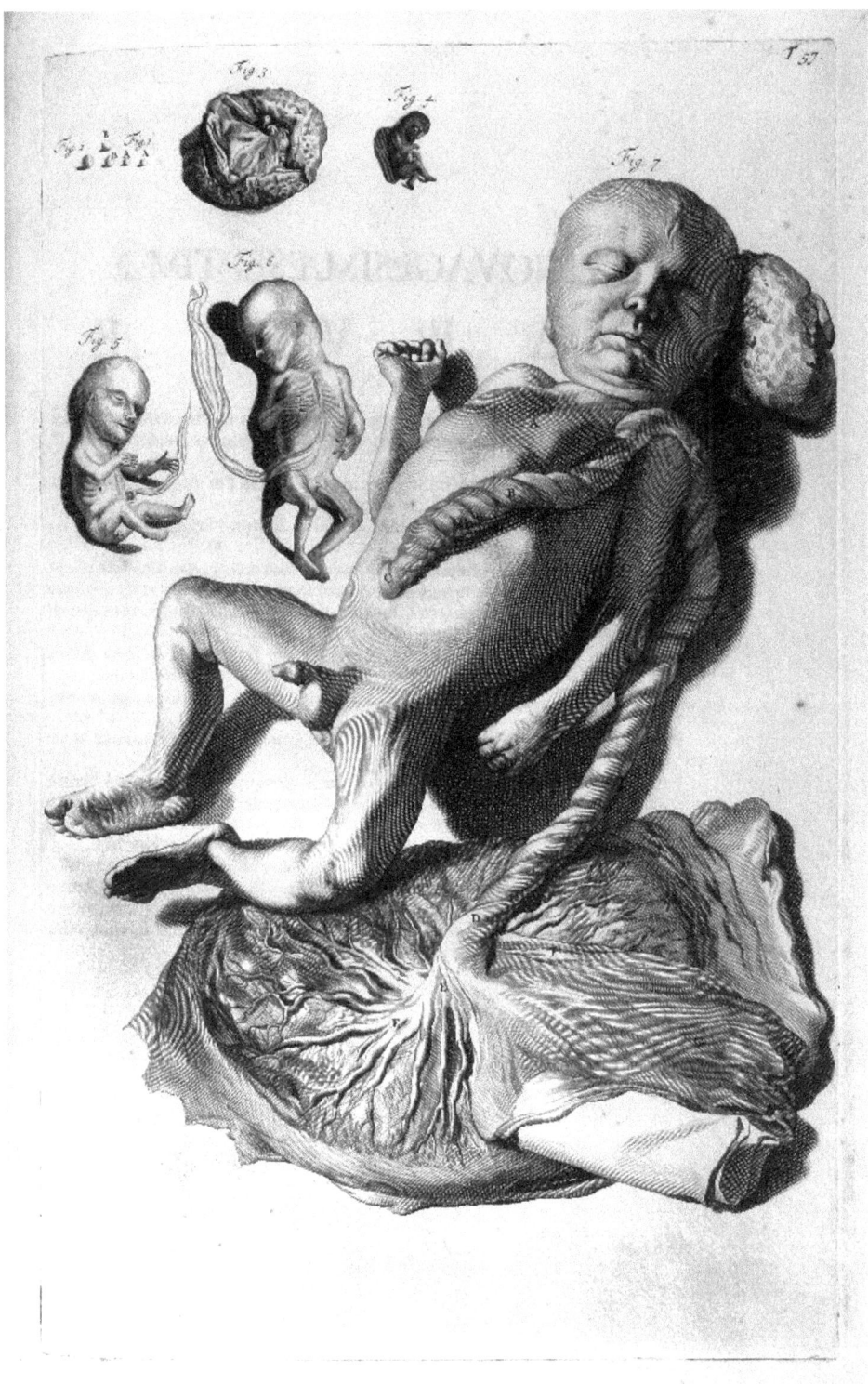

PLATE FROM BIDLOO, 1685

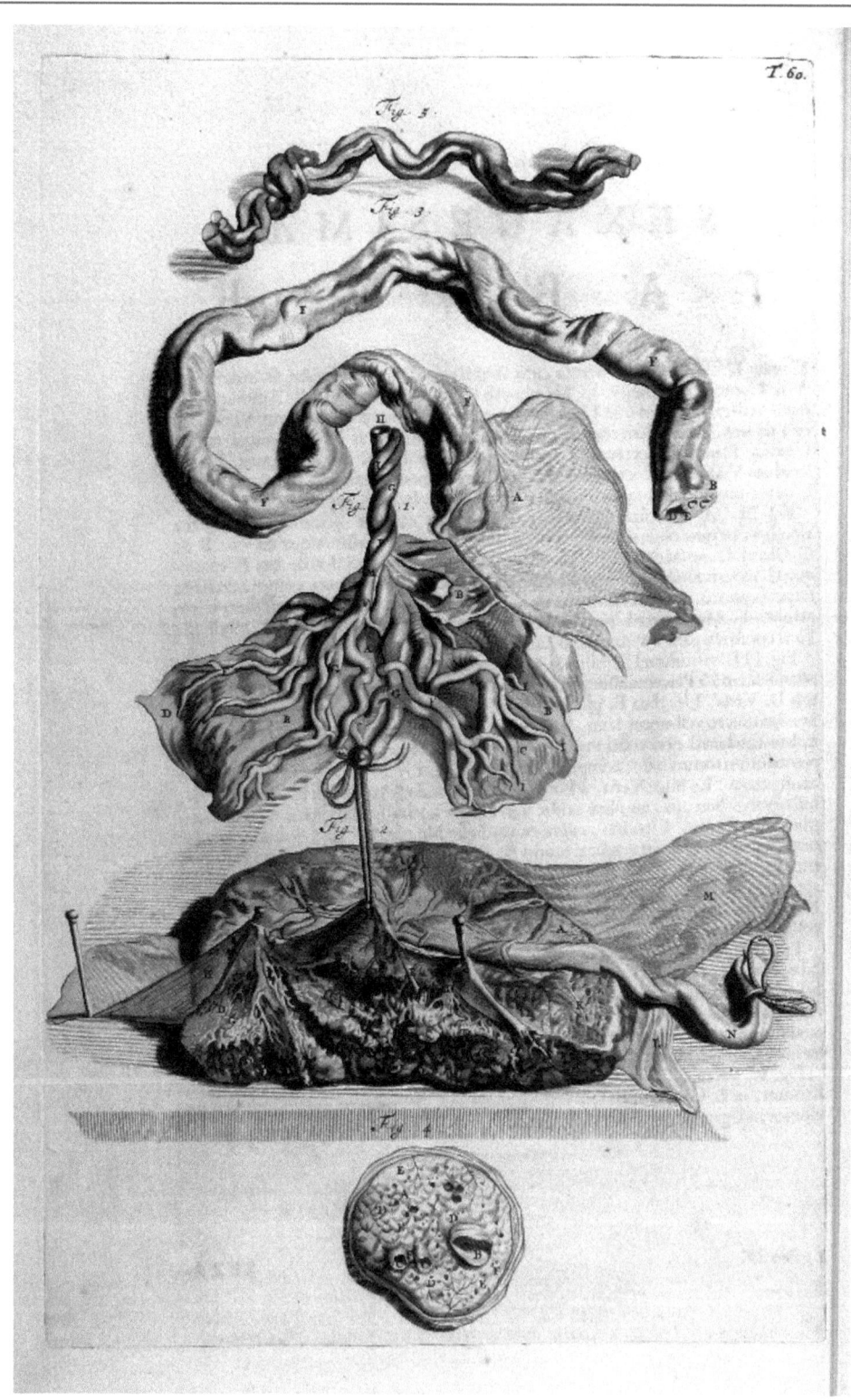

PLATE FROM BIDLOO, 1685

Philippe Peu

La pratique des accouchemens ... Paris, Chez Jean Boudot..., 1694.

In this volume, "The Practice of Childbirth", Peu (1623–ca. 1707) presented original views on many aspects of pregnancy and complications of childbirth. Chief *chirurgien accoucheur* [surgeon, man-midwife] at the *Hotel Dieu* in Paris, the present work is based on Peu's experience with over 4000 obstetrical cases. In seven full-page plates, Peu illustrates various entanglements of the umbilical cord with the fetal body or its limbs. An eighth plate depicts instruments, including two *crochets,* curved hooks used to pull a dead infant from the womb, or dismember an infant in cases of obstructed labor. Cautious in his approach to obstetrical problems, Peu advocated combined internal and external manipulation in cases of transverse lie. In cases of dystocia or obstructed labor he opposed trephination or crainiotomy of the fetal skull or cesarean section. He also advocated slowness in delivery of the placenta, as well as insistence upon its complete removal. The title page contains the aphorism *Sat cito, si sat bene* [set in montion is enough, if set in motion well/properly].

A student of François Mauriceau (1637–1709), Peu disagreed with his mentor over the optimal mode of vaginal delivery, and several other issues including use of the *crochet*. Peu was among the first of the seventeenth-century *accoucheurs* to describe septic inflammatory complications of the puerperium. He stated that in 1 year, 1664, a large number of parturients at the *Hotel Dieu* died, apparently of puerperal sepsis (pp. 268–269). This he ascribed to foul air rising to the lying-in ward from the ward directly below, which was overcrowded with sick soldiers, many of whom had infected wounds. As with Mauriceau's treatise (Mauriceau 1668), this work has been included among those of the early modern period in obstetrics that were replete with "blame narratives," accounts of cases in which responsibility for a patient's death was argued among the female midwives, male *chirurgien accoucheurs*, and physician consultants (McTavish 2006).

References

Cutter & Viets pp. 221–222; Krivatsy 8870; Ricci, p. 322.

Lemarié, E. *Etude sur le Traité d'Accouchement de Philippe Peu*. Dijon... 1891.

Mauriceau, F. *Des maladies des femmes grosses et accouchées....* Paris, Chez Jean Henault et al...., Imprimeries de Charles Coignara, 1668.

McTavish, L. Blame and vindication in the early modern birthing chamber. *Med Hist* 50: 447–464, 2006.

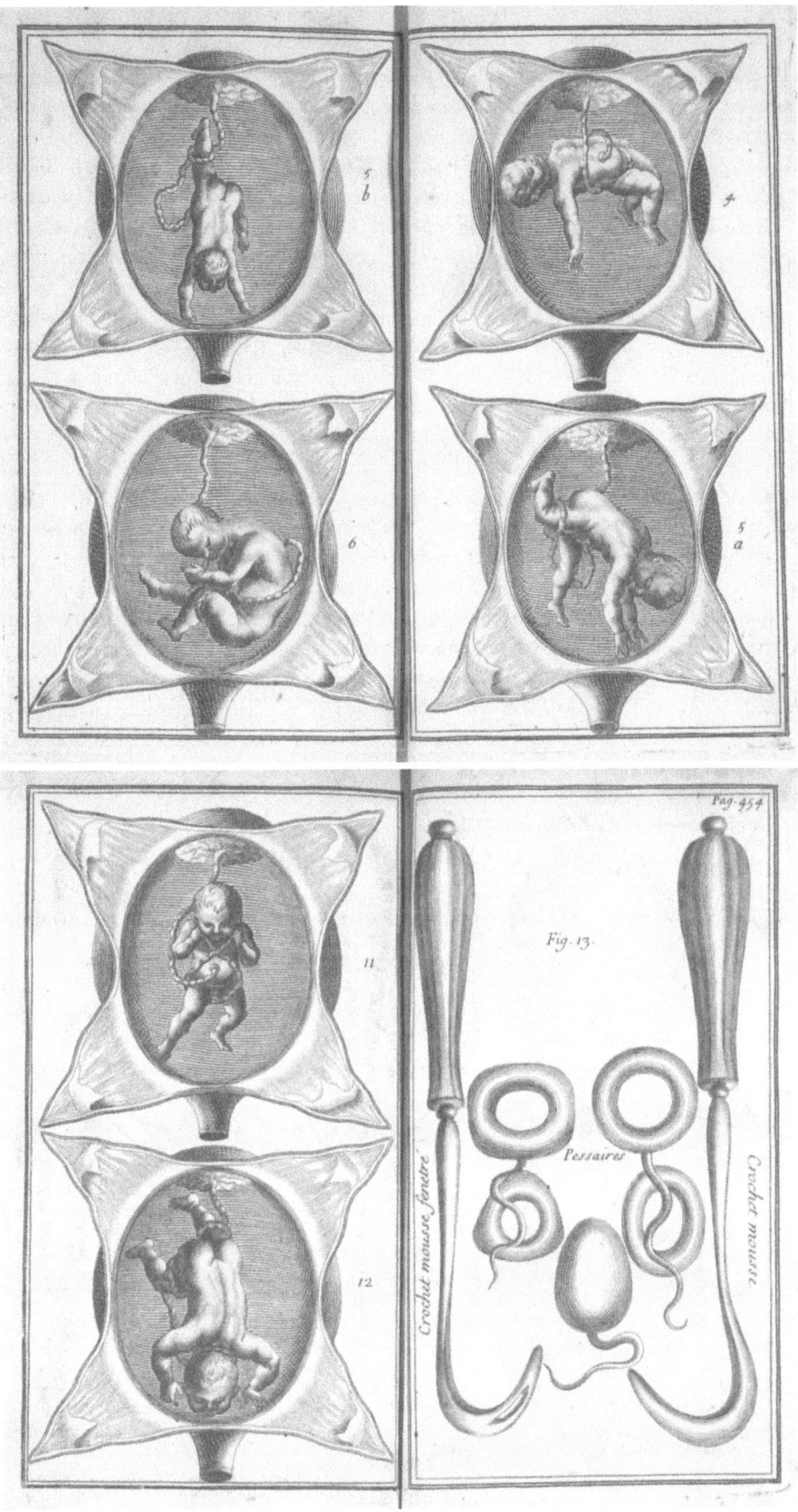

PLATES FROM PEU, 1694

Johan von Hoorn

Den swenska wäl-öfwade jord-gumman hwilken grundeligen underwijser huru med en hafwande handlas, en wåndande hielpas, en barna-qwinna handteras, och det nyfödda barnet skiötas skal. Mäst effter egen Förfarenhet, jemte wäl-öfwade personers Skriffer, all Läkare och Feldschere; men särdeles Barnmoderskor och Huus-Mödrar til Gagn och Nytta, med tienliga figurer författat ... på egen bekostning utgifwin. Stockholm, Tryckt uti Nathanael Goldenaus Tryckerij, 1697.

Von Hoorn's (1662–1724) "Well-trained Swedish midwife" is the first Swedish handbook for midwives, as well as the first complete medical textbook to be published in Sweden. Twenty-six years later and one year before his death, von Hoorn published a Swedish version of Paul Portal's (1630–1703) *La Pratique des Accouchemens* (1685) that he translated and [added, included] his own commentaries.

The first to teach midwifery to both midwives and medical students in Sweden, von Hoorn was a well-trained physician who upheld the highest standards of practice in his time. Born into an affluent family, von Hoorn studied at Leyden. In Amsterdam he studied under the anatomist-obstetrician Frederik Ruysch (1638–1731), and then to Paris where he studied obstetrics for over a year with François Mauriceau (1637–1709), Paul Portal (1630–1703), and Phillipe Peu (1623–ca. 1707). Much of his instruction came from Madame Allegrain, a midwife at the Hôtel-Dieu who practiced among the poor. As men were excluded from the hospital, von Hoorn delivered some of the midwife's patients, by offering bribes to some poor childbed women. Following receipt of his medical degree at Leyden (1690), in 1691, he returned to Stockholm. Here he gave private lessons to both midwives and physicians on the art of delivery, and in 1708 became City Physician. "The often barbarous conditions, ignorance and superstitions under which women were delivered during this period was alarming. The ... eighty midwives in Stockholm ... had no proper education and stood helpless when confronted with complicated deliveries."

Through the efforts of von Hoorn, official regulation of midwives in Stockholm came into force in 1711, which required the women to complete 2 years of education and practice, and pass an examination at the *Collegium Medicum* to obtain a license and membership of the guild of midwives. This allowed them to advertise by hanging a signboard with an illustration of a newborn infant. As William Smellie (1697–1763), half a century later (Smellie, 1752), von Hoorn permitted his students to observe difficult births, and used a manikin to teach the principles of delivery.

Many of the plates have been copied after those of Siegemundin's *Hoff-Wehe-Mutter* (1690). For instance, the title-page includes an illustration of the Rose of Jericho—a plant which was believed by ancient authorities to ease and aid delivery (not the placenta with umbilical cord as some have stated. The border with two peacocks and lilies symbolizes the attributes of Juno, the patroness of midwifery). Regarding breech delivery, von Hoorn observed that they seldom were delivered following a normal course of labor, but rather following the midwives "weary effort and [deft] manipulations". In 1715, von Hoorn published another textbook for midwives, *Siphra and Pua*, the names taken from the Torah of Moses, in which von Hoorn stressed the importance of vaginal examinations to assess progress of labor. He also described mouth-to-mouth resuscitation to revive an apparently dead newborn.

References

Blake p. 220, 359; Cutter & Viets pp. 222–223; Krivatsy 5962; Waller 4882 and 7577.

Hagelin, O. *The byrth of mankynde otherwyse named the womans booke; embryology, obstetrics, gynaecology through four centuries..., an illustrated and annotated catalogue of rare books in the library of the Swedish Academy of Medicine.* Stockholm, Svenska Läkaresällskapet, 1990, pp. 80–85.

von Hoorn, J. *The twenne gudfruchtige.... Siphra och Pua.* Stockholm, (Matthiae), 1715.

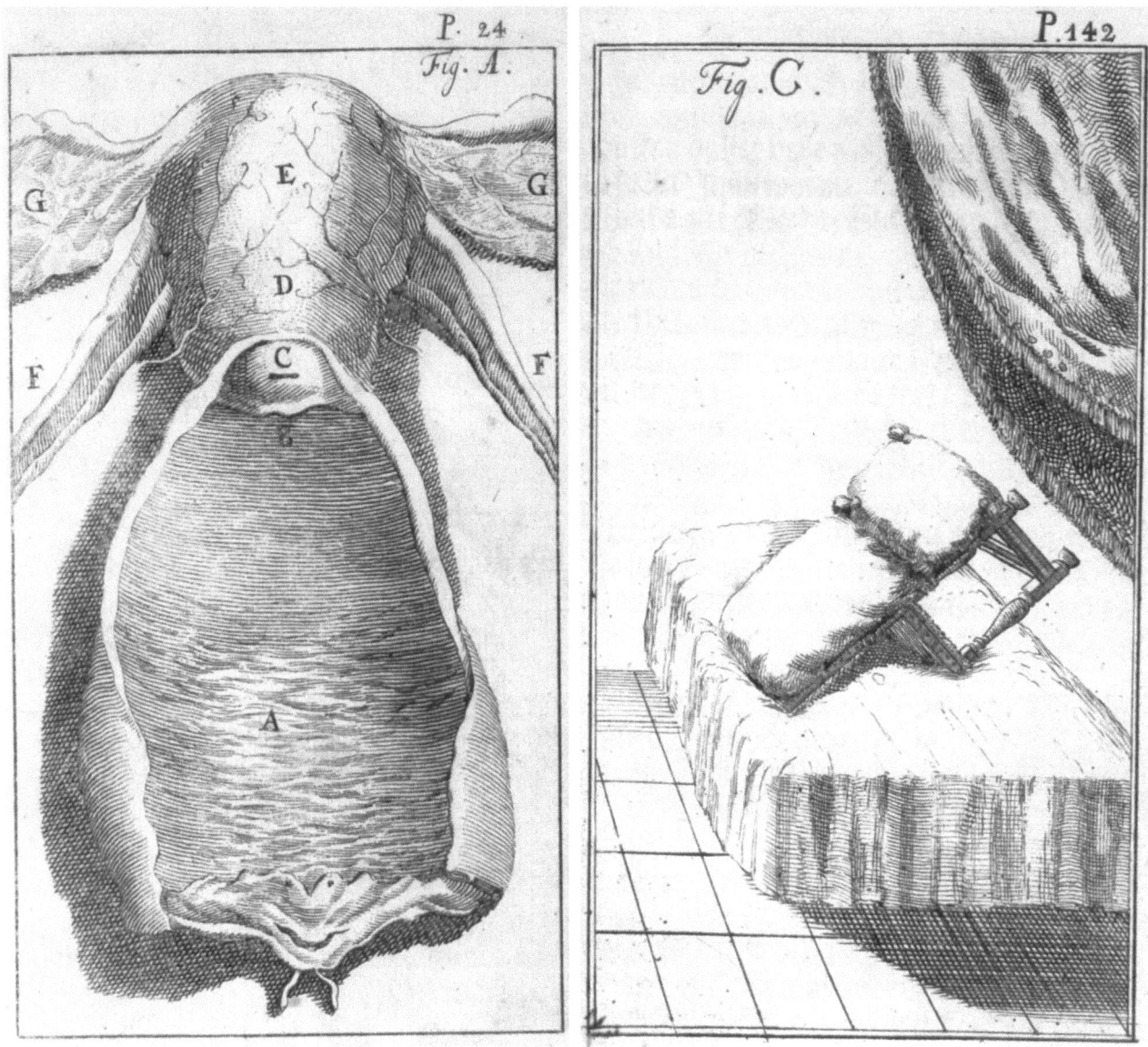

PLATES FROM VON HOORN OF FEMALE REPRODUCTIVE TRACT AND BIRTH CHAIR, 1697

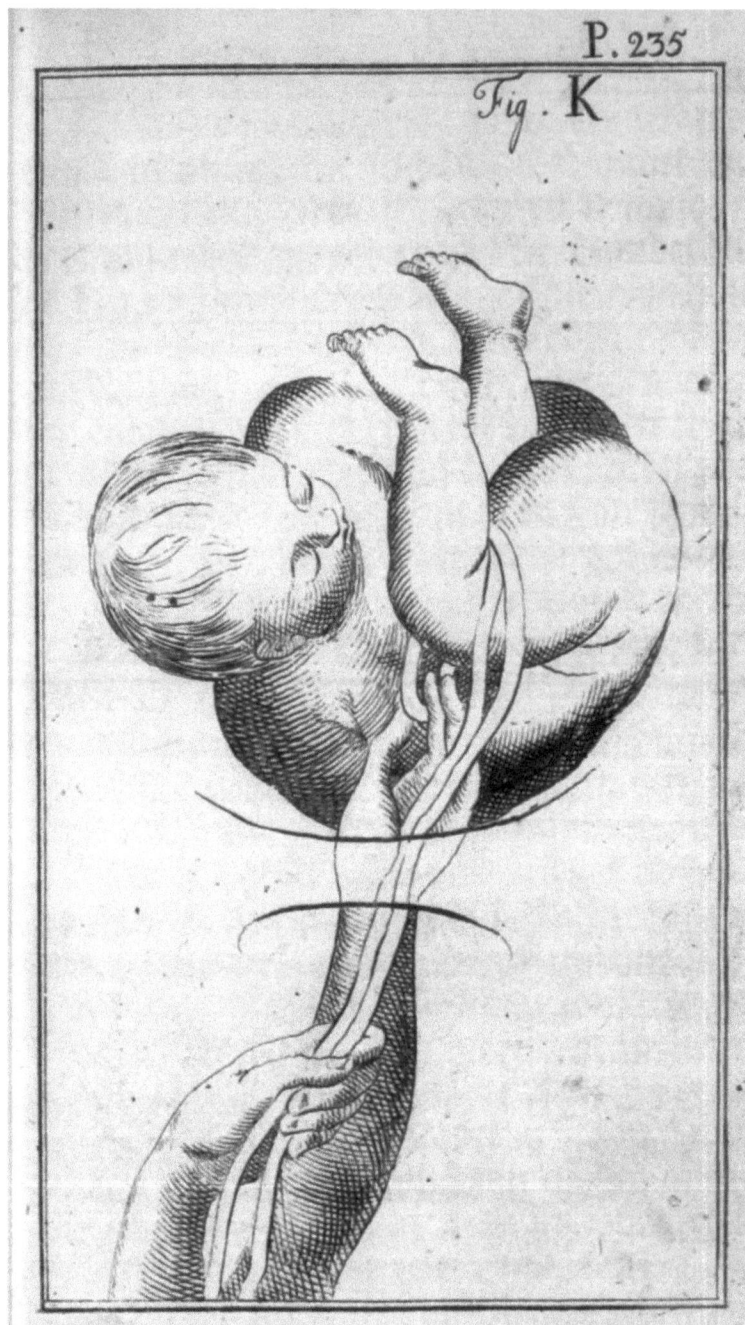

PLATE FROM VON HOORN, 1697

Additional Author(s) of Significance in the Seventeenth Century

Jacques Guillemeau

De l'heureux accouchement des femmes. Paris, Nicolas Buon, 1609.

Jacques Guillemeau (1550–1613), a pupil and son-in-law of Ambroise Paré (1510–1590) whose works he translated into Latin (Paré 1582), was surgeon at te Hôtel Dieu in Paris. During his lifetime he served as surgeon to three of France's Kings: Charles IX (1550–1574), Henry III (1551–1589), and Henry IV (Henry of Navarre; 1553–1610). The present work contains three sections: on pregnancy, delivery, and the diseases of women. In addition, are chapters on feeding and care of the newborn infant, diseases of children, and the diagnosis of pregnancy. Guillemeau first described what is called the Mauriceau-Smellie-Viet maneuver for management of the aftercoming head of a breech delivery. In this procedure, the after-coming head is delivered with the infant resting on the physician's forearm, and the operator's index finger introduced into the mouth to maintain flexion of the infant's head, while fingers of the other hand hooked over the shoulders exert traction, with the infant's body slowly being raised as the mouth, nose, brow, and occiput are successively brought over the perineum. Prior to the use of the obstetrical forceps or Caesarean section, this maneuver saved the lives of many infants. He also advocated podalic version in cases of placenta previa which, it is believed, he learned from Paré. Although he claimed to be conservative, he wrote about caesarean section. Guillemeau's son Charles (1588–1656) later edited several of his father's works.

References

Choulant-Frank p 213; Garrison Morton 6145.1

Guillemeau, J. *De la grossesse et accouchement des femmes....* Paris, Abraham Pacard, 1620.

Guillemeau, J. *Les oeuvres de chirurgie avec les portraits et figures....* Paris, Nicholas Buon, 1612.

Guillemeau, J. *Childbirth; or the happy deliverie of women.* Translated by A. Hatfield. London, A. Hatfield, 1612.

Paré, A. *Opera Ambrosii Parei Regis primarii Parisiensis chirurgi....* Paris, Apud Jacobum Du-Puys, 1582.

Caspar [Gaspard] Bauhin [Bauhinus]

Vivae imagines partium corporis humani aeneis formis expressae & ex Theatro Anatomico Caspari Bauhini... desumptae. Francofurti, Johann Theadori de Bry, 1620.

In his comprehensive atlas on anatomy, Bauhin (1560–1624) included a section on the female reproductive organs. For the most part, the 140 full-page engravings are copies from Vesalius (1543), and others, and according to Choulant, "have no artistic value".

Bauhin was professor of anatomy and of botany at the University of Basel, where he also served as rector. He wrote several textbooks, including *Theatrum anatomicum...* (1592). An enlarged edition of *Vivae imagines partium...,* but without illustrations, was published several decades later, the present volume appearing separately as an atlas. Bauhin published an early work on hermaphridites and other monstrous births, in which he included a lengthy tabulation of previous studies of teratological "monsters." Included are plates illustrating a pair of Siamese twins (male/female), a woman with a gigantic and deformed lower leg, and a monstrous fetus.

Bauhin also published a large botanical work, *Pinax theatri botanici...* [Illustrated Exposition of Plants] (1623), in which he described over 6,000 plants, a large number of which were from his herbarium. In this latter work, he introduced a system of nomenclature which distinguished genera and species, and grouped plants according to their similarities. This analysis had a great influence on the modern taxonomic classification of plants developed by the Swedish botanist, Carl von Linné [Linnaeus] (1707–1778) (1735). Bauhin with his brother Johann (1541–1613) are memorialized in the genus of plants *Bauhinia.*

References

Choulant-Frank p 229; Krivatsy 951.

Bauhin, C. *Gynaeciorum sive de mulierum affectibus commentarii* 4 vols, Basileae per T. Guarinum, 1586–1588. (GM 6012).

Bauhin, C. *Anatomica corporis virilie et muliebris historica,* Lugduni, J le Preux, 1597. (GM 437).

Bauhin, C. *Theatrum anatomicum infintis locis auctum, ad morbos accummodatum...*Basileae, S. Henri Petri, 1592. (GM 379).

Bauhin, C. *Pinax theatri botanici: sive index in Theophrasti Dioscoridis, Plinii et botanicorum qui a seculo scripserunt opera.* Basilae, 1628.

Bauhin, C. *Pinax theatri botanici:* sive index in Theophrasti Dioscoridis, Plinii et botanicorum qui a seculo scripserunt opera. 2nd Ed, Impensis Joannis Regis, 1671. Basileae.

Bauhin, C. *De hermaphroditorum monstrosorumque partum natura...Libri duo lactenus non editi...*Oppenheim, Hieronymus Galler, 1614.

Linné, C.V. *Systema naturae,* Lugduni Batovorum, apud Theodorum Haak, 1735. (GM 99).

Jan Swammerdam

Miraculum naturae; sive,uteri muliebris fabrica. Lugduni Batavorum, apud Severinum Mathaei, 1672.

Following the publication of Regner De Graaf's work on ovulation (1672), Swammerdam (1637–1680) proclaimed his priority. A graduate of the University of Leiden (and fellow student of De Graaf), Swammerdam became known for his expertise in microscopic dissection and wax injection studies. As with De Graaf, he mistook the ovarian follicles for the ovum, an error not rectified until the work of Karl Ernst von Baer (1827). Nonetheless, the concept of mammals having ovaries was important to a correct understanding of reproduction.

A naturalist with many interests, Swammerdam contributed to an understanding of nerve-muscle function, and devised methods of studying the circulatory system by means of injections. On the basis of extensive studies of over 3000 species of insects, and their developmental stages (1669) he formulated a classification that remained unsurpassed for decades. A supporter of the preformation doctrine, on the basis of his study of insect metamorphosis and seeing the folded butterfly perfectly formed within the cocoon, Swammerdam concluded that the butterfly had been hidden or "masked" in the caterpillar. Thus, it was no great leap in logic to regard the embryo in the egg in a similar light. "Each butterfly in each cocoon must contain eggs within it which in their turn must contain butterflies which in turn must contain eggs, and so on" (Needham 1934, pp 148–149).

Swammerdam's writings were characterized by a mystical sense of the beauty of life. He extended his ideas on preformation to man, "in nature… there is no generation but only propogation, the growth of parts. Thus original sin is explained, for all men were contained in the organs of Adam and Eve. When their stock of eggs is finished, the human race will cease to be" (Needham 1934, p 149). In 1675, he abandoned science for a life of religious contemplation. This lasted only a year, however, and he returned to Amsterdam to gather his writings into a "great work". Following his death, this was edited and published by Herman Boerhaave (1668–1742) (Swammerdam 1737–1738).

References

Blake p 440; Garrison Morton 1211; 292.1; Krivatsy 11603.

Baer, C.E.V. *De ovi mammalium et hominis genesi*. Lipsiae, L. Vossius, 1827 (GM 477).

Cobb, M. Reading and Writing. *The Book of Nature*: Jan Swammerdam (1637–1680). *Endeavor* 24:122–138, 2000.

Fournier, M. *Fabric of Life: Microscopy in the Seventeenth Century*. Baltimore, MD Johns Hopkins Press, 1996.

Needham, J. *A history of embryology*. Cambridge, at the University Press, 1934.

Schierbeck, A. *Jan Swammerdam*. Amsterdam, Swets & Zeitlinger, 1967.

Swammerdam, J. *Bybel der nature, sive historie insectorum….* 3 vols. Leipzig, Johann Friedrich Gleditschens Buchhandlung, 1752.

Swammerdam, J. *The book of nature or the history of insects: reduced to distinct classes, confirmed by particular instances… Thomas Flloyd, trans, with footnotes by John Hill*. London, C.G. Seyffert, 1758.

Swammerdam, J. *Historia insectorum generalis…*2 pts, Utrrecht [sic], M. van Dreunen, 1669. (GM 294).

Windsor, M.P. Jan Swammerdam. In: *Dictionary of Scientific Biography*. Vol XIII. Charles Coulston Gillispie (Ed.). New York, Charles Scribner's Sons, 1976, pp 168–175.

Bernardino Genga

Anatomia per uso et intelligenza del disegno ricercata non solo su gl'ossi, e muscoli del corpo humano… Roma, Domenico de Rossi, herede di Gio Jacomo, 1691.

In 56 cooper-plate engravings Genga (1620–1690), of Rome, presented one of the earliest anatomical works for artists.

References

Garrison Morton 386; Krivatsy 4655.

Genga, B. (English translation). London, Senex, 1723.

The Eighteenth Century

Exhausted by wars and the intolerance of previous years, the eighteenth century experienced continued advances in learning, with a deepening study of philosophy, economics, and expansion of colonial empires. The century also witnessed continued revolution, horrific wars, and social trauma. The revival of interest in classical aesthetics, "Neoclassicism," of the latter portion of the previous century, continued with the "Enlightenment," the critical examination of previously accepted doctrines and institutions from the point of view of rationalism. With ever increasing discoveries in science, "Materialism" followed, the philosophical opinion that matter in its movements and modifications is the only reality, and that everything in the Universe, including thought and feeling, can be explained in physical laws. With gradual abandonment of ancient traditions, the major intellectual developments, e.g., those of Renaissance Neoplatonism, Baconian experimental philosophy, Cartesian mechanism, and Newtonianism contributed to the scientific revolution. The "Restoration" of the second half of the seventeenth century was followed by the "Romantic" Period to the end of the eighteenth century. Political theory espoused the idea that the monarch derived his authority from God, and thus was perfectly suited to rule his people. The literature of the period was marked by reason, restraint, and good taste. Essayists promulgated the idea of rationalism and scientific reasonsing. Enticed by economic opportunities available in factories, shipyards, coalfields, and large conurbations, increasing industrialization prompted mass population movements. Major cities in Great Britain and the continent became multicultural melting pots. With its classicists, philosophers, and other scholars, Edinburgh became the "Athens of the North," helping to establish the Scottish Enlightenment. Despite the attack upon religious dogma by the new science, and with Enlightenment-Cartesian rationalistic philosophy and increasing secularization, both the Catholic and Protestant confessions flourished. The spread of literacy, education, and fora [forums] for public discussion with a culture of the public sphere were followed by major quests for social equality, the rise of the "common man," and control over nature. In terms of the present volume, it is important to situate the contributions within the larger history of ideas, and the wide range of literary, cultural, and scientific developments. This leads towards considerations of the relation of the body to the mind, the physical to the abstract, the empirical to the theoretical, and the concrete to the speculative.

Czar Peter I, "the Great" (1672–1725) founded Saint Petersburg (1703), the new capital of Russia, and continued the tradition of the previous century in courting scientific expertise and exchange. The Act of Union (1707) merged the Parliaments of England and Scotland, thus establishing the Kingdom of Great Britain. Louis XIV (1638–1715; the "Sun King" 1643–1715), died leaving France with monumental debt in the wake of wars and excesses of Royal expenditure. In

fact, some mark the century as those years from the death of Louis XIV to the beginning of the French Revolution in 1789. The British "South Sea Bubble," the economic hysteria that occurred from overheated speculation in shares of the South Sea Company, collapsed (1720). In 1727, George II (1683–1760) became King of Great Britain and Ireland, and Elector of Hanover. Several years later, in 1734, the mighty pillar of finance, the Bank of England, opened its doors. Frederick II (1712–1786), "The Great," was crowned King of Prussia (1740) and became patron of the arts and philosophy. In mid-century (1755) Lisbon experienced its great earthquake, and Samuel Johnson (1709–1784), lexicographer and conversationalist, introduced his dictionary of the English language.

The "Age of Reason" combined with the impact of the scientific revolution and social thought of the time, resulted in the idea of progress and a new humanism, with its extraordinary interest in the welfare of human kind. Democratic ideals, a period of unparalleled creativity with a rationalistic approach, substantially changed both popular and intellectual attitudes towards the sciences, society, religion, and government, all of which helped to enlarge the scope of medicine. The "First" industrial Revolution, with extensive mechanization of production, associated with a shift from home fabrication to large scale industrial manufacturing and production, brought extensive social and cultural change. Years of turbulence continued with the "Seven Years War" (1756–1763), fought among European powers. George III (1738–1820) became King of Great Britain and Ireland (1760), and lost the American Colonies. The American Revolutionary War (1775–1783) led to the Declaration of Independence (1776) with establishment of the United States of America. From 1762 until her death, Catherine II (1729–1796), "The Great," reigned as Empress of Russia. John Wesley (1703–1791) with his hymm writer brother Charles (1707–1788) founded Methodism (1784). The Scottish engineer James Watt (1736–1819) invented the modern condensing steam engine (1769), which helped to ignite the Industrial Revolution. The French Revolution (1789–1799) against the monarchy and aristocracy, with its bloodshed and chaos, continued for a decade until Napoleon I, Bonaparte (1769–1821) gained control and appointed himself Emperor of France (1804–1815). Pinckney's Treaty, also known as the Treaty of San Lorenzo, with Spain (1795) led to the Mississippi Territory being granted to the United States. The century ended with dissolution of the Dutch East India Company (1799). That same year, the Rosetta Stone was discovered, inscribed with a 196 BCE decree of Ptolemy V (204–181 BCE) in Egyptian hieroglyphic, Demotic, and Greek. Later, its deciphering (1822) by Jean François Champollion (1790–1832) opened the way for scholarly studies of Egyptology.

Notable individuals and thinkers during this century included Marie Antoinette (1755–1793), wife of Louis XIV; the philosophers and/or statesmen Edmund Burke (1729–1797), David Hume (1711–1776), George Washington

(1732–1799), first President of the United States (1789–1797), and both William Pitt "the Elder" (1708–1778) and "the Younger" (1759–1806), Thomas Jefferson (1745–1826) the third United States President (1801–1809), Immanuel Kant (1724–1804), Thomas Paine (1737–1809), Jean-Jacques Rousseau (1712–1778), Adam Smith (1723–1790), and Voltaire [François Marie Arouet] (1694–1778). In music, Isaac Watts (1674–1748) the English nonconformist minister, was writing memorable hymns, in addition to books on geography, grammar, and philosophy. Johann Sebastian Bach (1685–1770), Georg Frideric Handel (1685–1759), Franz Joseph Haydn (1732–1809), and Wolfgang Amadeus Mozart (1756–1791) created monumental works of genius.

In concert with these intellectual developments, scientists included Daniel Bernoulli (1700–1782), Karl Friedrich Gauss (1777–1855), Pierre Simon [Marquis de] Laplace (1749–1827), and Antoine Laurent Lavoisier (1743–1794). In the wake of the French philosopher and mathematician René Descartes (1596–1650), many scientists devoted themselves to understanding matter and its organization, the process of conception and development, the relation of the body to the mind, the brain to the soul, the physical to the abstract, the empirical to the theoretical, and the material to the speculative. Medicine moved from the "dark ages" of empiricism into the light of science, and some of the learned men of medicine acquired considerable power and prestige. A number of anatomical discoveries with the beginnings of physiology, combined with descriptions of specific diseases, helped to lay a foundation for rational therapy. As noted, medical theory was based on the Hippocratic/Galenic doctrine of the four humors: "blood, phlegm, yellow bile (choler), and black bile." These fluids were believed to determine character and general health, and to be associated with one's disposition, e.g., sanguine, phlegmatic, choleric, and melancholy, respectively, with disease a result of imbalance of these humors. The humors were believed to complement the four elements earth, air, fire, and water, and the four qualities hot, cold, moist, and dry. In Leiden, Herman Boerhave (1668–1738) established as a center of contemporary medical education. Teaching in Padua,

Giovanni Battista Morgagni (1682–1771) laid the foundation of pathological anatomy, and at Montpellier Paul Joseph Barthez (1734–1806) introduced the concept of a "vital principle" to explain the workings of the body. The Swedish naturalist Carl von Linné [Linnaeus] (1707–1778) developed the first modern classification of plants, animals, and minerals, while, in attempting to summarize all human knowledge of the natural world, the Frenchman Georges-Louis Leclerc, comte de Buffon (1707–1788), raised natural history to the status of a science. The philosopher-encyclopedist Denis Diderot (1713–1784) followed these workers in holding that natural history provides a key to understanding the world. In Paris, Franz Anton Mesmer (1734–1815) promoted his system of therapy based on his confused doctrine of a universal magnetic fluid that alike influenced ocean tides and men's minds.

In architecture, the rococo artistic style characterized by fanciful curved asymmetrical forms and elaborate ornamentation, originated in France and rapidly spread through Europe. In art, Thomas Gainsborough (1727–1788), William Hogarth (1697–1764), and Joshua Reynolds (1723–1792) created masterpieces. Illustrators sought new ways to introduce shades of gray into typically black and white prints. Mezzotint, discovered in the seventeenth century, (a method of engraving a cooper or steel plate by scraping and burnishing areas to produce effects of light and shadow) became especially popular. Other new techniques were developed to mimic the appearance of original drawings. These included aquatint, which approximated the appearance of wash drawings. As noted in the Introduction, this century saw the production of the great anatomical folios of William Cheselden (1688–1752) and Bernhard Siegfried Albinus (1697–1770). In these, particularly those of Albinus, landscape backgrounds enhanced the *chiaroscuro* of the figures to give them greater three-dimensional reality. The great obstetrical folio atlases published by William Smellie (1754), Charles Nicholas Jenty (1757), and William Hunter (1774) during the second half of the century depicted near-life-sized engravings and accompanying texts, presenting the gravid uterus and its contents in a new light.

Hendrik van Deventer

Manuale operatien, 1. deel zijnde een nieuw ligt voor vroed-meesters en vroed-vrouen. The Hage, The author, 1701.

In his "new light for midwives," van Deventer (1651–1724), gave the first accurate description of the female pelvis. A native of the Netherlands, Deventer obtained his medical degree from the University of Gröningen (1694), and then returned to den Hague to practice Obstetrics and Orthopedic surgery. With his interest in orthopedics, Deventer spent a great deal of time studying spinal deformity and pelvic abnormalities, conditions that were not uncommon in an age in which rickets, associated with vitamin D deficiency, was rampant. Importantly, he stressed the role of these disorders in complicating labor. It must be remembered that this was an age in which it was believed that the infant was delivered by its own efforts. Thus in instances of severe pelvic contraction with fetal demise, rather than the problem being the pelvic contracture *per se* preventing delivery, failure to progress was attributed to fetal death. Deventer was one of the first to study thoroughly the anatomy of the pelvis, and its deformities, and to give particular attention to the space within the pelvis required for the fetus to be born naturally. He emphasized the need of midwives' knowledge of pelvic anatomy, "...for without a clear knowledge of that matter they proceed uncertainly, and make use of their hands, like those that are blind" (p. 14). Addressing this deficiency in the literature, he wrote:

> As for the necessity of the knowledge of these bones, and their form and figure, I should take no notice of them, had I a mind to follow the method of other writers; or I should but slightly touch upon them, so that midwives would reap little advantage by it; but thinking the knowledge of these bones to be highly necessary to midwives, I thought it necessary also to represent their figures, as clearly as they could be represented by an expert painter, accurately to the life (p. 15).

In this volume, which he published at his own expense, Deventer included 38 figures to illustrate his text. These ranged from depictions of the pelvic bones to various fetal positions within the womb. With his great interest in the pelvic structure, his name is associated with the narrow and flat pelvis *pelvis plana Deventeri.* Figure 30 illustrates twins near term with the following annotations: aa, vertebral bodies; b,iliac bone; c, pubic bone; g, the uterus; h,h, the two infants; i, the umbilical cords; k,k, the twin placentas. As a practitioner in his native city, he directed his greatest efforts to pregnancy and to the study of its many unsolved problems. Deventer's wife was a midwife, and it is believed that much of his success was due to her assistance in making the book meet the needs of her fellow midwives. He was also the first to use the term "placenta previa" and to advise podalic version after piercing the placenta. He thus played a key role in founding modern obstetrics.

In considering the characteristics of midwives, Deventer observed, "Women fit for this employ ought to be inclinable to do good, given to hospitality and tender hearted, as ready to help the poor as the rich, ... and knowing that she stands in need of Divine Assistance, being ernest in Prayer, ... always relying on God, who graciously supplies with wisdom those that want it, and call upon him" (pp. 8–9). Possessing a global view of life, Deventer observed, "Not just one single person is capable of creating great changes. Obstetrics is a science, created by many physicians" (p.). "If we are unable to discover new facts, like our predecessors, science definitely will become extinct. Since we do not review our memory by self-study, we tend to forget small items. After a while our art and know-how will deteriorate and become obsolete. We must re-establish our facts and principles, in order to reach a new dimension" (p.). Deventer published a preliminary and much shorter version of this work 5 years earlier (Deventer 1696). A Latin translation of Deventer's work was published in Leiden the same year, and an English translation appeared in 1716. His work remained authoritative for over a century.

References

Blake 118 lists later editions; Cutter & Viets p. 13 & 180; Garrison Morton 6253.

Deventer, H.v. *Dageraet van vroet-vrouwen: ofte voorlooper van het tractate genaemt Nieuw ligt der vroet-vrouwen.* Leyden, 1696.

Deventer, H.v. *Operationes chirurgicae novum lumen exhibentes obstetricantibus, quo fideliter manifestatur ars obstetricandi,...* Lugduni Batarvorum, apud Andream Dyckhuisen, 1701.

Deventer, H.v. *The Art of Midwifery Improv'd. Fully and Plainly laying down Whatever Instructions are requisite to make a compleat Midwife. And the many Errors in all the Books hitherto written upon this Subject clearly refuted...Illustrated...To which is added, A Preface giving some Account of this Work, by an Eminent Physician.* London, Printed for E Curll et al., 1716.

Deventer, H.v. *Manuele operation, zynde een nieuw ligt voor vroed-meesters en vroed-vrouwen. Facsimile en commentaar, onder red. Van R. W. Bakker, M.A.C. Lubsen-Brandsma en A.Th.M. Verhoeven. (Dutch Classics on History of Science, XX).* Utrecht, Hes & De Graaf, 2001.

Houtzager, H.L. Henrik van Deventer. *Europ. J. Obstet, Gynec. & Reprod. Biol.* 21:263–270, 1986.

FIGURE 30 FROM VAN DEVENTER, 1701

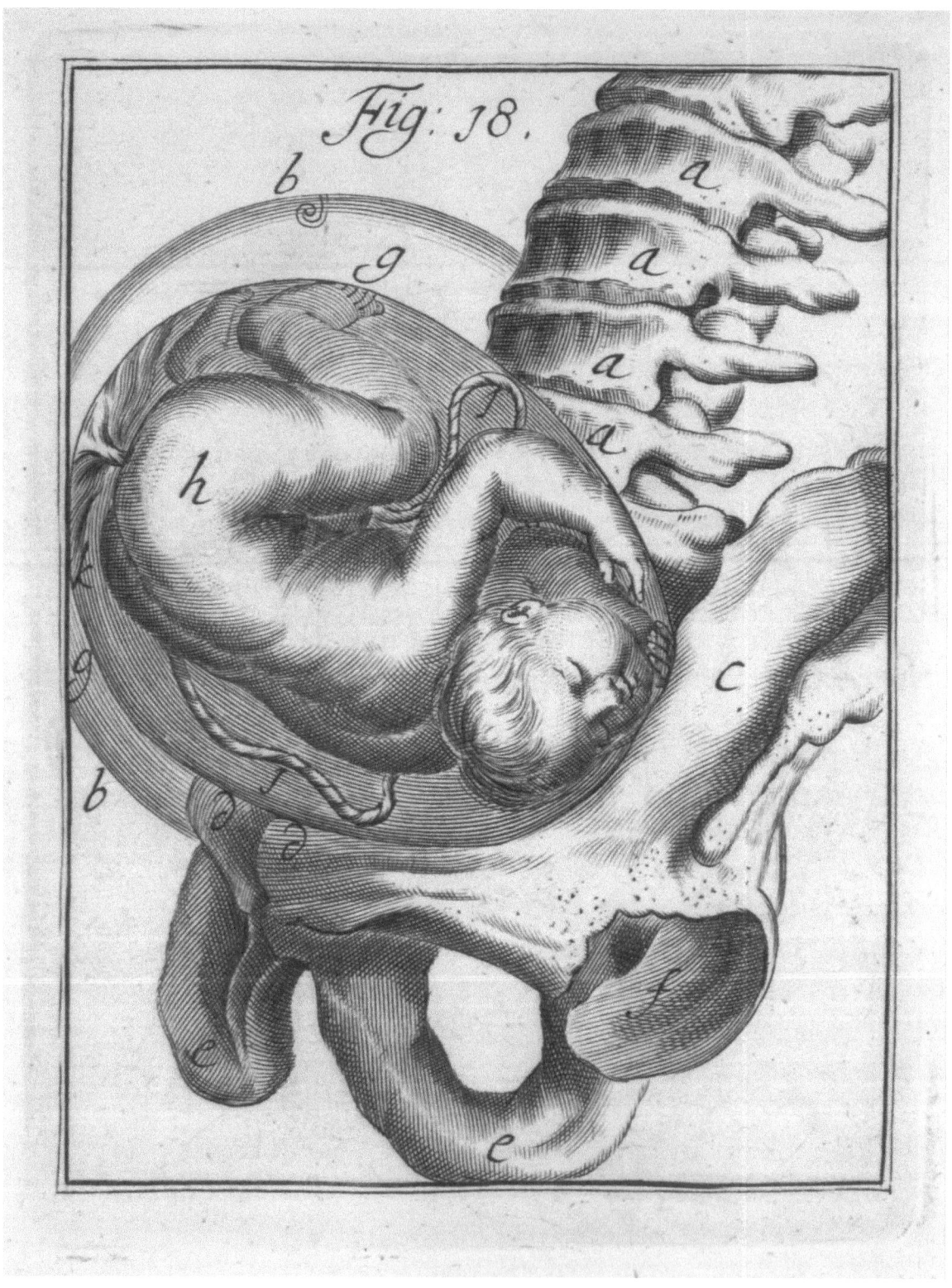

FIGURE 18 FROM VAN DEVENTER, 1701

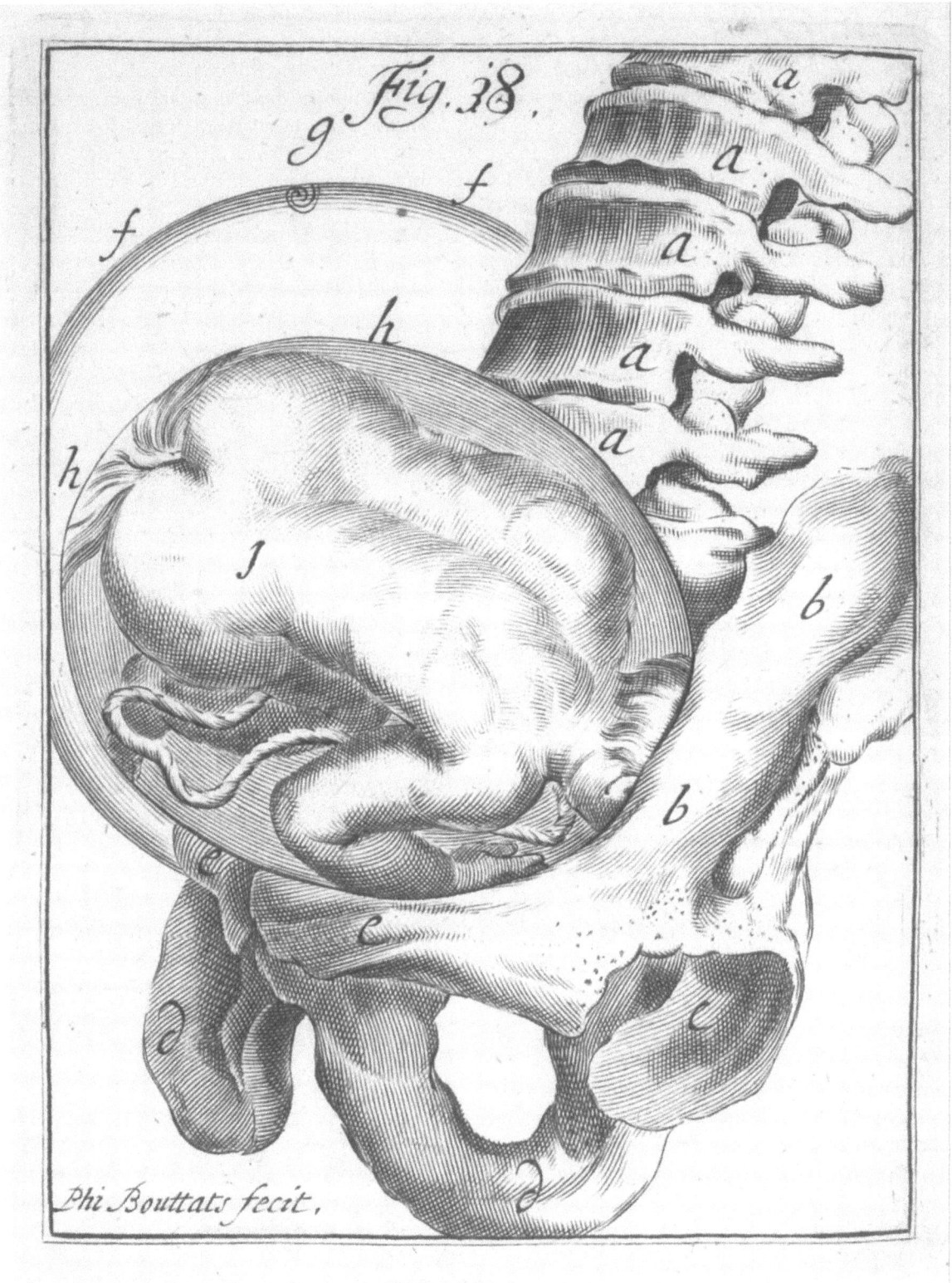

FIGURE 38 FROM VAN **DEVENTER, 1701**

Johann Christoph Ettner

Des getreuen Eckarths unvorsichtige Heb-Amme: in welcher wie eine Heb-Amme… Leipzig, Verlegts Johann Friedrich Braun, 1715.

In his *Trusty (or Faithful) Eckarth's careless midwife,* Ettner (1654–1724) presented a massive (944 page) compendium on the anatomy of female organs of reproduction, human reproduction, and the diseases of women and children. The engraved frontispiece caricaturizes a negligent midwife who stands before a table on which lie the bodies of two dismembered infants, one decapitated. In one hand, the midwife holds a piece of the torn placenta. In the background, the failing or dead mother sits in the birthing chair. Above, the omniscient eye of the almighty oversees the scene. An accompanying poem labels such midwives as murderers, subject to the wrath of God.

Rather than a novel with a plot in the contemporary sense, *The Careless Midwife…* records a lengthy and uneventful journey during which four men and Carilla, a midwife (who is not that depicted in the frontispiece), converse on various aspects of gynecologic midwifery, the medical care of women and children, quackery, and politics. Their discussion includes the requisites for a proper midwife: piety, caregiving, experienced, conscientious, good natured, and being the mother of several children. In this work, Ettner, speaking as Eckarth (a pseudonym), also described how a midwife who wished to preserve her good conscience should act.

It should be remembered that even in the eighteenth century the theory and practice of medicine was, from a contemporary standpoint, somewhat contradictory, being an amalgam of superstitions that dated from antiquity, arcane alchemical and astrological theory, and progressive ideas based on scientific observations of that time. A graduate of the University of Leipzig (1674), who lived in Gurau and later in Breslaw, Ettner developed a great interest in pharmaceuticals and alchemy, and sought to differentiate competing systems to discover the valid aspects of alchemical theory. *The Careless Midwife…* is the last of several novels Ettner wrote that consider medical issues (Ettner 1694, 1697, 1698, 1700, 1715). While this type of novel was written to proscribe proper social behaviour, it reflected the cultural prohibitions that governed midwives, surgeons, and physicians of that era. As in his other medical novels, the author promoted the use and effectiveness of Ettner's *Bezoardicum universale,* a pharmaceutical potion used as a universal remedy for numerous conditions.

References

Blake p. 139.

Ettner, J.C. *Des getreuen Eckharts [pseud.] medicinischen Maul-Affens erster Theil; oder, Der entlarvte Marcktschreyer, in welchen vornehmlich der Marcktschreyer und Quacksalber Bossheit und Betrugereyen, wie dieselben zu erkennen und zu meiden; hernach bewertheste Artzney-Mittel in allerhand Kranckheiten… zu gebrauchen… vorgestellet worden.* Franckfurt und Leipzig, Zu finden bey Michael Rorlachs sel. Erben, 1694.

Ettner, J.C. *Des getreuen Eckharts [pseud.] medicinischen Maul-Affens erster Theil; oder, Der entlarvte Marcktschreyer, in welchen vornehmlich der Marcktschreyer und Quacksalber Bossheit und Betrugereyen, wie dieselben zu erkennen und zu meiden; hernach bewertheste Artzney-Mittel in allerhand Kranckheiten… zu gebrauchen… vorgestellet worden.* Franckfurt und Leipzig, Zu finden bey Michael Rorlachs sel. Erben, 1694. (Krivatsy 3731).

Ettner, J.C. *Bezoardicum universale.* 1695.

Ettner, J.C. *Des getreuen Eckharts entlauffener Chymicus: in welchem vornemlich der Laboranten und Process-Krämer Bossheit und Betrügerey….* Augsburg; Leipzig, Lorentz Kroniger u. Gottlieb Göbels, 1696.

Ettner, J.C. *Desz getreuen Eckharts [pseud.] unwürdiger Doctor, in welchem wie ein Medicus, der rechtschaffen handeln will, beschaffen seyn soll; hernach bewährteste Artzney-Mittel in allerhand Kranckheiten… zu gebrauchen….* Augspurg und Leipzig, Bey Lorentz Kroniger und Gottlieb Gobels seel. Erben, 1697.

Ettner, J.C. *Desz getreun Eckardts [pseud.] verwegener Chirurgus, in welchem wie ein rechtschaffener Chirurgus beschaffen seyn sole, was er für Tugenden an sich nehmen, und welcherley Laster er zu fliehen; hernach bewährteste Artzney-Mittel in allerhand Kranckheiten… zugebrauchen….* Augspurg und Leipzig, Bey Lorentz Kroniger und Gorrlieb Gobels seel. Erben, 1698. (Krivatsy 3734).

Ettner, J.C. *Desz getreun Eckarths [pseud.] ungewissenhaffter Apotecker, in welchen wie ein rechtschaffener Apotecker beschaffen seyn, was er vor Tugenden an sich nehmen, und welcherley Laster er fliehen soll; hernach bewehrteste Artzeney-Mittel in allerhand Kranckheiten… zu gebrauchen….* Augspurg und Leipzig, Bey Lorentz Kroniger und Gottlieb Gobels seel. Erben, 1700. Krivatsy 3732).

Ettner, J.C. *Gründliche Beschreibung des egerischn Sauer-Brunns….* Nürnberg, Zieger, 1710. (Blake, p. 139).

Ettner, J.C. *Rosetum chymicum, oder, Chymischer Rosen-Garten: aus welchem der vorsichtige Kunst-Beflissene Voll-blühende Rosen….* Franckfurt; Leipzig, Bey Michael Rohrlachs Wittib und Erben, 1724.

Hardin, J. *Johann Christoph Ettner: eine beschreibende Bibliographie.* Bern, Francke, 1988.

Hardin, J. Johann Christoph Ettner: Eine beschreibende Bibliographie. *German Quarterly* 62: 527–529, 1989.

Hardin, J. Johann Christoph Ettner: physician, novelist, and alchemist. *Daphnis* 19:135–159, 1990.

Tatlock, L. Speculum feminarum: genedered perspectives on obstetrics and gynecology in early modern Germany. *Signs* 17:725–760, 1992.

FRONTISPIECE FROM ETTNER, 1715

Sebastiano Melli

La comare levatrice istruita nel suo ufizio. Secondo le regole piu certe, e gli ammaestramenti piu moderni... Venezia, Giovanni Battista Recurti, 1721.

Melli (fl. 1713–1750), professor of surgery at the University of Venice, based his obstetrical text on the 1596 work "Instructions for the office of midwife...", of Geronimo Scipione Mercurio (ca. 1550–1615). In four "books," Melli described the anatomy of the female reproductive system, serving for the "propagation of the species," the course of normal parturition, complicated childbirth and false pregnancy, and the management of the dead fetus, cesarean section, and other complications. In PlateI, Book III, Melli illustrated the mother positioned in childbirth allowing her legs to hang from the edge of the bed or table. He reported this to be particularly useful for patients who were obese, or in cases in which the fetus was in an abnormal or undesirable position. This posture for delivery also had been described and illustrated by Mercurio more than a century earlier as useful in cases of difficult birth (Mercurio 1596).

This dangling leg position was "rediscovered" another century and a half later by Gustav Adolf Walcher (1856–1935) (Walcher 1889), and carries the eponym Walcher's position. Figures ii and iii below show unborn infants in such abnormal positions with the head turned to the side and with the face presenting, respectively.

Melli also wrote on the practice of surgery (1713) and fistula of the lacrimal gland in the eye (1717).

References

Blake p. 300.

Melli, S. *Pratica chirurgica.* Venezia, 1713.

Melli, S. *Della fistole lacrimali....* Venezia, Gio. Battista Recurti, 1717.

Mercurio, G.S. *La commare o riccoglitrice....* Venettia, G.B. Ciotti, 1596.

Walcher, G. Die Conjugata eines Beckens ist kine konstante Grösse, Sondern, lässt sich durch Körperhaltung der Trägerin Verändern. *Centralb für Zbl Gynäk* 13: 892–893, 1889. (GM 6200).

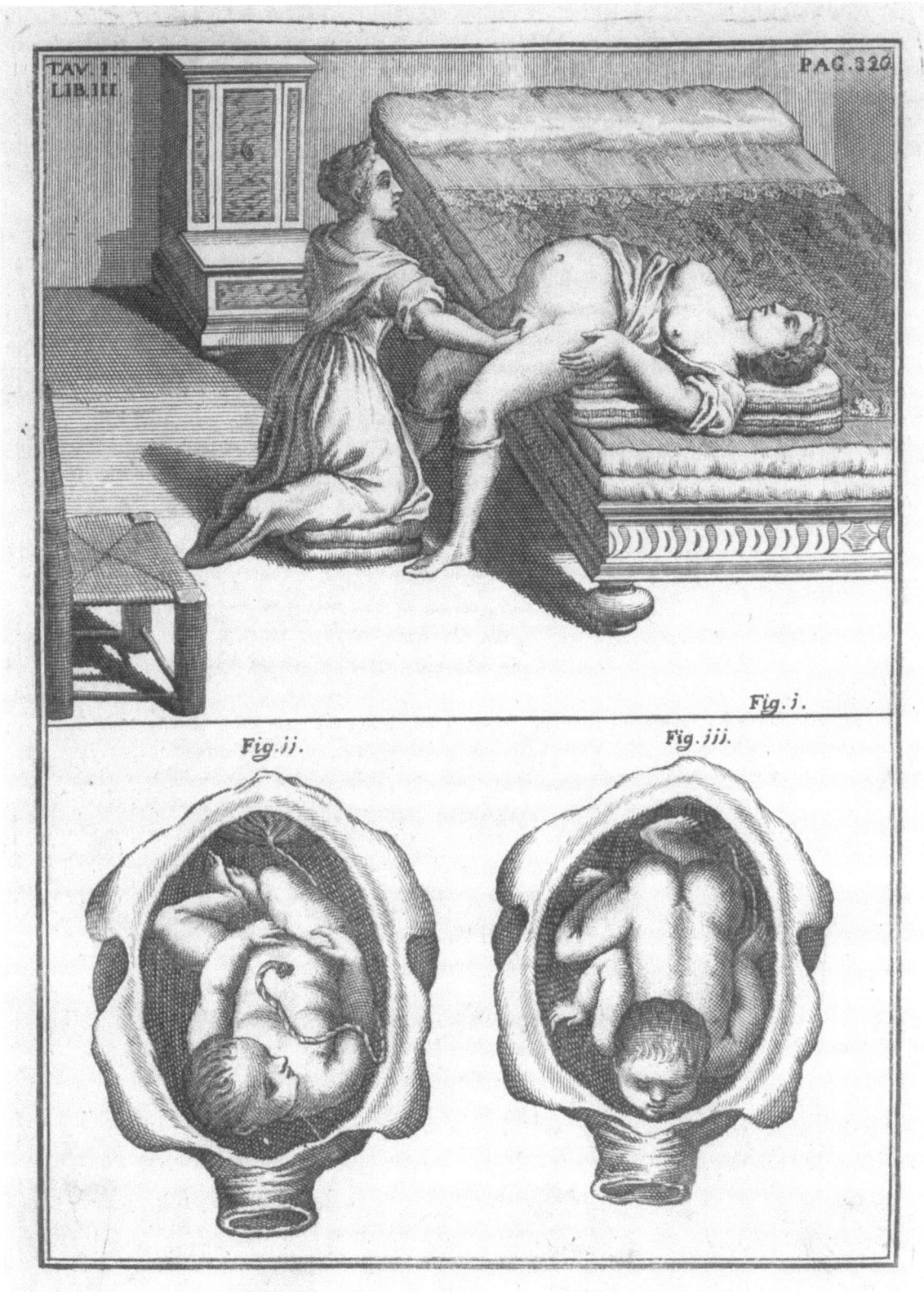

PLATE I, BOOK III FROM MELLI, 1721

Sir Richard Manningham

An exact diary of what was observ'd during a close attendance upon Mary Toft, the pretended rabbet-breeder of Godalming in Surrey, from Monday Nov. 28, to Wednesday Dec. 7 following. Together with an account of her confession of the fraud. London, Printed for Fletcher Gyles..., 1726.

Manningham (1690–1759), a prominent teacher in London, in 1726 was called into consultation in the case of Mary Toft (1703–1763) of Godalming (Godlyman), Surrey, who perpetuated the hoax of giving birth to rabbits. Initially, this was attested to by a local apothecary, John Howard, who had practiced midwifery for 30 years. Toft claimed that "As she was weeding in a Field, she saw a Rabbet spring up near her after which she ran with another Woman that was at work just by her: this set her a longing for Rabbets, being then, as she thought, 5 Weeks gone with Child; the other Woman perceiving she was uneasy, charged her with longing for the Rabbet they cou'd not catch, but she deny'd it: soon after another Rabbet sprung up near the same place, which she endeavour'd likewise to catch. The same Night she dreamt that she was in a Field with those two rabbets in her Lap and awaked with a sick Fit which lasted till Morning; from that time for above 3 Months, she had a constant and strong desire to eat Rabbets, but being very poor and indigent cou'd not procure any".

Later, Mary Toft was delivered in November, 1726 first of parts of a pig, and afterwards of a rabbit. She was attended by Howard, who over the next few days delivered Mary of nine rabbits. The amazed Howard sent word of the phenomenon to several people, including the information that he was still delivering more rabbits from Mary. This fiction even came to the attention of King George I (1660–1727; King 1698–1727), who commissioned his surgeon-anatomist Nathaniel St. André (1680–1776) and Samuel Molyneux (1689–1728), an astronomer to travel to Godalming to witness a delivery. St. André then delivered her of two rabbits, and was convinced that the births were genuine. This news spread like wildfire and Cyriacus Ahlers (fl 1727–1756), another surgeon to the King, was sent to Guilford to investigate. The rabbits appeared after Mrs. Toft had dreamed of them and continued to come over a number of days finally ending in a total of 18 rabbits.

Because matters were somewhat confused, with those who believed and those who disbelieves the veracity of these reports, the King dispatched Sir Richard Manningham, the leading man-midwife of his day, who recorded that he, with St. André, set out for Godalming "at about four in the morning" Sunday 28 November ... arriving there "a little after twelve at noon" (p. 7).

Manningham examined Mary Toft, noting that the uterus was contracted and cervix closed with no evidence of a pregnancy (p. 8). Some time later, when she was in pain he examined her and removed what appeared to be a piece of hog's bladder from the vagina. Sir Richard informed those present that it was his opinion "that Membrane never came out of the *uterus*," after which Mary Toft cried, supposing he "thought her a Cheat" (p. 14). Manningham stated that after he with colleagues retired to the *White Hart* pub, "very warm disputes arose amonst us" as to the veracity of her accounts. The following day, Mary Toft was taken to London. Over the next several days, the subject experienced pains with "motion on the right side of her Belly" (p. 21), and several apparent epileptic seizures. Later, on Sunday 4 December, both the porter and Mary Toft's sister acknowledged that they had procured a "Rabbet in a clandestine manner," but that it was for her to eat as she had "long'd for it" (p. 26). Upon being brought to London, her fraud was discovered after a porter admitted smuggling a rabbit into her quarters. Finally, on Wednesday 7 December 1776, she admitted that she had stuffed dead rabbits into her vagina, and then pretended to give them birth.

This affair, which attracted wide publicity, led eighteenth century English pamphleteers to a heyday in controversy and discussion, which kept them busy publishing pamphlets and broadsides for a year. Subsequently, Mary Toft was imprisoned in Bridewell as "an infamous Cheat and Imposter". Several months later she was released without being prosecuted. One explanation for the widespread acceptance of this hoax was that the notion of a human giving birth to rabbits fit well with the belief held by many eminent individuals of that time, the notion of "maternal impressions"—in other words, that a pregnant woman's experiences are directly imprinted on her unborn child. This theory was used to explain birth defects and monstrosities. For example, a child born deaf was believed due to the mother having been shocked by a loud sound during pregnancy, or a child born blind might be due to the pregnant mother having looked at a blind person. Thus, Toft's tale about her desires, dreams, and garden exploits fit well with the hypothesis of maternal impressions and lent plausibility to her account.

Following this deception, the medical profession was mocked and derided widely for its gullibility. The English caricaturist, engraver and critic William Hogarth (1697–1764) published the print, *The Cunicularii [rabbit warren] or the Wise Men of Godliman in Consultation* showing her in the process of giving birth. Later, he published the print *Credulity, Superstition, and Fanaticism,* (shown on the previous page) in which Manningham with St. André observe her delivery. The English satirist, and Dean of St. Patrick's Cathedral, Dublin, Jonathan Swift (1667–1745), using the pseudonym Lemuel Gulliver, is believed to have been the

Credulity, Superstition, and Fanaticism by William Hogarth

author of a tract ridiculing the deception that had so captured the imagination of "the Minds of the People of this Island… so as scarce to leave them any leisure for things of a more sublime nature and of greater consequence and importance" (1727, p. 3). The poet and satirist Alexander Pope (1688–1744), with William Pulteney Earl of Bath (1684–1764), wrote the satire "The discovery: or, the squire turn'd ferret…" (Pope and Pulteney 1727). Another satire was attributed to "Mary Tuft" (1727).

Manningham, a graduate of Cambridge University received his medical degree in 1725, after being elected a Fellow of the Royal Society and licentiate of the Royal College of Physicians (both 1719). In 1722, he was knighted by George I. One of Manningham's major contributions was the establishment with the aid of Public Subscriptions, in 1739, of a lying-in residence for women next to his own home. This was, in effect, the first maternity hospital in London. In his *Artis obstetricariae compendium…* (1739), he presented precepts in the form of aphorisms on normal and complicated deliveries, abnormal fetal presentations, and the diseases to which pregnant women were susceptible. In his *An abstract of midwifry…* of 1744 he described a "glass machine" or manakin that he created for the education of students in the art of midwifery (Manningham 1744).

References

Blake, p. 286; Cutter and Viets, pp. 15–16, 180–181.

Ahlers, C. *Some observations concerning the woman of Godlyman in Surrey, made at Guilford on Sunday, Nov. 20. 1726. Tending to prove her extraordinary deliveries to be a cheat and imposture.* London, J. Roberts, 1726. (Blake, p. 6).

Brathwaite, T. *Remarks on a short narrative of an extraordinary delivery of rabbets, perform'd by Mr. John Howard, Surgeon at Guilford, as publish'd by Mr. St. Andre, anatomist to his Majesty. With a proper regard to his intended recantation.* London, N. Blandford, 1726. (Blake, p. 64).

Douglas, J. *An advertisement occasion'd by some passages in Sir R. Manningham's diary lately publish'd.* London, Printed for J. Roberts… and J. Pemberton, 1727. (Blake, p. 125).

Graham, H. *Eternal Eve.* London, Heinemann, 1950, pp. 334–345.

Gulliver, L. [Pseudonym] *The anatomist dissected: or the man-midwife finely brought to bed. Being an examination of the conduct of Mr. St. Andre: touching the late pretended rabbit-bearer: as it appears from his own narrative. By Lemuel Gulliver, surgeon and anatomist to the Kings of Lilliput and Blefuscu.* Westminster, A. Campbell, 1727. (Blake, p. 190).

Manningham, R. *Artis obstetricariae compendium tam theoriam quam praxin spectans: morborum omnium qui foeminis inter gestandum in utero et in puerperio….* Halae Magdeburgicae, Luderwaldian, 1739. (Blake, p. 286).

Manningham, R. *An abstract of midwifry, for the use of the Lying-in Infirmary….* London, T. Gardner, 1744.

Manningham, R. *Aphorismata medica….* Londini, Apud J. Robinson…, 1756.

Pope, A. & W. Pulteney. *The discovery: or, the squire turn'd ferret. An excellent new ballad, to the tune of high boys! Up go we….* Westminster, A. Campbell, for T. Warner, 1727.

Rhodes, P. Sir Richard Manningham. In: *Oxford Dictionary of National Biography.* Vol. 36. Oxford, The University Press, 2004, pp. 514–515.

Robertson Aitchison, W.E. Extraordinary delivery of rabbits. *Ann Med Hist* 10: 368–375, 1928.

St. André, [N]. *A short narrative of an extraordinary delivery of rabbets, perform'd by Mr. John Howard surgeon at Guilford.* London, John Clarke, 1727. (Blake, p. 397)

Tuft, M. *Much ado about nothing; or, a plain refutation of all that has been written or said concerning the rabbit-woman of Godalming: being a full and impartial confession from her own mouth, and under her own hand, of the whole affair from the beginning to the end.* London, A. Moore, 1727.

Edmund Chapman

An essay on the improvement of midwifery; chiefly with regard to the operation. To which are added fifty cases, selected from upwards of twenty-five years practice. London, A. Blackwell et al, 1733.

In his *Essay...* with fifty case reports, Chapman (ca. 1680–1756) a surgeon-man midwife, first published an account of the Chamberlens' "secret instrument", the obstetrical forceps which had been invented about one-half a century earlier. Chapman recorded his experience as a man-midwife and noted that he had used the forceps as early as 1723. He gave sound directions for their use. "As to the *Forceps*, which, I think, no Person has yet any more than barely mentioned, it is a noble Instrument, to which many now living owe their Lives, as I can assert from my own Knowledge and Practice. This Instrument, ...must yet be used with great Caution; You are first to pass one part thereof above, gently introducing it, and guarding and directing the *Bow* as far as you can, with all the fingers of the Left Hand (the Instrument lying in the Hollow of the Hand,) carefully preventing any Valve [fold] of the *Vagina* from getting between the Instrument and the Head of the Child, which would at once prevent any Hold of the Head, (and consequently foil you in the Attempt,) and bruise the Part that intervenes, so as to engender an *Excoriation, Inflammation*, or, perhaps, a *Mortification*... One Part thus passed over the Head, the other is then to be passed underneath, exactly opposite to the former, and this with the Care before mentioned. When this is passed, observing that it be done on the *right* Side, the *Handle* Part on the other, they are then to be brought close together ... and thus you may extract the Head, by drawing gently down ..." (pp. 13–14).

Chapman presented reports of several successful cases in which he had personally delivered. He also described his modifications of the instrument, which was substantially the same as that used by the Chamberlens. Chapman confessed that he came by his design "by mere *Accident*, as I believe, is often done. For many Years my *Forceps* happen'd to be made of so soft a Metal as to bend or give way, or suffer some Alteration of their *Curve*. They were made, as usual, with the *Screw* fixed to one Part or Side of them. These I used for some Years, but they often happening to slip off sideways (as before mentioned,) it lessened my Opinion of the Instrument so much, for several Years after, I used it but seldom,.... At length I caused another Pair to be made me, of better Metal, some other Improvements" (pp. 15–16). Chapman described removing a screw that he had placed to assist in holding the handles together. He noted, however,

that in one instance he lost the screw in the bedcloths at the time of delivery, and then discovered that without it, "The Instrument did its Office much better without the *Screw*, or the two Parts being fixed" (pp. 15–16). Chapman observed that the forceps were well known to the principal men of the profession, and apparently, for this reason, he neglected to include an illustration in his original account. He corrected this oversight in the book's second edition published in 1735. Here he shows a pair of obstetrical forceps with fenestrated blades, both a cephalic and pelvic curve, and inturned handles (see Illustration).

In the Preface, Chapman observed, "My greatest Aim... is... to instruct, not such as are quite ignorant, but such as have already made some Progress in the Science, and to point out to them those dangerous *Rocks* on which many have been cast away....". He concluded, "If by submitting my self to the *Censure* of others, I should have the good fortune to excite some more masterly Pen to improve this *Art*, it would much augment my Pleasure and Satisfaction, as I shall thereby be instrumental it [sic] conveying a greater Good to my Fellow-Creatures" (pages un-numbered). Chapman, a "country practitioner", worked in South Halstead, Essex. This was only about 40 Km from Woodham Mortimer, where almost a century later the obstetric forceps of the Chamberlen family were discovered. As an aside, William Giffard who also described the "Secret Instrument" (Giffard 1734), lived in the same region. Chapman met with such success that, in about 1733, he was induced to move to London and teach man-midwifery. Lorenz Heister (1683–1758) first published an illustration of a single blade of the obstetric forceps ("*les mains de fer*" of Jean Palfyn (1650–1730) in 1724 (Heister 1724); however, he failed to include an adequate description of the instrument and later wrote that he was unable to demonstrate its utility. William Giffard's posthumous *Cases in Midwifery...*(Giffard 1734) contains a description and illustration of the complete obstetric forceps. Almost simultaneously, Alexander Butter of Edinburgh published an illustration of Dusée's (fl. 1730) instrument (Butter 1735). John Burton (1697–1771) also described several early varieties of the forceps (Burton 1751).

References

Blake p. 84; Cutter and Viets pp. 22–23, 61–63, 181–182; Garrison Morton 6156.3, Spencer p. 27.

Burton, J. *An essay towards a complete new system of midwifery, theoretical and practical.* London, J. Hodges, 1751. (GM 6268).

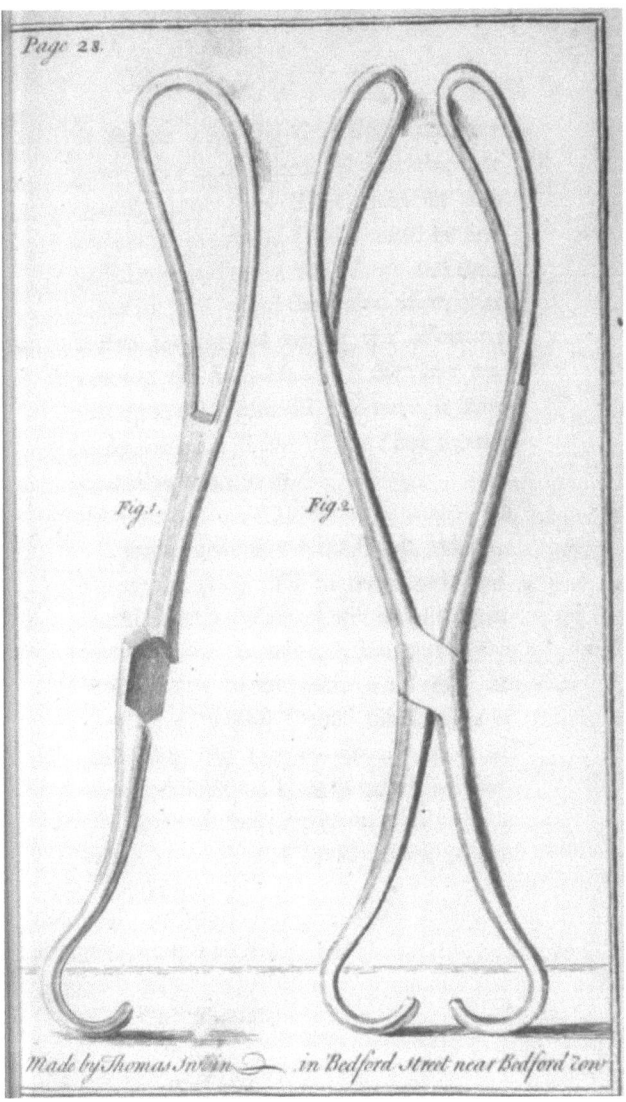

Page 28.

Fig.1.   Fig.2.

Made by Thomas Irwin⏝ .in Bedford Street near Bedford row

FIGURE FROM CHAPMAN, 1733

Butter, A. The description of a forceps for extracting children by the head when lodged low in the pelvis of the mother. *Med Essays Observ (Edinburgh)* 3: 320–, 1735.

Chapman, E. *A treatise on the improvement of midwifery, chiefly with regard to the operation. To which are added fifty-seven cases, selected from upwards of twenty-seven years practice.... 2nd Ed.* London, John Brindley et al., 1735.

Chapman, E. *A reply to Mr. Douglass's short account of the state of midwifery in London and Westminster....* London, Printed for T. Cooper..., 1737.

Chapman, E. *A treatise on the improvement of midwifery; chiefly with regard to the operation.... 3rd Ed.* Ed. By L.D. Longo. Birmingham, AL, The Classics of Obstetrics and Gynecology Library, Gryphon Ed, 1981.

Doran, A. Dusée: his forceps and his contemporaries. *J Obstet Gynaec Brit Emp* 22: 119–143, 1912.

Doran, A. Dusée, De Wind and Smellie: an addendum. *J Obstet Gynaec Brit Emp* 22: 203–207, 1912.

Giffard, W. *Cases in midwifery... Revised and published by Edward Hody....* London, B. Motte and T. Wotton, 1734.

Heister, L. *Chirurgie, in welcher alles was zur Wund-Artzney gehöret....* Nurnberg, J. Hoffman, 1724. (GM 5576 is 1718 ed.).

Radcliffe, W. *The secret instrument. The birth of the midwifery forceps.* London, W. Heinemann, 1947. (GM 6311).

Rhodes, P. Edmund Chapman (fl. 1735). *J Obstet Gynaec Brit Comm* 75: 793–799, 1968.1735.

William Giffard

Cases in midwifry. Written by the late Mr William Giffard, surgeon and man-midwife. Revis'd and publish'd by Edward Hody, M.D. and Fellow of the Royal Society, London, Printed for B. Motte, T. Wotton, and L. Gilliver… and J. Nourse…, 1734.

Giffard (birth date unknown–1731), a London surgeon and man-midwife, recorded 225 cases of complications of pregnancy over the almost 8 year period from January 1724 until October/November 1731. He is considered the first English obstetrician to have published substantial contributions to clinical midwifery. He described extrauterine pregnancy (Case CLVII), the case being that of a fetus at about 6 months gestation, which terminated by rupture into the rectum. Giffard reported that he had seen the woman about 3 months previously, when it appeared that she was having a miscarriage. Sometime thereafter, her husband reported that she believed she was pregnant, sensing "quickening." 6 or 7 weeks later, she experienced severe abdominal and back pain. Several weeks later, she bled and passed a fetus per rectum. The following week the woman died. At autopsy, Giffard found the right ovary to be "…dilated into a large *Sacculus* of an irregular form, extending itself behind the *Uterus*, (to the back part of which it adhered) and…was connected to that part of the *Colon* that terminates in the *Rectum*… In this *Sacculus* we found part of a *Placenta*, and the remains of the lacerated Membranes; and besides the Aperture of the *Fallopian Tube* mentioned before, there was another about four inches in *Diameter*, into the middle of the *Rectum*" (p. 380). Page 380 (Figs. 1 and 2) illustrates a case of ectopic pregnancy. The legends read: "Fig. 1. Shews the Uterus with the Sacculus behind it, part of the Colon and the Rectum; the Fallopian Tubes, Ovary on the left side, Ligamenta rotunda, and the vagina laid open to the Os Tincae; A, The Uterus; B, The Fallopian Tube; C, The Ovary; D, The Ligamenta rotunda; E, The Vagina laid open; F, That part of the Colon that terminates in the Rectum; G, The Rectum continued to the Anus under the Vagina; H, The Fallopian Tube on the right side, whose extremity opens into the Sacculus formed from the Ovary; I, The Sacculus extending itself behind the Uterus, wherein we found part of the Placenta, several lacerated membranes, and from whence there was a large opening into the Rectum; Fig. 2 shows the Inside of the Sacculus, and its Aperature into the Rectum: A, The Intestine; B, The Sacculus adhering to it; C, The Opening from the Sacculus into the Rectum; D, The Membranes found withing the Sacculus; E, The Vagina turned to the right."

In addition, Giffard was among the first, after the Chamberlens, to use the forceps, which he called an "extractor", using the instrument in April 1726. Three years following his death, his series of cases was edited by Edward Hody (1698–1759), who recorded in the Dedication that these 225 women, "for the most part were attended with a great deal of danger and difficulty". In regards to Giffard's writing style, Hody stated further that "with Books, as it is with Men, we ought principally to regard the *Use* they are of to Mankind: and … whoever shall pursue these Cases with an intent to *learn* the Practice of Midwifry, will not think his time ill spent" (p. v–vi).

Before presenting the cases, the book illustrates in three plates the "extractors" designed by Giffard and John Freke (1688–1756). The latter, a surgeon at St. Bartholomew's Hospital, is particularly known for his contributions to ophthalmic surgery. This was the first time that the obstetrical forceps were illustrated, invented by and previously kept secret by the Chamberlen family, report of a case (number XIV) in which they were used. Edmund Chapman had published the first account of the use of foreceps in his monograph of the previous year (Chapman 1733).

The "End of Book" plate shows the uterus of a patient who died with a "retained placenta" (Case CLXXXVI). Following her delivery, she had continued to bleed intermittently, and the day prior to her demise, the patient had delivered "several large substances, form'd from a great number

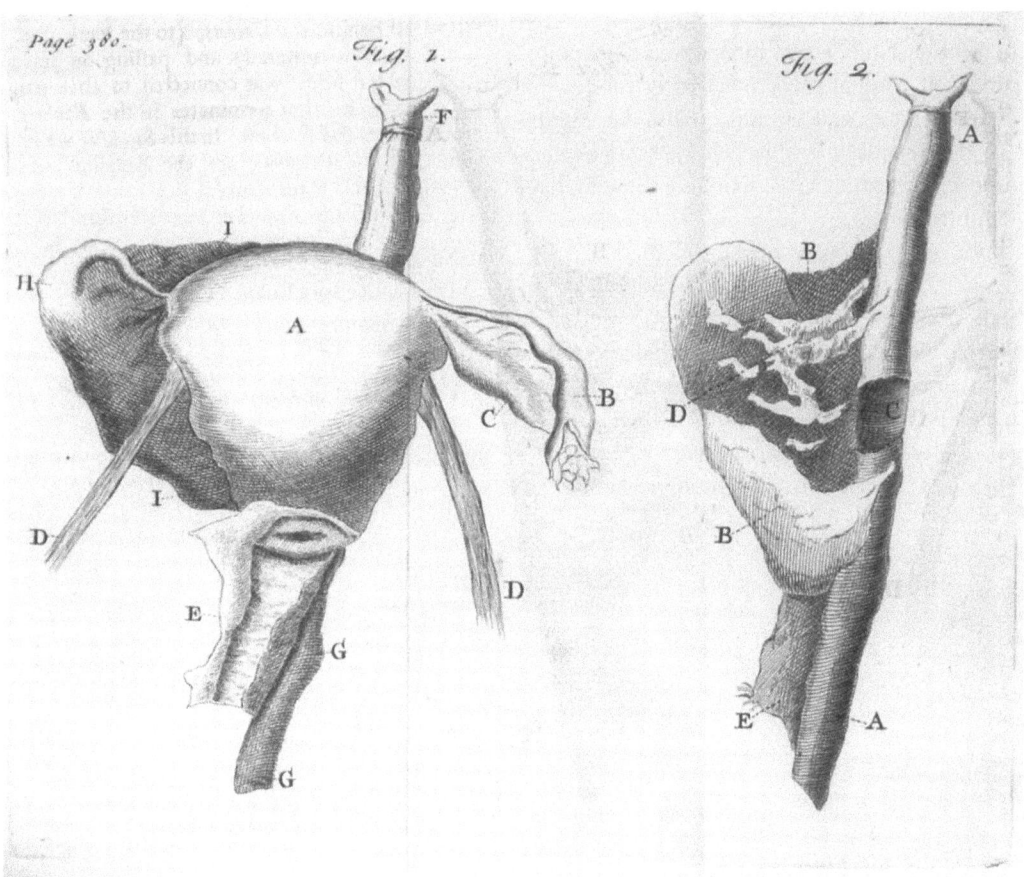

PAGE 380 FROM GIFFARD, 1734

of Hydatides…. From that time … she was subject to sudden and pretty frequent gushings out of blood from the womb." She also experienced "violent pains in her belly", following which she "died of a Fever, and Mortification in the Womb" (pp. 518–519). This probably was, in fact, a chorionepithelioma, which developed over the 12 months following passage of a hydatiform mole.

Giffard's work is also of interest for his method of delivering the after-coming head in breech presentation (Case XIII) [the Smellie maneuver], which preceded its description by William Smellie (1697–1763) by several decades (Smellie 1752). Giffard also described the Ritgen maneuver to preserve the patients' perineum at the time of delivery, almost a century before Ferdinand August Marie Franz Ritgen (1787–1867) (Ritgen 1828). Unfortunately, little is known of Giffard's life.

References

Blake p. 175; Cutter and Viets pp. 18–22; Garrison Morton 6156.3; Spencer, pp. 18–22.

Chapman, E. *An essay on the improvement of midwifery.* London, A. Bettsworth, 1733.

Giffard, W. *Cases in midwifry… Revis'd and publish'd by Edward Hody, M.D.…* Ed by L.D. Longo. New York, NY, The Classics of Obstetrics and Gynecology Library, Division of Gryphon Editions, 1995.

Ritgen, F.A.M.F. Geburtshülfliche Erfahrungen und Bemerkungen. *Gemein deutsch Ztschr für Geburtsk* 3: 147–169, 1828.

Smellie, W. *A treatise on the theory and practice of midwifery.* London, D. Wilson, 1752. (GM 6154)

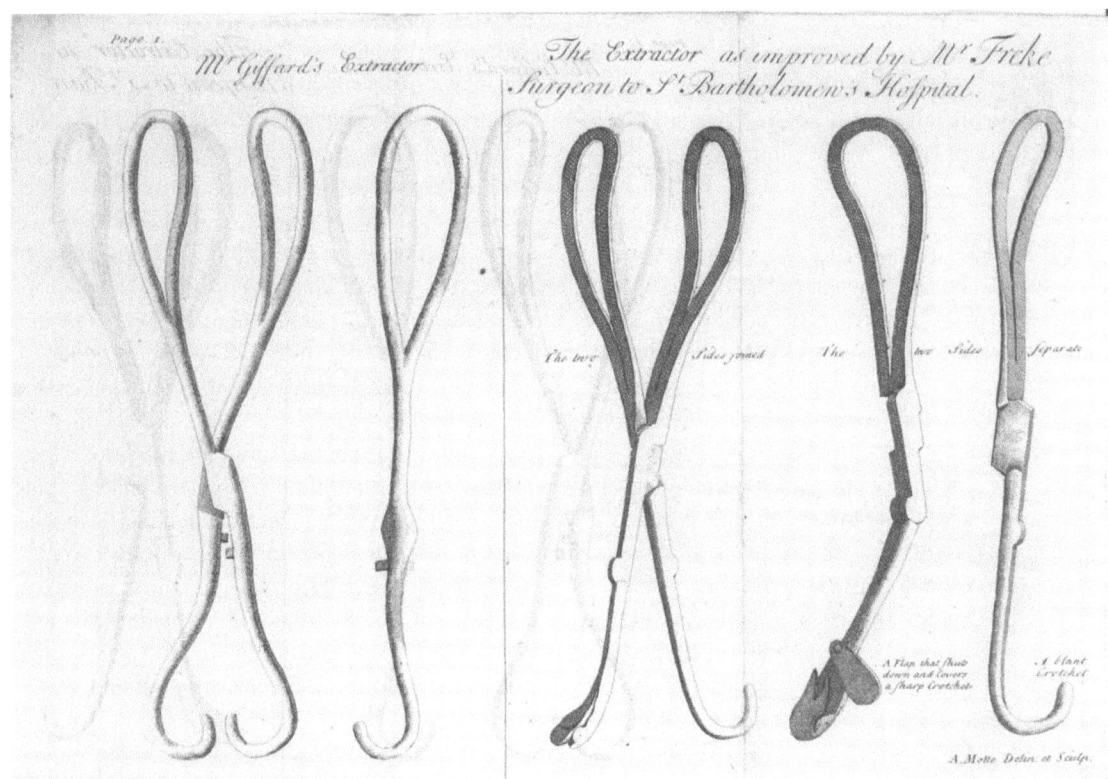

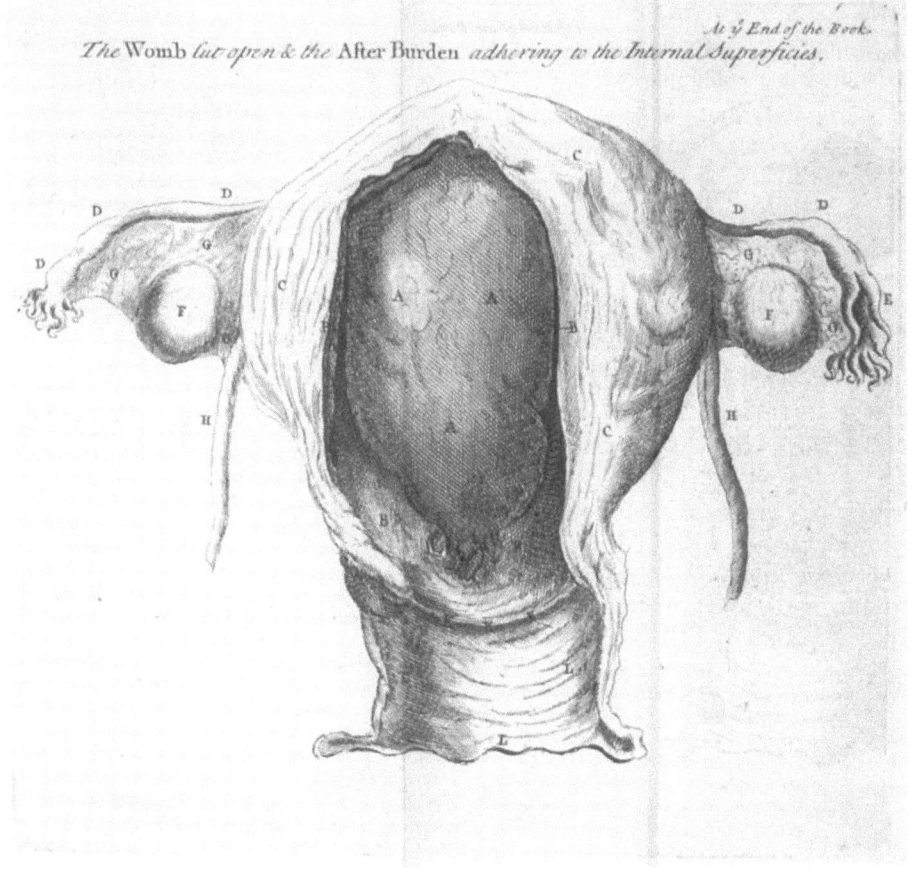

PLATES FROM GIFFARD, 1734

Albertus Seba

Locupletissimi rerum naturalium thesauri accurata descriptio, et iconibus artificiosissimis.... Amsterlaedami, Apud Janssonio-Waesbergios, & J. Wetstenium, & Guil. Smith, 1734–1765.

In his "Accurate description of the very rich thesaurus of [the most important] natural objects" Seba (1665–1736) displayed the astounding collection of natural curiosities: plants, animals, snakes, and insects that he had collected from around the world. Commencing in 1731, Seba commissioned illustrations of his specimens, which he arranged in four volumes with 446 copperplates, 175 of which were double paged. The volumes were published independently over three decades, with the commentary on the plates in a Latin-French and a Latin-Dutch edition, in an effort to reach a broad readership. Volume I (1734) depicts plants and animals from South America and Asia. Volume II (1735) is dedicated to reptiles and snakes. Volume III (1758) concentrates on marine life and sea creatures, and Volume IV (1765), on insects, fossils, and minerals. An impressive example of a Baroque book, the detailed drawings and engravings by a number of artists are quite remarkable. Of these Seba wrote, "all sorts of exquisite pieces from the East and West Indies," among which were about "700 jars containing the rarest exotic animals and many particularly rare snakes. Also brought together thus are every exceptional sort of beautiful and rare conch, the finest and most complete butterflies from the 4 corners of the Earth [and samples] of all the plants, some familiar pieces, but unfamiliar ones too." In addition, Seba included many monstrosities both real, and some mythical such as dragons.

In this reproduction from an original hand painted edition in the Koninklijke Bibliotheek, Den Haag, The Netherlands, Tomus I, Tabula 111 illustrates several fetuses from various species: Fig. 1, Asiatic elephant; Fig. 2, human at about 18 weeks gestation; Fig. 3, sheep; Fig. 4, pig; Figs. 5 and 6, mice.

Beginning in the sixteenth century, volumes on the natural world contained detailed and realistic descriptions. An early example of this genre was the epochal herbal *De historia stirpium...* by Leonhart Fuchs (1501–1566), which included over 500 woodcuts (Fuchs 1542). It is believed that

Seba, an Amsterdam pharmacist, commenced his collecting with plants and herbs of medicinal value, then extending his collection of specimens in the seventeenth and eighteenth century tradition of John Ray (1628–1705), Hans Sloane (1660–1753), and John Hunter (1728–1793) of London, and other physician—or apothecary—naturalists of this period. In amassing such cabinets of curiosities, these collectors strove to obtain a more complete understanding of the natural world. The "cabinet" upon which Seba's *Thesaurus* is based was actually his second. In 1717, the Russian czar, Peter I, "the Great" (1672–1725), who was visiting the Dutch Republic to shop for his own cabinet of wonders from around the world, purchased Seba's first collection (as he did the collection of Frederik Ruysch (1638–1731) see Ruysch 1701–1716). In addition to commissioning some travelers and seafarers to bring him specimens from distant lands, Seba would purchase specimens from sailors upon their return to the port city of Amsterdam from trips abroad. On the basis of his contributions to natural history, Seba was elected to Fellowship in the Royal Society of London, as well as the Accademia Scienzia de Bologna. Perhaps surprisingly, many of Seba's specimens have survived in the collections of the zoological and natural history museums of Amsterdam, Stockholm, and St. Petersburg, as well as at the British Museum in London. The taxonomist Carl von Linné (Linnaeus; 1707–1778) used and made considerable reference to Seba's collection and the *Thesaurus* in formulating his classification system of biology (Linné 1735, 1737).

References

Fuchs, L. *De historia stirpium commentarii insignes....* Basileae, In officina Isingriniana, anno Christi MDXLII, 1542. (GM 1808).

Linné, C.V. *Systema naturae.* Lugduni Batavorum, apud Theodorum Haak, 1735. (GM 99).

Linné, C.V. *Genera plantarum.* Lugduno Batavorum, apud Conradum Wishoff, 1737. (GM 1829).

Seba, A. Cabinet of natural curiosities, Das Naturalien-Kabinett, Le cabinet das curiosites naturellis, Locupletissimi rerum naturalium thesauri, 1734–1765. Ed by I. Musch, R. Willmann, and J. Rust, Köln..., Taschen GmbH, 2001.

FIGURE FROM SEBA, TOMUS I, TABULA 111, 1734.
Courtesy Internet Archive (http://archive.org)

Pietro Berrettini Da Cortona

Tabulae anatomicae. Rome, Fausti Amidei, 1741.

In about 1618, da Cortona (1596–1669), a prominent artist of the Italian High Baroque period, presented in 27 drawings elegant representations of human anatomy, chiefly muscles, nerves, and blood vessels. These were prepared from dissections made at the *Santo Spirito* hospital, in Rome. The drawings were executed by the use of several techniques: black chalk, pen and ink, sepia wash and white heightening on buff paper tinted with a gray wash. The copperplate engravings are believed to have been prepared by Luca Ciamberlano (ca. 1580–1641). It is unclear for what volume or work these figures were prepared, or why it was that the drawings were not published for over a century (da Cortona 1986). Some of the figures in the plates hold a medallion depicting some anatomical region in greater detail (some have stated these are mirrors in which are reflected other images). Tab XXVII, Figure I of a standing woman illustrates, within her opened abdomen, the uterus (K), Fallopian tubes (N), and ovaries (L), as well as the posterior aspect of the abdominal cavity. Figure II shows the opened uterus, within which is a stylized fetus.

The editor, Gaetano [Cajetano] Petrioli, surgeon to Victor Amadeus II (1666–1732) of Sardinia, prepared the text and added small numbered anatomical figures to the margins of the plates. A note in Italian, which accompanies the drawings, states that they were prepared with the aid of the surgeon Nicholas Larchée [Larche] (1602–1665). In 1772, the original drawings for the plates were obtained by the anatomist–obstetrician William Hunter (1718–1783), and according to Hunter, the plates were intended to teach neurology. The drawings now are preserved in the Hunterian Library at the University of Glasgow (Kemp 1976). In addition to his gifts as an artist, da Cortona, who took the name of the city in Tuscany in which he was born, was influential as an architect, decorator, and designer of tombs. In particular, he is noted as the creator of large integrated compositions in ceiling frescos, as opposed to the practice of preparing small individual compositions. He was involved in the reconstruction and decoration of several major churches including *Santa Bibiana* and *Santi* [Saint] *Luca* [Luke] *e Martina,* the latter of which was built originally in the seventh century. He also was responsible for the Gran Salone of the *Palazzo Barberini*, the latter of which contains his frescos *Allegory of Divine Providence* and *Berberini Power*. da Cortona also helped to plan the *Louvre* in Paris and the *Palazzo Pitti* in Florence. Over 150 of his frescos and paintings have been documented (Briganti 1962).

References

Blake p. 42; Choulant, pp. 235–239; Garrison Morton 395.2.

Briganti, G. *Pietro da Cortona o dela pittura barocca.* Florence, Sansoni, 1962.

Da Cortona, P. *Tabulae anatomicae ex archetypis egregii pictoris Petri Berrettini Cortomensis expressae... Alteram hanc editionem recensuit, nothas iconas expunxit, perpetuas explications adjecit Franciscus Petraglia....* Rome, Impensis Venantii Monaldini... excud. Johann. Zempel., 1788.

[Da Cortona] *The anatomical plates of Pietro da Cortona. 27 Baroque masterpieces. With a new introduction by Jeremy M. Norman.* New York, Dover, 1986.

Duhme, L. *Die tabulae anatomicae des Pietro Berrettini da Cortona.* Cologne, Institut für Geschichte der Medizin, 1980.

Kemp, M. Dr. William Hunter on the Windsor Leonardos and his Volume of Drawings attributed to Pietro da Cortona. *Burlington Magazine* 118: 144–148, 1976.

Olry, R. & K. Motomiya. Baroque anatomy masterpieces for plastinated specimens. *J Int Soc Plastination* 12: 18–22, 1997.

Wittkower, R. *Art and architecture in Italy. 1600 to 1750, 2nd rev. ed.* London, Penguin Books, 1965.

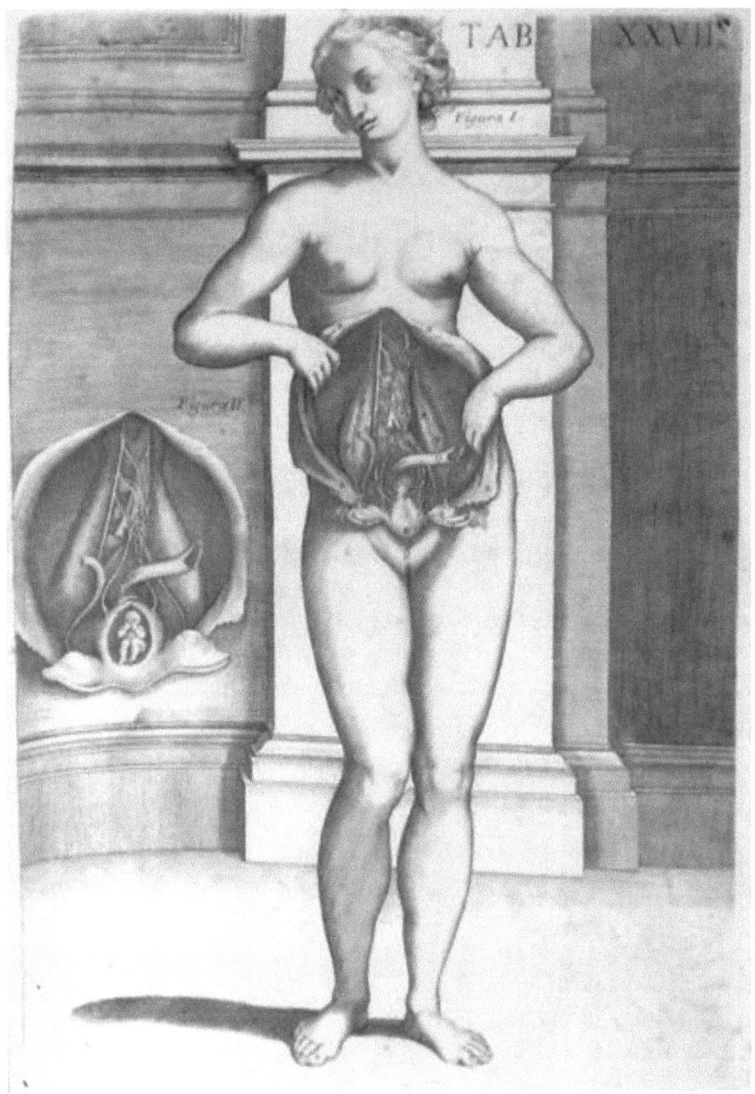

TAB XXVII, FIGURES I AND II, FROM DA CORTONA, 1741.
Courtesy J. Norman's History of Science Online Bookshop (http://www.jnorman.com)

Sir Fielding Ould

A treatise of midwifery. In three parts. Dublin, Oli Nelson & Charles Conner, 1742.

In the early to mid-eighteenth century most deliveries were performed by midwives without formal training. The majority of the few surgeons called into consultation were equally ignorant of the process of childbirth, and physicians were not allowed to practice midwifery. Also during this period, few legitimate textbooks of midwifery/obstetrics were available. A pivotal figure in the evolution of unskilled midwifery into modern obstetrics was Fielding Ould (1710–1789), a "man-midwife" of Dublin. His *Treatise* is considered the first textbook of importance in the English language.

To place this work in perspective, it is useful to consider obstetrics at this time. Prior to the establishment of lying-in hospitals, many pregnant women, particularly the poor and those in large cities, brought their children into the world under the most wretched of circumstances. Destitute of proper care, warmth, or food, the mortality rate has been estimated at about 3 % (Le Fanu 1972). The deplorable circumstances for pregnant women in the city of Dublin have been eloquently described (Kirkpatrick 1912, Kirkpatrick and Jellett 1913). Also, at this time, midwifery was considered the lowest form of medical practice. Occupying a higher rung on the therapeutic ladder were apothecaries and barber surgeons, some of whom dabbled in midwifery. Physicians, who stood at the pinnacle of this medical pyramid, were the only practitioners with a university education; however they did not practice obstetrics. In Dublin, although several hospitals were established during the early eighteenth century, none gave clinical teaching. At the University of Dublin, also called Trinity College, rather than a clinical curriculum, the Medical Faculty (founded in 1711) offered only courses in botany, chemistry, and natural philosophy. By and large, physicians had to travel to London or the continent to complete their education.

In the wake of the Irish potato famine of 1740, Dublin and other cities were flooded with the poor and destitute. One sixth to one fifth of the population are believed to have died as famine-related epidemics raged. In an effort to alleviate some of the suffering by women in childbirth, and inspired by his experience at seeing the maternity facilities in Paris, in 1745, the surgeon and man-midwife Bartholomew Mosse (1712–1759) established the first bona fide maternity hospital in the British Isles, Dublin's Hospital for Poor Lying-In Women. In December 1757, this facility was moved, and opened as The New Lying-In Hospital complex, later called the "Rotunda".

As its title suggests, *A Treatise...*, consists of three sections: normal parturition, complicated obstetrics, and the use of instruments for delivery. The work was the first to consider the mechanism of normal labor, to recommend the use of opiates and rest in prolonged labor, to describe episiotomy, to urge delivery of the second twin shortly following that of the first, and to caution against premature extraction of the placenta. In addition, Ould was one of the earliest *accoucheurs* to consider ethical issues and the need to consult with colleagues.

In the Preface Ould expressed wonder, "Certainly the wisdom of providence is very fully displayed in the whole scene of procreation; and the concurring circumstances which contribute to parturition are surprizingly beautiful..." (p. viii). Committed to improving the lot of women in childbirth, he declared that one should "contribute his mite... towards the perfection of the art of midwifery." Ould criticized several authorities, stating that "many of their schemes are like those of some navigators and geographers, who never made use of a compass, but in their closet" (pp. viii-ix). He affirmed that for himself, the facts advanced had "truth and demonstration on their side, being confirmed by practice", rather than arising from his imagination (p. x). He concluded the preface with the caveat that the *Treatise* did not pretend to present a "compleat system...", by which that art may be perfectly learned", or that it would make one adept "without ocular demonstration, and a long continued exercise of the hands in this, as well as other chirurgical operations" (p. xxv).

In Part I, Ould began with "anatomy... as the surest foundation whereon to build the rest" (p. 5). He expressed his belief in the importance in communicating whatever one can about normal pregnancy and delivery to relieve "distressed women, from extraordinary pain and torture, innumerable disorders and death, the consequence of bad practice; from misapply'd and ill contrived instruments; and even from the injudicious management of the hands" of the *accoucheur* (p. 2). He referred to previous works, giving his motive for writing: "Should I be happy enough to strike... the least spark of light, by the help of which, others more accomplished may illustrate this subject, my ends are answered" (p. 5). Nonetheless, he acknowledged being "thoroughly conscious of my inequality to the task" (p. 4).

Undoubtedly Ould's greatest scientific contribution was to consider for the first time the normal rotation of the fetal head in the pelvic cavity, i.e., the mechanism of labor. In presenting this discovery, he claimed that it deserved "more circumspection" than it hitherto had received, as "hereon depends in a great measure, either the happiness or misery both of mother and child" (p. 28). He recounted how he had first discovered the mechanism of normal labor in a patient in whom fetal "progress grew tedious." This stimulated him to perform thereafter the "strictest examination" upon every woman he delivered (pp. xvi–xvii). A decade later William Smellie

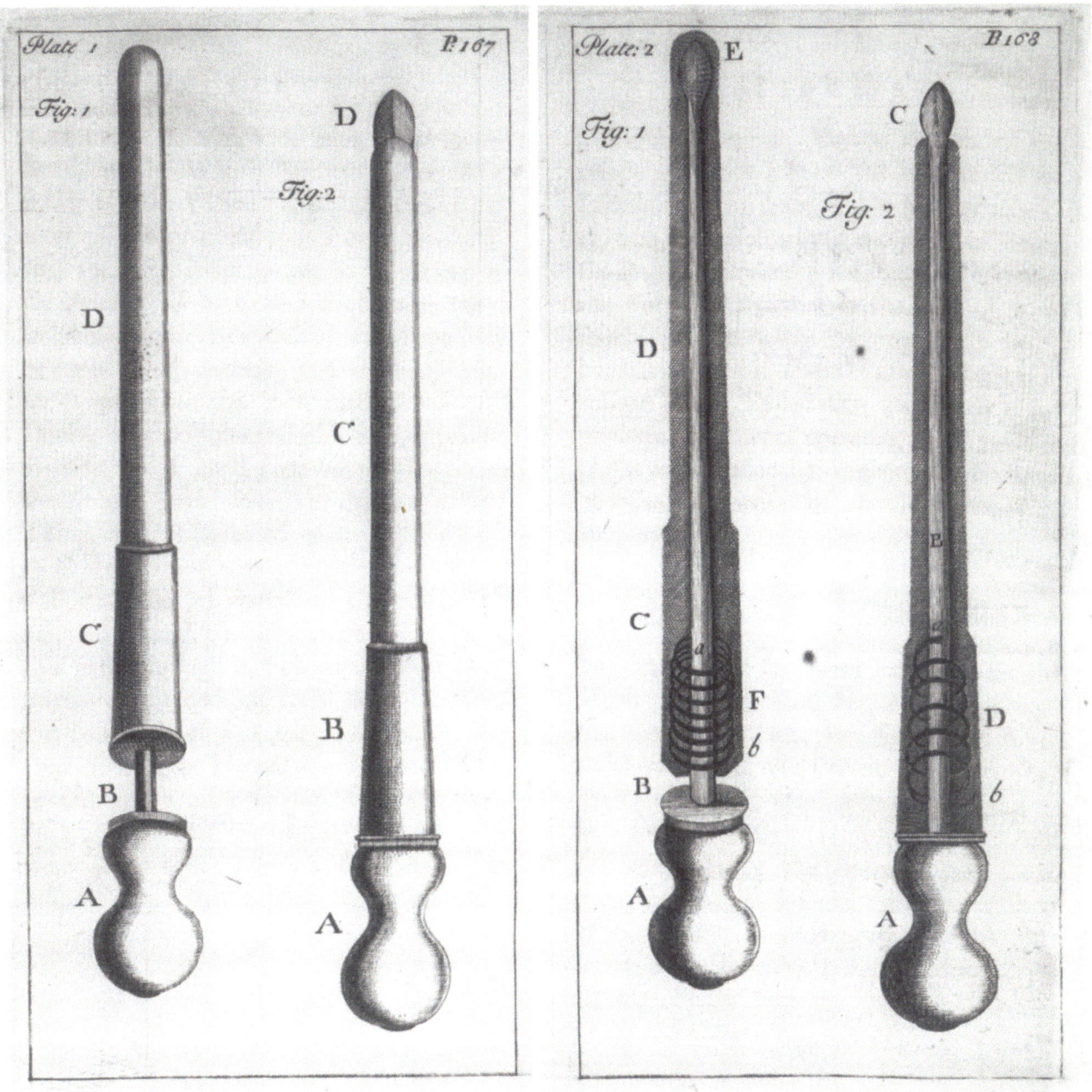

FIGURES FROM OULD, 1742

(1697–1763) expanded on and amplified details of the mechanism of normal labor (Smellie 1752). In complicated cases, Ould also advocated consultation among doctors.

In the text's third section, on instrumental delivery, Ould first described episiotomy. Although used secretly by the Chamberlen family for over 100 years, it was not until a decade earlier that the obstetrical forceps had been first described (Chapman 1733) and illustrated (Giffard 1734). In cautioning against the use of forceps and other dangerous "foreign materials" except when "absolutely unavoidable," Ould stated: "Though I have gone through the foregoing part of this treatise with great pleasure, yet what is to come strikes me with horror" (p. 140). Also as an ardent foe of cesarean section, Ould rejected the favorable recommendations of previous writers for the procedure. He argued that it represented a "piece of cruelty"… "repugnant, not only to all rules of theory or practice, but even of humanity" (pp. xxiii–xxiv), "this unparalleled piece of barbarity,… this detestable, barbarous, illegal piece of inhumanity." He reasoned that because "this operation must necessarily destroy the mother; the question must arise, whether there be… any case [in which its] performance… is warrantable, while the mother is yet living" (pp. 198–203). In an era in which few mothers survived this surgery, perhaps such a view was understandable.

In place of the abdominal approach for obstructed labors, Ould devised what he referred to as a "*terebra occulta*" (hidden piercer) for perforating the fetal skull, thorax, or abdomen, and which he illustrated in two plates (see Figs). Although *accoucheurs* of this era employed a veritable multitude of instruments of torture for this purpose—the crotchet, bistoury, fillet, vectis, various perforators, and scissors—Ould held that his trocar was less likely to injure the mother, and was more efficient in performing "the business required" (p. 169). He recounted several cases of severely contracted pelvis, with obstructed labor and fetal demise, in which it had proved useful. Regarding one such case he recorded, "This was the most laborious operation I ever performed." Following the procedure his left hand was so swollen it could not be used for 10 days (p. 182).

Born in Galway, Ireland, at about age 19 Ould commenced working as a prosector in the Trinity College Medical School's department of anatomy. During this time he attended the medical school lectures, although he was not officially matriculated as a student. After 5 years at Trinity, Ould moved to Paris where for 2 years he studied midwifery, probably under the *accoucheur* Grégoire the Elder (fl. 1730) who gave private lessons at the *Hôtel Dieu*. In regard to that experience, he could not "…help declaring the necessity of being indebted to France for the true knowledge of practical midwifry; for the opportunities which are there met with, are no where else to be found…

namely, those of occular demonstration of women being delivered, both in natural and preternatural labours" (pp. 71–72). He also recalled that while studying in Paris, "I made the strictest examination of every woman, which I either delivered, or saw delivered" (p. xvii). In 1738, Ould returned to Dublin and shortly thereafter was licensed to practice midwifery by the King's and Queen's College of Physicians of Ireland. In 1745, he was appointed assistant master of the newly founded Lying-In Hospital, and in 1759 following the death of Mosse became master. In 1759, he was knighted by the Lord Lieutenant, Duke of Bedford and Viceroy of Ireland, although the reasons for this honor remain unclear. It was said of him, "Sir Fielding Ould is made a knight; He should have been a Lord by right. For then the ladies' cry would be, O Lord, good Lord deliver me." Among royalty that Ould delivered was Arthur Wellesley (1769–1852), the first Duke of Wellington.

References

Blake p. 334; Cutter and Viets pp. 24–26, 182; Garrison Morton 6151.

Brody, S.A. The life and times of Sir Fielding Ould: man-midwife and master physician. *Bull Hist Med* 52: 228–250, 1978.

Browne, A. (Ed.) *Masters, Midwives, and Ladies-in-waiting: the Rotunda Hospital, 1745–1995… with a foreword by Mary Robinson.* Dublin, A&A Farmar, 1995.

Chapman, E. *An essay on the improvement of midwifery; chiefly with regard to the operation. To which are added fifty cases, selected from upwards of twenty-five years practice.* London, A. Blackwell, 1733.

Giffard, W. *Cases in Midwifery….* Edited by Edward Hody. London, B. Motte, T. Wotton, and L. Gittiver, 1734.

Kirkpatrick, T.P.C. *History of the medical teaching in Trinity College Dublin and of the School of Physic in Ireland.* Dublin, Hanna and Neale, 1912.

Kirkpatrick, T.P.C. and H. Jellett. *The Book of the Rotunda Hospital. An illustrated history of the Dublin Lying-In Hospital from its foundation in 1745 to the present time.* London, Adlard & Son, Bartholomew Press, 1913, pp. 48–61.

Le Fanu, W. The lost half-century in English medicine, 1700–1750. *Bull Hist Med* 46: 319–348, 1972.

McClintock, A. On the rise of the Dublin school of midwifery; with memoirs of Sir Fielding Ould, and Dr. J.C. Fleury. *Dublin Quart J Med Sci* 25: 1–20, 1858.

Ould, F. *A treatise of midwifry in three parts, with an introduction by L.D. Longo.* Birmingham, AL. Classics in Obstetrics & Gynecology Library, Gryphon Ed, 1990.

Wilhelm Noortwyk

Uteri humani gravidi anatome et historia. Lugduni Batavorum, Joan et Herm Verbeek, 1743.

In this work of the anatomy of the gravid uterus, Noortwyk (c. 1712–1778) presented his investigations on the corpse of a young woman who died at 6 months gestation. Noortwyk obtained permission of the husband to excise the pregnant uterus from the corpse, and took it home for dissection. Tabula I of four copper engraved plates illustrates the opened uterus with external membranes (B), uterine wall (G, g), left Fallopian tube (A), cervix (L and M), uterine artery (O), and posterior uterine wall with orifice of vessels and decidua (which Noortwyk mistakenly believed was the adherent placenta; (P)). Tabula III illustrates the 24 week fetus as it would lie *in utero*. Drawings for the four plates were made by J Wandelaar and engraved by Jan van der Spyk. In this description, Noortwyk asserted correctly that the maternal and fetal circulations were separate.

Noortwyk, a native of Leiden and graduate of that university's medical school (1735), was a pupil of the anatomist Bernhard Siegfried Albinus (1697–1770) who created impressive preparations of injected organs (Albinus 1737–1747), apparently was the first to inject the uterine vessels of a young woman who had died near term. In this preparation, he believed incorrectly, that some of the vessels were fetal in origin, and joined those of the uterus.

References

Blake p. 326; Cole 1477; Waller 6894.

Albinus, B.S. *Tibulae sceleti et musculorum corporis humani.* Lugduni Batavorum, J&H Verbeek, [1737–1747]. (GM 399).

Needham, J. *A history of embryology.* Cambridge, at the University Press, 1934. p. 174.

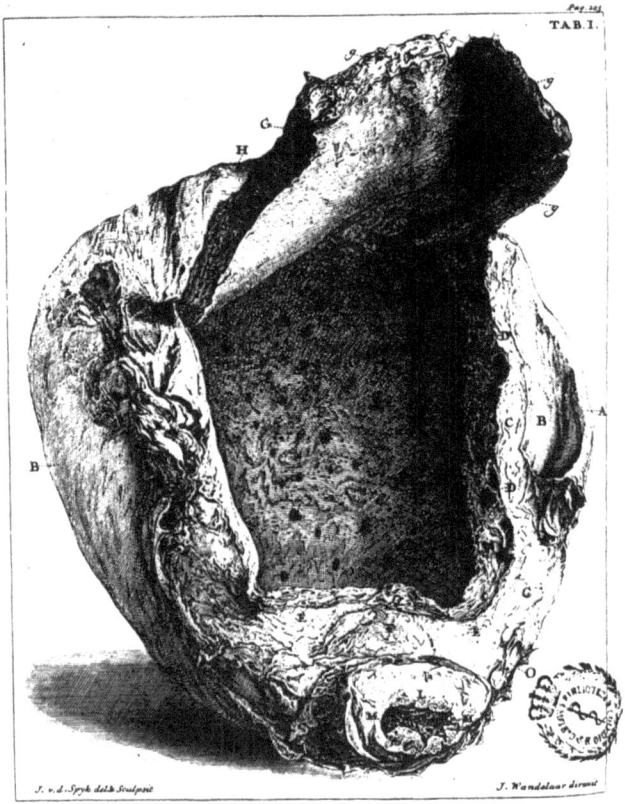

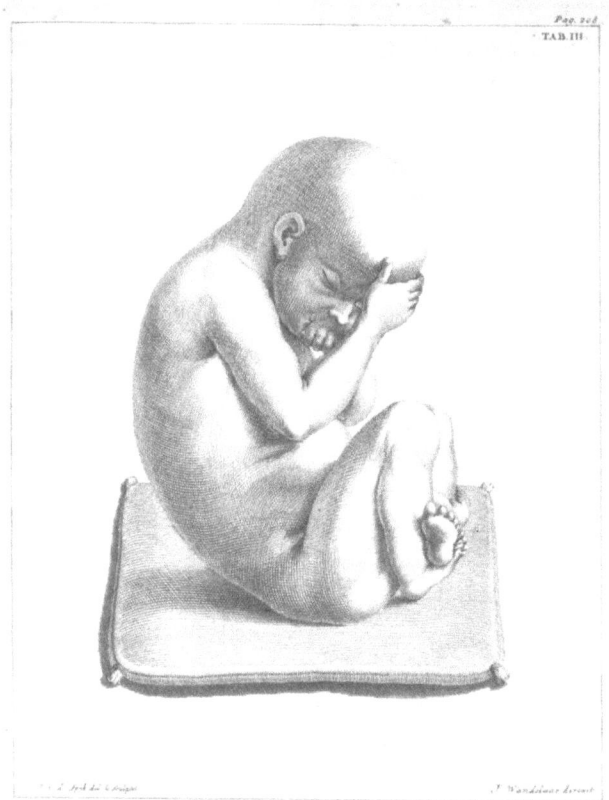

Tab I and III from Noortwyk, 1743.
Tab I courtesy Google Books (http://books.google.com)

Bernhard [Bernard] Siegfried Albinus

Tabulae VII uteri mulieris gravidae cum jam parturiret mortuae, [with] Tabularum uteri mulieris gravidae appendix. Lugdini Batavorum, Apud J & H Verbeek, 1748, 1751.

In his "illustrations of the gravid uterus after death of the parturient..." Albinus (1697–1770), presented a series of eight large copperplate illustrations (including one in the appendix). Albinus was born at Frankfort am Oder. He commenced his medical studies at the University Leiden, and then transferred to Paris. In 1719, prior to receiving his medical degree, he began teaching at Leiden. Two years later, following the death of his father Bernard (1653–1721) who was a professor of medicine, at age 24 Bernhard was appointed professor of anatomy and surgery at the University. Albinus set a high standard in anatomical study by producing some of the most beautiful plates in the history of engraving. This magnificent work, with that of his *Tabulae sceleti et musculorum corporis humani*, [Illustrations of the skeleton and of muscles of the human body] (1737–1747), is the apogee of a remarkable collaboration between Albinus and the artist Jan [Jean] Wandelaar (1690–1759) (who both drew the figures and engraved the copper plates). This collaboration lasted over 30 years, and was supported by the publishers Johannes and Hermanus Verbeek. Albinus supervised each drawing, insuring that the smallest details were depicted with great accuracy. Regarding the artistic merits of Wandelaar, Albinus recorded that he produced everything with truth and accuracy and with marvelous refinement.

In the present volume, seven plates depict life-size the uterus far advanced in pregnancy, and one shows the fetus *in utero*. Tab V illustrates a fetus attached by the umbilical cord (p-u) to the placenta, with amnion (a, b) and chorion (e, f, g)

membranes. The umbilical arteries (r, s, t, u) and vein (p, q) are clearly delineated, as are their distribution on the chorionic plate of the placenta.

For their major anatomical folio (Albinus 1737–1747), Wandelaar placed each figure in a carefully chosen landscape setting, with results so successful that the figures appear to be actually stepping out of the picture. The historian, Johann Ludwig Choulant (1791–1861), stated that Albinus' work with Wandelaar commenced "...an epoch of high perfection during which the mere outward appearance, superficial investigations, or the mere copying of subjects observed prove insufficient. Artistic and faithful representations of the true form and connection of anatomic structures, discovered through repeated comparative studies, are now demanded" (Choulant-Frank 1920).

References

Blake p. 9; Choulant-Frank 1920, p. 276, ff; Garrison Morton 399; Russell 5, 6.

Albinus, B.S. *Historia musculorum hominis.* Leidae Batavorum, Apud Theodorum Haak & Henricum Mulhovium, 1734.

Albinus, B.S. *Tabulae sceleti et musculorum corporis humani.* Lugduni Batavorum, J & H Verbeek, [1737–1747].

Albinus, B.S. *Icones ossium foetus humani, accedit osteogeniae brevis historia.* Lugduni Batavorum, Apud. J & H Verbeek, 1737.

Albinus, B.S. *Tables of the skeleton and muscles of the human body.* London, J & P Knapton, 1749.

van der Pas, P.W. Bernard Siegfried Albinus. In: *Dictionary of Scientific Biography.* Vol XV, Suppl. I. Charles Coulston Gillispie (Ed). New York, Charles Scribner's Sons, 1978, pp. 4–6.

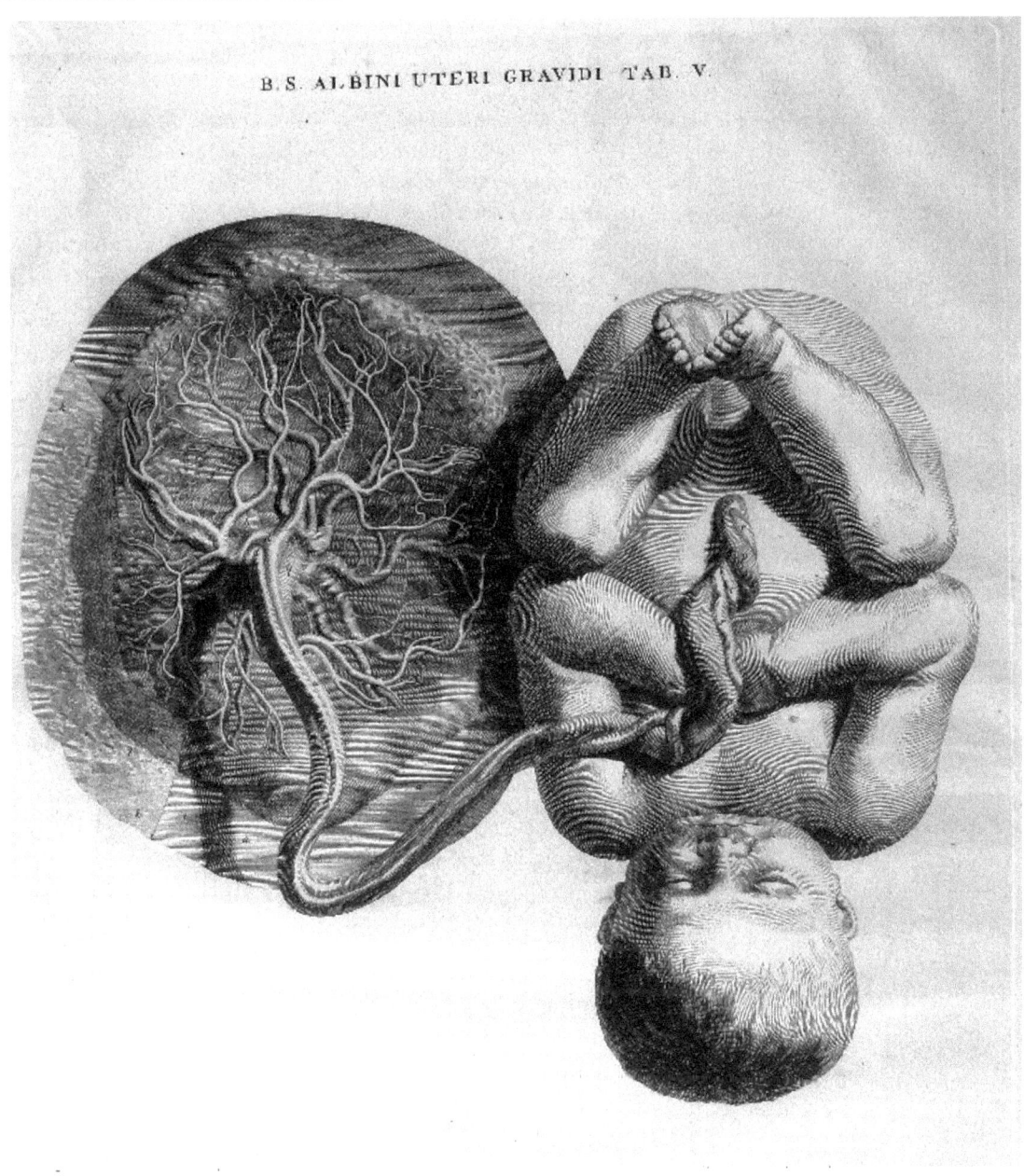

TAB V FROM ALBINUS, 1748–1751

John Burton

An essay towards a complete new system of midwifry, theoretical and practical... London, James Hodges, 1751.

In his *Essay...* Burton (ca. 1710–1771), of York, in describing the placenta wrote, "it is evident, that the Circulation is not carried on from the Mother to the *Foetus*, nor from the *Foetus* to the mother, by continued Canals, but by the Extremities of the Umbilical Vein taking up Liquors by Absorption, in the same way as the Lacteal Vessels do in the Guts.... By the Means the *Foetus* is solely nourished, and thus the Circulation and Communication of the Humours betwixt Mother and *Foetus* are performed ..." (see Table 6; pp. 62–63).

Burton was a strong advocate that physicians should "apply themselves a little more to that practice" [e.g., midwifery], as in "bad Cases" where their assistance is needed, but no one can be found "... both Woman and Child frequently perish" (p. 291). In the eighteenth century, he was also one of the first to present a detailed discussion of cesarean section, advocating its performance both in the living woman when the fetus could not be delivered vaginally, and as soon as possible in the instance of a woman who died but the fetus was still alive (pp. 71, 91–92). Burton included detailed instructions for its performance (pp. 261 ff).

In the Preface, Burton extolled the writings of several of his predecessors, but stated that for the most part, the authors published, "from no other Motive than to let the World know there were such Persons in Being... the Public, in the mean Time, not reaping the least Benefit by this Content". He continued, stressing the importance of understanding reproduction for "the Preservation of our Species", and "in bring Children alive into the World, and in preserving the Lives of the Mothers...." (p. ix) Further in the Preface, Burton states of some critics, "*But for those People, who like* Birds of the Night scream in the Dark, *when none can see them; and like* cowardly Enemies, unseen, *shoot their invenomed Darts at me, in* secret Whispers, *or* anonymous Papers, *such* Creatures *may spit their* malignant Cholor, *till it consume* Themselves, *before I shall regard them in the least*" (p. xv).

Apparently, Burton also was the first to suggest that puerperal fever was contagious. He also described the female pelvis, the uterus and placenta, instruments used in delivery. In addition, he showed the fetus in various positions (see plates), some of which are similar to the "babies in a bottle" first illustrated in Eucharius Rösslin (ca. 1470–1526) (Rösslin 1513), and Jacob Rueff (1500–1558) (Rueff 1554). The 18 plates in this volume were by the later to be celebrated artist George Stubbs (1724–1806), who lived in York at this time. These were Stubbs' first published anatomical illustrations, and were based on his own dissections. However, he apparently was not satisfied with their quality, as he is not acknowledged as the artist. Stubbs' illustrations for his celebrated the *Anatomy of the Horse* (Stubbs 1766) were his only other anatomical studies published during his lifetime. Drawings for another volume that remained unpublished during his lifetime, *Comparative Anatomical Illustration,* were only discovered in the mid-twentieth century (Doherty 1974).

In his *Essay*, Burton violently attacked William Smellie (1697–1763), accusing Smellie of being purely a theorist and poor practitioner. He again returned to this theme in a later work (Burton 1753). Burton concluded his *Essay* stating, "I flatter myself that the Improvements which I have made in the Method of *Practice*, for the Preservation of both Mother and Child, and the several vulgar Errors which I have refuted, will sufficiently atone for the Size of the *Book*" (p. 390).

Burton studied at the Universities of Cambridge, Leiden, Paris, and Rheims, the latter institution at which he received his medical degree. After practicing in Wakefield for several years, he moved to York, where he played an important role in founding the York Hospital (Infirmary). In 1745, at the time of the Jacobite movement to restore Prince Charles Edward Stuart (1720–1788) who had been exiled in France, Burton's sympathies resulted in his being imprisoned for almost a year and one-half. In his novel *Tristram Shandy*, Laurence Sterne (1713–1768) satirized Burton as "Dr Slop" (Sterne 1759–1767), a vulgar, man-midwife who during the delivery of Tristram, had crushed the bridge of his nose with his newly-invented forceps. Sterne's satire is generally recognized as an unfair and cruel travesty of the real gentleman and scholar. Burton also contributed an important work on the antiquities of Yorkshire (Burton 1758).

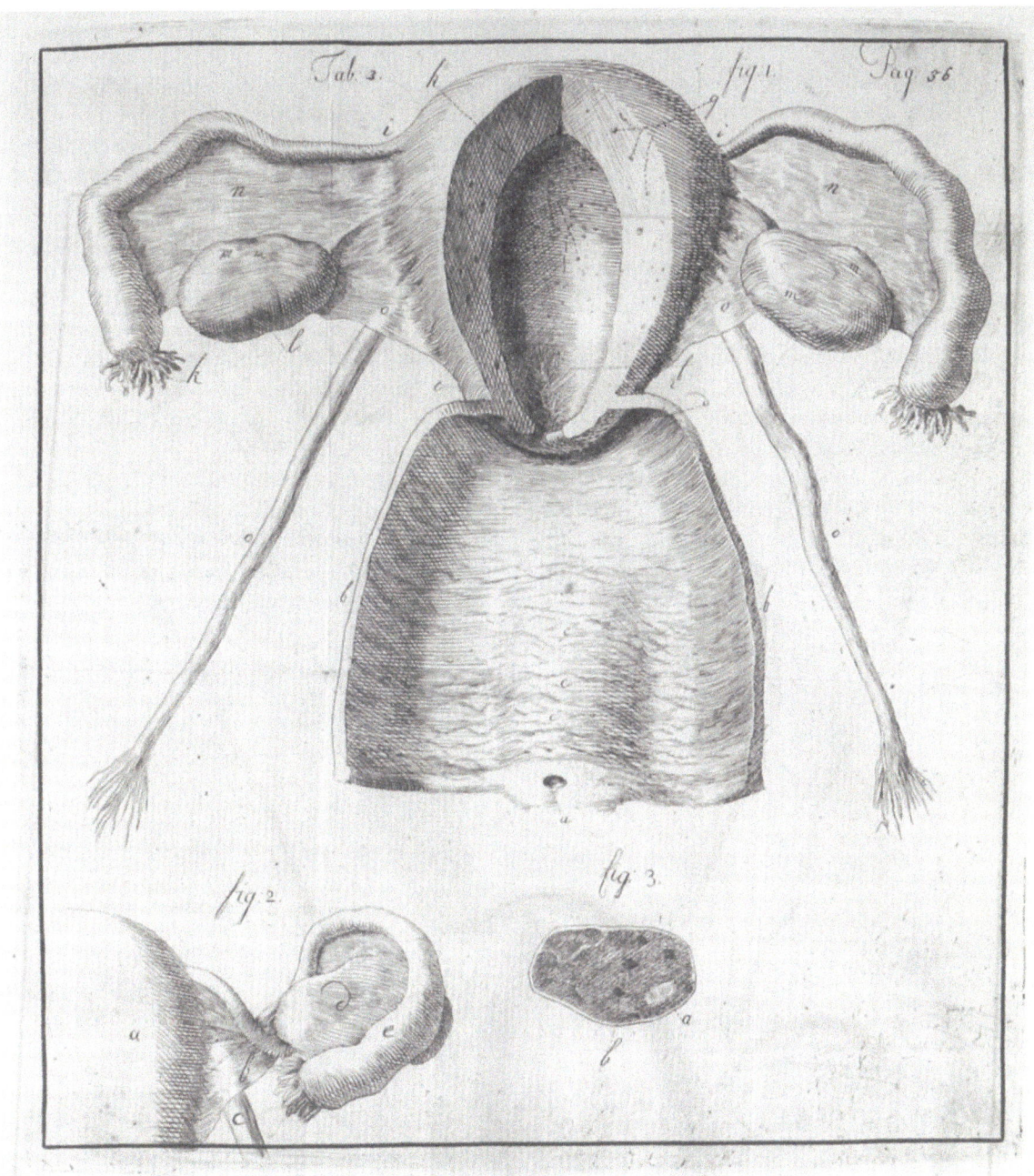

Plate from Burton, 1751

References

Blake, p. 72; Cutter and Viets pp. 28–31, 67–68, and 182–183; Garrison Morton 6268.

Allport, W.H. Tristram Shandy and obstetrics. *Am J Obstet Gynec* 65: 612–617, 1912.

Brady, F. Tristram Shandy: Sexuality, Morality, and Sensibility. *Eighteenth-Century Studies* 4: 41–56, 1970.

Burton, J. *An essay towards a complete new system of midwifry, theoretical and practical....* Edited by L.D. Longo. Birmingham, AL. The Classics in Obstetrics and Gynecology Library, Gryphon Editions, 1995.

Burton, J. *A letter to William Smellie, M.D., Containing critical and practical remarks upon his Treatise on the theory and practice of midwifery.* London, W. Owen, 1753.

Burton, J. *Monasticon Eboracense and the ecclesiastical history of Yorkshire....* York, printed for the author by N. Nickson, 1758.

Cross, W.L. *The life and times of Laurence Sterne.* New York, Macmillan, 1909.

Doherty, T. *The anatomical works of George Stubbs.* London, Secker and Warburg, 1974 (American Edition; Boston, David R. Godine, 1975), pp. 2–4 and 25–36.

Doran, A. Burton ("Dr. Slop"): his forceps and his foes. *J Obstet Gynaec Brit Emp* 23: 3–24 and 65–86, 1913.

Dunn, P.M. Dr. John Burton (1710–1771) of York and his obstetric treatise. *Arch Dis Child Fetal Neonatal Ed* 84: F74-F76, 2001.

Osler, W. Dr. Slop. In: *Men and Books by Sir William Osler, Ed by E.F. Nation.* Durham, N.C., Sacrum Press, 1987.

Rösslin, E. *Der schwangern Frauwen und Hebammen Roszgarten.* Strassburg, Martinus Flach, Junior, 1513. (GM 6138).

Rueff, J. *Ein schön lustig Trostbüchle von den Empfengknussen und Geburten der Menschen.* Tiguri, Apud Frosch [overum], 1554.. (GM 6141).

Sterne, L. *The life and opinions of Tristram Shandy, gentleman....* 9 vols. York, 1759–1767.

Stubbs, G. *The anatomy of the horse; including a particular description of the bones, cartilages, muscles,....* London, J. Purser for the author, 1766.

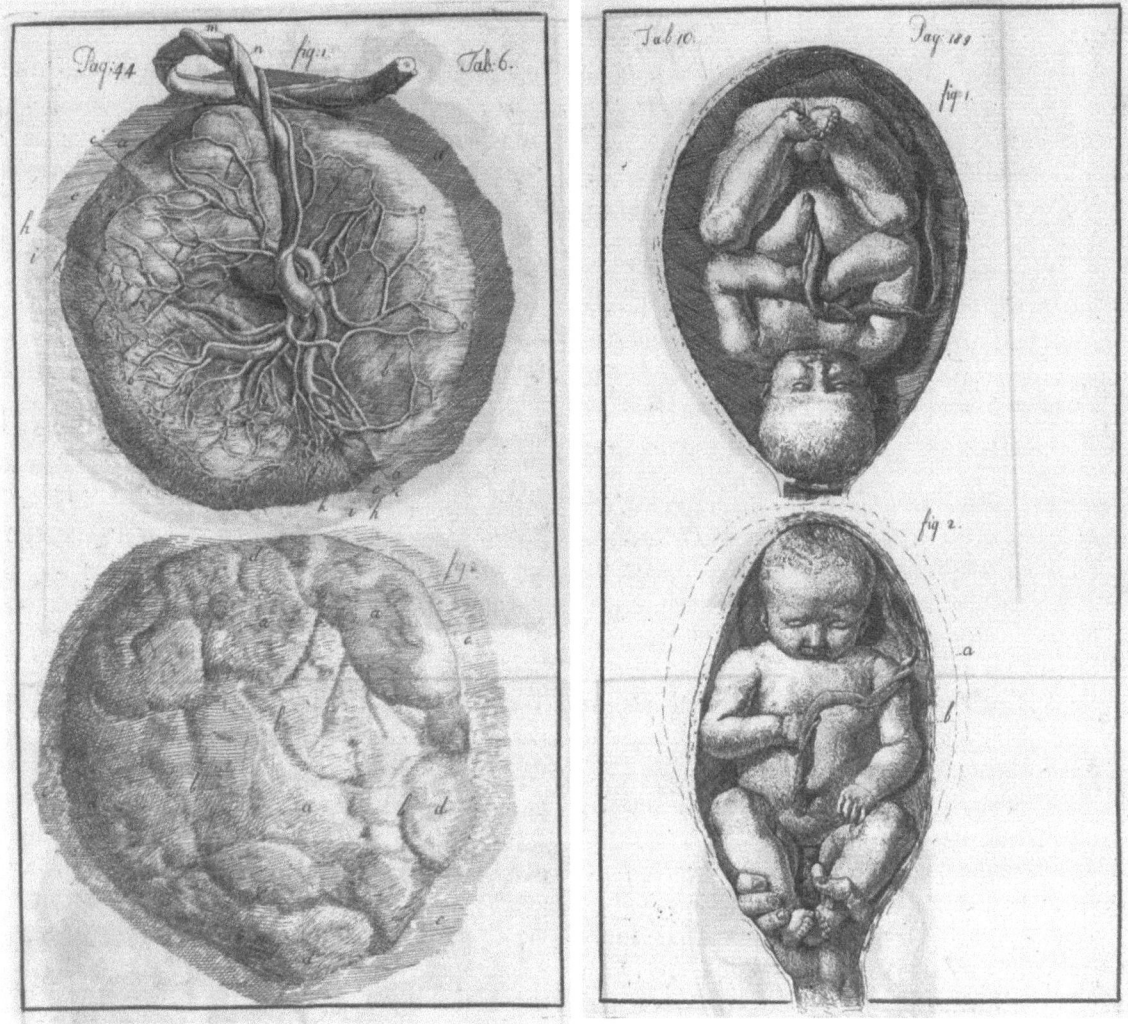

PLATES FROM BURTON, 1751

Johannes Fatio

Helvetisch-Vernunftige Wehe-Mutter, oder Gruendlicher Unterricht, wie mit den Schwangern, Gebahrenden, Kindbetterinnen und neugebohrnen Kindern umzugehen… Samt einer ausfeuhrlichen Beschreibung von Fortpflanzung des menschlichen Geschlechts, und aller weiblichen Leibes-Theilen, auch der Empfangniss, Formir- und Bildungn der Frucht in Mutterliebe. Nebst des Verfassers curiosen Anmerckungen, selbst bewahrten Handgriffen, Curen und dazu dienlichen Arzney-Mitteln. Dem Loeblichen Frauenzimmer, geschwohrnen Weibern, und andern ehrbaren Frauen zu Nutz. Basel, verlegts Johann Rudolph Imhof, 1752.

The *Helvetic Reasonable Midwife*, by the Basel surgeon and obstetrician Fatio (1649–1691), was published 61 years after the author's death. The work is divided into five parts: the anatomy of women and on generation, the pregnant woman and her diseases, natural and complicated deliveries, the childbed woman, her disease, food and drink, and the care of newborn children and their diseases. In the Socratic manner, Fatio's printed marginal notes pose a series of questions as to how, why, and when a certain condition, complaint, or disease arises. Fatio then provided the response in the main body of the text, frequently including prescriptions for a variety of pills, salves, tonics, dressings, purgatives, and sedatives.

Believed to be the first surgeon to write in detail on many aspects of surgery on children, of particular note is Fatio's successful separation of a pair of conjoined twins. This is the first such account in recorded history. In Chapter 20, after asking in what way there can be the anomalies at birth and in what way they can be separated, he presents the case with a discussion and illustrations. The twins Elisabet and Catharina, born 23 November 1689 (Figs. I and II), were conjoined at the lower sternum (xiphoid process) and umbilicus (e.g., xypho-omphalopagos twins). The mother, Clemtia Meijerin [or Meinin], was 42 years of age, and previously had given birth to two children. The first twin was born in the occiput presentation, the second in a *conduplicatio corpore* [body double or embraced]. Fatio was called to see the infants the following day. Then, with an audience that included the two Lord Mayors of Basel and a group of physicians, Fatio proceeded to separate the twins, dissecting the vessels of the umbilical cord, which, fortunately the midwife had cut at some distance from the infant, to their entry into the abdominal wall. Fatio ligated these with a silk cord, and cut the bridge of tissue from the umbilicus to the xiphoid. Figure I presents the twins *in utero*, while Fig. II shows them in a cradle.

Illustrated are the bridge of connecting tissue (C), the umbilical cord (D, E), and the two individual sets of one umbilical vein (F) and two umbilical arteries (G). Figure III shows the separated infants shortly following their baptism, while Fig. IV shows an enlarged view of the anatomy of the umbilical connection tissue of each twin (a, b). Fatio also provided an illustration of the two placentas from the conjoined twins (next page), which appear dichorionic and diamniotic.

Fatio records that on the ninth day postoperatively the ligated band of connecting tissue separated spontaneously from the abdominal wall of the infants. He debrided the oozing stumps with a scalpel, treating these with a truss and small cushion soaked in red wine. The following day reasonably complete healing had occurred and the infants were breast fed. Both girls recovered. This, the successful separation of conjoined twins, was not repeated until early twentieth century. Unethical by present day standards, Emanuel Koenig (1658–1731), a young physician who was an observer of Fatio's success, first published the case including illustrations, without crediting Fatio (Koenig 1689 or 1690). Theodor Zwinger III (1658–1724), another witness, also published an account with illustrations (Zwinger 1690). For the present work, the anonymous editors rearranged and reproduced Zwinger's plate as shown here.

In this work, a number of representations of the fetus *in utero* clearly were modeled after those of François Mauriceau (1637–1709) (1668). This volume, which provides a fascinating guide into midwifery practices at the end of the seventeenth century, is based largely upon Fatio's own practice and experiences, most of which were believed to be valid some 60 years later. The frontispiece presents a fine engraving of Fatio by Georg Daniel Heuman (1691–1759). Below his image is the motto Labore Laurem Lego [A legacy of work crowned with laurel].

Because of this and his other contributions to surgery in children; repair of peno-scrotal hypospadiac boys, hydrometrocolpos, altresia of the terminal colon with imperforate anus, hemangiaoma, and exomphalos, Fatio is sometimes referred to as the "father of pediatric surgery." In addition, Fatio twice, in 1675 and 1684, operated successfully to repair vesico-vaginal fistula. Prior to the mid-nineteenth century and the work of James Marion Sims (1813–1883), the relief of this condition was a major stumbling block of gynaecology (Sims 1852). Although many surgeons had made attempts to repair such lesions, they failed to be successful. With the patient in the lithotomy position, Fatio freshened the fistula margins with fine scissors, passed several sharpened quills through the edges of the fistula, and wound threads snugly about them to

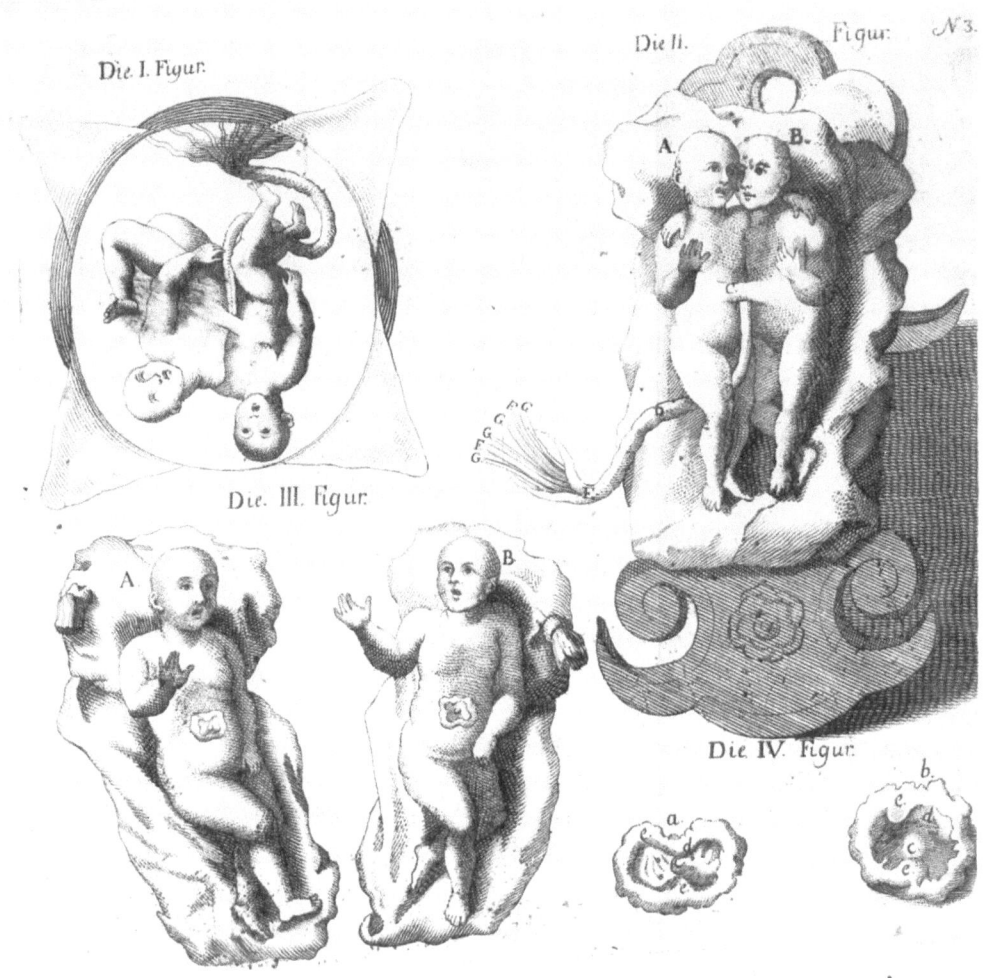

FIGURES I–IV FROM FATIO, 1752

approximate the fistula margins. These successes are believed to be the first for this terrible scourge (Wangenstein and Wangenstein 1978).

Fatio, had a large practice in Basel and won high repute among his patients. A Master-Surgeon at the Guild of Barbers, and believed to have received a doctor of medicine degree at Valence, he was, however, treated as an outcast and never accepted by the University Faculty of Medicine. At this time the Canton of Basel had not joined the Swiss federation. To aid in the midst of considerable political turmoil, Fatio had helped to write a new constitution. Nonetheless, in 1691 he was arrested for "political conspiracy," labeled a rebel as leader of an insurrection against the current city government, and was tortured repeatedly. Finally, in the marketplace in Basel with a large gathering of soldiers and onlookers, he was beheaded. According to some authors, his head was then used in a game of "football". Although Fatio is alleged to have written several other volumes on surgery the manuscripts for these are believed to have been burned by the authorities. His *Reasonable Midwife* is the only work that survived (Rickham 1986).

References

Blake p 144; Garrison Morton 6357.51; Hirsch II 483; Waller 2956.

Bondeson, J. The Biddenden Maids: a curious chapter in the history of conjoined twins. *J Roy Soc Med* 85: 217–221, 1992.

Fatio, J. *Der Arzney Doctor, Helvetisch-Vernünftige Wehe-Mutter*. Basel, Switzerland, Johann Rudolph Imhof, 1752.

Gerster, J. *Johannes Fatio ein Basler Chirurg und Geburts Helfer des XVII. Jahrhunderts....* Basel, B. Schwabe & Co., 1917.

Hagelin, O. *The byrth of mankynde otherwise named the womans booke; embryology, obstetrics, gynaecology through four centuries, an illustrated and annotated catalogue of rare books in the library of the Swedish Academy of Medicine.* Stockholm, Svenska Läkaresällskapet, 1990, pp 78–79.

Kompanje, E.J. The first successful separation of conjoined twins in 1689: some additions and corrections. *Twin Res* 7: 537–541, 2004.

König, E. Gemelli invicem adnati feliciter separati. *Miscellanea Curiosa sive Ephemeridum Medico-Physicarum Germanicarum Academiae Imperialis Leopoldinae Naturae Curiosorum,* 305–307, 1689.

Mauriceau, F. *Des maladies des femmes grosses et accouchees.* Paris, Jean Henault, 1668. (GM 6147).

Rickham, P.P. The dawn of paediatric surgery: Johannes Fatio (1649–1691) – his life, his work and his horrible end. *Prog Pediatr Surg* 20:94–105, 1986.

Rickham, P.P. Historical aspects of pediatric surgery. *Prog Pediatr Surg* Vol. 20, 1986. (GM 6359.90).

Rintelen, Von.F. Der Basler Chirurg und Rebell Johannes Fatio 1649–1691. *Gesnerus* 40: 149–158, 1983.

Sims, J.M. On the treatment of vesico-vaginal fistula. *Amer J Med Sci* 23: 59–82, 1852. (GM 6037).

Spencer, R. Anatomic description of conjoined twins: a plea for standardized terminology. *J Pediatr Surg* 31: 941–944, 1996.

Spencer, R. Theoretical and analytical embryology of conjoined twins: Part I: Embryogenesis. *Clinical Anatomy* 13: 36–53, 2000.

Spencer, R. *Conjoined twins, developmental malformations and clinical implications.* Baltimore, Johns Hopkins University Press, 2003.

Spitz L. Hunterian Lecture. Surgery for conjoined twins. *Ann R Coll Surg Engl* 85: 230–235, 2003.

Spitz, L. & Kiely, E.M. Conjoined twins. *JAMA* 289:1307–1310, 2003.

Wangensteen, O.H. & Wangensteen S.D. *The rise of surgery, from empiric craft to scientific discipline.* Minneapolis, University of Minnesota Press, 1978, pp. 244–245.

Weiden, R.M.F. van der. The first successful separation of conjoined twins (1689). *Twin Res* 7: 125–127, 2004.

Zwinger, D.T. Historia Admirandi Partus Gemellarum Vivarum, umbilicotenus sibi invicem connatarum. *Miscellanea Curiosa sive Ephemeridum Medico-Physicarum Germanicarum Academiae Imperialis Leopoldinae Naturae Curiosorum, Annus Nonus,* 229–232, 1690.

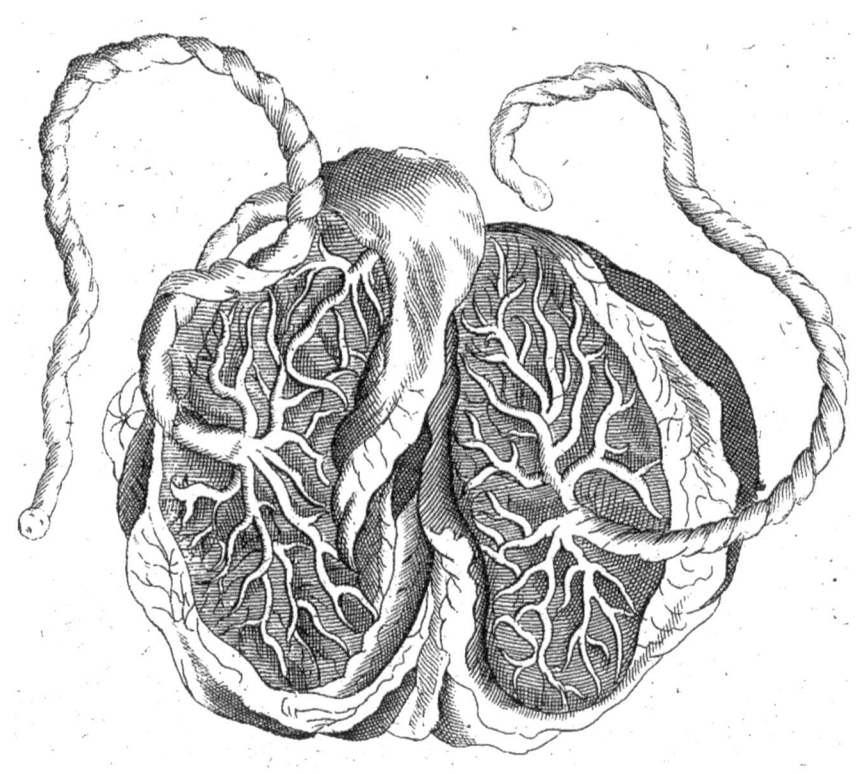

FIGURE OF TWIN PLACENTAS FROM FATIO, 1752

André Levret

L'art des accouchemens, démontré par des principes de physique et de mécanique: pour servir de base et de fondement à des leçons particuliéres. Paris, Delaguette, 1753.

Levret (1703–1780) presented a schematic illustration of coronal sections of the uterus, with adenexa and vagina, at various stages of gestation. A major contribution was his attempt to apply the principles of physics and mechanics to the course of labor, and his emphasis on the concordance, or lack of it, in the relation of the size of the fetal head to the pelvic diameters in determining the outcome of labor. These contributions helped to earn him the distinction of the "founder of rational operative obstetrics." He also illustrated the normal and rachitic female pelvis. In addition, he diagrammed the [variations] in the axis of the near-term uterus, and placentation with multiple pregnancy. In this work he also considered diseases such as syphilis, rickets, puerperal convulsions, and other complications. Moreover, he invented other obstetric instruments and made important observations on pelvic anomalies, and improved the operation of version with extraction of the fetus. In the second edition of this work published almost a decade later, Levret described the pelvic planes and axes, diagramming the parabolic curvilinear axis of fetal descent (Levret, 1761). This parturient axis was further described and clarified by Carl Gustav Carus (1789–1869, Carus 1820).

One of the most important French obstetricians of the eighteenth century, Levret published on many aspects of obstetrics and gynecology. With William Smellie (1697–1763, Smellie 1754) and Benjamin Pugh (1715–1798; Pugh 1754), Levret is credited with introduction of the pelvic, maternal curve with fenestrated blades of the obstetrical forceps (Levret 1747). He noted that he developed these to assist in disengaging the fetal head "empacted in the pelvis."

Among his distinguished patients was the Dauphiness, mother of Louise XVI (1754–1793). A skillful surgeon-accoucheur, he was elected to membership in the Royal Academy of Surgery of Paris.

References

Blake 269; Cutter Viets pp. 63–64, 89–90; Garrison pp. 338, 340; Garrison Morton 6153.

Carus, C.G. *Lehrbuch der Gynakologie, oder systematische Darstellung der Lehren von Erkenntnis....* Leipzig, G. Fliescher, 1820.

Levret, A. *Observations sur les Causes et les Accidens de Plusieurs Accouchemens laborieux avec des Remarques sur ce qui a été proposé ou mis en usage pour les terminer;....* Paris, 1747.

Levret, A. *Observations sur la cure radicale de plusieurs polypes de la matrice, de la gorge et du nez opérée par de nouveaux moyens inventés....* Paris, Chez Delaguette, 1749. (Blake, p. 270).

Levret, A. L'art des accouchemens, *démontré par des principes de physique et de mécanique: pour servir de base et de fondement à des leçons particuliéres.* 2nd Ed. Paris, Le Prieur, 1761. (Blake, p. 269).

Levret, A. Essai sur les abus des règles *générales et contre les préjugés qui s'opposent aux progès de l'art des accouchemens.* Paris, Prault & P.Fr. Didot le jeune, 1766. (Blake, p. 269).

Pugh, B. *A treatise of midwifery, chiefly with regard to the operation. With several improvements in that art....* London, J. Buckland, 1754. (Blake, p. 365).

Smellie, W. *A Sett of Anatomical Tables, with Explanations, and an Abridgment of the Practice of Midwifery, with a view to illustrate a Treatise on that Subject and Collection of Cases.* London, 1754.

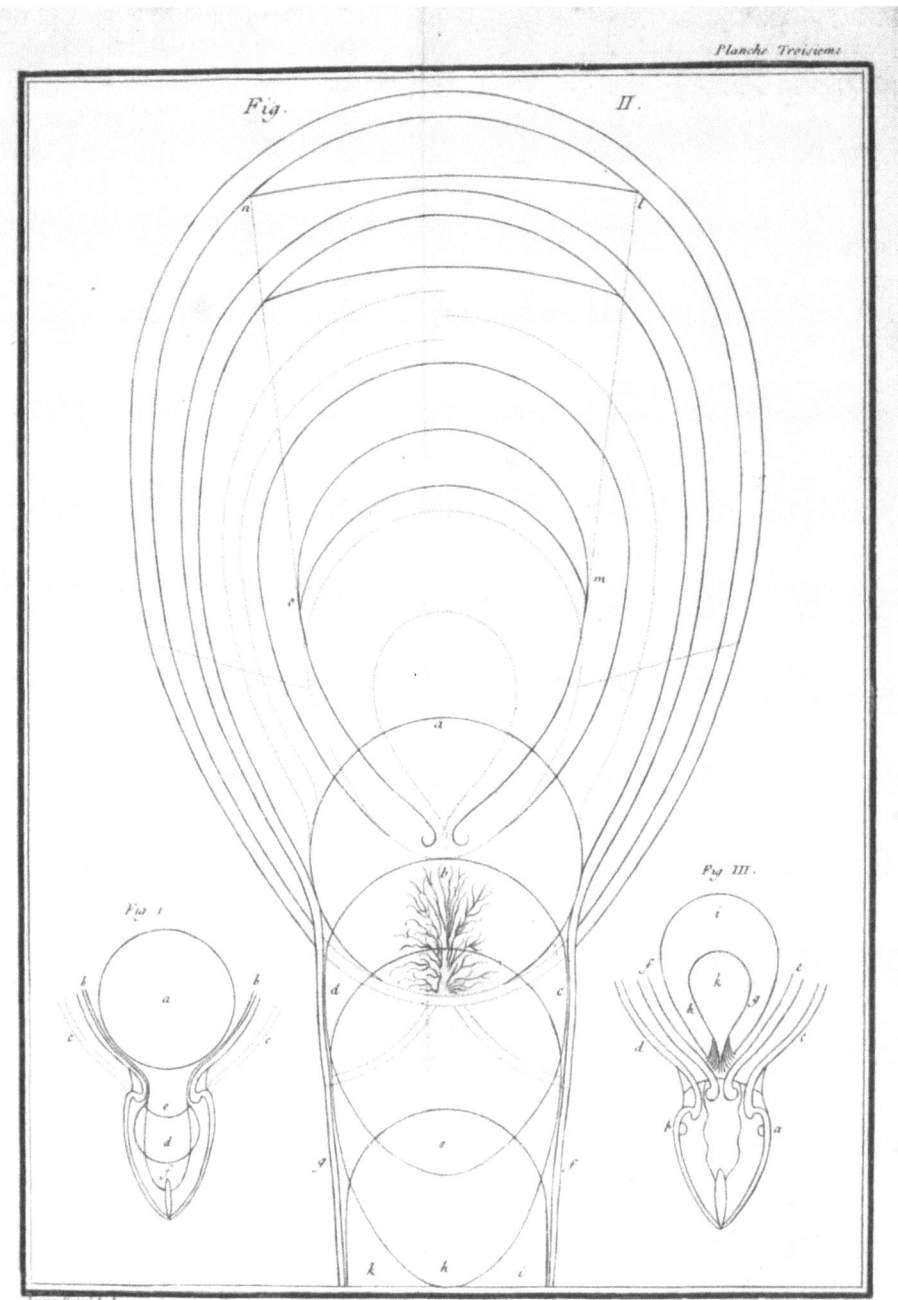

FIGURES I–III OF UTERINE MECHANICS FROM LEVRET, 1753

William Smellie

A Sett of Anatomical Tables, with Explanations, and an Abridgment of the Practice of Midwifery, with a view to illustrate a Treatise on that Subject and Collection of Cases. London, 1754.

Smellie's (1697–1763) *Sett of Anatomical Tables...,* was published as a supplement to his *Treatise ...* of 1752, and was designed to illustrate as accurately as possible the female pelvis and fetus. Smellie, the "master of British Midwifery," is stated to have been the foremost obstetrician of the eighteenth century, and made a number of contributions to the field. In the Preface, Smellie gave his rationale for this work. "As in a long course of teaching and practice in Midwifery, I hope I may without vanity say, that I have done something towards reducing that Art, into a more simple and mechanical method than has hitherto been done. I have attempted to explain the same in my Treatise of the Theory and Practice of Midwifery...." He continued, that he was induced to prepare this volume as, "... most of the representations hitherto given ... were in many respects deficient." Smellie assured the reader, "the greatest part of the figures were taken from Subjects prepared on purpose, to show everything that might conduce to the improvement of the young Practitioner ..." (page unnumbered).

Smellie described more accurately than any previous writer the mechanical relations of the fetal head to that of the mother's pelvis during parturition, i.e., the mechanism of labor. He also stressed the importance of precise pelvic measurements, and was the first to measure the internal or diagonal conjugate. Importantly for the *accoucher*, he laid down rules regarding the safe use of obstetrical forceps which remain valid today. He introduced the steel "English" lock on the forceps and, coincidentally with André Levret (1703–1780) of Paris (1753), added the pelvic curve. He apparently was the first to use forceps to rotate the fetal head, and to use forceps on the aftercoming head of a breech delivery. He also invented several important obstetric instruments.

The anatomic plates of this elephant folio, and classic of illustrated obstetrics, are far superior to any that had appeared previously, and present masterly representations of the relations of the parts of the mother and child at several stages of gestation. The *Sett of Anatomical Tables...,* (the *Sett* of which was a misprint, corrected in later editions) may have achieved more in the spread of correct ideas of labor and parturition than all of the works previously written on the subject. Undoubtedly, it also played an important role in Smellie's rise to fame. Of the 39 plates, 25 or 26 are by the Dutch artist and comparative anatomist Johannes van Riemsdijk (also Jan van Riemsdyk; van Rymsdyk; fl. 1750–1788). In fact, this was the first of the three great eighteenth century obstetric folios illustrated by van Riemsdijk.

Smellie's decision to produce this atlas from anatomic dissections could not have been undertaken lightly, and the expense must have been great. The artist later created the illustrations for Charles Nicholas Jenty's (fl. 1720–1770) *Uteri praegnantis un ad partum* (1757), William Hunter's (1718–1783) *Gravid Uterus* (1774), as well as for a quarto, of Thomas Denman (1733–1815). Eleven of the plates were by Pieter Camper (1722–1789), one of his students from Holland and anatomist at Franeker. Smellie himself is believed to have drawn the illustrations for two plates (37 and 39). All were beautifully engraved by Charles Grignon (1714–1810), a French artist who lived in London. Among the plates is the first illustration of a rachitic pelvis, and several on the mechanism of normal labor and illustrations of abnormal presentations. Plate VIII illustrates the fetus at 6 or 7 months of gestation, with its placenta (K) and surrounding amniotic membranes (L), within the uterine wall (A) and bones of the pelvis (B, C, D). In Plate X, Smellie depicts near-term twins, with their individual placentas (K) and membranes (L), within the uterus (A, I) and the bones of the pelvis (B, C, D). Many editions were published subsequently, including those in French and German (Russell 1963, 1987).

Smellie was born in Lanarkshire, Scotland. Following an apprenticeship with a local apothecary-physician, naval service, and general practice, he studied for a year in Paris. In 1741, he moved to London where he taught obstetrics in his home. To teach obstetrics on a scientific basis, Smellie used a mannequin fabricated from pelvic and fetal bones covered with leather. Practicing among the poor, Smellie became a popular teacher. In his Preface to "A Treatise..." during one decade, he gave over 280 courses of midwifery "... for the instruction of more than 900 pupils, exclusive of female students: and in that series of courses, 1150 poor women have been delivered in presence of those who attended me; and supported during their lying-in, by the stated collection of my pupils: over and above those difficult cases to which we were often called by midwives, for the relief of the indigent. These considerations, together

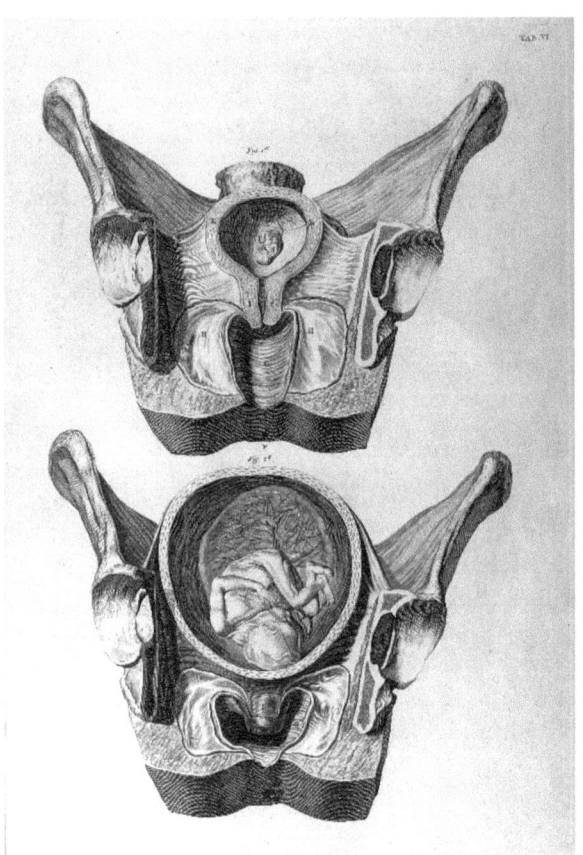

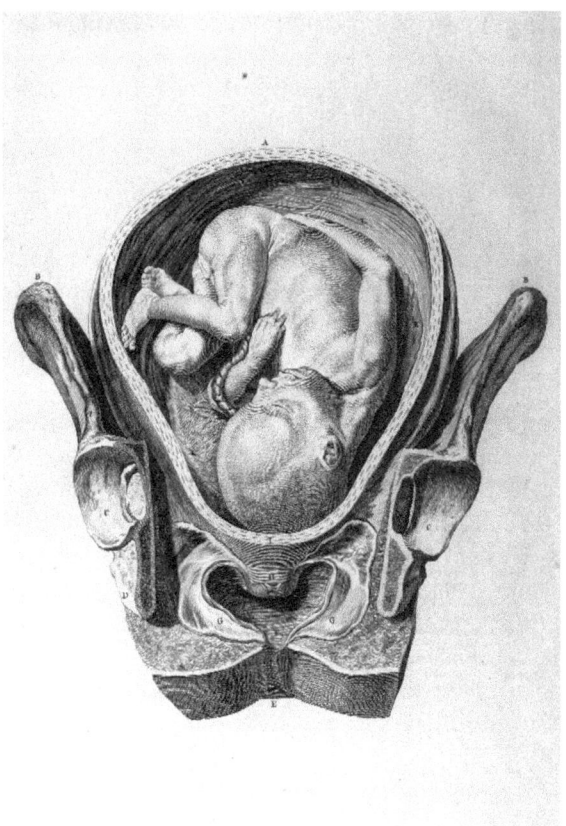

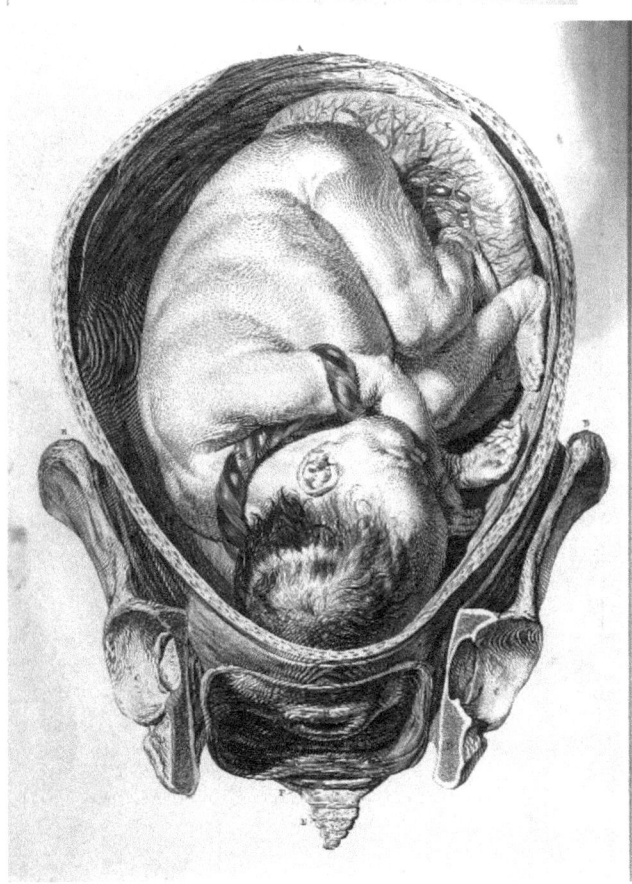

FIGURES FROM SMELLIE, 1754

with that of my own private practice, which hath been pretty extensive, will, I hope, screen me from the imputation of arrogance, with regard to the task I have undertaken; and I flatter myself, that the Performance will not be unserviceable to mankind" (Smellie 1752, pp. v–vi). Among his pupils was the celebrated William Hunter (Smellie 1752).

Volume one of his *Treatise...*, is a general discussion of normal obstetrics, as well as pathological problems, and was the first important textbook of obstetrics written by a Briton. In agreement with several previous writers on the weight of the newborn infant, Smellie wrote that "… at 9 months [the infant would weigh] from ten to twelve, and sometimes sixteen pounds" (Smellie 1752, p. 121). As noted earlier, it was Johann Georg Roederer (1727–1763) of Göttingen who first recorded the correct weight and length of the newborn infant (Roederer 1753). The second and third volumes of the *Treatise...*, published in 1754 and 1764, respectively, included 531 case histories collected from Smellie's extensive practice. Smellie began compiling volume three in his retirement at his home in Lanark, Scotland. Following his death, this was completed by his friend the novelist Tobias George Smollett (1721–1771) who also practiced midwifery (Huffman 1969, 1970; Thornton and Want 1979).

References

Blake 420; Garrison Morton 6154 and 6154.1; Choulant-Frank, p. 75; Norman 43; Russell 753 (atlas); Thoms, pp. 123–129.

Christie FGS. William Smellie, 1697–1763, the master of British midwifery. *Calgary Assoc Clin Hist Bull* 22: 189–200, 1957.

Denman, T. *A collection of engravings, tending to illustrate the generation and parturition of animals, and of the human species*. London, Sold by J. Johnson …, 1787.

Huffman, J.W. Jan van Riemsdyk. Medical illustrator extraordinary. *JAMA* 208:121–124, 1969.

Huffman, JW. The great eighteenth century obstetric atlases and their illustrator. *Obstet Gynecol* 35:971–976, 1970.

Hunter, W. *Anatomia uteri humani gravid tabulis Illustrata* …. Birmingham, John Baskerville, 1774.

Jenty, C.N. *The demonstrations of a pregnant uterus of a woman at her full term*. London, Printed for … the author, 1757.

Johnson RW. *William Smellie*. Edinburgh, E. & S. Livingston Ltd., 1952.

Knight E. William Smellie. *St. Bartholomews Hosp J* 64:282–285, 312–316, 1960.

Levret, A. *L'art des accouchemens*. Paris, Delaguette, 1753.

Oxorn H. William Smellie and the mechanism of labor. *Am J Obstet Gynecol* 77:41–49, 1959.

Roederer, J.G. De pondere et longitudine infantum recens natorum. *Comment Roy Soc Göttingen*, 1753, p. 140.

Russell, K.F. *British anatomy, 1525–1800. A bibliography*. Melbourne, Melbourne University Press, 1963.

Russell, K.F. *British anatomy 1525–1800. A bibliography of works published in Britain, America and on the Continent*. 2nd Ed. Winchester, St. Paul's bibliographies, 1987.

Rydberg E. The 200 years' teaching of the mechanisms of labor. *Am J Obstet Gynecol* 68:236–244, 1945.

Smellie, W. *A Treatise on the Theory and Practice of Midwifery*. London, D. Wilson 1752.

Smellie, W. *A Collection of Cases and Observations in Midwifery* …. London, D. Wilson & T. Burham, 1754.

Smellie, W. *A Collection of Preternatural Cases and Observations in Midwifery*. London, D. Wilson & T. Durham, 1764.

Spencer HR. *History of British Midwifery, From 1650 to 1800* …. London, J. Bale & Sons & Danielsson, Ltd., 1927, pp. 43-60.

Thornton, J.L. and P.C. Want. Jan van Rymsdyk's illustrations of the gravid uterus drawn for Hunter, Smellie, Jenty and Denman. *J Audiov Media Med* 2:11–15, 1979.

Wall LL. William Smellie (1697–1763): The father of scientific obstetrics. *Med Heritage* 2:158–167, 1986.

Willocks J. William Smellie and the birth of modern obstetrics. *Surgo* 33:4–12, 1966.

Charles Nicholas Jenty

The demonstrations of a pregnant uterus of a woman at her full term. London, Printed for…the author, 1757.

The London surgeon Jenty (fl. 1720–1770) presented an atlas of six life-sized mezzotint plates of a near term pregnant woman and her fetus. In the mezzotint process, sometimes called "the black art" (De Lint 1916), the surface of a metal plate is stippled or roughened all over. Then lighter shades are produced by smoothing down the appropriate areas. Rather than sharp lines as in an engraving, the resultant prints show a smooth transition between the light and darker areas. Jenty justified this method which was not common for medical illustrations, noting,

> If it should be asked, Why, in these Plates, I chose Mezzotinto, instead of Engraving? I answer, that not only the difficulty and Length of time requisite to have executed these TABLES, by able persons, nor the expence, which would have been considerable, prevented my Determination to Engraving; but the Engraving itself, how well soever performed, would not have answered my intention for Colouring, so well as Mezzotinto; as this method is softer, and capable of exhibiting a nearer imitation of Nature than Engraving, as Artists themselves acknowledge that Nature may admit of light and shades, well blended and softened, but never did of a harsh outline: So it must be confessed, that these Prints may want the Smartness which Engraving might have contributed; but the Softness which they possess, may approach nearer to the Imitation of Nature, when coloured, than any engraving possibly could, merely thro' the unavoidable Delineation of the Outline.

These drawings were made by Johannes van Riemsdijk (also Jean or Jan van Riemsdyk; van Rymsdyk; fl. 1750–1788), while several engravers prepared the mezzotint plates (Russell 1963, 1987). van Riemsdijk also produced the illustrations for both William Smellie's *Sett of Anatomical Tables … (1754)* and William Hunter's *The Anatomy of the Human Gravid Uterus Explained by Figures* (1774). With the illustrations of Jenty's classic work, a separate text was published simultaneously. Because the plates probably were used for teaching, being hung in dissecting rooms, few copies have survived. In addition to the original English edition, Latin, French, German, and Dutch editions of Jenty's major work were published.

Jenty also dissected other anatomical specimens. A number of drawings (probably 16 of these drawn by van Riemsdijk) with several others were acquired by the London physician John Fothergill (1712–1780). A friend of the American William Shippen, Jr. (1736–1808), later Professor of Anatomy and Surgery and co-founder in 1765 of the College of Philadelphia (now University of Pennsylvania), Fothergill sent 18 anatomical drawings with several casts to Shippen at the Pennsylvania Hospital in Philadelphia. These

were used by the medical students and staff and where they remain today (Packard 1938, Scott 1904).

Despite the fame of these plates, little is known of Jenty and his life. Based on his writing, he is believed to have been born and educated in France. Listing himself as Professor of Anatomy and Surgery in London, he is thought to have practiced there for a decade or more. Listed as a British Surgeon's Mate, he served in Portugal for several years of the Seven Year's War (1756–1763), which pitted the great powers of Europe against one another (Dobson 1954). Following the cessation of hostilities, he remained in Lisbon for several years and later moved to Madrid (Thornton and Want 1978).

References

Blake 235 lists other editions; Garrison Morton 6156.4.

De Lint, J.G. The plates of Jenty. *Janus* 21:129–135, 1916.

Dobson, J. The army nursing service in the eighteenth century. *Ann R Coll Surg Engl* 14:417–419, 1954.

Hunter, W. *Anatomia uteri humani gravidi tabulis Illustrata… The Anatomy of the Human Gravid Uterus Explained by Figures.* Birmingham, John Baskerville, 1774.

Jenty, C.N. *A course of anatomico-physiological lectures on the human structure and animal oeconomy… 2 vols.* London, J. Rivington & J. Fletcher, 1757.

Jenty, N. A remarkable case of cohesions of all the intestines, &c. In a man of about 34 years of age, who died some time last summer, and afterwards fell under the inspection of Mr. Nicholas Jenty. *Phil Trans* 50:550–552, 1757.

Krumbhaar, E.B. The early history of anatomy in the United States. *Ann Med Hist* 4:271–286, 1922.

Packard, F.R. *Some account of the Pennsylvania Hospital from its rise to the beginning of the year 1938.* Philadelphia, Pennsylvania Hospital, 1938.

Russell, K.F. *British anatomy, 1525–1800; a bibliography.* [Melbourne], Melbourne University Press, 1963.

Russell, K.F. *British anatomy, 1525–1800. A bibliography of works published in Britain, America and on the Continent.* 2nd Ed. Winchester, St. Paul's bibliographies, 1987.

Scott, J.A. Concerning the Fothergill pictures at the Pennsylvania Hospital. *Univ Pennsylvania Med Bull* 16:388–393, 1904.

Smellie, W. *A Sett of Anatomical Tables, with Explanations, and an Abridgment of the Practice of Midwifery, with a view to illustrate a Treatise on that Subject and Collection of Cases.* London, 1754.

Thornton, J.L. & P.C. Want. C.N. Jenty and the Mezzotint plates in his "Demonstrations of a pregnant uterus", 1757. *J Audiov Media Med* 1:113–115, 1978.

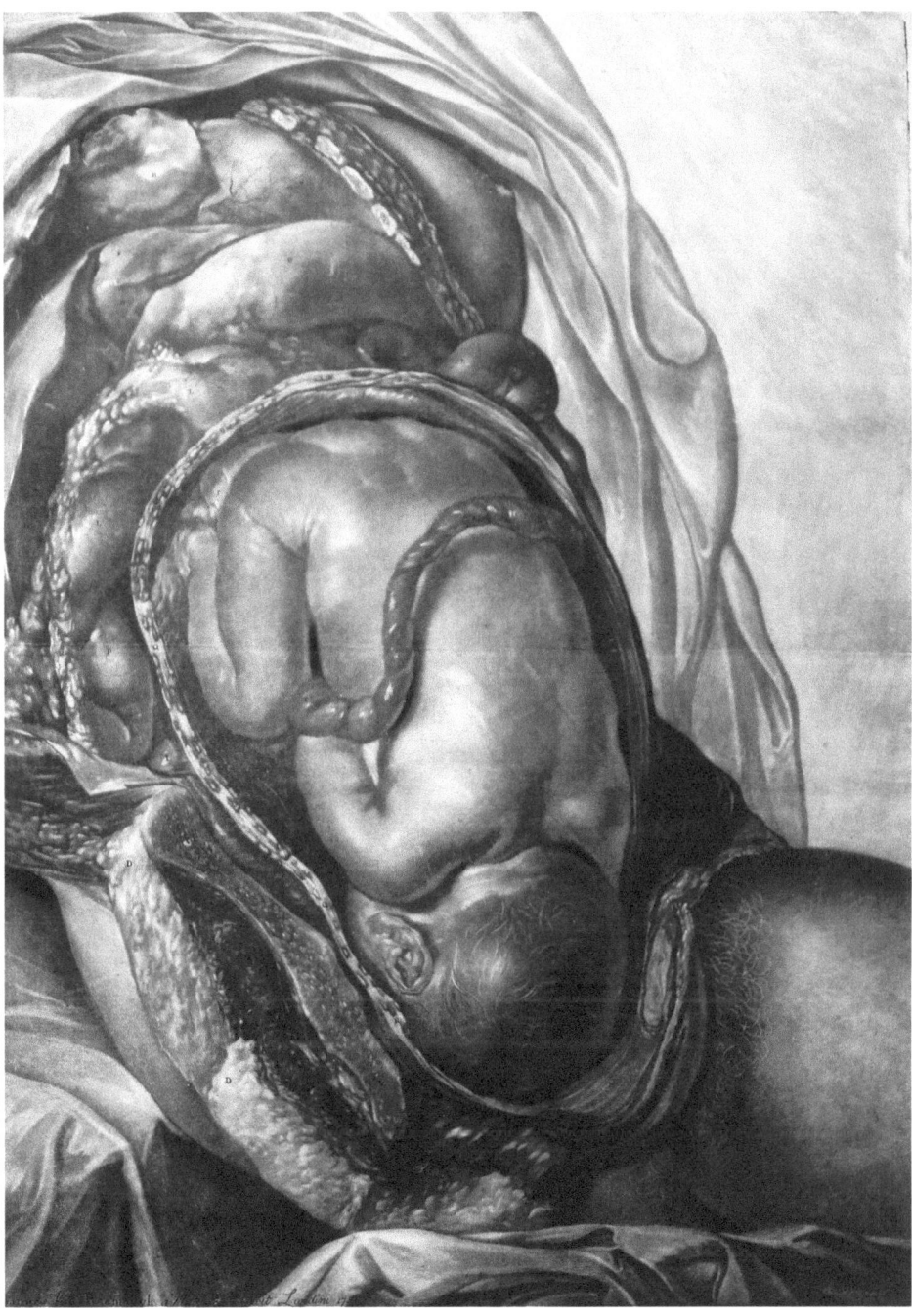

FIGURE FROM JENTY, 1757.
FETUS IN UTERO

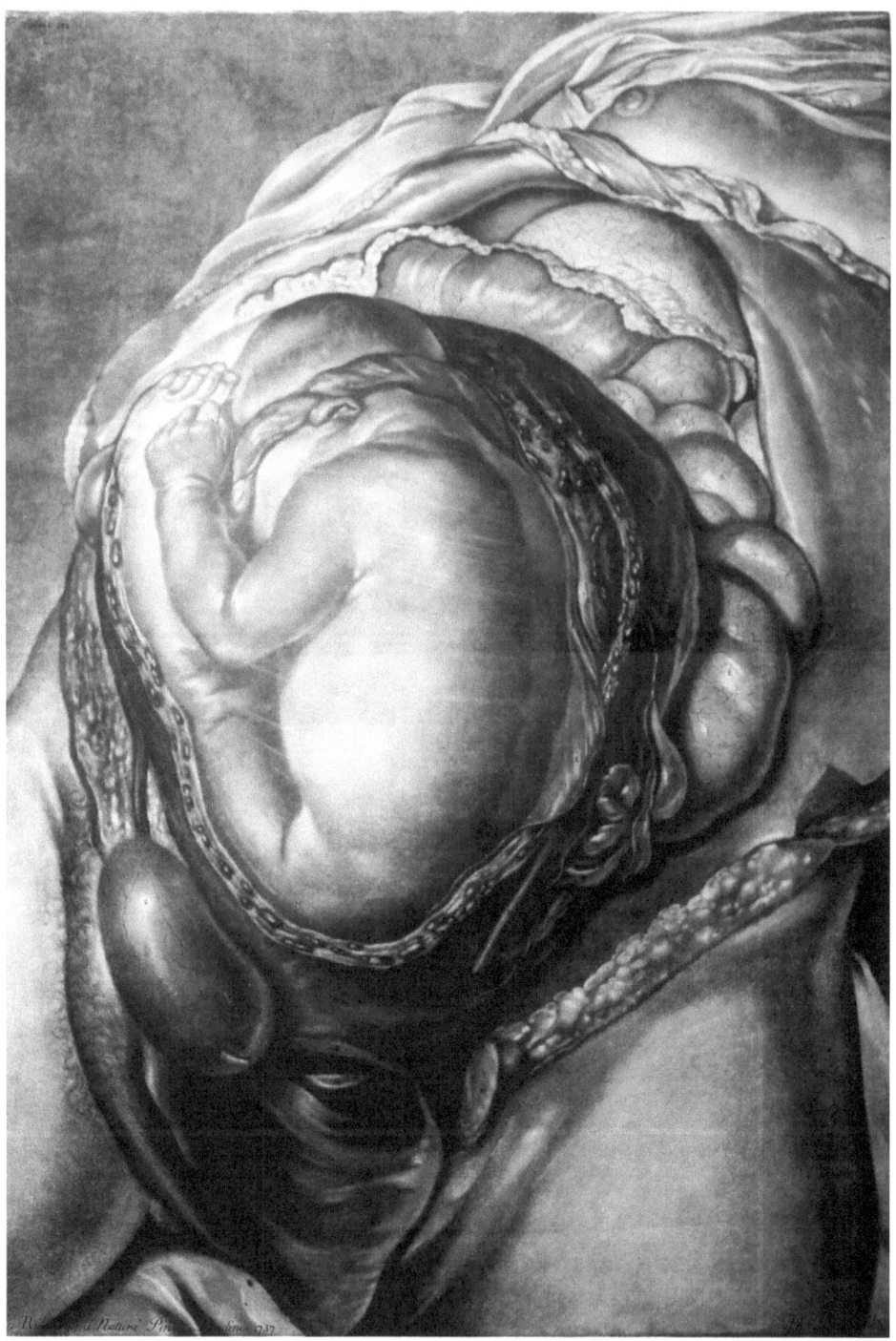

FIGURE FROM JENTY, 1757.
FETUS IN UTERO (BREECH)

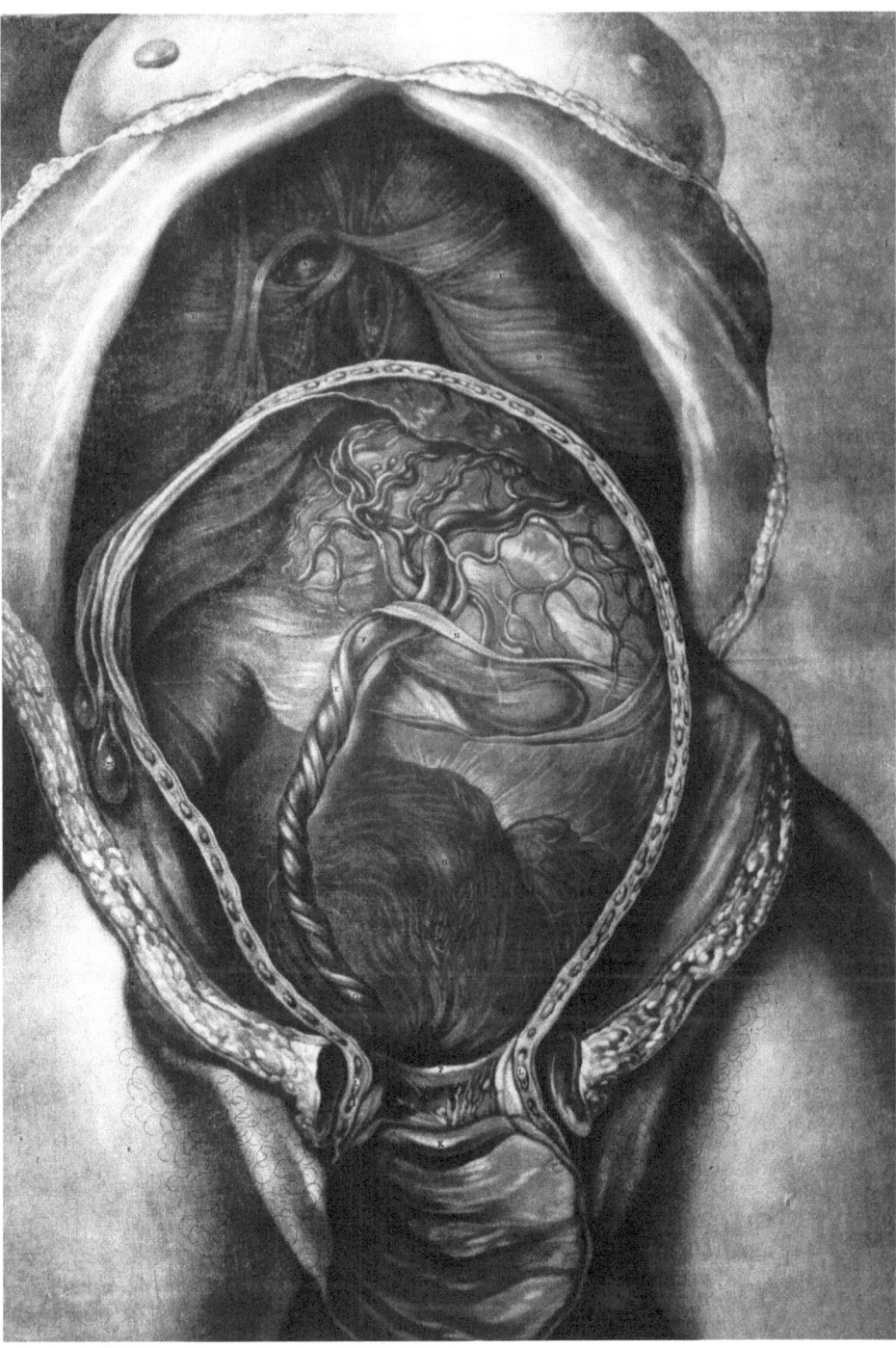

FIGURE FROM JENTY, 1757.
THE UTERUS AFTER BIRTH WITH PLACENTA

Johann Georgii [Georg] Roederer

Icones uteri humani observatinibus illustratae. Gottingae [Göttingen], sump Vandenhoeckianis, 1759.

Originally from Strassburg, in 1748 Roederer (1726–1763) went to London to study anatomy with William Hunter (1718–1783) and midwifery with William Smellie (1697–1763). With Bernard Siegfried Albinus (1697–1770) at the University of Leyden, and Pieter Camper (1722–1789) at the University of Groningen, he explored anatomy further before returning to Strassburg where he graduated from medicine and published his dissertation, *De foetu perfecto* [Formation of the Fetus] (1750). Roederer then was recruited to the University of Göttingen by Albrecht von Haller (1708–1777), where he was appointed as Germany's first Professor of Obstetrics, and founded the first lying-in hospital for the training of male *Geburtschelfer* [obstetricians]. Among his contributions were studies on fetal positions, and the mechanism of delivery.

Based on his autopsy studies at the lying-in hospital at Göttingen, Roederer prepared his *Icones...*, with seven folio plates, drawn and engraved by Joel Paul Kältenhofer (1716–1777), one of Germany's finest copper-plate engravers. Tabula III illustrates the placenta *in situ* and attached to the caudo-lateral uterine wall after delivery. Seen are the opened wall of the uterus (A to C), with cross-sections of the uterine arterial supply (F) and veins (D, E), the umbilical vein (H) and its branches (L, l), the umbilical arteries (I, K) and their branches (M, m), placental membranes (N, P, Q), the cervical os (S), and the right Fallopian tube (T). Tabula V illustrates a term fetus in utero.

Roederer also published on the mechanism of delivery of the fetus, and gave the first clear account of the changes in the breast during pregnancy (Roederer 1753), and also of cancer of the uterus (Roederer 1756). With Carl Gottlieb Wagler (1731–1778), he wrote an extensive study of typhoid fever, a disease not uncommon among mid-eighteenth century obstetrical patients; howver, which the authors confused with dysentery and relapsing f ever.

References

Blake 385; Cutter & Viets, p 202.

Roederer, J.G. *De foetu perfecto,* 1750.

Roederer, J.G. *Elemento artis obstetrical in usum praelectionum academicarum,* Göttingae, 1753. Roederer, J.G. *De uteri schirrho commentio medica,* Göttingen, V. Bossiegel, 1756.

Roederer, J.G. and C.G. Wagler. *De morbo mucoso.* Göttingae, V. Bossiegel, 1762 (GM 5021).

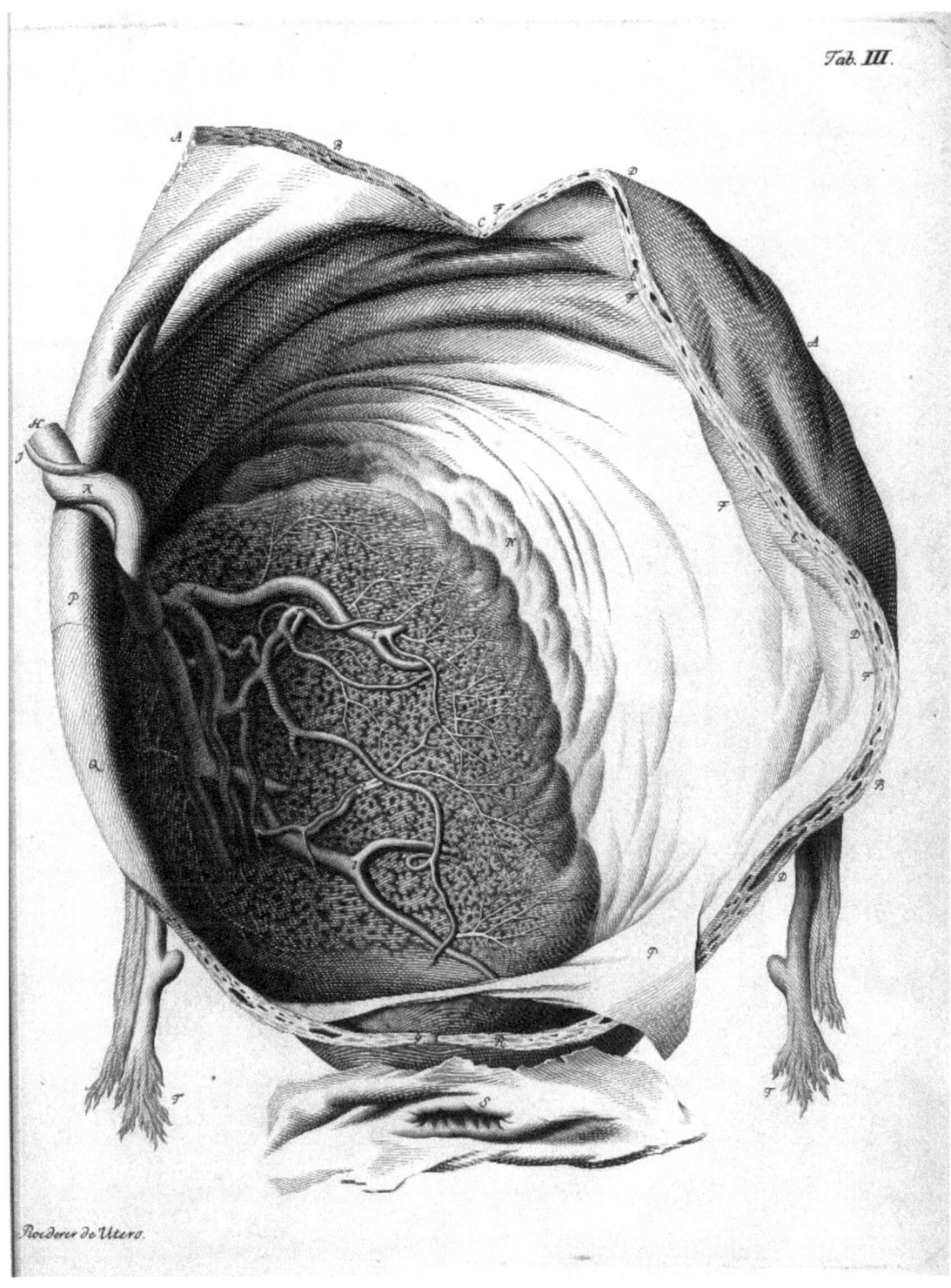

TAB. III FROM ROEDERER, 1759

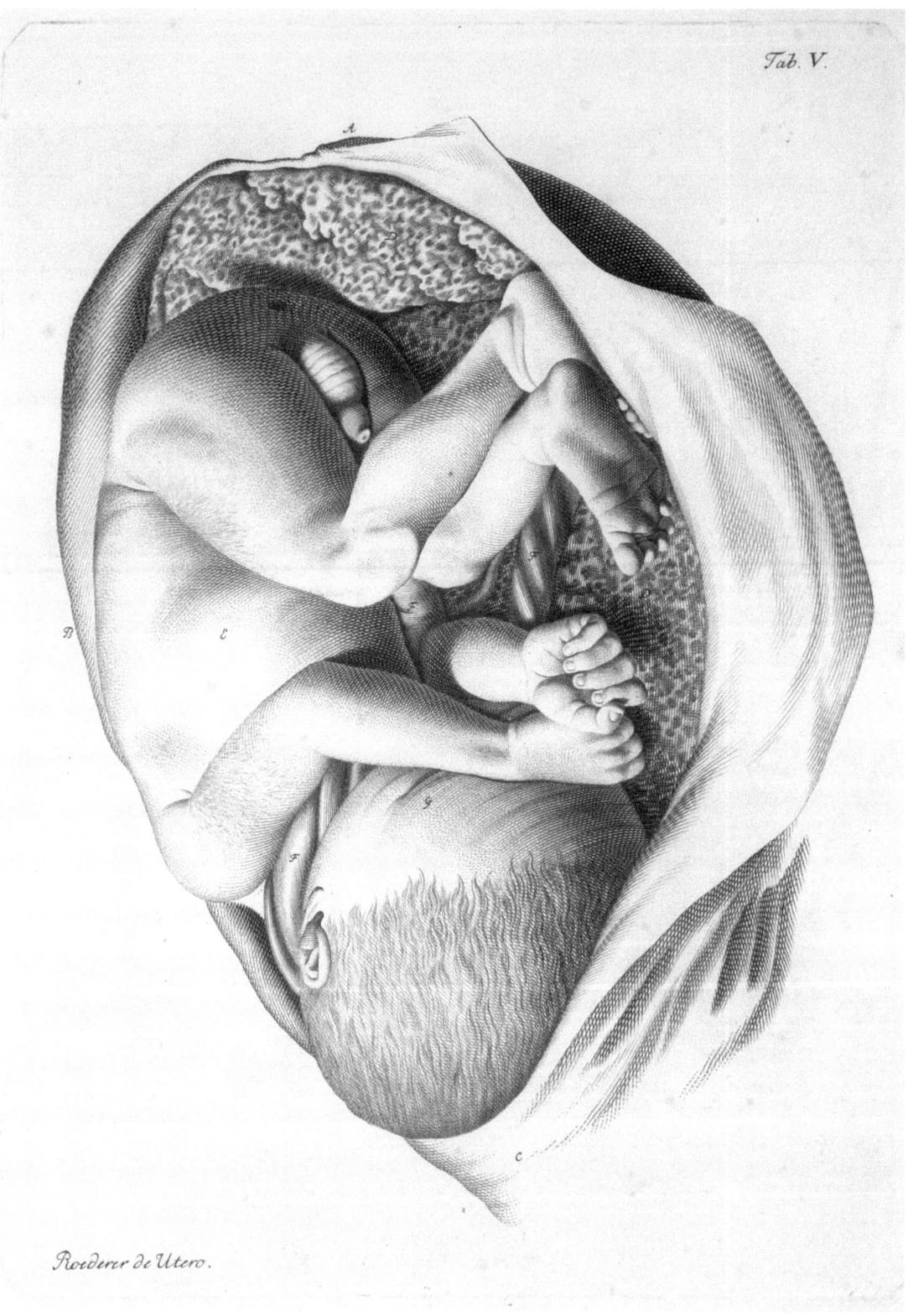

Tab. V from Roederer, 1759

Johann Friedrich Schütz

Gründliche Anweisung zur Hebammenkunst… mit kupfern.
Hildburghausen, Johann Gottfried Hanisch, 1770.

Schütz of Sonnenberg dedicated this well illustrated work for midwives to duchess Charlotte Amalie of Saxony. It is divided into three parts, dealing with the periods before, during, and after birth. The first part is introduced with a description of the profession and the duties of midwives, followed by a section on anatomy, the examining of a pregnant woman, virginity, the signs of pregnancy, and the conduct of pregnant women including advice on adequate clothing. The second discusses childbirth in general, labor, and various positions for delivery of the infant. This is followed by chapters on natural births, turning, and "unnatural births", that require the intervention of a physician or midwife, and the occasional use of instrumens. Included is advice in complicated births caused by a narrow pelvis, a misshapen uterus, premature births, the still-born infant, and extra-uterine pregnancy. The third portion of the text concerns postnatal care, delivery of the placenta, removal of the umbilical cord, conduct of the lying-in woman, and the care of the newborn infant. The last chapter is dedicated to the duties of wet-nurses. The 19-folded plates contain 37 engraved illustrations of the female pelvis and uterus, various fetal positions, and the manipulations required to effect delivery. Figures V and VI illustrate infants in an oblique lie, that require extraction of the legs and breech delivery. Figures VII and VIII demonstrate a method of placing a lasso around the ankles to bring the legs down for extraction. Figures XXIX and XXX illustrate the manner in which a midwife might perform assisted breech delivery of the fetus.

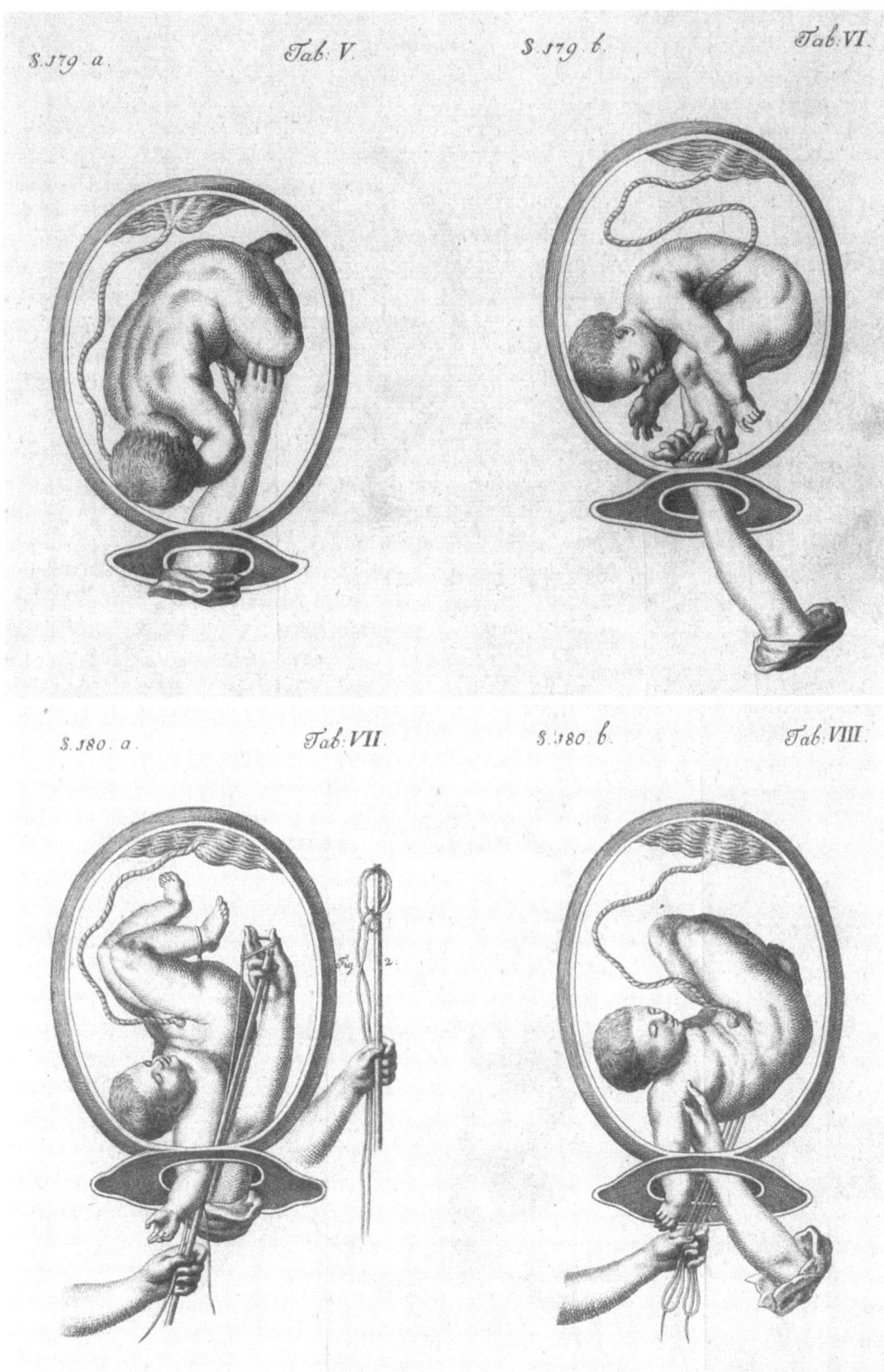

Figures from Schutz, 1770

Jacques Fabien Gautier D'Agoty

Anatomie des parties, de la géneration de la femme, représentées avec leurs couleurs naturelles, se lon le nouvel art. Jointe a l'angeligi de tout le corps human, et a ce qui concern la grosserse et bs accouchemens. Paris, chez J. B. Brunet, 1773.

In his *Anatomie des parties de la generation…*, Gautier D'Agoty (1717–1785) published the first such work on female anatomy produced by printing in four colors rather than by hand coloring. Although not always anatomically exact, the magnificent mezzotint representations are striking, and are of interest in the history of art and of anatomic illustration.

Gautier was born at Marseilles. Here, he became the assistant of the printer Jacob Christoph le Blon (1667–1741), who invented the technique of colored mezzotinting by first applying Isaac Newton's (1642–1727) theory of colors (Newton 1704) to printing. This involved making three separate copper plate impressions (with yellow, red, and blue inks). After Le Blon's death, Gautier claimed to have invented color plate printing, although his contribution was only to add a fourth black plate to Le Blon's three color plates. Nonetheless, his representations are striking, and, in the opinion of Choulant, "… will always retain their value in the history of art and especially in the history of anatomic illustrations." (Choulant 1962, p. 270). Although drawn from dissected bodies, the features are shown lifelike. Plate V, a colored mezzotint, illustrates the upper body of a near-term pregnant woman, showing the fetus *in utero* with placenta (a to h), as well as her exposed superficial musculature (A to H). The subject, looking back at the artist,

presents a characteristic pose of eighteenth-century French portraiture.

He also published *Essai d l'anatomie en tableaux imprimés* (Paris, 1745), which contained two life-sized anatomical figures printed in four colours, one of a man, the other of a woman, each figure being made up of three plates. Later, Gautier D'Agoty associated himself with the surgeon Jacques-François-Marie Duverney (1661–1748), and together they produced several other large colored anatomical atlases including *Anatomie gènèrale des viscère …* (Paris, 1752). Gautier's works were printed on large size paper (French *columbier*) folio. In the text, Gautier gives not only description of the plates, but information on the history of the publication.

References

Blake 169 lists another printing; Garrison Morton 398; Choulant-Frank 270–275.

Choulant, L. History and bibliography of anatomical illustration. Translated and annotated by Mortimer Frank, New York, Hafner, 1962, pp. 265–274.

Gautier D'Agoty, J.F. *Eassi d l'anatomie en tableaux imprimé.* Paris, Chez Gautier, 1745.

Gautier D'Agoty, J.F., Mertrud, A., and Duverney J.F.M. *Anatomie générale des viscére…* Paris, Gautier, 1752.

Newton, I. *Opticks, or, A treatise of the reflexions, refractions, inflexions and colours of light; Also two treatises of the species and magnitude of curvilinear figures.* London, Printed for Samuel Smith and Benjamin Walford, printers to the Royal Society…, 1704.

PLATE V FROM D'AGOTY, 1773

William Hunter

Anatomia uteri humani gravidi tabulis Illustrata...The Anatomy of the Human Gravid Uterus Explained by Figures. Birmingham, John Baskerville, 1774.

One of the most magnificent obstetric atlases ever published, an Elephant Folio with life sized figures, "anatomically exact and artistically perfect," this work illustrates the gravid uteri of three women who died during the last third of gestation, and several others that succumbed earlier in pregnancy. Hunter (1718–1783), a Scotsman who worked in London, commenced working on the atlas in 1751, when it was expected to contain ten plates illustrating the anatomy of a single subject parturient. With the addition of the illustrations of a dozen other subjects, the number of plates rose to 36, two of which were discarded, leaving the present 34 Thus, the completion of this folio required nearly 23 years. Hunter spared no expense in having one of the leading anatomical artists of the time, Johannes van Riemsdijk (also Jan van Riemsdyk or van Rymsdyk) execute 31 of the drawings in red chalk, the plates of which were engraved by over a dozen engravers. Strangely, Hunter failed to credit van Riemsdijk by name, referring only to "the ingenious artists." Some problems raiseds by this affront have been reviewed (Corner 1951).

As noted earlier van Riemsdijk had also executed the plates for the folio atlases of both William Smellie (1754, pg. 184) and Charles Nicholas Jenty (1757, pg. 188) (Huffman 1969, 1970, Thornton and Want 1979). This also is one of only two medical books to have been printed by the finest printer of that age, John Baskerville (1706–1775) of Birmingham, who not only created his unique type fonts, but made his own fine paper and ink (Ollerenshaw 1974, Thornton and Want 1974a). The life-size line engravings achieve effects of depth and contrast without losing sharpness of detail, and present a wide range of normal and pathologic conditions of the womb and fetus. Each page of text is printed in two columns, one in Latin and the other in English. John Hammond Teacher (1869–1930) of Glasgow has presented an account of the preparation of the plates (Teacher 1900, pp. xxi–xxvii), and several writers have provided considerable detail on the individual artists and engravers who prepared the 34 plates (Ollerenshaw 1974, Thornton and Want 1974a). Bibliographical details of later editions of this classic also have been given (Goodall 1958, Le Fanu 1958).

Regarding this work Hunter stated in the Preface, "The art of engraving supplies us... with what has been the great desideratum of the lovers of science, an universal language. Nay, it conveys clearer ideas of most natural objects than words can express; makes stronger impressions upon the mind; and to every person conversant with the subject, gives an immediate comprehension of what it represents" (p. i).

Later he expanded upon this theme, "Anatomical figures are made in two very different ways; one is the simple portrait, in which the object is represented exactly as it was seen; the other is a representation of the object under such circumstances as were not actually seen, but conceived in the imagination" (p. ii). He also noted that the illustrations represent only "...what was actually seen," that they will carry "...the mark of truth... [and be]...almost as infallible as the object itself" (p. iii).

Regarding Plate VI, the opened abdomen exposing the fetus of a woman near term, Hunter wrote, "Its body was covered with a white, greasy mucus, which is commonly seen on children at their birth. This is represented at the upper part of its back, where it was intersected with lines, from the wrinkles and motion of the child's body. Every part is represented just as it was found; not so much as one joint of a finger having been moved to show any part more distinctly, or to give a more picturesque effect." This plate was engraved by Sir Robert Strange (1721–1792), whom Hunter praised in the preface "for having by his hand secured a sort of immortality for two of the plates (the other being plate IV), but for having given his advice and assistance in every part with a steady and disinterested friendship" (p. iv). Plate XIII depicts the gravid uterus and fetus in breech presentation at 9 months of gestation. Note the exquisite detail and the umbilical vessels as they insert on the placental disc (top of figure). It has been suggested that the plates for *The Anatomy of the Human Gravid Uterus ...* originally were intended as illustrations to *An Anatomic Description of the Human Gravid Uterus and its Contents,* which Hunter left unfinished at his death. His nephew, Mathew Baillie (1761–1823) published this latter work a decade later with a preface and several inaccurate additions (Hunter 1794). A second edition prepared by Edward Rigby (1747–1821) appeared in 1843 (Hunter 1843).

William, with his brother John (1728–1793), also explored the anatomy of the placenta, and described the separate circulatory systems of the mother and the fetus (J. Hunter 1786, Hunter 1794). Four decades earlier, Alexander Monro *primus* (1697–1767) of Edinburgh, in his consideration of fetal nutrition had denied that any connection between the maternal and fetal circulations could exist. He recounted that in careful examination of the bodies of five women who had died in labor he identified, "... a thick, fungus, succulent, cellular substance ..." [the decidua] between the muscular myometrium and the villous coat in which were the uterine venous sinuses and through which numerous thin-coated vessels passed, and in this cellular substance the sinuses were. Excepting its sinuses it resembled the internal coat of the intestine ... Munro concluded that the tips of the fetal vessels pass through this decidua to approximate the maternal blood (Monro 1734). In his careful description of the

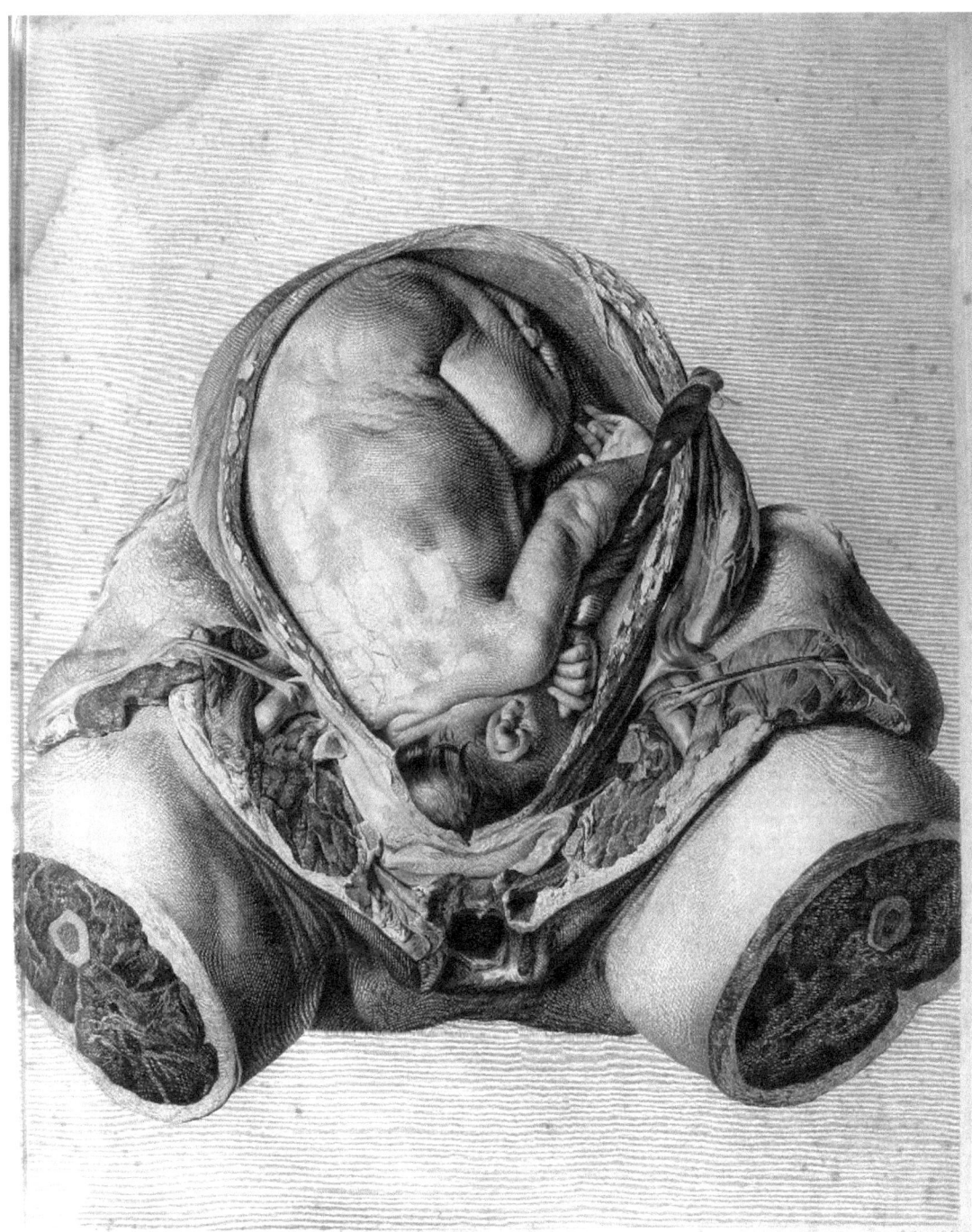

TAB. VI. *Fœtus in utero, prout a naturâ positus, recisa omnine parte uteri anteriori, ac Placenta ei adhærente.*

TAB. VI FROM HUNTER, 1774

decidua, Hunter distinguished between the parietal lining (*membrana decidua vera*) and that of the capsule (*membrana decidua reflexa*).

William Hunter had been a student of Monro, and no doubt had read his 1734 essay. It was the Hunter brothers, however, with their careful injections of the vasculature, who definitively demonstrated the maternal and fetal circulations to be separate. In Plate XXIV of *The Gravid Uterus*, Hunter illustrates the opened uterus of a woman at 6 months gestation. The uterine arteries and veins had been injected with red and blue wax, respectively, and these with umbilical vessels on the fetal placental surface, which contained no wax, are observed. Here and elsewhere, Hunter noted the "convoluted" (e.g., spiral) arteries of the uterine vasculature, and the separate circulation of the mother from that of the fetus (Hunter 1774). In his full *Anatomic Description*, Hunter described the placenta as a blending of two components—a continuation of the fetal umbilical vessels and the uterine (decidual) aspect. He described in some detail his experiments of filling the uterine arteries and veins with liquid wax of different colors (Hunter 1794, p. 47), and also of injecting the umbilical vessels (Hunter 1794, p. 48), concluding that the circulations were quite separate. He summarized his experience:

> From all these experiments and observations which have been often repeated and diligently attended to, with no other desire than to discover truth, it seems incontestable that the human placenta, ... is composed of two distinct parts, though blended together, *viz*, an umbilical, which may be considered as part of the foetus, and an uterine, which belongs to the mother; that each of these parts has its peculiar system of arteries and veins, and its peculiar circulation, receiving blood by its arteries, and returning it by its veins; that the circulation through these two parts of the placenta differs in the following manner: in the umbilical portion the arteries terminate in the veins by a continuity of canal, whereas in the uterine portion there are intermediate cells into which the arteries terminate, and from which the veins begin. Though the placenta be completely filled with any injection thrown into the uterine vessels, none of the wax finds its way into any of the umbilical vessels; and in the same manner fluids injected into the umbilical vessels never can be pushed into the uterine, except by rupture or transudation (Hunter 1794, p. 48).

In his account, William Hunter noted the occasional absence of one umbilical artery (Hunter 1794, p. 33), a condition now appreciated to be associated frequently with fetal malformation (Benirschke and Brown 1955, Benirschke and Dodds 1967).

In addition to demonstrating that the maternal and fetal vessels are completely separate, the brothers also clearly demonstrated maternal blood in the intervillous space. These two observations have led to our modern view of transplacental exchange between the fetal and maternal circulations in humans. Plate X, Fig. I (lower right) illustrates the exterior of the postpartum uterus over the site of placental implantation, with both arteries and veins filled with wax. Figure II (lower left) shows the fetal surface of the placenta, with the umbilical vein and two umbilical arteries injected. Figure III (upper panel) depicts the inner surface of the postpartum uterus at the site of placental implantation. A, B. and C show the uterine wall, and D, E, and F show the inner uterine surface with large venous orifices.

An alternative view on resolving the problem of the placental circulation has credited John Hunter (1728–1793) with that discovery. According to his account, in 1754 Colin Mackenzie (d. 1775), an assistant to the London obstetrician/anatomist William Smellie (1697–1763), had made an unusually careful dissection of the gravid uterus after injecting melted wax into the uterine arteries and veins of a pregnant woman who had died, this apparently in an attempt to confirm or deny the report of Wilhelm Noortwyk (ca. 1712–1778). John Hunter recorded that he was asked by Mackenzie to examine the dissection, and wrote, "The facts being now ascertained and universally acknowledged, I consider myself as having a just claim to the discovery of the structure of the placenta, and to its communicate with the uterus, together with the use arising from such structure and communication, and of having first demonstrated the vasculatiry of the spongy chorion" (J. Hunter 1780). As a consequence, the brothers William and John disagreed on priority of this discovery. They went their separate ways, and it was not until near William's death in 1783 that they were reconciled (Simmons 1783).

Hunter originally studied at the University of Glasgow for a career in the ministry. Soon, however, he discovered his love for medicine. After an apprenticeship with William Cullen (1710–1790) and a year in Edinburgh with Monro, he moved to London to study under William Smellie (1697–1763) and James Douglas (1675–1742). Beginning in October 1746, Hunter offered courses in anatomy from his home. These courses offered, "the opportunity of gentlemen learning the art of dissecting during the whole winter season, in the same manner as at Paris." From Hunter's "Great Windmill Street School of Anatomy" came many of the leading anatomists and surgeons of the day, including his brother John. Hunter's *Anatomia uteri humani gravidi...* stimulated the publication of several related works, including John Burns' (1774–1850) *The anatomy of the gravid uterus* (Burns 1799), and Samuel Thomas Sömmerring's (1755–1830) *Icones embryonum humanorum* (Sömmerring 1799).

In 1762, Hunter attended Queen Charlotte (1738–1820) during her first confinement in which she was delivered of the Duke of Cornwall (1762–1830). In 1767 he was elected to fellowship in the Royal Society, and the following year King George III (1738–1820) appointed Hunter the first professor of anatomy at the Royal Academy of Arts. In addition to this work as a leading anatomist and reformer of the practice of midwifery, Hunter also is remembered for his original description of arteriovenous aneurysm, and for clarifying the

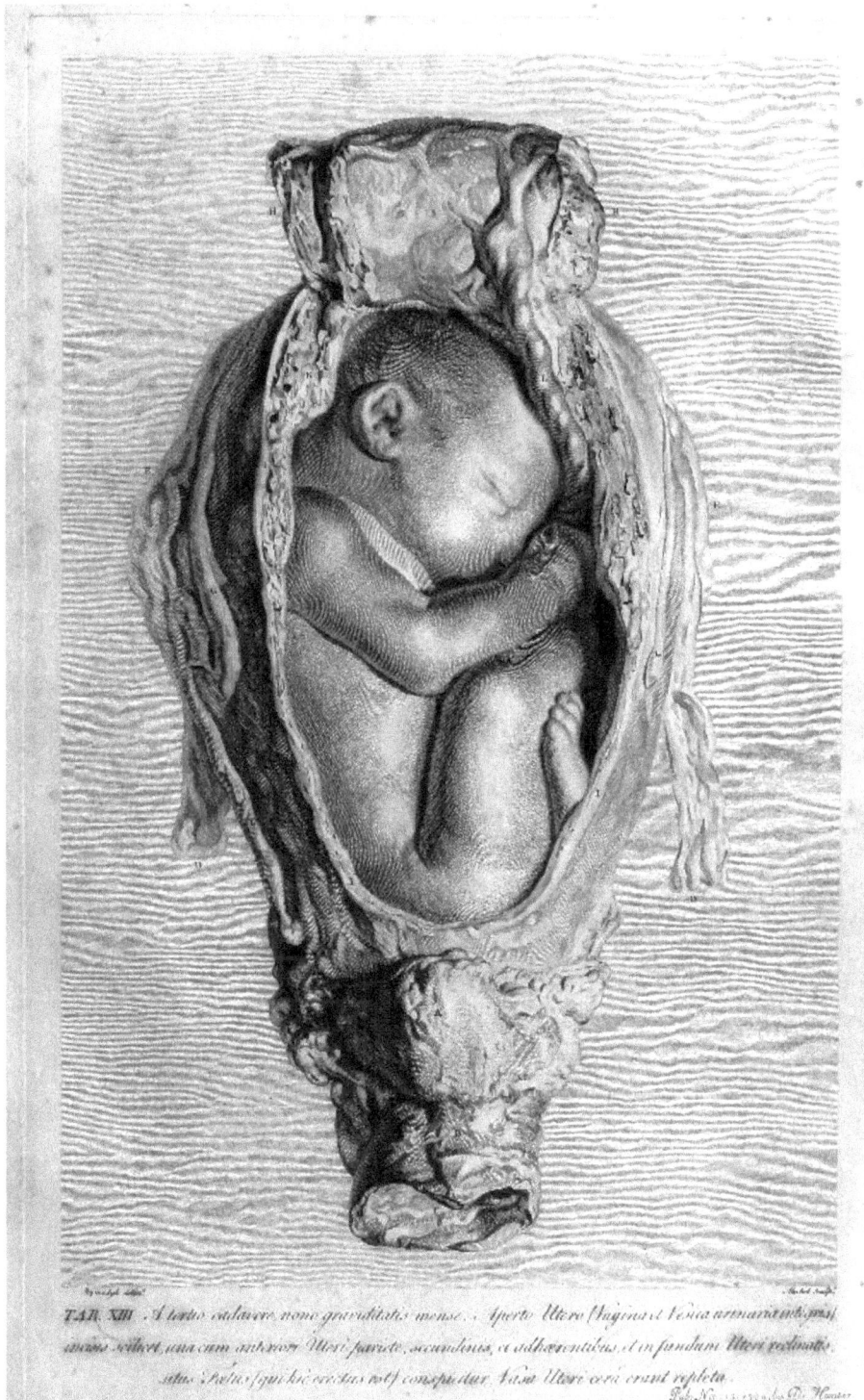

Tab. XIII from **Hunter, 1774**

anatomy and physiology of the lymphatics. Hunter served as President of the Medical Society of London (1780), and received numerous honors. Following his death, his magnificent collection of books, manuscripts, and anatomical and pathological specimens were bequeathed to the University of Glasgow, where they are still archived to this day. As may be evident from the list of references, in contrast to many works presented in this volume, after almost two and a half centuries Hunter's The Anatomy of the Human Gravid Uterus … continues to be of relevance and to attract widespread interest.

References

Blake 226; Garrison Morton 6167; Heirs of Hippocrates 591, Le Fanu 133; Osler 3026; Russell 452.

Andrews, H.R. William Hunter and his work in midwifery. *Br Med J* 1: 277–282, 1915.

Benirschke, K. and W.H. Brown. A vascular anomaly of the umbilical cord; the absence of one umbilical artery in the umbilical cords of normal and abnormal fetuses. *Obstet Gynecol* 6:399–404, 1955.

Benirschke, K. and J.P. Dodds. Angiomyxoma of the umbilical cord with atrophy of an umbilical artery. *Obstet Gynecol* 30:99–102, 1967.

Burns, J. *The anatomy of the gravid uterus. With practical inferences relative to pregnancy and labour*. Glasgow, At the University Press, 1799.

Choulant, L. *Geschichte und bibliographie der anatomischen abbildung nach ihrer beziehung auf anatomische wissenschaft und bildende kunst*. Leipzig, R. Weigel, 1852.

Choulant, L. *History and bibliography of anatomic illustration in its relation to anatomic science and the graphic arts*. Translated by M. Frank. Chicago, University Press, 1920.

Corner, B.C. Dr. Ibis and the artists: a sidelight upon Hunter's Atlas, The Gravid Uterus. *J Hist Med Allied Sci* 1:1–21, 1951.

Dennistoun, J. *Memoirs of Sir Robert Strange, knt., engraver and of his brother-in-law, Andrew Lumisden*. London, Longman, Brown, Green, and Longmans, 1855.

De Witt, F. An historical study of theories of the placenta to 1900. *J Hist Med Allied Sci* 14:360–374, 1959.

Dobson, J. William Hunter. In: *Dictionary of Scientific Biography*. Vol VI. Charles Coulston Gillispie (Ed.). New York, Charles Scribner's Sons, 1972, pp. 568–570.

Fox, R.H. *William Hunter, Anatomist, Physician, Obstetrician (1718–1783)* …. London, H.K. Lewis, 1901.

Goodall, A.L. The writings of William Hunter, F.R.S. *Bibliotheck* 1:46–47, 1958.

Huffman, J.W. Jan van Riemsdyk. Medical illustrator extraordinary. *JAMA* 208:121–124, 1969.

Huffman, J.W. The great eighteenth century obstetric atlases and their illustrator. *Obstet Gynecol* 35:971–976, 1970.

Hunter, J. *On the structure of the placenta*. Communicated to the Royal Society, 1780.

Hunter, J. On the structure of the placenta. In: *Observations on certain parts of the animal anatomy*. London, 1786, pp. 127–139.

Hunter, W. *An anatomic description of the human gravid uterus and its contents*. Edited by M. Baillie. London, Johnston & Nichol, 1794.

Hunter, W. *Anatomia uteri humani gravidi tabulis Illustrata…The Anatomy of the Human Gravid Uterus Explained by Figures. Second Edition*. London, Messrs. Cox, 1815.

Hunter, W. *An anatomical description of the human gravid uterus and its contents*. Edited by Edward Rigby. London, Renshaw, 1843.

Illingworth, Sir Charles. *The Story of William Hunter*. Edinburgh, E. & S. Livingstone, Ltd., 1967.

Jenty, C.N. *The demonstrations of a pregnant uterus of a woman at her full term*. London, Printed for…the author, 1757.

Mather, G.R. *Two great Scottssmen, the brothers William and John Hunter*, Glascow, Pages 6, 9, 24, 1893. Monro, A. An essay on the nutrition of the fetuses. *Edinburgh Med Ess Observ* 2:121–224, 1734.

Morris, W.I. Brotherly love; an essay on the personal relations between William Hunter and his brother John. *Med Hist* 3:20–32, 1959.

Ollerenshaw, R. Dr. Hunter's "Gravid uterus" – a bicentenary note. *Med Biol Illus* 24:43–57, 1974. Peachey, G.C. *A memoir of William & John Hunter*. Plymouth, Brendon, 1924.

Russell, K.F. *British Anatomy 1525–1800. A bibliography*. Melbourne, Melbourne University Press, 1963.

Russell, K.F. *British Anatomy 1525–1800. A bibliography of works published in Britain, America and on the Continent*. 2nd Ed. Winchester, St. Paul's bibliographies, 1987.

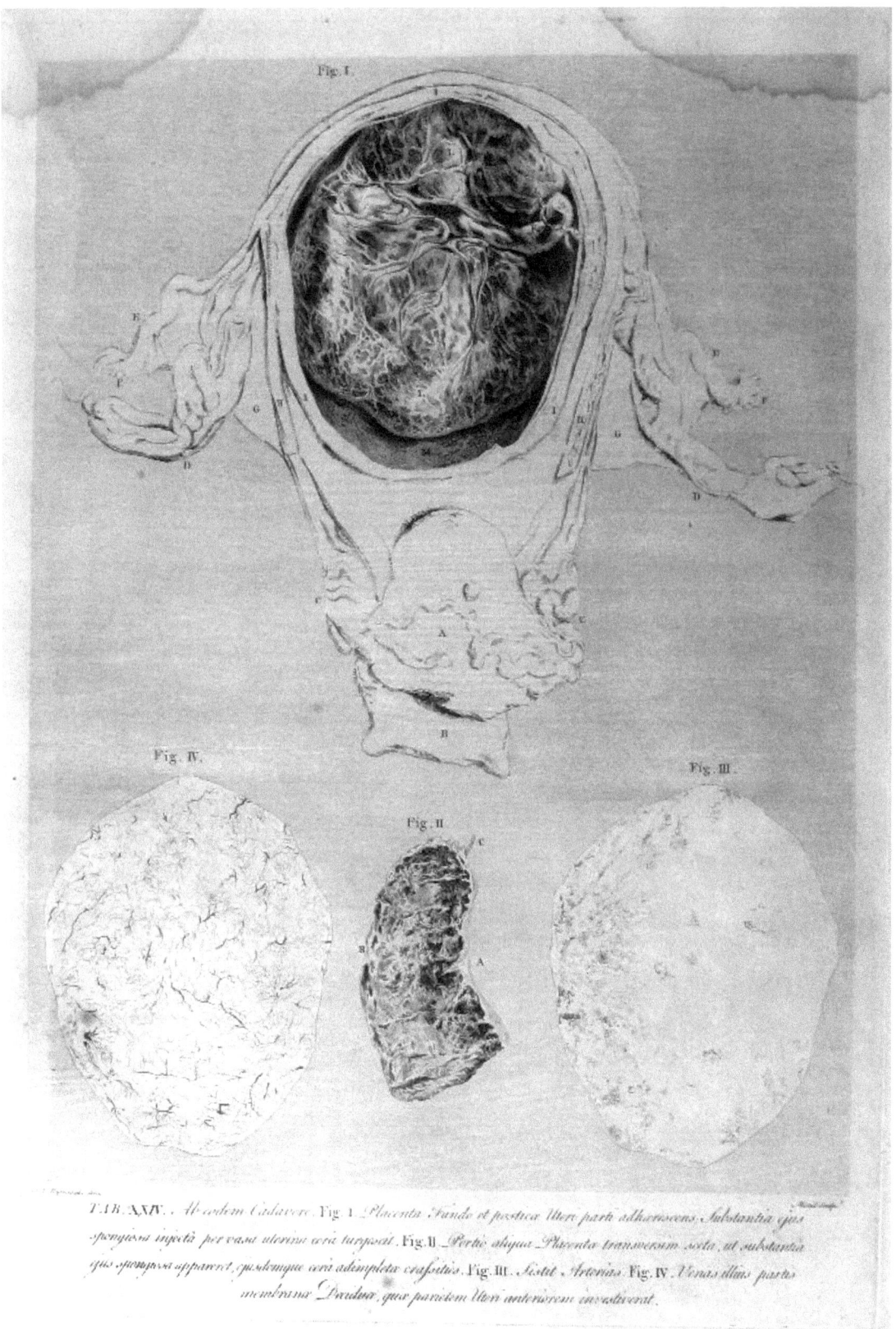

TAB. XXIV FROM HUNTER, 1774

Schumann, E.A. William Hunter lecturing on obstetrics and infant care. *Trans Stud Coll Physicians Phila* 9:155–183, 1941.

Simmons, S.F. *An account of the life and writings of the late William Hunter....* London, Printed for the author by W. Richardson, 1783.

Smellie, W. *A Collection of Cases and Observations in Midwifery* .... London, D. Wilson & T. Burham, 1754.

Sömmerring, S.T. *Icones embryonum humanorum.* Frankfurt am Main, Varrentrapp & Wenner, 1799. Teacher, J.H. *Catalogue of the anatomical and pathological preparations of Dr. William Hunter in the Hunterian Museum, University of Glasgow* .... 2 vols. Glasgow, 1900. (Note: see Vol II, pp. 707–710 for placenta and independence of maternal and fetal care).

Thornton, J.L. and P.C. Want. William Hunter's "The anatomy of the human gravid uterus" 1774–1974. *J Obstet Gynaecol Br Commonw* 81:1–10, 1974a.

Thornton, J.L. and P.C. Want. Artist versus engraver in William Hunter's 'Anatomy of the human gravid uterus', 1774. *Med Biol Illus* 24:137–139, 1974b.

Thornton, J.L. and P.C. Want. Jan van Rymsdyk's illustrations of the gravid uterus drawn for Hunter, Smellie, Jenty and Denman. *J Audiov Media Med* 2:11–15, 1979.

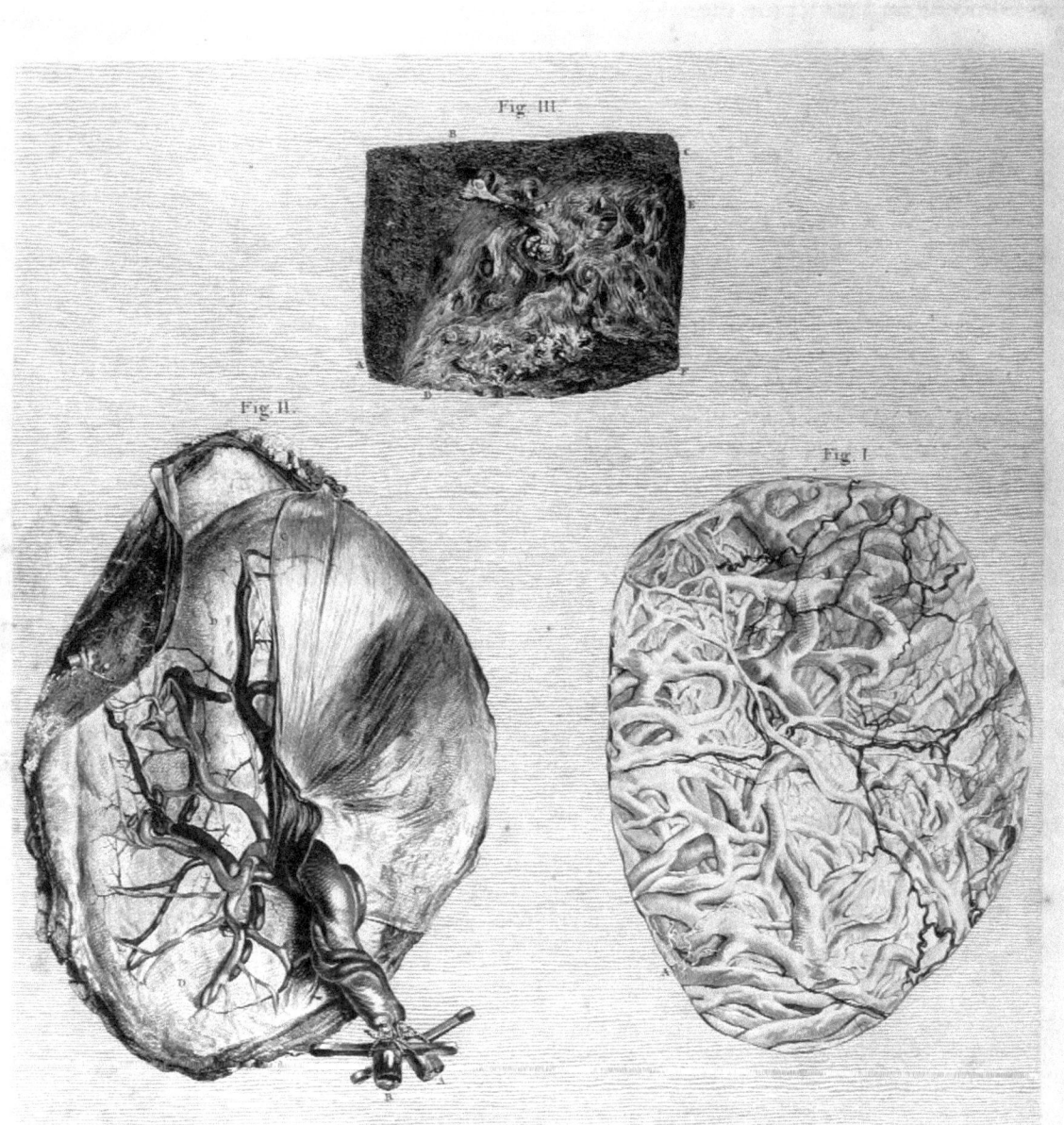

Fig. III.

Fig. II.

Fig. I.

TAB. X. Fig. I. *Uteri pars anterior et externa prout se præbuit omnino siccata, exhibens faciem vasorum uterinorum, qualem præ se ferunt eo loci, ubi Placenta utero adhæret.* Fig. II. *Facies interna Placenta, cujus vasa per funiculum umbilicalem cera sunt repleta.* Fig. III. *Ab alia muliere bidua post partum extincta, exhibet partem superficia intimæ uteri.*

TAB. X FROM HUNTER, 1774

Joseph Jakob Ritter Plenk [von Plenck]

Anfangsgründe der Geburtshülfe. Zwote verbesserte Auflage. Wien, Gräfferischen Buchhandlung, 1774.

Plenk (1738–1807), a renowned Viennese physician and polymath, first published this obstetrical work, "Elements of Childbirth" in 1768. With engraved vignette on title-page, decorative ornaments and two fine copperplates of instruments and obstetrical chair, this enlarged and corrected second edition shows in Plate I views of Der Geburtstuhl [obstetrical chair] that can folded back to make a bed (Figs. I and II), a blunt spatula or lever for aiding in delivery (Fig. III), a hook for reducing (amputating) a fetal part (Fig. V), a tape to affix to a fetal limb so that it might be delivered (Fig. IX), and a perforator (Fig. X) with the needle extended (a). Plate II depicts obstetrical forceps (Fig. VI) with the two blades separated (Fig. VI, LA and LB), a wide pincer or perforator (Fig. VII) and serrated pliers (Fig. VIII).

Plenk was a graduate of the University of Vienna (1763), and disciple of Heinrich Johann Nepomuk von Crantz (1722–1799) the first professor of obstetrics at that University (Crantz 1756a; 1756b). For a short period, Plenk served as professor of anatomy, surgery and obstetrics at the University of Basel (1770). Soon, however, he was summoned by Emprèss Maria Theresia (1717–1780) archduchess of Austria and Queen of Hungary and Bohemia (1740–1780) to teach at the University of Tyrnau in Hungary. He also taught at the Universities of Buda and Pest. In 1783, he became professor of chemistry and botany at the Josephinum, the military medico-chirurgical academy at Vienna for training army officers, founded by Emperor Joseph II (1741–1790), the Holy Roman Emperor (1765–1790). A prolific author, Plenk wrote over 30 works on surgery (Plenk 1777–1778), toxicology, forensic medicine (Plenk 1781), venereal disease (Plenk 1766, 1793), and botany (Plenk 1788, 1798). An admirer and student of Carl von Linne [Linnaeus] (1707–1778) (Linne 1735), he employed classification by genus and species to define various entities. For instance, for diseases of the skin, he pioneered a classification based on their clinical appearance rather than anatomical location, arranging over 100 clinical conditions into ten major groups (Plenk 1776). Although having shortcomings, this classification served as a basis for the development of contemporary dermatology (Jackson 1977, Lane 1933). He also authored a major review of the advances in ophthalmology during the eighteenth century (Plenk 1777), and one on diseases of the teeth and gingiva (Plenk 1778). Although his works passed through many editions and translations, some have argued that aside from his contributions to systems of classification, the works presented little that was innovative or original. Because of his many contributions, in 1798 he was elevated to Hungarian nobility and received the appellation, *von* Plenk.

References

Hirsch IV, pp. 590–591.

Aliotta, G., G. Capasso, A. Pollio, S. Strumia, and N.G. De Santo. Joseph Jacob Plenck (1735–1807). *Am J Nephrol* 14:377–382, 1994.

Bauer, G. Austrian forensic medicine. *Forensic Sci Int* 144:143–149, 2004.

Crantz, H.J.N.von. *Commentarius de rupto in partus doloribus a foetu utero*. Lipsiae, Impensis Iohannis Pavli Kravsii, 1756a. (Blake, p. 102).

Crantz, H.J.N.von. *Einleitung in eine wahre und gegründete Hebammenkunst*. Wien, Gedruckt bey Johan Thomas edlen von Trattnern, 1756b. (Blake, p. 102 in 1768 ed.)

Holubar, K. and J. Frankl. Joseph Plenck (1735–1807). A forerunner of modern European dermatology. *J Am Acad Dermatol* 10:326–332, 1984.

Lane, J.E. Joseph Jacob Plenk 1738?–1807. *Arch Dermatol Syph* 28:193–214, 1933.

Linne, C. von. *Systema naturae*. Lugduni Batavorum, apud Theodorum Haak, 1735. (GM 99)

Plenk, J.J. *Schreiben an Herrn Rumpelt von der Wirksamkeit des Quecksilbers und Schirlings*. Viennae, 1766.

Plenk, J.J. *Anfangsgründe der Geburtshülfe*. Wien, In der Gräfferischen Buchhandlung, 1768.

Plenk, J.J. *Doctrina de morbis cutaneis: qua hi morbid in suas classes, genera and species rediguntur*. Viennae, Apud Rudolphum Graeffer, 1776.

Plenk, J.J. *Doctrina de morbis oculorum*. Viennae, Rudolphum Graeffer, 1777.

Plenk, J.J. *Anfangsgründe der chirurgischen Vorbereitungswissenschaften*. 3 vol. Wien, 1777–1778.

Plenk, J.J. *Doctrina de morbis dentium ac gingivarum*. Viennae, Apud Rudolphum Graeffer, 1778. (Blake, p. 355).

Plenk, J.J. *Dottrina de'morbi degli occhi*. Venezia, Appresso Francesco di Niccolò Pezzana, 1781. Plenk, J.J. *Elementa medicinae et chirurgiae forensis*. Viennae, Apud Rudolphum Graeffer, 1781. (Blake, p. 356).

Plenk, J.J. *Icones plantarum medicinalium secundum systema Linnaei digestarum cum enumeratione virium et usus medici chirurgic atque diaetetici....* 8 vols. (The eighth volume was edited by Joseph Kerndl). Vienna, Apud R. Graeffer et soc., 1788–1812.

Plenk, J.J. *De morbi venerei*. Venezia, Pezzana, 1793. (Blake, p. 355).

Plenk, J.J. *Anfangsgründe der botanischen Terminologie, und des Geschlechtssystems der Pflanzen*. Wien, 1798.

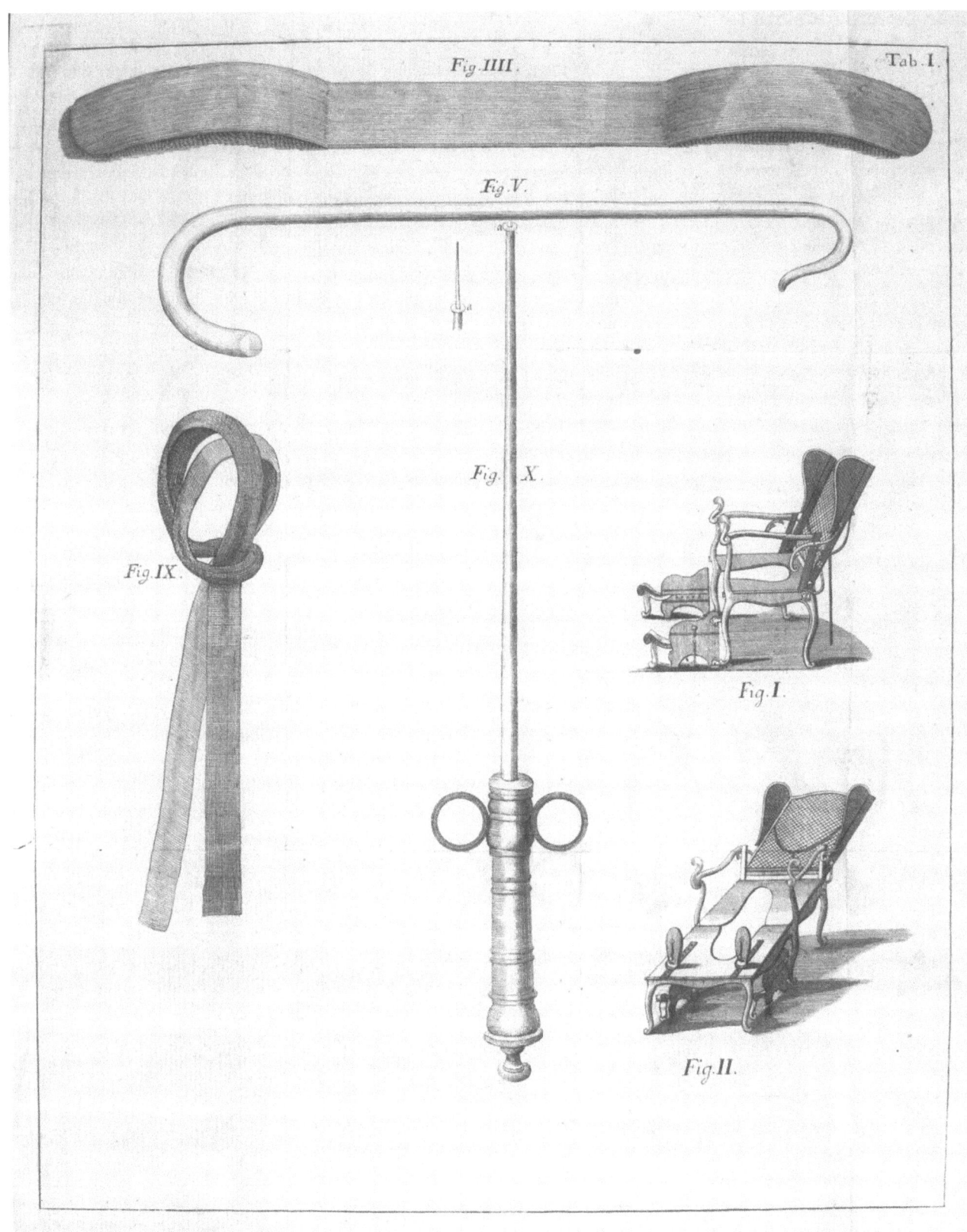

Tab. I from Plenk, 1774

Jean-Louis Baudelocque

L'art des accouchemens 2 vols. Paris, Mequignon, 1781.

Baudelocque (1746–1810), chirurgien accoucheur of Paris, contributed to an understanding of the anatomy of the female pelvis, and possible abnormalities. In that era, rickets, a disease now known to be caused by vitamin D deficiency characterized by defective bone growth, was particularly common. In women this resulted in pelvic deformity, dystocia, and difficulty in childbirth. Baudelocque invented a pelvimeter and established the importance of external pelvic menstruation in assessing the probability of a woman being able to deliver an infant vaginally. By his careful descriptions of the female pelvis, he helped to advance knowledge of pelvimetry and an understanding of the mechanisms of labor. The external conjugate diameter, that of "the thickness of the woman from the middle of the pubis to the tip of the spine of the last lumber vertebra…" is known as Baudelocque's diameter.

Baudelocque's treatise contained little that was original. His objective, he wrote, was to prepare an authoritative manual for the edification of students. Regarding vaginal delivery he wrote, "this operation, which is entirely mechanical, and subject to the laws of motion, is most frequently performed by the natural force of the organs of the woman…" (p. 1). In this two volume work, Baudelocque considers normal labor as well as essentially every complication of which one can conceive. The work includes 14 large plates by the Parisian artist and engraver Jean Jacques Avril the Elder (1744–1831).

Plate VI, Volume II, illustrates the *accoucheur* using outlet forceps to assist with the delivery of the head in an otherwise normal vaginal delivery: a, a, lumbar vertebrae; b, b, sacral vertebrae;… h, the pubis; *et cetera*. The first five plates included in volume I, present details of normal pelvic anatomy and menstruation, as well as that of two deformed pelves. Plates VII and onward (Volume II) illustrate the use of forceps in abnormal presentations such as face, the aftercoming head of a breech delivery, and so forth.

In 1798, following the French Revolution, Baudelocque was appointed professor of obstetrics at the *École de Sauté* and director of the Maternité. *Le grand* Baudelocque became the acknowledged master *accoucheur* in France, attending Empress Marie-Louise (1791–1847), second wife to Napoleon I [Bonaparte] (1769–1821), in her first confinement.

The first work to describe in detaile the female pelvis, was that by Hendrick van Deventer (1651–1724) of Den Hague (Deventer 1701). In addition, André Levret (1703–1780) (Levret 1747), William Smellie (1697–1763) (Smellie 1752) and others contributed to an understanding of the anatomy of the female pelvis, and the various abnormalities to which it is subject.

References

Blake p. 34; Cutter & Viets, pp. 90–93; Garrison Morton 6255.

Baudelocque, J.L. *Principes sur l'art d'accoucher*.... Paris, Didot, 1775.

Baudelocque, J.L. *A system of midwifery, Translated by John Heath, 3 vols*. London, the author, 1790.

Baudelocque, J.L. *Two memoirs on the Cesarean operation.... Translated from French ... by John Hull*. Manchester, Sowler and Russell, 1801.

[Baudelocque, J.L.] *An abridgement of Mr. Heath's translation of Baudelocque's midwifery. With notes by William P. Dewees*.... Philadelphia, Bartram and Reynolds, 1807.

Deventer, H.v. *Manuale operation, I. deel zijnde een nieuw ligt voor vroed-meesters en vroed-vrouwen*.... Den Hague, the author, 1701. (GM 6253)

Dunn, P.M. Jean-Louis Baudelocque (1746–1810) of Paris and *L'art des accouchemens. Arch Dis Child Fetal Neonatal Ed* 89: F370-F372, 2004.

Levret, A. *Observations sur les causes et les accidens de plusieurs accouchemens laborieux*. Paris, C. Osmont, 1747. (GM 6152)

Smellie, W. *A treatise on the theory and practice of midwifery*. London, D. Wilson and T. Durham, 1752. (GM 6154)

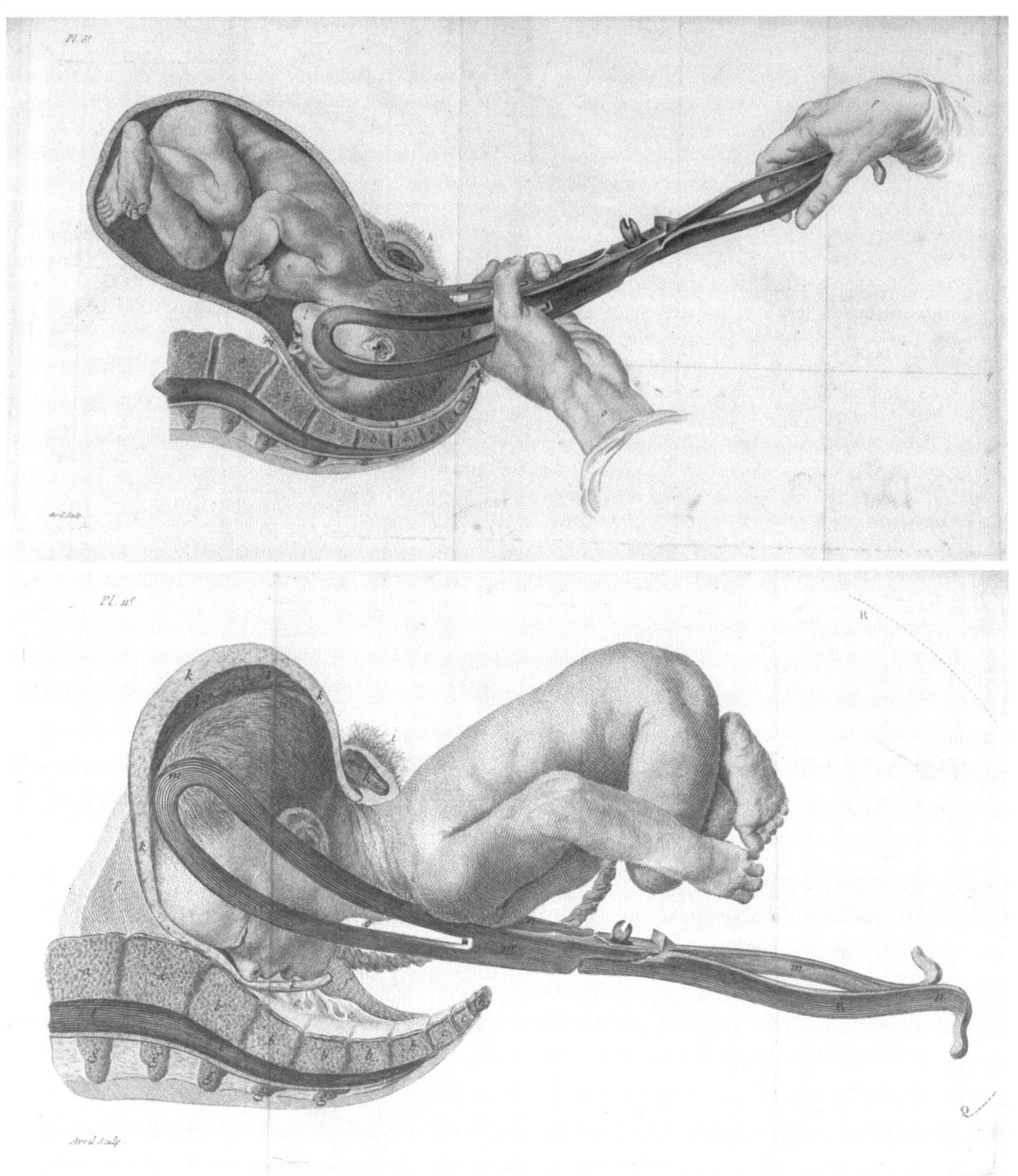

PLATES 6 (*TOP*) AND 11 (*BOTTOM*) FROM BAUDELOCQUE, 1781

George Herbiniaux

Traité sur divers accouchemens laborieux, et sur les polypes de la matrice; ouvrage dans lequel on trouve la déscription d'un nouveau levier, imité de celui de Roonhuysen, & mis en parallel avec le forceps: ainsi que d'un nouvel instrument, proper à la ligature des polypes, approuvé pa l'Académie Royale de Chirurgie de Paris. Brussels, chez J.L. De Boubers, 1782.

In his obstetrics manual, Herbiniaux (ca. 1740–date of death unknown) of Brussels, described a lever-syringe, an instrument like a single blade of a pair of forceps with a cloth band attached, and tube for baptizing the infant in utero in instances of impending fetal death. As noted in the work of Francesco Emmanuele Cangiamila (1702–1763), because of the belief that the unbaptized fetus was condemned to damnation and eternal torment in hell, the instance in which death of the fetus or the pregnant woman was imminent, many advocated baptism in utero by means of a syringe (Cangiamila 1758). The fine plates show details with cutaway views and transverse sections of the instruments. Plate II, Fig. I depicts the spatula and lever in profile with a sturdy linen cord tied in place. Figure II shows a similar lever with a more lengthy curvature. Figure III illustrates the hollow cylinder or syringe to be used for baptism, while Fig. IV shows an alternate nipple that can be used, and Fig. V presents the syringe plunger for the several conformations of this apparatus. Figures VI and VII show a birth chair, the "bed of misery" as Herbiniaux called it, that is separated to illustrate parts such as the coarse hair mattress (q), treadle or stirrup (i), and associated attachments. Figure VIII, and IX show the forceps of André Levret (1703–1780) for comparison (Levret 1747). Figure X depicts the forceps used to extract a dead fetus, with "j" illustrating the several hooks to be used with this instrument. Figure XI illustrates the wrench used with the screw for the forceps. The instruments illustrated are one-half normal size. Plate III depicts the several instruments in use.

In this volume, Herbiniaux included a lengthy history of the lever, as described by Hendrik van Roonhuyze (1622–1672) (Roonhuyze 1663). Herbiniaux first described an osseous prominence anterior to the sacrum that caused narrowing of the pelvic inlet and birth canal. This condition *spondylolisthesis* [Greek, slipped vertebrae], was a result of anterior subluxation or displacement of the fifth lumbar vertebrae over the sacral vertebrae. Hermann Friedrich Kilian (1800–1863), first used the term *spondylolisthetic pelvis*, as *pelvis obtecta* [covered or obstructed pelvis] in 1854 (Kilian 1854a, b). Currently spondylolisthesis is classified into several different types. Herbiniaux also invented several instruments to avulse polyps from the uterine cavity (1771). An illustration of their use is depicted in Plate IV of the present volume.

References

Blake, p. 208; Ricci, p. 394.

Cangiamila, F.E. Embryologie sacra, sive, de officio sacerdotum, medicorum, et aliorum circa aeternam parvulorum in utero existentium salutem. Libri quatuor. [Palermo], Panormi, 1758.

Herbiniaux, G. Parallèle de différens instrumens, avec les méthodes de s'en server pour pratiquer la ligature des polypes dans la matrice; en forme de letter à Mr. Roux.... À La Haye, chez P. Gosse & D. Pinet, 1771.

Kilian, H.F. *Schilderungen neuer Beckenformen and ihres Verhaltens im Leben.* Mannheim, Verlag von Bassermann und Mathy, 1854. (GM 6261).

Kilian, H.F. De spondylolisthesi gravissimae pelvangustiae causa nuper detecta. Bonnae, C. George, [1854]. (GM 6262).

Levret, A. *Observations sur les causes et les accidens de plusieurs accouchemens labourieux.* Paris, C. Osmont, 1747. (GM 6152).

McPhee, B. Spondylolisthesis and spondylolysis. In: *Neurological Surgery: a comprehensive reference guide to the diagnosis and management of neurosurgical problems.* Third Edition, Vol. 4. Philadelphia, W.B. Saunders Company, 1990, pp. 2749–2784.

Roonhuyze, H.van. *Heel-konstige aanmerkkingen betreffende de gebreeken der vrouwen.* Amsterdam, weduwe van T. Jacobsz, 1663. (GM 6015).

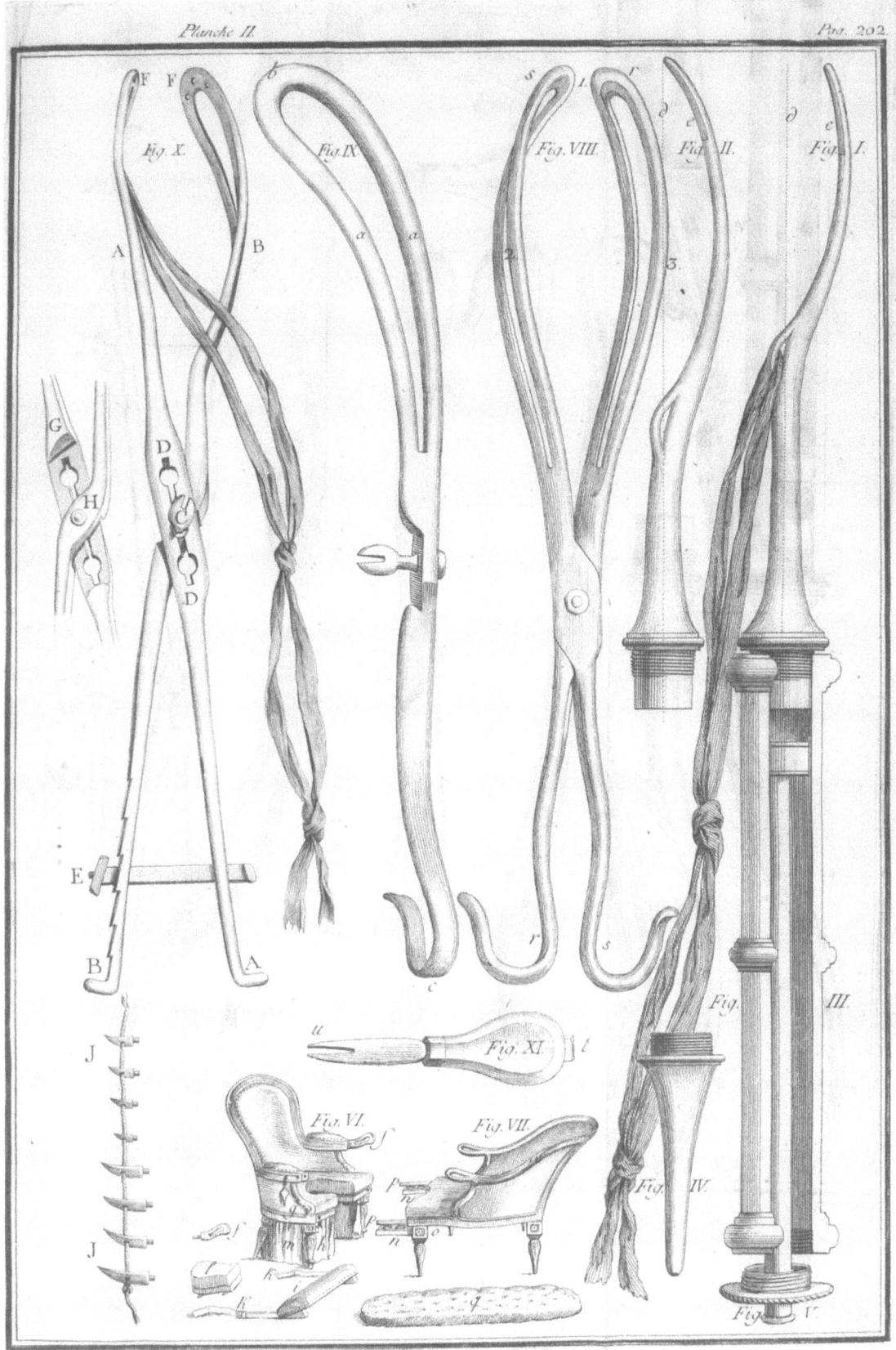

PLATE II FROM HERBINAUX, 1782

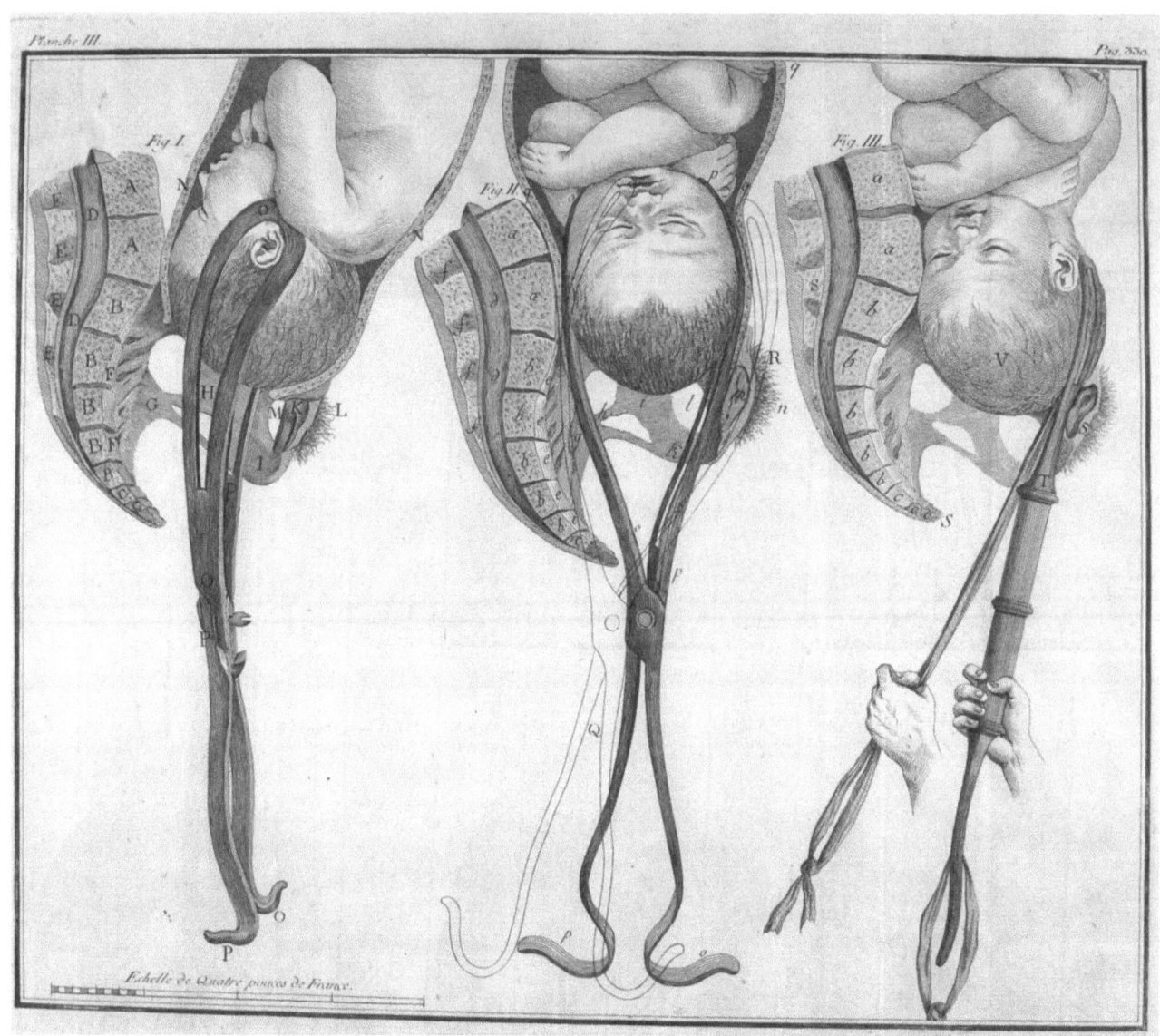

PLATE III FROM HERBINIAUX, 782

Jean Bernard Jacobs

Vroedkundige Oeffenschool, vervattende in een klein bestek meerendeels alles, 't geen tot deze konst eenige betrekking heeft, de geneeskunde uitgezonderd, uitgegeeven bij wijze van lessen met kopere plaeten. Gent, Vanderschueren, 1784.

In his "Midwifery Practice School..." of over 60 chapters, which was the most complete compendium for midwives at the time, Jacobs (1734–1790) provided instruction on essentially every aspect of female reproductive anatomy, fetal development, and the complications of labor and delivery. Twenty-one finely engraved plates illustrate his teachings, of helping women during pregnancy and childbirth. On the title page is the injunction, "Safe, good Godess of Birth! The pregnant woman: Protect the Fruit, and that it will be ripe [and] safely reborn." Following a laudatory introduction in which he acknowledged the support of administrative and political leaders of the region, in his Preface Jacobs stressed that his treatise was written for students and midwives, rather than for masters of the art. Also, rather than a claim of originality, he credited notable authors of the past for their many contributions. He observed that while the illustrations were intended to be of value, he believed that practice on the obstetrical manaquin was of the greatest aid in learning.

Plate V, Fig. 1 depicts the spinal column, ribs, and pelvic bones, with outline of a near-term fetus. Figure 2 shows the delivered fetus with placenta attached to the posterior uterine wall. Figures 3 and 5 present cross sections of the umbilical cord. Plate VIII illustrates a recently delivered stillborn infant (Fig. 1) with the abdominal viscera exposed to view. Shown are the umbilical cord (A), umbilical vein (B), and both umbilical arteries (C). Figure 2 depicts the opened uterus with placenta attached to the posterior uterine wall, Fig. 3 illustrates the manner of extracting the placenta with the membranes, while Fig. 4 shows the extracted placenta in the hand of the midwife.

In addition to learning the science and art of safe obstetrics, one of his goals was to avoid the need for the deadly caesarean operation. Jacobs noted, "… I was able to accelerate birth in favour of the mother and without having to inconvenience her … after an hour of intense labour I granted the mother a pause, which would be good for her. This way she could regain some strength again …." Nonetheless, he admitted that the "frightful and terrifying" procedure that exposed the mother to "great peril and pain" in an attempt to save the child, was necessary when no other means of delivery was possible. Jacobs then listed six conditions requiring caesarean section, including: severely contracted pelvis, uterine rupture, extra uterine pregnancy, and in instances in which the woman dies before delivery, so that the infant might be baptized and if possible, kept alive. He concluded by recounting several cases in which the operation had been successful (pp. 400–401). Jacobs urged his colleagues to act always in the best interest of the women, admonishing them to operate as swiftly and softly as possible, in order not to prolong the pains of labour.

From a well to do family in a small town near Ghent, the capital of East Flanders, Jacobs trained in surgery in Prussia and then worked in hospitals in Berlin and Magdeburg. In 1758, he returned to Ghent where he became a master in the surgeon's guild. In 1775, Jacobs traveled to Ypres, where he attended the class of the celebrated French midwife Angélique Marguerite le Boursier du Coudray (1714/5–1794). He apparently purchased one of her life sized phantom "machines" with a fetal doll, and then returned to Ghent to offer midwifery classes of 6 weeks in duration, based on her methods (du Coudray 1769). He thus introduced the teaching of obstetrics with the aid of a life-sized anatomical figure in Flanders. For part of the course, the apprentice midwives and *accoucheurs* were taught practical obstetrics in the home of a midwife. To become recognized officially, the students had to attend the course twice during a 2 year period.

At this time, a significant public health problem was the prevalence of individuals dying by drowning in canals or rivers, as few people knew how to swim. Works such as *Colymbetes sive de arte natandi...* by Nicolaus Wynman (fl. 1540) (Wynman 1538), *De arte natandi...* by Everard Digby (fl. 1590) (Digby 1587) translated by Christopher Middleton (ca. 1560–1628) as *A short introduction for to learne to swimme* (Middleton 1595), both of which contain lovely woodcut illustrations, championed the importance of knowing how to swim. However, for the general public this was not common knowledge. Thus, the common view was that drowning victims were beyond hope. In major cities, societies were formed for the rescue and resuscitation of drowning victims. The first of these was the "Society for Preservation of Life from Accidents in Water" in 1767 in Amsterdam. This soon was followed by the "Society for the Recovery of

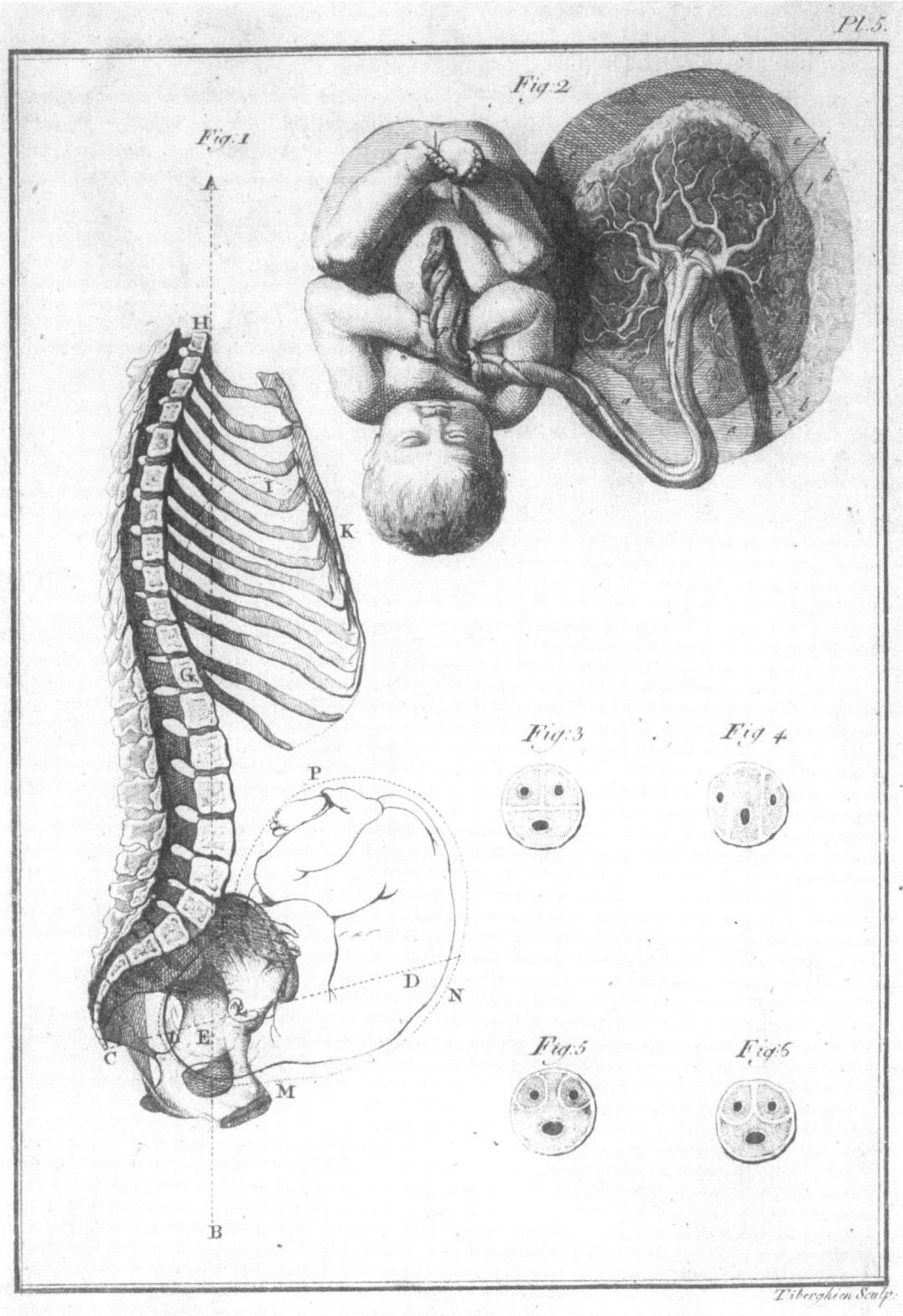

Plate 5 from Jacobs, 1784

Persons Apparently Drowned" in 1774 in London, which in 1776 became the "Humane Society," and in 1787 the "Royal Humane Society." Its motto was Latcal scivlillul forsan [a small spark may perhaps lie hid]. Initially, acts of courage and success in saving the life of such unfortunate beings were rewarded with money. Later, the rescuers received certificates, a gold medal, and registration in the annual reports of the Society (Bishop 1974, Mundell 1895, Thomson 1963). In 1775, Jacobs translated an early work on this subject by Joseph Jacques de Gardane (fl. 1770) of Paris (Gardane 1774). In his translation, Jacobs augmented the text with his own experience (Jacobs 1775). He also described and illustrated a portable "smoke box" for resuscitation, which he stated should be in the possession of every pastor and Lord, and in the home of those with a large household, and that classes should be organized to teach lay people its use. In addition, Jacobs promoted free swimming classes to minimize the need of resuscitating the asphyxiated (Jacobs 1775).

In 1788, Jacobs was appointed to teach surgery at the University of Leuven. The following year, during the Revolution in the Province of Brabant (surrounding Brussels, a former duchy of the Netherlands, now divided between Netherlands and Belgium), he joined the Austrian forces apparently because of his loyalty to the Habsberg regime. He died later that year of typhus. After Jacobs' death, his extensive library and collection of artifacts including rare fossils, minerals, shells, stones, and so forth were sold at an auction that lasted for more than a week. The present work was translated into French (Jacobs 1785) and German (Jacobs 1787), and remained a classic for many years.

References

Blake, p. 232; Hirsch III, p. 363; Lindeboom, Dutch Medical Biography, I, pp. 964–965.

Bishop, P. *A short history of the Royal Humane Society to marks its 200th anniversary*. London, Royal Humane Society, 1974.

Digby, E. *De arte natandi libri duo, quorum prior regulas ipsius artis, posterior vero praxin demonstrationemque continet*. Londini, Excudebat Thomas Dawson, 1587.

Digby, E. *A short introduction for to learne to swimme. Gathered out of Master Digbies Booke of the Art of Swimming. And translated into English for the better instruction of those who understand not the Latine tongue. By Christofer Middleton*. At London, Printed by James Roberts for Edward White, and are to be sold at the little North doore of Paules Church, at the signe of the Gun, 1595.

du Coudray, A.M.le B. *Abbrege de l'art des accouchemens, dans lequel on donne les precepts necessaries….* A Saintes, Chez Pierre Toussaints, libraire, imprimeur…, 1769.

Gardane, J.J.de. *Avis au people, sur les asphyxies ou morts apparentes et subites….* Paris, Ruault, 1774. (Blake, p. 166).

Gardane, J.J.de. *Bericht aen het volk, aengaende de asphyxia ofte schynbaere ende schielyke dood….* Gend, Judocus Begyn…, 1775.

Jacobs, J.B. *Ecole prastique des accouchemens*. Ghent, Chez J.F. Vander Schueren, 1785. Jacobs, J.B. *Praktischer Unterricht der Entbindungskunst*. Marburg, 1787.

Mundell, F. *Stories of the Royal Humane Society*. London, Sunday School Union, 1895.

Schechter, D.C. Role of the humane societies in the history of resuscitation. *Surg Gyn Obstet* 129:811–815, 1969.

Trubuhovich, R.V. History of mouth-to-mouth rescue breathing Part 1. *Crit Care Resusc* 7:250–257, 2005.

Trubuhovich, R.V. History of mouth-to-mouth rescue breathing Part 2. *Crit Care Resusc* 8:157–171, 2006.

Thomson, E.H. The role of physicians in the humane societies of the eighteenth century. *Bull Hist Med* 37:43–51, 1963.

Van Bortel, T. Climbing up the medical ladder, the surgeon-accoucheur Jan Bernard Jacobs (1734–1790). *Acta Chir Belg* 107:228–236, 2007.

Wynman, N. *Colymbetes sive de arte natandi dialogus*. Augustae Vindelicorum, excudebat Henricus Steyner, 1538.

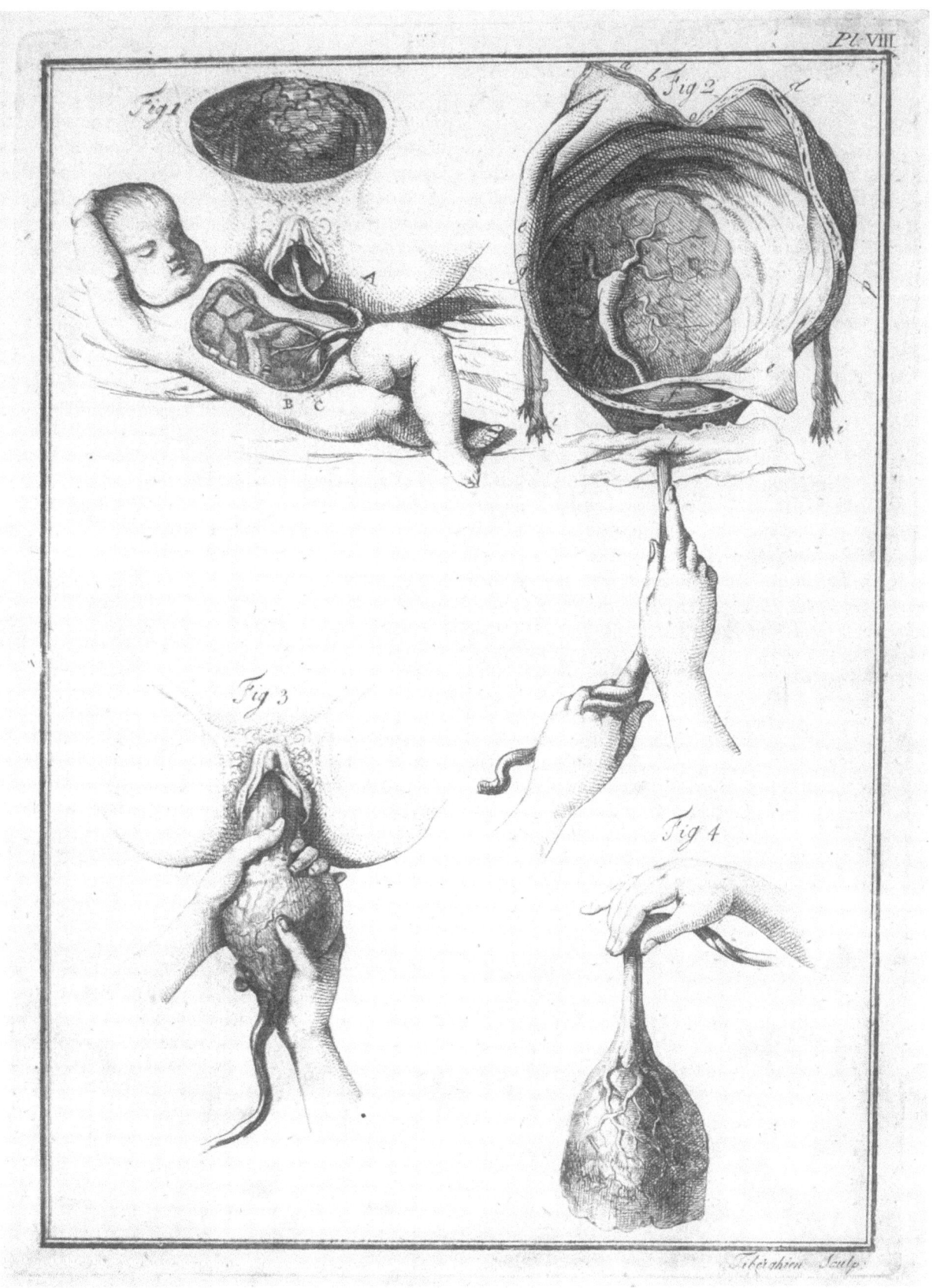

PLATE VIII FROM JACOBS, 1784

Thomas Denman

A collection of engravings, tending to illustrate the generation and parturition of animals, and of the human species. London, Sold by J. Johnson ..., 1787.

In his folio *Collection of engravings...* Thomas Denman (1733–1815) Physician-Accoucher at the Middlesex Hospital, London presented 15 plates by Johannes van Riemsdijk (also Jan van Riemsdyk; van Rymsdyk; fl. 1750–1788), who had prepared the plates for the great folios by William Smellie (Smellie 1754), Charles Nicholas Jenty (Jenty 1757), and William Hunter (Hunter 1774) (Thornton and Want 1979). Included is a lovely representation of the fetus at 3 months gestation, and a plate depicting rupture of the uterus. The 1815 edition of Denman's work included 17 plates (Denman 1815, Russell 1987).

One of Denman's chief claims to fame was his *magnum opus, An introduction to the practice of midwifery* published in two volumes (Denman 1794/1795). Appearing when he was 61 years of age, a wide range of experience from his years in practice and teaching were displayed in its pages. In a 42 page Preface, he surveyed some historical aspects of the field to his day. In the eight chapters of volume I Denman reviewed the bones of the pelvis, the organs of generation, menstruation, conception, diseases of pregnancy, and the course of normal labor. In the seven chapters of volume II (two are misnumbered), he considered difficult labor, the use of forceps and vectis, craniotomy of the fetal head, cesarean section, other aspects of complex labor, those labors attended by convulsions (probably eclampsia), multiple gestation, prolapse of the umbilical cord, and postpartum management of the mother. This work included several of his previously published essays (Denman 1794/1795). In discussing the third (1815) edition of this work in his *History of British Midwifery ...*, Herbert Ritchie Spencer (1860–1940) referred to Denman's volume as "... perhaps the most splendid work on midwifery in the English language, whether regarded from the point of view of the format, paper, printing and illustrations of the work; the learning and knowledge it exhibits; or the ordered, lucid, and judicial manner in which that knowledge is presented" (Spencer 1927, p. 138). During Denman's lifetime, five editions of his *Introduction to the practice of midwifery* were published.

Denman is credited with popularizing the induction of premature labor in patients with a small or distorted pelvis, as early as the seventh month of pregnancy. Although this practice was known, it was not widely performed (Dunn 1992). Following delivery, he also advocated delay in cutting the umbilical cord until its pulsations ceased. Overall, Denman practiced so-called "conservative" obstetrics, relying on the forces of nature, and cautioning against the use of forceps unless absolutely necessary to avoid new evils or aggravate those that may be existing. Denman also was opposed to cesarean section except in the case of a woman with extreme pelvic contraction (Denman 1783). His ultra conservatism probably influenced his son-in-law Sir Richard Croft (1762-1818), Surgeon-in-Ordinary to Princess Charlotte of Wales (1796–1817) and her husband Prince Leopold (1790–1865), whose refusal to expedite delivery led to the Princess' death in childbirth in 1817. (Because of remorse and a threat of public inquiry into the cause of the Princess' death, the following year Croft committed suicide).

After spending 8–10 years (different sources give different numbers), as a Naval ship's surgeon, Denman returned to London where at St. Georges Hospital he attended the midwifery lectures of William Smellie (1697–1763). In conjunction with a longtime friend, William Osborne (also Osborn; 1736–1808), he commenced lecturing on midwifery using manikins, skeletons, and other teaching aids. Continuing for a decade and a half, this helped to expand his influence and reputation. In 1783, following the death of William Hunter, Denman's influence was expanded further and he assumed the care of many of the "first families" of the city (Cutter and Viets 1964). That same year Denman was admitted as a Licentiate in Midwifery by the College of Physicians, one of the rare persons to achieve this distinction when the college first recognized midwifery as a legitimate form of medical practice. Denman wrote on many other aspects of obstetrics, including natural (Denman 1786c) and preternatural (Denman 1786b; 1787) labours. He is credited with first suggesting that puerperal fever was contagious, and may be caused by the attending man-midwife (Denman 1768). He also wrote on rupture of the uterus (Denman 1810), and described membranous dysmenorrhea (Denman 1794).

— — del!

C. Knight sculp.

London Publifh'd Dec.ʳ 22. 1783 by Dʳ. Thomas Denman.

FIGURE FROM DENMAN, 1787.
FETUS AT 3 MONTHS OF GESTATION.
COURTESY GOOGLE BOOKS

References

Cianfrani, T. *A short history of obstetrics and gynecology.* Springfield, IL, Thomas, 1960.

Cutter, I.S. and H.R. Viets. *A short history of midwifery.* Philadelphia, W.B. Saunders Company, 1964. Denman, T. *Essays on the puerperal fever, and on puerperal convulsions.* London, Printed for J. Walter ..., 1768.

Denman, T. *Aphorisms on the application and use of forceps: on praeternatural labours, and on labours attended with hemorrhage.* London, [s.n.], 1783.

Denman, T. *An essay on uterine hemorrhages depending on pregnancy and parturition.* London, Printed for J. Johnson ..., 1786a.

Denman, T. *An essay on preternatural labours.* London, J. Johnson, 1786b.

Denman, T. *An essay on natural labours.* London, Printed for J. Johnson, 1786c.

Denman, T. *An essay on difficult labours.* London, Printed for J. Johnson, 1787.

Denman, T. *An introduction to the practice of midwifery.* 2 vols. London, Printed for J. Johnson ..., 1794–1795.

Denman, T. *Observations on the rupture of the uterus, on the snuffles in infants, and on mania lactea.* London, J. Johnson, 1810a.

Denman, T. *Observations on the cure of cancer.* London, J. Johnson, 1810b.

Denman, T. *Engravings representing the generations of some animals; some circumstances attending parturition in the human species; and a few of the diseases to which the sex is liable.* London, Printed by T. Bensley for E. Cox and Son, 1815.

Dunn, P.M. Dr. Thomas Denman of London (1733–1815): rupture of the membranes and management of the cord. *Arch Dis Child* 67:882–884, 1992.

Munk, W. *The roll of the Royal College of Physicians of London, compiled from the annals of the College and from other authentic sources.* Vol. II. London, Longman, Green, Longman, and Roberts, 1856 Russell, K.F. *British anatomy, 1525–1800. A bibliography.* Melbourne, Melbourne University Press, 1963.

Russell, K.F. *British anatomy 1525–1800. A bibliography of works published in Britain, America and on the Continent.* 2nd Ed. Winchester, St. Paul's bibliographies, 1987.

Spencer HR. *History of British Midwifery, From 1650 to 1800. The Fitz-Patrick Lectures for 1927 delivered before the Royal College of Physicians of London.* London, J. Bale & Sons & Danielsson, Ltd., 1927.

Thoms, H. Classical contributions to obstetrics and gynecology. Springfield, IL, Charles C. Thomas, 1935.

Thornton, J.L. and P.C. Want. Jan van Rymsdyk's illustrations of the gravid uterus drawn for Hunter, Smellie, Jenty and Denman. *J Audiov Media Med* 2:11–15, 1979.

Wilson, A. *The making of man-midwifery: childbirth in England, 1660–1770.* Cambridge, MA, Harvard University Press, 1995.

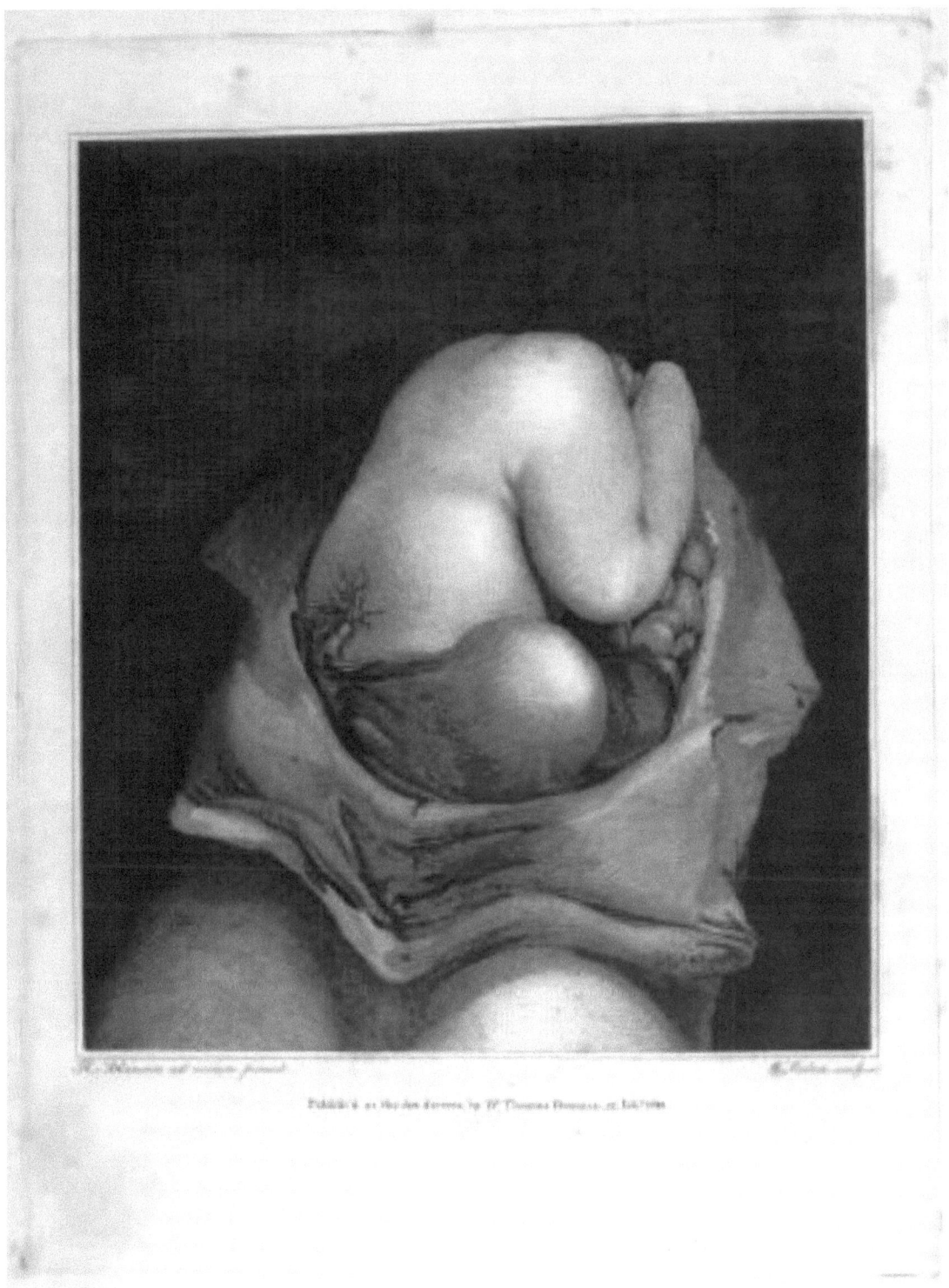

FIGURE FROM DENMAN, 1787.
RUPTURED UTERUS, WITH FETUS, SHOWN IN SITU. ANTERIOR ABDOMINAL WALL DIVIDED AND REFLECTED TO SHOW BODY OF FETUS IN THE ABDOMEN,
FETAL HEAD REMAINING WITHIN THE UTERUS. ANTERIOR VIEW.
COURTESY UNIVERSITY OF TORONTO, FISHER LIBRARY DIGITAL COLLECTIONS

Matthias Saxtorph

Nyeste Udtog af Fødsels-videnskaben, til brug for jordenmødrene ... Kiøbenhaven, Sælges i F. C. Pelts Boghandling paa Børsen ..., 1790.

In his "New excerpt for birth science for midwives..." the prominent eighteenth century Danish obstetrician Matthias Saxtorph (1740–1800) of Copenhagen detailed many aspects of the physiology of pregnancy and mechanics of labour. Saxtorph divided his study into 19 chapters dealing with several aspects of childbirth, including the mechanism of normal labor, premature birth, difficulties with labour, the birth of twins, and false labour. With a background in mathematics and physics, his contributions were noted for their thoroughness and clarity. A major contribution of this work is his description of the mechanism of labor, with the relation of the fetal head to the maternal pelvis during the course of its descent. The English man-midwife William Smellie (1697–1763) originally had described this mechanism (Smellie 1752), and Saxtorph helped to make it more widely understood in northern Europe. In a 1772 monograph, he also described these interrelations, noting that the fetal head enters the pelvic brim in the oblique rather than the transverse diameter.

In sixfolded, engraved plates, Saxtorph illustrated in detail various views of the fetal head and the maternal pelvic birth canal. Plate IV illustrates several of these, including: Fig. 1, the near-term fetus deep within the maternal pelvis; Figs. 2 and 3, the fetal head internally rotated immediately prior to delivery; Fig. 4, the maternal perineum as the head commences to emerge; and Fig. 5, a pair of dichorionic, fraternal (non identical) twins. Saxtorph's admiration for Smellie is reflected in many of his figures, which are essentially copies from Smellie's *Sett of anatomical tables...* (Smellie 1754), and much of the text of this work clearly is derived from Smellie's *A treatise ...* (Smellie 1752).

A graduate of the University of Copenhagen and, in 1785, a founding member of the *Societas Disputatoria Medical Hauniensis*, Saxtorph introduced the phrase *uterus reflexus* for retroversion of the uterus. Saxtorph also differentiated uterine prolapse from an inverted uterus.

References

Hirsch V p. 193; Blake p. 403.

Leishman, W. On the mechanism of parturition. *Glasgow Med J* 11:181–215, 276–309, and 432–459, 1863.

Saxtorph, M. *Theoria de diverso partu ob diversam capitis ad pelvim relationem mutuam experientia fundata et figures aeneis illustrata.* Havniae & Lipsiae, Apud Frid. Christ. Pelt ..., 1772.

Smellie, W. *A Treatise on the Theory and Practice of Midwifery.* London, D. Wilson 1752.

Smellie, W. *A Sett of Anatomical Tables, with Explanations, and an Abridgment of the Practice of Midwifery, with a view to illustrate a Treatise on that Subject and Collection of Cases.* London, 1754.

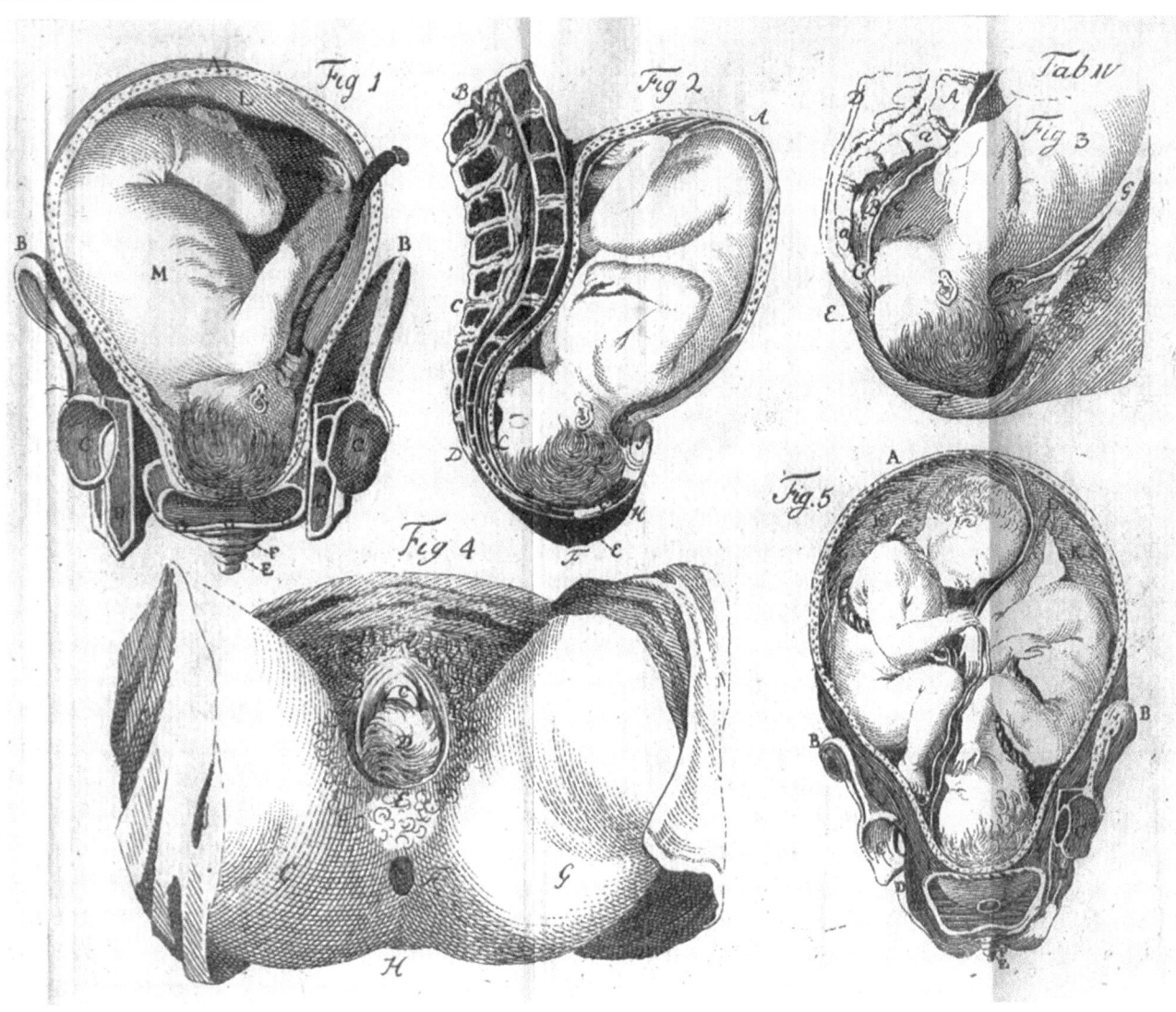

TAB. IV FROM SAXTORPH, 1790

Henrico [Henry] Krohn

Foetus extra uterum historia. Cum inductionibus quaestionibusque aliquot subnexis.... The history of a case of extrauterine conception. With inductions and queries .... Londini, Gulielmi Bulmer, 1791.

In this folio, Krohn (ca. 1736–1816) presented the case report of a woman in the seventh month of gestation, who died suddenly from hemorrhage in association with ovarian pregnancy. The patient was "upwards of 30 years of age,... [and] of a somewhat delicate constitution" (p. 1). In June 1790, she had been hospitalized for 2 weeks for "poor health with pain and chilliness" in association with early pregnancy. Krohn stated that upon admission to the Middlesex Hospital 2 months later, the patient believed she was in the seventh month of pregnancy, and complained of pain with fullness of the lower abdomen. She had a fever with rapid pulse. Physical examination revealed a large pelvic mass. However, because the usual signs of pregnancy were absent, Krohn believed that rather than being pregnant, the patient had a retroverted uterus. Ten days later, her condition worsened. The patient experienced severe pressure in her perineal area, her pulse was 150 per minute. Two days later she expired. Autopsy disclosed a large blood-filled sac that was separate from the uterus, containing a female fetus at 7 months gestation, lying transversely with its head on the left side (see illustration). The palcenta was "uncommonly large," and filled the cavity between the rectum and the sacrum. Kohn noted in particular that the vessels supplying the blood from the mother to the placenta "were so exceedingly small and few in number, as to excite very great surprise, how an adequate share of nourishment could be conveyed to enable the foetus to acquire such a size" (p. 5).

At the end of his case report, Krohn presented six Queries, among which were: to what extent similar cases occur in animals; has an ovarian pregnancy with a fetus of this size been reported previously; because such cases must invariably result in death of both mother and fetus, should not one consider surgery to extract the fetus; based on this case should not we believe that conception occurs in the ovary, with the impregnated ovum then moving to the uterus for "proper warmth and security;" should we not conclude that despite an unfavorable environment, "Nature... is enabled to accommodate the foetus to the local inconveniences of its situation," analogous to a plant that can "take root in a soil, which is ill adapted to its vegetation?;" and, finally, does not this case "indicate a propensity in Nature to promote fecundity," by affording the impregnated ovum or embryo by some "extraordinary resource" the means of nourishment (pp. 6–7).

The text of this volume was written in both Latin and English. The figures in four plates were drawn from the dissected body by Henry Bernard Chalon (1770–1849), and then engraved by Duterran. Each plate is accompanied by a figure in outline (in sepia) with labeling of the various specific parts.

Little is known of the life of Krohn, a native of Hamburg, who qualified in medicine at the University of Utrecht (1762). Shortly thereafter he moved to London, where he was appointed "physician man-midwife" at the Middlesex Hospital, and where he remained for nearly three decades. After retiring in 1798, he moved from London to Huntingdonshire.

References

Blake p. 248; Cutter and Viets p. 228.

Munk, W. *The roll of the Royal College of Physicians of London....* Vol II, 1701 to 1800. London, Published by the College, 1878.

Wilson, E. *The history of the Middlesex Hospital during the first century of its existence. Compiled from the Hospital records.* London, John Churchill, 1845, pp. 209–212.

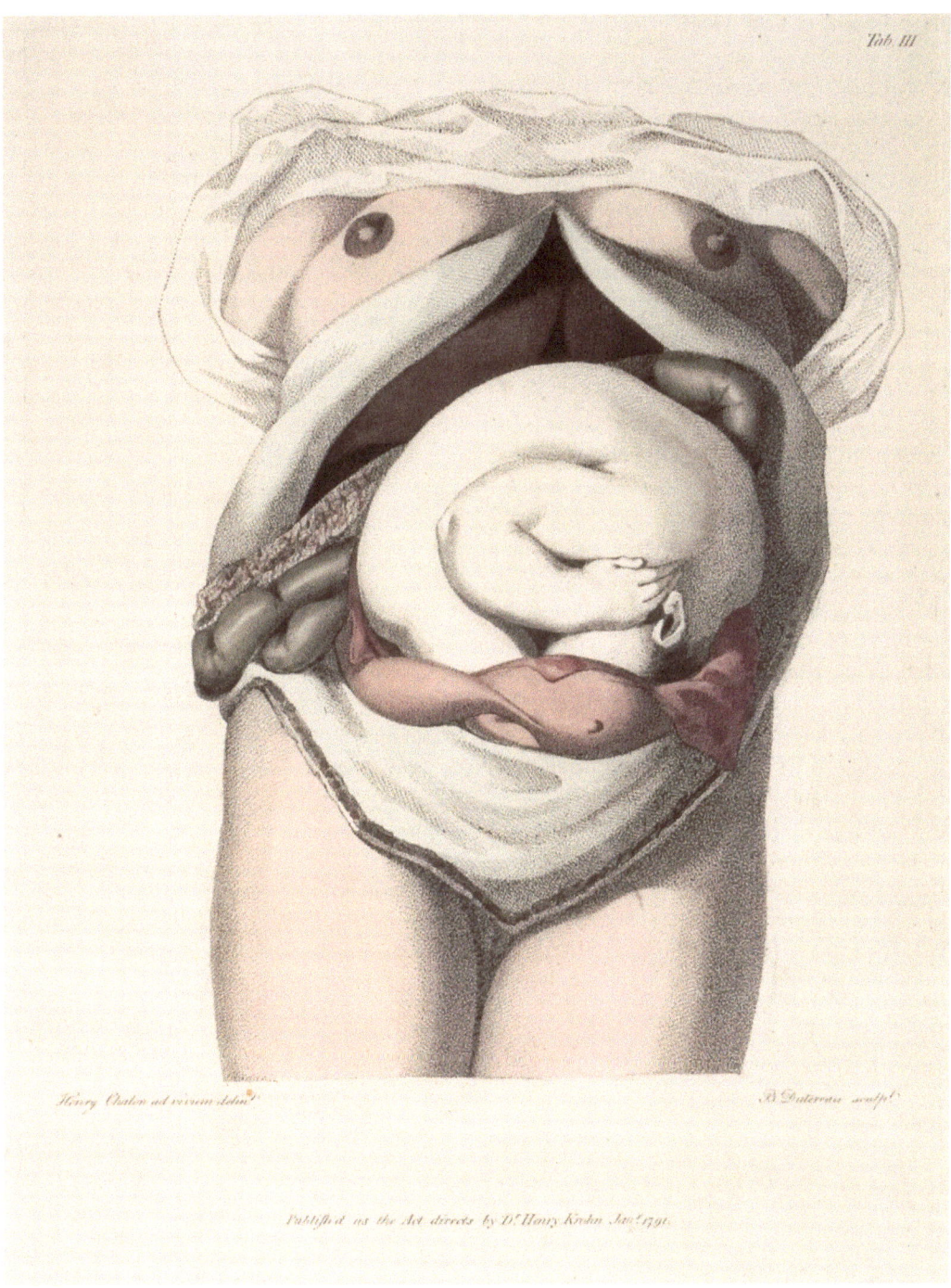

TAB. III FROM KROHN, 1791

Joseph Jacob Freiherr von Mohrenheim

Abhandlung über die Entbindungskunst. St. Petersburg, Kaiserliche Akademie der Wissenschaften, 1791.

In this sumptuous two volume work, "Treatise on Obstetrics/ Midwifery", Baron Mohrenheim (1759–1799) presented splendid life-sized engravings of the gravid uterus and its contents. Some of the plates are by the author, and are original. However, many were taken from William Smellie's, *A Sett of Anatomical Tables* (1754), and Roederer et al. (1697– 1763). They represent embryos, the foetus in utero, deliveries, instruments, and so forth. The engravings were by Russian artists. The work is divided into two sections: I. *Den leichten natuerlichen Geburten* [natural childbirth], and II. *Den schweren, widernatuerlichen, und gefahrlichen Geburten* [difficult, unnatural, and dangerous childbirth]. Volume I includes a brief literary history of obstetrics. Some of the plates are by the author, and many of the engravings are by Johann Christian Mayr (1764–1812) of St. Petersburg.

Copper plate XV illustrates the fetal (Fig. I) and maternal (Fig. II) surfaces of the near-term placenta. Figure I shows the smooth, flat, inner, fetal aspect with branches of the umbilical vessels (a), the severed umbilical cord (b), the membranes at the edge of the placenta (c), and the amniotic membrane (d). Figure II illustrates the maternal placental surface following its detachment from the uterine wall, with attached membranes and umbilical cord (a).

An Austrian by birth, Mohrenheim graduated from medicine in Vienna, where he practiced general and ophthalmologic surgery (Mohrenheim 1780–1783), and obstetrics. In 1783, Mohrenheim was summoned to St. Petersburg by Catherine II "The Great" (1729–1796) Empress of Russia, to be Professor of Surgery and Obstetrics, and to serve as court physician as well as he *accoucheur*. Born a German Princess, in addition to her wide influence in Russian political administration, the Empress was responsible for several innovations in "westernizing" Russian life and medicine, including the practice of obstetrics. Catherine commissioned the present volume, the second Russian book on the subject, underwriting the considerable expense of its production and publication. Because of its rarity, this work has escaped the notice of many historians. Mohrenheim is eponymized in Mohrenheim's fossa, a triangular space between the pectoralis major and deltoid muscles beneath the clavicle.

References

Blake p. 307; Garrison Morton 6161; Cutter and Viets pp. 228–229.

Mohrenheim, J.J. *Beobachtungen verschiedener chirurgischer Vorfälle*. Vienna, R. Graffer, 1780 (vol. 2: Dessau: Kosten, 1783).

Mohrenheim, J.J. *Dissertatio inauguralis medica sistens novam conceptionis atque generationis theoriam....* Königsberg, Regiomonti, Typis G.L. Hartungii, 1789.

Mohrenheim, J.J. *Ueber die Krankheiten der Solönen.* Berlin, 1789.

Mohrenheim, J.J. *Abhandlung über die Entbündungskunst... Erster Theil. Von der Entbindungskunst uberhaupt, und insbesondere von den leichten naturlichen Geburten....* Leipzig, H. Gräff, 1803.

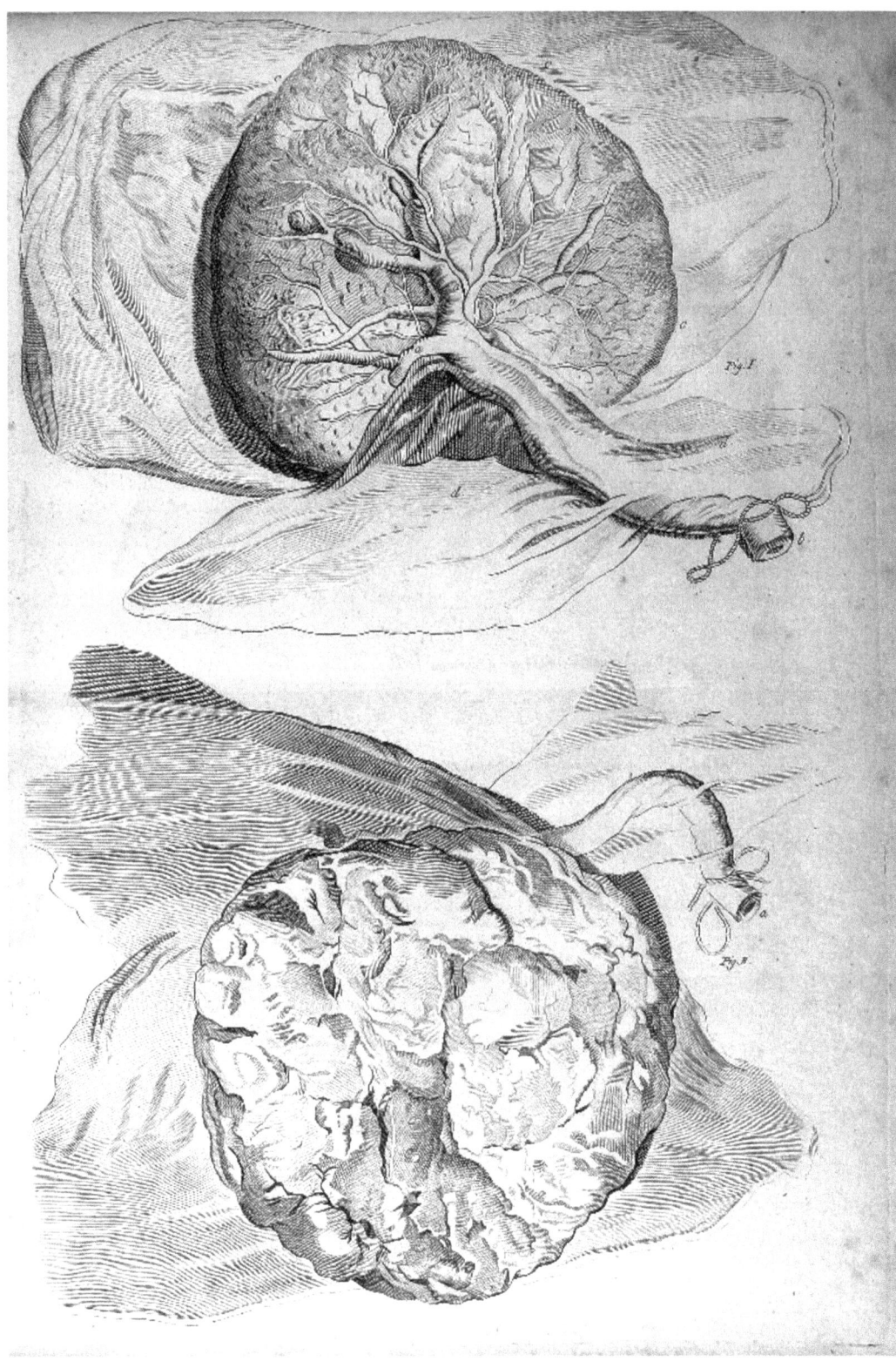

PLATE XV FROM MOHRENHEIM, 1791

[Samuel William Fores]

Man-midwifery dissected: or, The obstetric family-instructor; for the use of married couples, and single adults of both sexes …. London, Published for the author by S.W. Fores…, 1793.

Before the eighteenth century, almost all infants were delivered by midwives, although in difficult cases a doctor/surgeon might be called in to assist. These midwives, who usually were older and had children, often were familiar with the parturient, and brought considerable comfort and support to the mother and her family. Early in the eighteenth century, following disclosure of the obstetrical forceps invented by the Chamberlens, the employment of male midwives gained increasing acceptance. Nonetheless, the usurping by men of what previously had been the province of women was much criticized by many. Man-midwives were accused of indiscriminate use of the forceps, and many women objected to the presence of men at the time of delivery, as did their families. This frontispiece illustrates the controversy.

The author and publisher Fores (1761–1838), who used the pseudonym John Blunt, was a native of Scotland. Fores was, in fact, a dealer in prints and books, with his shop in Piccadilly in London's Westend. Here, he established himself as "Caricaturist to the First Counsel" (a group of lawyers giving professional advice). Isaac Cruikshank (ca. 1756–1811), the contemporary self-taught caricaturist (not to be confused with his more distinguished sons George (1792–1878) and Robert (1789–1856)), prepared the frontispiece. To the right is the midwife who offers a feeding cup and is featured in a domestic setting. In contrast, the man-midwife to the left holds forceps (lever) and on the shelves are more horrific instruments as well as love potions, allegedly to stimulate sexual desire in patients, "for my own use." The contrast and conclusions are clear. The text on the lower portion of the plate reads, "A man-*mid*-wife or a newly discovered animal, not known in Buffon's time; for a more description of this *Monster*, see, an ingenious book … Man-midwifery dissected, containing a variety of well authenticated cases elucidating this animals Propensities to cruelty and indecency …."

Fores described himself as, "a student under different teachers, but not a practitioner of the art." This statement may imply that he studied medicine for a year or more. In his 14 "letters" addressed to Alexander Hamilton (1739–1802), an influential professor of midwifery in Edinburgh (see Hamilton 1775, 1780, 1784), Fores argued that male midwives perpetuated "mischief done by ignorant and cruel male operators, and likewise of the new-fangled obstetric butchery lately invented at Paris" (p. xiii).

Fores also criticized "Those pusillanimous husbands who feel themselves overborne by custom, and cannot muster up resolution enough to protect their wives' persons from injury and insult, [who] may be compared to a captain who quits his ship … out of compliment to his pilot" (p. xv). He asserted that the husband should remain with his wife to insure against *accoucheurs* "making too free with women's persons, manually, ocularly, and instrumentally" (p. 143). He argued of great mischief being done, "owing to the ignorance and impatience of those professors who erroneously imagined, their instruments *must* be used on *all* occasions, whether the labours were natural or difficult" (p. viii). Fores enforced this argument, stating that the decision to use obstetrical forceps was "for the convenience and profit for the operator, rather than the comfort and safety of mother and child" (p. 57). He asserted further that not only were men unnecessary and their practice "indecent and cruel," but that "much more mischief has been done by the instruments of *skillful* men, than by the hands, or by the omissions of *ignorant* women; and I shall also take notice of the ignorance of *men* midwives, and *their* blunders" (p. 52).

In the first of his 14 letters to Hamilton, Fores outlined six objectives of this essay; e.g., to describe the anatomy of the pelvis and its contents, classify labor, to inquire whether Hamilton's instruction was "consistent with decency," demonstrate the effects of such instruction, ask in what manner the practice of male-midwives was consistent with safety, and explore reasons for the relative lack of "good midwives and present a plan for their instruction." In conjunction with the last of these objectives, he set out to "prove that man-midwifery is a personal, a domestic, and a national evil" (p. 14).

In arguing this thesis, Fores considered the ignorance and cruelty of some accoucheurs, and the impropriety of teaching midwifery to men. As an example, he recited the instance of a male midwifery student "caught in bed with his patient the day before her delivery" (p. 87). In letter XIV to Hamilton, Fores explored the sixth of his propositions, that of a plan to improve midwives' education so that they might become better qualified. He proposed establishment of a midwifery school "as near the centre of London as possible" (p. 182) to teach anatomy, and held that midwives be required to pass an examination for certification. To avoid conflict of interest by having male midwives do any of the teaching, he advocated "The gentleman employed to deliver these lectures shall not be a man-midwife by profession, lest his *own interest* should cause him to with-hold *necessary* instructions from the female pupils" (p. 185). With the proper education of midwives, Fores predicted that "the great *obstetric idol*, MAN, will fall to the ground by its own ponderous weight in a few years" (p. 187). Many of his arguments were along the line of those put forth three decades previously by the English midwife Elizabeth Nihell (Nihell 1760).

To a certain extent seeds of this controversy continue to the present, although in Europe and many other parts of the world midwives have retained their important role in society. For instance in Britain, state certification was introduced in 1902, and their status further enhanced by the 1936 Midwives

FRONTISPIECE FROM FORES, 1793

Act and the establishment of the Royal College of Midwives in 1947. In the state of California practicing midwives are classified as "lay," those who lack formal training and who practice without a license; "licensed," those who have graduated from a 3-year training program, provide prenatal care, and deliver low-risk pregnant women in their home; and "certified nurse midwives," those who are registered nurses, have several years additional training in the care of pregnant women, and are licensed to practice in a hospital setting under protocols established by licensed physicians, and who are under the authority of the California Board of Nursing.

References

Blake p. 151.

Aveling, J.H. *English midwives: their history and prospects.* London, J.A. Churchill, 1872.

Emery, A.E.H. and M.L.H. Emery. *Mother and child care in art. Foreword by Sir John Hanson.* London, Royal Society of Medicine Press, 2007, pp. 42–43.

Hamilton, A. *Elements of the practice of midwifery.* London, Printed for J. Murray, 1775. (Blake, pp. 196–197).

Folkert [Volkert] Snip

Vroedkundige Aanmerkingen en Afbeeling eener Bezwangerde Baarmaeder. Amsterdam, J.B. Elwe, 1793.

Snip (1733–1771), a pupil of Pieter Camper (1722–1789) and graduate of the University of Gröningen, The Netherlands, became professor of anatomy and surgery at the University of Amsterdam. His atlas, "Obstetric observations and illustration of a pregnant womb," a life-size, elephant folio illustrated by Maarten Petersz Houttuyn (1770–1798), a physician and naturalist, was published posthumously. The four plates, dated 1767, show dissections of the pregnant uterus, each from a different perspective.

Plate II (illustrated) shows the fetus encased in the placental membranes. Other plates show the fetus in cephalic presentation with tightened umbilical cord wrapped around its neck (III, next page) and depict the placenta attached to the posterior uterine wall after the removal of the fetus (IV; two pages hence). Snip commissioned the engraver Frans de Bakker (fl 1736–1767) to engrave the drawings on copper plates.

Originally, Houttuyn had prepared a fifth plate. Because Snip was dissatisfied with this plate, he and the anticipated publisher quarreled and the project was abandoned. Later, the engraved plates and manuscript text were acquired by the publisher Jan Barend Elwe (1746–1816), who prepared the work for publication with the help of Snip's pupil David van Gesscher (1736–1810).

References

Blake p. 422.

Bower, F. Early Dutch obstetric atlas, *Austral & New Zealand J Obstet Gynaec* 44:86–87, 2004.

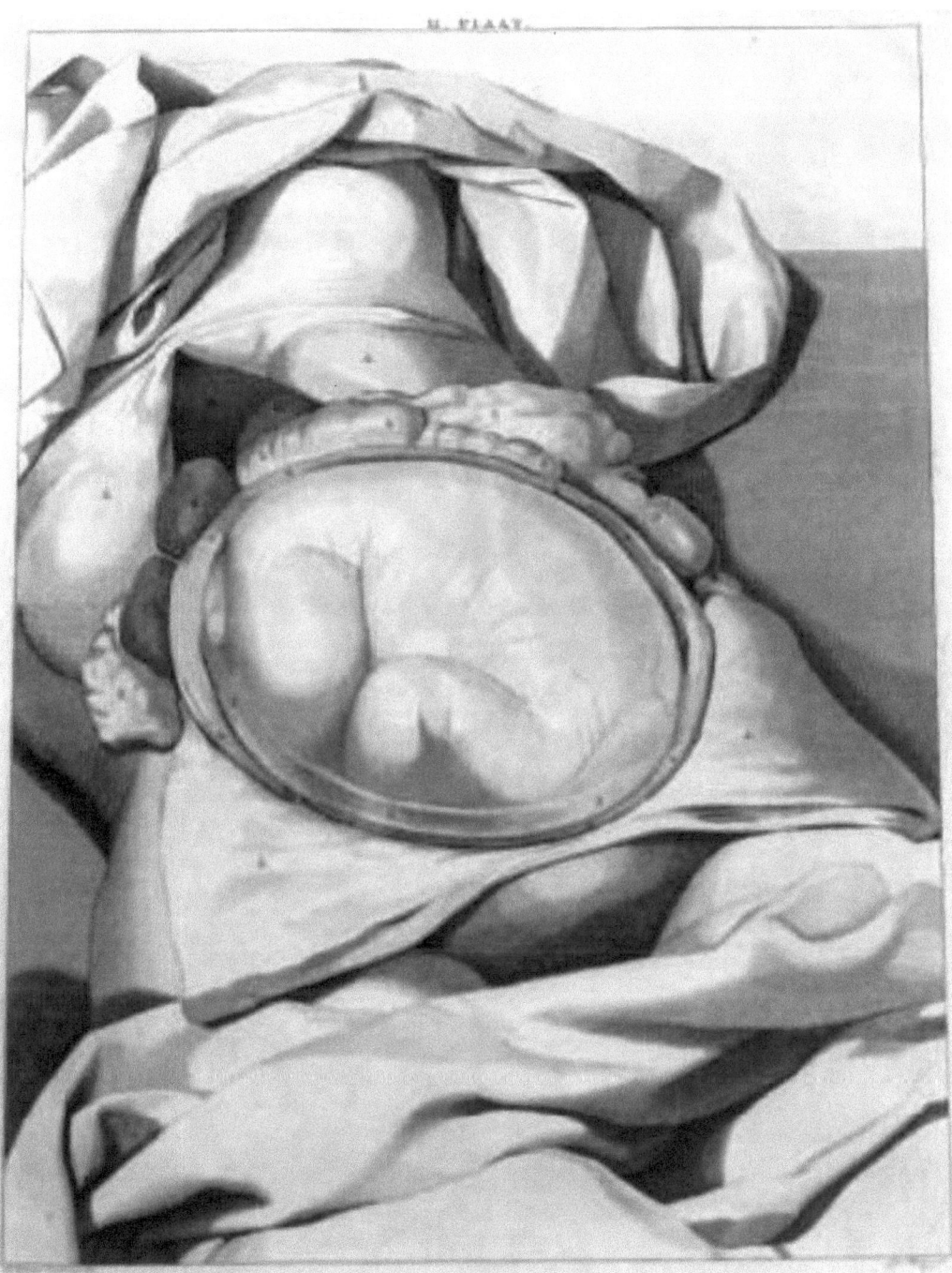

PLATE II FROM SNIP, 1793

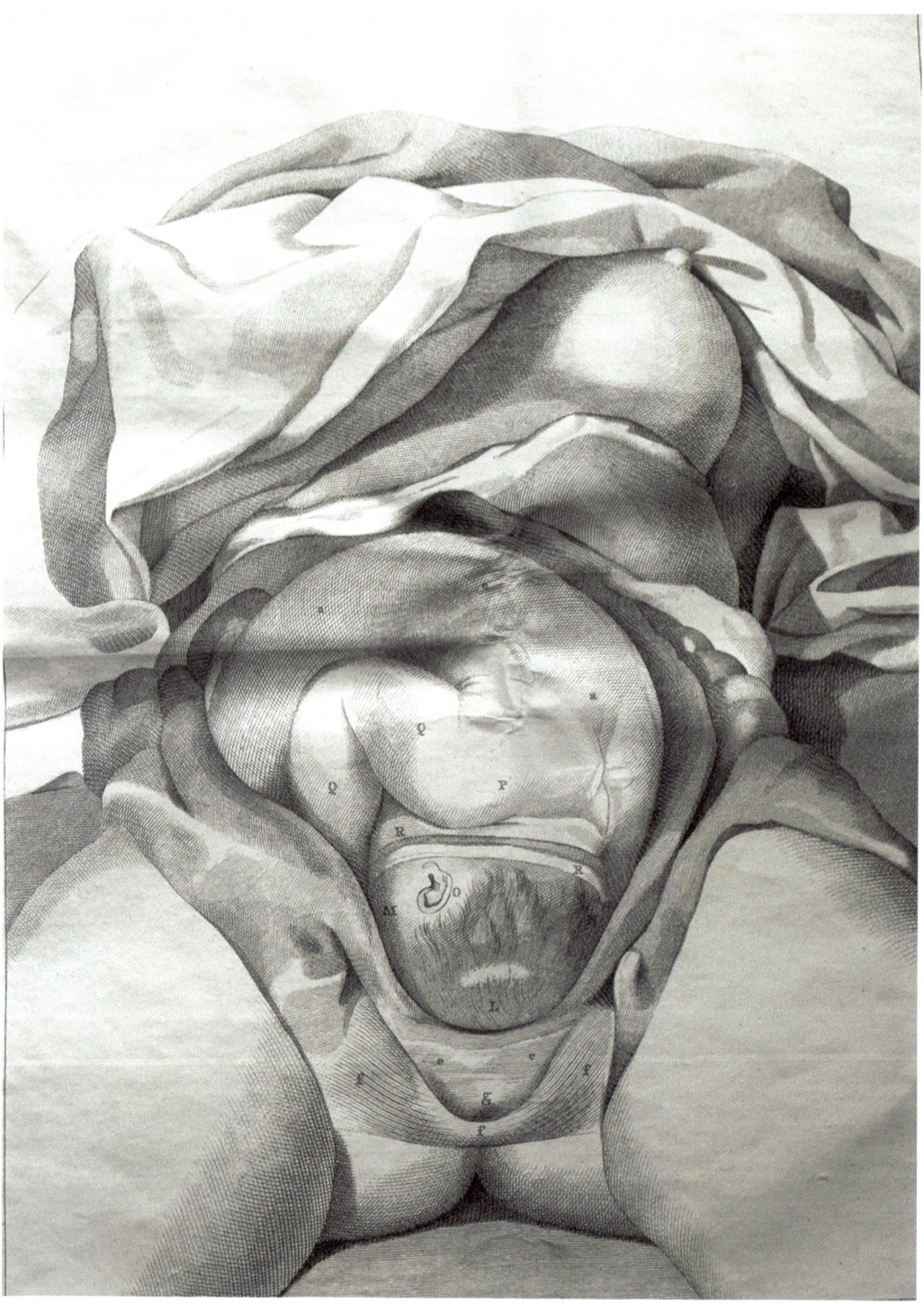

PLATE III FROM SNIP, 1793

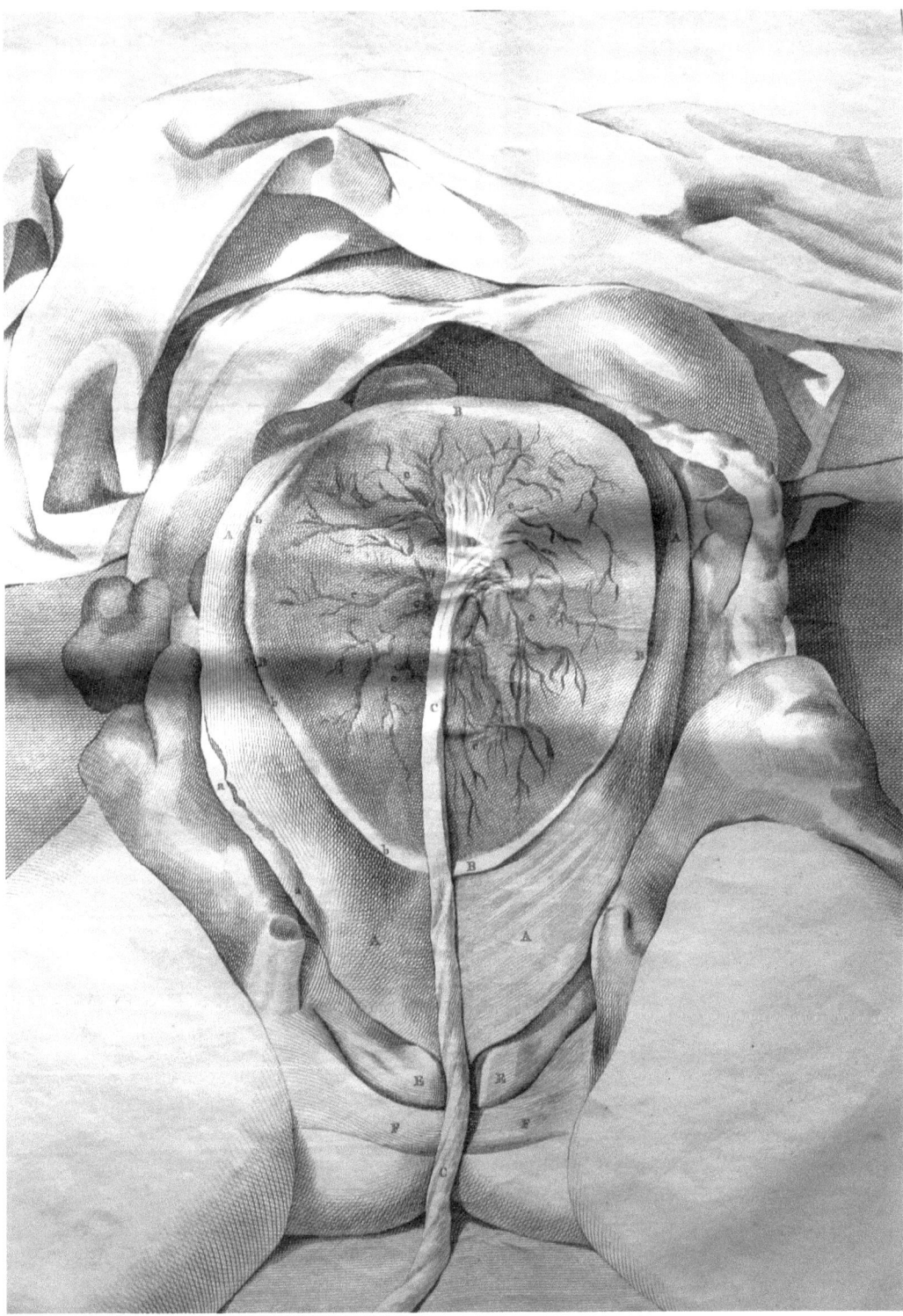

PLATE IV FROM SNIP, 1793

Additional Author(s) of Significance in the Eighteenth Century

Frederik Ruysch [Ruijsch]

Thesaurus anatomicus, i-x. Amstelaedami, J. Wolters, 1701–1716.

Ruysch (1638–1731), professor of anatomy at the Universities of Leiden and Amsterdam. A native of Den Hague, Ruysch originally apprenticed as an apothecary and had his own shop. He then took medicine, graduating at the University of Leiden in 1664. Two years later, he became *praelector anatomiae* [reader in anatomy] at the Guild of Surgeons in Amsterdam. In addition to teaching anatomy to the surgeons, he performed public dissections during the winter. Following the death of Hendrik van Roonhuyze (1622–1672), Ruysch was appointed city obstetrician, presenting lectures and demonstrations to the midwives in an effort to improve their skills. He also became physician to the Court of Justice (1679), and professor of botany and supervisor of the botanical garden (1685).

Primarily, however, Ruysch considered himself an anatomist. He developed a unique technique of injection into blood vessels of a compound(s) that solidified. This allowed preservation of his specimens for many years. Ruysch's *Wunderkammern* [cabinet of anatomical specimens] attracted a number of visitors, and in 1691 he published a catalogue of the *Musaeum Ruijschianum Anatomicum*. By 1710 the museum displayed more than 1,300 anatomical preparations. When Peter I, "the Great" Czar of Russia (1672–1725), visited the Netherlands in 1717 on a collecting trip, he purchased Ruysch's collection (as well as that of Albertus Seba (1665–1736) (Baljet and Oostra 1998, Franke 1971, Seba 1734–1765). Despite his advanced years (79), Ruysch then commenced amassing a new collection, some specimens of which are in the Anatomical Museum of the Faculty of Medicine, University of Leiden. In addition to his unrivaled preparations of so called "Ruijschian Art", he made a number of important anatomical discoveries, including description of the arterial vasculature of the bronchial tree and the heart, a rich capillary plexus, the *lamina choriocapillaris*, an inner thin tissue layer behind the retina of the eye, "Ruijsch's membrane", and the circular musculature of the uterine fundus "Ruijsch's muscle". He also first described valves in the lymphatic vessels (Ruysch 1665). Ruysch became a Fellow of the Royal Society of London (1720), the Academie des Sciences in Paris (1727), and received other honors. The 1683 painting "The Anatomy Lesson of Frederik Ruijsch" by Jan van Neck (1635–1714) is displayed in the Historical Museum of Amsterdam.

References

Blake 395; Garrison Morton 389.

Baljet, B. Frederik Ruijsch (1638–1739?), obstetrician, gynecologist and examiner of midwives. In: *Historical perspectives of obstetrics and gynaecology in the low countries*. Houtzager, H.L., Lammes, F.B., Eds. Zeist, Medical Forum International, 1997, pp. 37–49.

Baljet, B. & R-J. Oostra. Historical aspects of the study of malformations in the Netherlands. *Am J Med Genetics* 77: 91–99, 1998.

Cole, F.J. *A history of comparative anatomy: from Aristotle to the Eighteenth Century.* London, McMillan, 1944.

Franke, H. Peter der Grosse und die anatomo-pathologische Präparate sammlung von F. Ruijsch. *Med Wochenschr* 133: 488–491, 1971.

Lindeboom, G.A. Frederik Ruysch. In: *Dictionary of Scientific Biography*. Vol XII. Charles Coulston Gillispie (Ed.). New York, Charles Scribner's Sons, 1975, pp. 39–42.

Ruijsch, F. *Observationem Anatomico-Chirurgicorum Centuria*. Amsterdam, Müller, 1691.

Ruijsch, F. *Opera omnia anatomico-medico-chirurgica*. Amsterdam, Janssoons van Waesberge, 1724. Ruysch, F. *Dilucidatio valvularum in vasis lymphaticis et lacteis*. Hagae-Comitiae, ex officina H. Gael, 1665. (GM 1099).

Guillaume Mauquest de La Motte

Triaté complet des accouchemens. Paris, Chez Laurent d'Houry, 1721.

In contrast to many of the *accoucheurs* of early eighteenth century France, Mauquest de La Motte (1655–1737) was not associated with any of the Paris hospitals. In his treatise, which was one of the most important obstetrical works of the early eighteenth century, Mauquest de La Motte presented lucid descriptions of clinical cases from contemporary obstetrical practice of more than three decades (1690–1720) in Valognes, a village in Normandy. He appended comments or "reflections" to almost all of his 411 cases, noting, "I have set down my thoughts and observations in the best manner I could, having less pretence to learning then experience. I hope [trust] that this confession will not make me loose the esteem of my reader, but will engage him to mind the subject itself before its regularity, or the choice of words". The frankness and candor of Mauquest de La Motte's descriptions, their detail, the record of failures as well as successes, stamp this work as one of the most valuable contributions to the science of midwifery of the period. Although he was a disciple of Hendrik van Deventer (1651–

1724) and François Mauriceau (1637–1709), he developed his own views. For instance, he challenged Deventer's idea that the coccyx was an important bone in obstructing labor, establishing priority for his conclusion that rather it was the pelvic inlet often obstructed descent of the fetus within the pelvis.

An example of his practical turn of mind is seen in one case (Observation 321) in which a large femoral hernia threatened to interfere with labor. Mauquest de La Motte placed the woman in the "Trendelenburg" (head down) position (which was not yet described) using gravity to reduce the hernia until the child was delivered. "When she got up I persuaded her to wear a truss, which prevented its coming down, and made her life much easier". He observed further: "A proper situation is certainly of great service in a difficult and lingering labor, but when the pains are brisk, and the child strong and vigorous, a woman would be brought to bed, tho' her head hung down, and her feet were up".

In another case (Observation 223), he described a patient whose labor was progressing unusually slowly, when he discovered that although the bladder was full, urine could not be passed voluntarily. Mauquest de La Motte introduced a catheter but this met with resistance. He then pushed the infant's head up, "as high as I could, which was no sooner done, than the water gushed forth in such quantity that it is hardly credible how the bladder could distend so much without breaking. This gave her ease immediately, and she enjoyed her health to her *delivery*, having shown her how to ease herself".

He also described many cases of eclampsia, and although he made no pretense of understanding its etiology, he advised delivery as promptly as possible after first emptying the bladder and rectum. Mauquest de La Motte disagreed with both Mauriceau and Ambrose Pare (1510–1590) who followed the Hippocratic writers in believing that an infant born during the seventh month had a better chance of survival than one born during the eighth month. He wrote "I have proved by my practice that children born at 7 and at 8 months may live, but better at the latter than at the former". On the other hand, he advanced the theory that the fetus changed its position in-utero "each time that he feels the need to do so", and that the "spirals of the cord have a great influence on the change of position, or, on the other hand, on maintaining it". Mauquest de La opposed the ancient theory that menstruation is regulated by the moon, and ardently advocated podalic version and extraction, particularly in instances of prolapsed umbilical cord, placenta previa, transverse lie, and contracted pelvis. In the management of twin pregnancy, he extracted the second twin without delay by version if necessary, and although he never performed cesarean section in a living patient, he admitted of possible circumstances that would justify the procedure.

Mauquest de La Motte recalled the horror the rough and inhuman practice of the unskilled among the profession, who by use of a *crotchet* and instruments to dismember the child, had the pregnancy terminated with a dead mother as well as infant. "I have always conformed myself to nature, which sometimes by a sudden happy change, brings to good issue a labor that was just before desperate; the contrary too often happens, and a labor that gave the best prospect in its beginning, may prove to last a very laborious one, nothing being so variable and uncertain as labors."

Regarding the Chamberlen family of England retaining the secret of the obstetrical forceps for a century or more, Mauquest de La Motte commented: "he who keeps secret so beneficial an instrument as the harmless obstetrical forceps deserves to have a worm devour his vitals for all eternity" (Levert 1753).

The work abounds with euphemisms such as: "I delivered as quickly and easily as taking a handkerchief from my pocket… the child escaped from the vulva as readily as an eel slips through your hands… in less time than needed to recite a *Pater* and an *Ave*".

Mauquest de La Motte's work was recognized quickly and several editions and translations appeared during the following half century. In the preface to the English translation, Thomas Tompkyns (_____) states that he undertook the work at the suggestion of William Smellie (1698–1763), "…a gentleman, who is not satisfied with being serviceable to mankind by his own labors, but with indefatigable industry studies to enable others to be as serviceable as himself, and… whose excellent lectures… will soon cause France to cease being our rival in this branch of surgery, as it has long ceased being so in all the other branches of it".

In his essay on Mauquest de La Motte, Theophilus Parvin (1829–1898) concluded "La Motte's *Traite des accouchemens* is one of the professional treasures the well educated physician will not neglect. This work is a monument to the industry, the knowledge and skill of a practitioner who for more than half a century, in a comparatively obscure part of the country, faithfully toiled, not to get riches or fame, or to secure a place in hospital or college, but for the glory of the Creator and the relief of man's estate" (Parvin 1892, p. 18).

References

Blake p. 293; Cutter & Viets, p. 197; Garrison Morton 6150.

Cumston, C.G. Mauquest de La Motte and his treatise on obstetrics. *Am J Obstet Dis Women Child* 52: 508–526, 1905.

Mauquest de La Motte, G. *Dissertation sur la generation, sur la superfetation, et la response au livre intitule de l'indecence aux homes d'accoucher les femmes, et sur l'obligation aux meres de nourrir leurs enfans de leur proper lait*. Paris, 1718.

Mauquest de La Motte, G. *Vollkommener Tractat von Kranckheiten Schwangerer und Gebährender Weibs-Persohnen, in welchem gehandelt wird, wie denenselben so wohl bey natürlicher als nicht natürlicher Gebährung beyzuspringen... translated from the French by J.G. Sheid.* Strasburg, Johannes Beck, 1732–1734.

Mauquest de La Motte, G. *A general treatise of midwifery: illustrated with upwards of four hundred curious observations and reflexions concerning that art... Translated by Thomas Tomkyns.....* London, James Waugh, 1746.

Parvin, T. A famous country obstetrician two centuries ago. Lecture delivered at the Philadelphia Hospital. *Phila Hosp Reports* pp. 1–18, 1892.

Albrecht von Haller

Sur la formation du coeur dans le poulet...2 vols. Lausanne, Bousquet, 1758.

Among his many contributions to anatomy and embryology, von Haller (1708–1777) devised a numerical method to describe the rate of fetal growth. He demonstrated the relatively rapid growth of the embryo/fetus during the first trimester, with the rate slowing in the latter part of gestation. He also calculated the rates of growth of both the chick and human embryos.

Haller, a native of Berne, commenced the study of medicine at Tübingen, and then transferred to Leiden where he studied under Herman Boerhaave (1668–1738) and Bernhard Siegfried Albinus (1697–1770). Following his graduation in 1727, he spent a year visiting academic centers in England, France, and Germany. Soon thereafter, he made an alpine tour and commenced the botanical collection that was to be the basis of his massive work on Swiss flora (Haller 1768b). Following a decade of medical practice and anatomical studies, he became professor at the new University of Göttingen where he remained for almost two decades. In 1753, he returned to Berne where he entered a life of combined academic studies and public service.

A major figure in eighteenth century medicine, Haller worked on a number of problems in anatomy and physiology. These include contractility of the heart, respiration, and function of the nervous system. He considered anatomy and physiology a unit, *anatomia animate*. By comparing many anatomical specimens and tabulating the frequency of variants, he established the principle of anatomic "norm." Haller's investigations of congenital malformations and "monstrosities" led him to acknowledge the influence of external factors such as disease of the developing fetus. As opposed to the contemporary view of epigenetics, that embryonic development involved formation of new structures, Haller's investigations convinced him of the erroneous "preformation" theory; e.g., that the organism was already formed in the egg or the sperm. In a series of letters and publications, he disputed the views of Caspar Friedrich Wolff (1733–1794) and others regarding epigenesis (see Wolff 1759). Rather, Haller claimed that development is a process of preformed folded structures becoming visible as fluids are pumped through them by the beating heart, itself initially invisible. Haller also believed, incorrectly, that in the chick the membranes that surround the yolk were continuous with its intestinal membranes, so that the embryo existed in a miniscule invisible state, connected to the yolk sac, prior to the beginning of development.

In addition to other accomplishments, Haller made numerous contributions to the new science of bibliography in the fields of botany, fields of anatomy, medicine, and surgery (Haller 1772, 1775, 1777, 1776–1788).

References

Blake 196; Garrison Morton 469.2.

Haller, A.v. *Icones anatomicae.* 8 pts. Göttingae, A. Vandenboeck, 1743–1756. (GM 397).

Haller, A.v. *Elementa physiologiae corporis humani.* 8 vols. Lausannae, Sumptibus Marci-Michael. Bousquet, 1757–1766. (GM 588).

Haller, A.v. *Operum anatomici argumenti minorum tomus tertius de monstris.* Lausanne, François Grasset, 1768a. (GM 534.54).

Haller, A.v. *Historia stirpium indigenarum Helvetiae inchoata....* 2 vols. Bernae, Sumptibus Societatis Typographicae, 1768b.

Haller, A.v. *Bibliotheca botanica.* 2 vols. Tiguri, apud Orell, Gessner, Fuessli, et al., 1771–1772. (GM 1833).

Haller, A.v. *Bibliotheca chirurgia.* 2 vols. Bernae & Basiteae, Haller & Schweighauser, 1774–1775. (GM 5789).

Haller, A.v. *Bibliotheca anatomica.* 2 vols. Tiguri, apud Orell, Gessner, Fuessli et al., 1774–1777. (GM438).

Haller, A. v. *Bibliotheca medicinae practicae.* 4 vols. Basle, J. Schweighauser, Berne, E. Haller, 1776–1788. (GM 6747).

Hintzsche, E. (Victor) Albrecht von Haller. In: *Dictionary of Scientific Biography.* Volume VI. Charles Coulston Gillispie (Ed.). New York, Charles Scribner's Sons, 1972, pp. 61–67.

Wolff, C.F. *Theoria generationis.* Halae ad Salam, lit. Hendelianis, 1759. (GM 470).

Caspar Friedrich Wolff

Theoria generationis. Halae ad Salam, lit. Hendelianis, 1759.

By his careful observations of early embryonic development in his "Theories of Generation," Wolff (1733–1794) described the formation of organs from leaf-like blastodermic or germ layers into their mature form. He thus revived and supported William Harvey's doctrine of epigenesis (1651), the gradual building up of parts from undifferentiated cells. Wolff's studies in both plants and animals established the nature of early development, as opposed to the erroneous idea of preformation in which the embryo was a well formed *homunculus* [little man] encased in the sperm. In opposition to the preformist theory of his contemporary Albrecht von Haller (1708–1777) (von Haller 1758), Wolff argued that, if in the embryo the various organs exist in their mature form and shape, they should be visible with a high-power microscope at early stages of growth. On the contrary, with development of the embryo one sees a series of advancing structures, each differing from its previous appearance. He used as an example, development of the vasculature, tracing the formation of "blood islets", and development of the heart and blood vessels in the embryonic chicken. Lacking detailed knowledge of cellular structure, he asserted that growth and development are a consequence of *vis essentialis* [an essential force]. Several years later, he demonstrated that the chick intestines develop from a ventral fold into a closed tube of flat membrane. He applied his observations on the folding and fusion of tissues to the theory of epigenesis and to the embryological development of other organs (Wolff 1768). Wolff thus helped lay to rest the preformation theory, and became a founder of observational embryology. Joseph Needham (1900-1995) recorded, "…the facts brought forward by Wolff have never been contradicted, but have been used as a foundation to which numberless morphological embryologists have added facts discovered by themselves" (Needham 1934).

This work, his graduation thesis from the University of Halle (1759), consists of three parts devoted to: the development of plants, the development of animals, and theoretical considerations. Because of his then considered heretical views, he failed to obtain a teaching post in Germany. Following service as a field doctor in the Prussian army, which was at war with Russia, in 1767, at the invitation of Empress Catherine II, "the Great" (1729–1796), Wolff moved to St. Petersberg to become head of the department of anatomy at the St. Petersberg Academy of Sciences (now Russian Academy of Sciences).

Wolff is eponymically remembered for his discovery of the "Wolffian bodies," (the elongated abdominal masses that comprise the mesonephros or primitive kidney), and their excretory ducts, the "Wolffian ducts," which he discovered in embryos of the chick (see Rathke 1832–1833). These paired organs are found during embryogenesis in mammals including humans. The duct connects the primitive kidney or Wolffian body (or mesonephros) to the cloaca, and serves as the anlage for certain male reproductive organs. In the male, it develops into a system of connected organs between the testis and the prostate, namely the *rete testis*, the efferent ducts, the epididymis, the vas deferens, the seminal vesicle, and the prostate. For this development to proceed in a normal manner, it is critical that the ducts are exposed to testosterone during embryogenesis. In the mature male, the function of this system is to store and mature sperm, and provide accessory semenal fluid. In the female, in the absence of testosterone support, the Wolffian ducts fail to develop and regress. In the early nineteenth century, the comparative anatomist Johann Friedrich Meckel the younger (1781–1833) translated many of Wolff's reports into German, and revived interest in his work.

Wolff's surviving manuscripts include a treatise on the "theory of monsters," in which he attempted to synthesize his ideas on epigenesis. Unfortunately, his sudden, untimely death from a massive stroke prevented completion of this project.

References

Blake p 494 is later edition; Garrison Morton 470; Needham p. 197 ff.

Dye, F.J. *Dictionary of Developmental Biology and Embryology*. New York, Wiley-Liss, 2002.

Gaissinovitch, A.E. Caspar Friedrich Wolff. In: *Dictionary of Scientific Biography*. Vol XV, Suppl. I. Charles Coulston Gillispie (Ed). New York, Charles Scribner's Sons, 1978, pp. 524–526.

Haller, A.v. *Sur la formation du coeur dans le poulet...2 vols*. Lausanne, Bousquet, 1758. (Blake, 196; GM 469.2)

Harvey, W. *Exercitationes de generatione animalium. Quibus accedunt quaedam de partude membranis ac humoribus uteri: & de conceptione*. Londoni, Octavian Pulleyn, 1651.

Kirchhoff, A. *Die Idee der Pflanzen-Metamorphose bei Wolff und bei Göthe*. Berlin, Gaertner, 1867. Locy, W.A. *Biology and its Makers*. New York, Henry Holt & Co., 1908.

Meckel, J.F. *Beyträge zur vergleichenden Anatomie*. 2 vols. Leipzig, C.H. Reclam, 1808–1811. (GM 314)

Meckel, J.F. *System der vergleichenden Anatomie*. Halle, Renger, 1821.

Needham, J. *A history of embryology*. Cambridge, At the University Press, 1934, p. 199.

Rathke, M.H. *Ablandlungen zur Bildungs – und Entwicklungs – Geschichte der Menschen und der Thiere*, 2 pts. Leipziq, F.C.W. Vogel, 1832–1833. (GM 480)

Wheeler, W.M. C*asper Friedrich Wolff and the Theoria Generationis*. In: *Biological Lectures, Marine Biological Laboratory, Wood's Hole, Massachusetts for 1898*. Boston, Ginn & Co., 1899, pp. 265–284.

Wolff, C.F. *Theorie von der Generation in zwo Abhandlungen erklärt und bewiesen*. Berlin, Birnstiel, 1764.

Wolff, C.F. De formatione intestinorum praecipue.... *Novi Comment Acad Sci Petropoli* 12:43–47, 40 507, 1768; 13:478–530, 1769. (GM 471).

Wolff, C.F. *De formatione intestinorum/La formation des intestins (1768–1769). Translated by Michel Jean-Louis Perrin. Introduction and notes by Jean-Claude Dupont. (De Diversis Artibus: Collection of Studies from the International Academy of the history of Science, 68)*. Turnhout, Belgium, Brepols, 2003.

Wolff, C.F. *Theoria generationis, edito nova, aucta et emendate*. Halle Ad Salam, Typis et sumtu Io. Christ. Hendel, 1774.

Wolff, C.F. *Über die Bildung des Darmkanals im bebrüteten Hühnchen*. Halle, Rengersche Buchhandlung, 1812.

Wolff, C.F. *Theorie von der Generation in zwei Abhandlungen erklärt und bewiesen (Berlin 1764)/Theoria Generationis (Halle 1759), with an introduction by R. Herrlinger*. Hildesheim, Olms, 1966.

Jean Astruc

Tracatus de morbis mulierum (pars altera ... Ars obstetricia ad sua principia redacta, 2 vols. [Venetiis], J.A. Pezzana, 1763–1767.

His comprehensive "treatise on disease incident to women", developed from his lecture courses, Astruc (1684–1766), professor of medicine at the University of Paris, was the most ambitious of its kind in the eighteenth century. The work includes chapters on conditions related to menstruation, pregnancy, sterility, miscarriage, cancers, and disorders of the breast. Volume II contains the author's *Ars obstetrician ...*, including chapters on "natural" and "unnatural" positions of fetal presentation, on protracted or difficult labour, and on further complications arising during or after childbirth.

Astruc, a graduate of the University of Montpellier (1703), taught there and held several other positions before, in 1743, being appointed Professor of Medicine at the University of Paris. In addition to presenting a 6-year series of courses with lectures on essentially every phase of medicine, he gave the first academic course for midwives in France. In his *Traité sur les maladies des femmes...* (1761), Astruc described puerperal fever, septicemia, pregnancy in

the Fallopian tubes and abdomen, and other complications of pregnancy. He was an advocate of surgery for extrauterine pregnancy, and limited use of Caesarean section. Astruc also wrote a classic treatise on the history of venereal diseases (1736).

Among his many activities, Astruc served as physician to Augustus II of Poland (1670–1733), municipal magistrate of Toulouse, and counselor and physician to the King of France, Louis XV (1710–1774). Astruc's name is well known among theologians. In 1753, he published anonymously what is regarded as the first work on biblical criticism, focusing on the book of Genesis and the first chapters of Exodus in the Pentateuch. His exegesis with analysis and interpretation focused on variant accounts of the creation story and other aspects of early history as recorded by Moses (Astruc 1753; see Osler 1929).

References

Blake 22.

Astruc, J. *De morbis veneris libri sex*, Lutetiae Parisorum, G. Cavelier, 1736. (GM 5195; see Osler 1851) Astruc, J. *A Treatise on the Venereal Disease, in six books; containing an account of the original, propagation, and contagion of this distemper in general. As also of the nature, cause, and cure of all venereal disorders in particular, whether local or universal.... Written originally in Latin ... And now translated into English by William Barrowby, M.B.* London: Printed for W. Innys and R. Manby ... C. Davis ... and J. Clarke..., 1737.

Astruc, J. *A treatise on all the disease of women. Containing an account of their causes, differences, symptoms, diagnostics, prognostics, and care.... Translated from a manuscript copy of the author's lectures read at Paris, 1740.* London, M. Cooper, 1743.

[Astruc, J.] *Conjectures sur les mémoires originaux dont il paroit que Moyse s'est servi pour composer le Livre de la Genèse....* Bruxelles, Chez Fricx, 1753. (Osler No 11)

Astruc, J. *Traité des maladies des femmes*. 6 vol, Paris, P.G. Cavelier, 1761–1765. (GM 6019) Astruc, J. *L'art d'accoucher réduit a ses principes....* Paris, P. Guillaume Cavelier, 1766.

Astruc, J. *A treatise on the disease of women, in which it is attempted to join a just theory to the most safe and approved practice... Translated from the French original....* 2 vols. London, J Nourse, 1762.

Astruc, J., and A.C. Lorry. *Mémoires pour servir a l'histoire de la Faculté de médicine de Montpellier....* Paris, P.G. Cavelier, 1767. (Osler 6219)

Huard, P. Jean Astruc. In: *Dictionary of Scientific Biography*. Vol I. Charles Coulston Gillispie (Ed.). New York, Charles Scribner's Sons, 1970, pp. 322–324.

Osler, W. *Men and books by Sir William Osler*. Collected and reprinted from the Canadian Medical Association Journal, with an introduction by Earl F. Nation, MD. Pasadena, CA., Privately printed at the Castle Press, 1959, p. 7-8.

James Hamilton

A collection of engravings designed to facilitate the study of midwifery, explained and illustrated. London, Robinson, 1796.

James Hamilton, the younger (1767–1839) was the son of Alexander Hamilton (1739–1802), professor of midwifery at the University of Edinburgh, who in 1791 had founded the Edinburgh lying-in hospital, the first of its kind in Scotland. James succeeded his father in that position. He was designated "the younger" to distinguish him from another James Hamilton, the elder (1749–1835), a contemporary physician and who did not practice or write on midwifery.

References

Blake, p. 197; Cutter and Viets, pp. 229–230.
Hamilton, J. *Select cases in midwifery; extracted from the records of the Edinburgh General Lying-In Hospital, With remarks*. Edinburgh, Printed for the benefit of the Hospital…, 1795.

Hamilton, J. *Hints for the treatment of the principal disease of infancy and childhood adapted to the use of parents*. London, 1809.
Hamilton, J. *Practical observations on various subjects relating to midwifery. E.H.* 2 Vols. Edinburgh, Bell & Bradfute; London, Longman, 1836.
Young, J.H. James Hamilton (1767–1839) obstetrician and controversialist. *Med Hist* 7:62–73, 1963.

John Bell and Sir Charles Bell

The anatomy of the human body. 4 vols. Edinburgh, Cadell & Davies, 1797–1804.

John Bell (1763–1820) and his brother Charles (1774–1842) contributed the most important work in anatomy in Great Britain during the late eighteenth and early nineteenth centuries, stressing the important of this [field] for surgeons. John Bell also wrote on cesarean section.

References

Blake p. 39; Garrison Morton 401.3, see also Garrison Morton 402; Russell 461.

The Nineteenth Century

Historians sometimes define this century as commencing with the 1815 Congress of Vienna, at which, following the exile of Napoleon I, the major European powers redrew the map of the continent and restored monarchies, and extending to 1914 with the outbreak of World War I. While this era saw the creation of the modern age and the nation state, internal strife rather than invasion characterized much of this period. During the century, the Spanish and Portuguese empires began to dissolve; and in 1806 the Holy Roman Empire ceased to exist. Following the Napoleonic wars, Great Britain became the world's leading power enforcing the *Pax Britannica*, and, with its dominance of the seas, encouraging world trade. With a rise in the production of petroleum, electricity, and steel, the "Second" Industrial Revolution with its factory system allowed Germany, the United States, and Japan to become more powerful, each racing to create empires of their own. In Europe and the United States toward the end of the eighteenth and continuing into the nineteenth century, the wide ranging intellectual, literary, and artistic movement "Romanticism" sought to assert the validity of human emotion and subjective experience, and to eschew the prevailing rationalism, with its subordination of content and feeling to more classical forms.

At the beginning of the century (1801), Ireland merged with Great Britain to create the United Kingdom. With the Louisiana Purchase (1803), the United States ended France's territorial claims in North America. Napoleon Bonaparte (1769–1821) was crowned Emperor of France (1804), and that year the steam railway locomotive was developed. The war of 1812 between the United States and Great Britain extended from 1812 to 1815. In Belgium, Napoleon was defeated at the Battle of Waterloo (1815). Under the reign of Victoria Alexandria (1819–1901), Queen of the United Kingdom (1837–1901), industries and railways expanded, in London the underground was opened (1863) and incandescent electric lights illuminated the streets, and the British Empire encompassed the globe. In 1834, the British Houses of Parliament were burnt, and Mt. Vesuvius erupted again to spread devastation. During this time, Andrew Jackson (1767–1845), seventh President of the United States (1829–1837) worked against high finance and for populism.

In the humanities, the English romantic poet-artist William Blake (1757–1827) declared "I must create a system or be enslaved by another man's." Percy Bysshe Shelley (1792–1822) and John Keats (1795–1821) penned lyric poetry, while Honorè de Balzac (Balssa; 1799–1850), with Victor Marie Hugo (1802–1885), Nathaniel Hawthorne (1804–1864), Edgar Allen Poe (1809–1849), Herman Melville (1819–1891), and other authors were writing substantive novels and plays. The creativity of Wilhelm

Richard Wagner (1813–1883) contributed much to the musical répertoire.

Karl Marx (1818–1883), German philosopher and political economist, with his socialist colleague Friedrich Engels (1820–1895), published the *Communist Manifesto* (1848). The Gold Rushes in California (1848–1858) and Australia (1851–1860s) inflamed men's minds with the dream of wealth. The Crimean War (1854–1856) between Great Britain and France and the Ottoman Empire and Russia hemorrhaged these countries of much of their youth. With the *Risorgimento* the republic of Italy was founded (1861). The Civil War (1861–1865) between the North and the Confederacy wasted over one half million American lives. With the Thirteenth Amendment (1865), slavery was abolished in America, as it was in Brazil (1888). In 1869, completion of both the Suez Canal to link the Mediterranean to the Red Sea, and the transcontinental railroad to span America, opened new possibilities for trade and commerce. Under Otto Eduard Leopold von Bismarck (1815–1898) the German Empire was formed (1871), and for several decades many European powers struggled to gain colonies in Africa. Near the end of the century, revival of the Olympic Games in Greece, and the Klondike Gold Rush in Alaska (both 1896), with the Unites States gaining control of Cuba, Puerto Rico, and the Philippines following the Spanish American War (1898), presented contrasting scenarios for an era of progress.

The nineteenth century has been called the "age of science." In science and education, the founding of the University of Berlin (1810) marked the establishment of the research-based university, which became a model for institutions of higher education in both Europe and America. Innovators who helped to revolutionize scientific thought included the British physicist and chemist John Dalton (1766–1844) who formulated atomic theory, the French mathematician and physicist André Marie Ampère (1775–1836), the Italian physicist Amedeo Avogadro (1776–1856), the British chemist and physicist Michael Faraday (1791–1867), and physician-naturalist Charles Robert Darwin (1809–1882) who expounded the theory of evolution by natural selection.

Medicine was not exempt from the profound and far-reaching cultural, political, and economic transformations that were occurring. The controversy concerning the role of "vitalism," the doctrine that life processes possess a unique character radically different from physiochemical phenomena, played a significant role in the establishment of the experimental life sciences. As noted earlier, health and disease were regarded as representing a balance or imbalance of the four humors; blood, phlegm, choler (yellow bile), and black bile. Based on this concept, the therapy of bloodletting, and the use of emetics or laxatives, were directed toward reestablishing the equilibrium of these humors. The "Heroic" medicine, expounded by Benjamin Rush (1745–

1813), saw extreme use and abuse of these therapies. Nonetheless, many healers began to see patients with their maladies, as not only having humoral disequilibrium, but as being afflicted with distinct diseases. Beginning in mid-century, a virtual revolution redefined medical doctrine and the understanding of the causes of disease, anatomic pathology, medical practice, and institutions. Marie Francois Xavier Bichat (1771–1802) contributed to descriptive anatomy and tissue biology, Johannes Müller (1801–1858) made numerous contributions to biology, anatomy, and physiology, Theodor Schwann (1810–1882) viewed cells as elemental structures of animal tissues and organs. From London's Broad Street water pump, John Snow (1813–1858) tracked the epidemiology of the infectious disease cholera. In addition, James Marion Sims (1813–1891) successfully repaired vesico-vaginal fistula, Ignaz Philipp Semmelweis (1818–1865) elucidated the etiology of puerperal sepsis, Henry Jacob Bigelow (1818–1890) with William Thomas Green Morton (1819–1868) invented surgical anesthesia, Rudolph Ludwig Karl Virchow (1821–1902) defined cellular pathology, Joseph Lister (1827–1912) surgical antisepsis, Louis Pasteur (1822–1895) microbiology and immunology, Theodore Albrecht Edwin Klebs (1834–1913) pathology-microbiology, John Hughlings Jackson (1835–1911) neurology and ophthalmology, Robert Koch (1843–1910) microbiology. Many others contributed to the striking advances in preventive and curative medicine. With these startling advances in microbiology and cellular pathology, the concept of illness changed from that of an imbalance of the humors to one of the pathogens and risk factors. Near the end of the century (1895), Wilhelm Conrad Röntgen (1845–1923) discovered X-Rays, although this did not come into widespread use for diagnosis and therapy until a decade or two later. In part, due to the vision of Baron Friedrich Heinrich Alexander von Humboldt (1769–1859) the University became an institution in which research and teaching were inseparable, with freedom of scholarly inquiry being guaranteed by law.

In art, many fine artists introduced color into their prints. In addition, lithography was introduced, a process in which the image is rendered on a flat stone (or later sheet zinc or aluminum) and treated to retain ink while non-image areas are treated to repel ink. This allowed artists to create multiple images from a given drawing. Realism and Romanticism in early nineteenth century art gave way to Impressionism and Post-impressionism in the later part of the century. With the advent of steam-driven presses (~1814), printing was transformed as was the distribution of periodicals and books.

Samuel Bard

A compendium of the theory and practice of midwifery, containing practical instructions for the management of women during pregnancy, in labour, and in child-bed; calculated to correct the errors, and to improve the practice of midwives; as well as to serve as an introduction to the study of this art, for students, and young practitioners. New York, Collins & Perkins, 1807.

Samuel Bard (1742–1821) published the first American textbook of midwifery. In the introduction, Bard decried the "pecuniary circumstances" and "deficiency of education", which prevented midwives from obtaining the knowledge for their calling. He noted, "I have thought that a concise, cheap book, containing a set of plain but correct directions for their practice in natural labours, and for the relief of such complaints, as frequently accompany pregnancy and labour, or which follow after delivery, would in the present state of this country prove an useful work" (p. 3).

Bard expressed a philosophy that might be adopted by all teachers of obstetrics, "...it has been my object to be useful, rather than to appear learned, ...and to detail such facts and observations as have been long known, and have received the stamp of time and experience rather than to offer new opinions" (p. 3). Later, he included medical students and physicians among those who required instruction, noting that he had given additional information "...with a view to render the work acceptable... as an introduction to the theory and practice of their art; and to furnish them in one small volume, and at little expense, with copies of the most useful plates, which are to be procured only by the purchase of many expensive works". In concluding his introduction, Bard admonished "every young person engaging in the study of midwifery, not to trust wholly to the information he may derive from books, in an art, in which so much depends on that which is to be obtained only from practice... (p. 10). Anticipating our present system of postgraduate specialty training, Bard advised the student "...to spend at least one or two seasons, under the professor of this branch of medical learning, at one of the colleges... where he may have an opportunity to add experience to theory, and while he is learning the rules, see their application in actual practice: for as is the case in every other mechanical operation...: although the manner of performing them may be described in words, and the principles on which that depends may be acquired by study, practice alone can give that coolness and dexterity which are necessary to ensure success" (p. 11).

In 239 pages Bard's text considered normal pelvic anatomy, the deformed pelvis, the reproductive organs, fetal development, menstruation, conception, abortion, the stages of labor, management of the complications of labor and the puerperium, including puerperal fever, and care of the newborn infant and its common diseases. Plate VI shows "... a child's head passing through a pelvis." Plate XII (next page) shows: "A the sides of the womb; B decidua vera, the deciduous membrane; C decidua reflexa, or deciduous membrane reflected over the ovum; D the chorion; E the amnion; F the placenta; and G the fetus; all at "... about the eighth week."

Bard's observation, "Hope and confidence increase the action of the womb while fear and dread retard it", presented the concept of childbirth without fear a century before this concept became fashionable. An appendix included recipes for therapeutic remedies and a glossary of medical terms. He described toxemia of pregnancy, recommending repeated copious bleeding in line with the "heroic" treatment of that era. Bard eschewed the use of forceps, believing that intervention by unskilled operators was more dangerous than the most desperate case left to nature. He wrote, "I confess, not without severe regret, that towards the latter end of 30 years of practice, I found much less occasion for the use of instruments than I did in the beginning... and I believe we may certainly conclude that the person who, in proportion to the extent of his practice, meets with most occasions for the use of instruments, knows least of the powers of nature, and that he who boasts of his skill in their application, is a very dangerous man". In the case of newborn infants that apparently were dead, Bard outlined appropriate resuscitation, "let the infant be laid in an easy posture, ... with a finger clear the child's mouth and throat of any mucous that may clog them;... close the nostrils, and blow forcibly into the mouth; then removing your own mouth, press the chest gently down, so as to force out the air, and in this way initiate breathing...." He continued, "A faint sighing, or a feeble pulsation of the heart, are the first signs of returning life, which are to be encouraged by gentle perseverance; but carefully avoid all sudden and rude motion, by which it may be extinguished as easily as a candle just beginning to flame" (pp. 207–208).

To a great extent, Bard's concepts reflected the contemporary views of the best European writers. In essence, this work represents a product of its time, neither better nor worse than other books of the period. One of its virtues was its size, for it was a *vade mecum* "pocket edition" similar to handbooks and outlines used by students and young practitioners today. Bard's text met with immediate success. Subsequent editions appeared in 1812 and again in 1815. To the fourth edition (1817) Bard added about 100 pages, focusing the presentation more completely to the require-

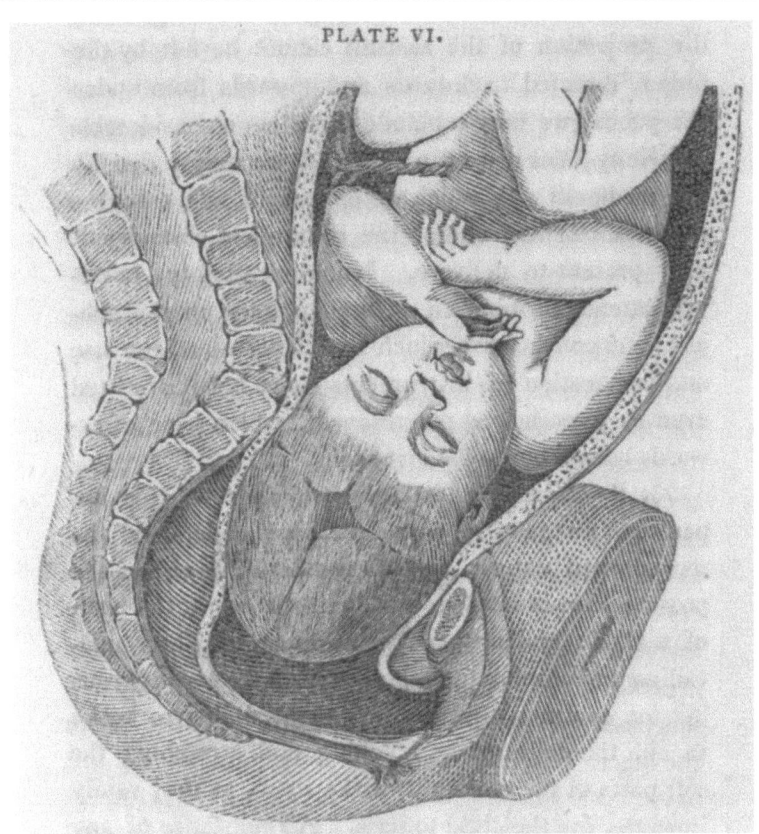

PLATE VI FROM BARD, 1807

ments of medical students and physicians. With each edition he included additional case reports to illustrate his teachings, until with the fifth edition (1819) these numbered 150. A sixth edition was in preparation at the time of his death. Bard also wrote several noteworthy essays on medical ethics and the responsibilities of the physician (Bard 1769) and medical education (Bard 1819, 1921). Through his writings Bard instructed a large number of American physicians in the art of midwifery. A further, rather intriguing problem with the *Compendium* regards the delineator of the wood engravings which illustrate the volume. These display a fine hand and highly accomplished artistic ability; however, the artist is unknown. It has been suggested that the artist was Alexander Anderson (1775–1810), a noteworthy New York illustrator of this period; but, alas, that must remain a surmise.

Bard, a native of Philadelphia and graduate of the University of Edinburgh (1765), was one of the founders of King's College medical faculty in New York (1767). Following the American Revolution (1775–1783) this was renamed Columbia College. Here, Bard served as professor of the Theory and Practice of Medicine, and later was dean of the faculty (1787–1804) and as president (1811–1820), when the school was reorganized into the College of Physicians and Surgeons of Columbia University. Bard also played a key role in founding the New York Hospital (1773). Bard cared for many notables, including President George Washington (1732–1799) for whom he lanced a deep-seated carbuncle shortly following his inauguration (1789). In addition to his writings on medical ethics (1769) and education (1819), Bard was an authority in veterinary medicine, preparing a handbook on raising sheep (1811).

References

Cutter and Viets, pp. 160–164, 213–214; Garrison Morton 6163.1.

Bard, S. *A discourse upon the duties of the physician,* New York, A. & J. Robertson, 1769. (GM 1763).

Bard, S. *A discourse into the nature, cause, and cure of the angina suffocative....* New York, S. Inslee & A. Car, 1771.

Bard, S. *A guide for young shephers....* New York, Collins & Co., 1811.

Bard, S. *A discourse on medical education....* New York, C.S. Van Winkle, 1819.

Bard, S. *Two discourses dealing with medical education in early New York.* New York, Columbia Univ Press, 1921.

Editorial. Samuel Bard (1742–1821) Colonial Physician. *JAMA* 205: 114–115, 1968.

Humphrey, D.C. The King's College Medical School and the professionalization of medicine in pre-revolutionary New York. *Bull Hist Med* 49: 206–234, 1975.

Kelly, H.A. & W.L. Burrage. *Dictionary of American Medical Biography, Lives of Eminent Physicians of the United States and Canada, from the Earliest Times.* New York, D. Appleton and company, 1928, pp. 58–60.

Langstaff, J.B. *Doctor Bard of Hyde Park. The Famous Physician of Revolutionary Times, The Man Who Saved Washington's Life...Introduction by Nicholas Murray Butler.* New York, E.P. Dutton & Co., 1942.

McVickar, J. *A domestic narrative of the life of Samuel Bard, M.D., LL.D....* New York, The literary rooms [Columbia College], A. Paul, printer, 1822.

Williams, P.F. A book review: Samuel Bard's "A compendium of the theory and practice of midwifery". *Am J Obstet Gynec* 70: 701–710, 1955.

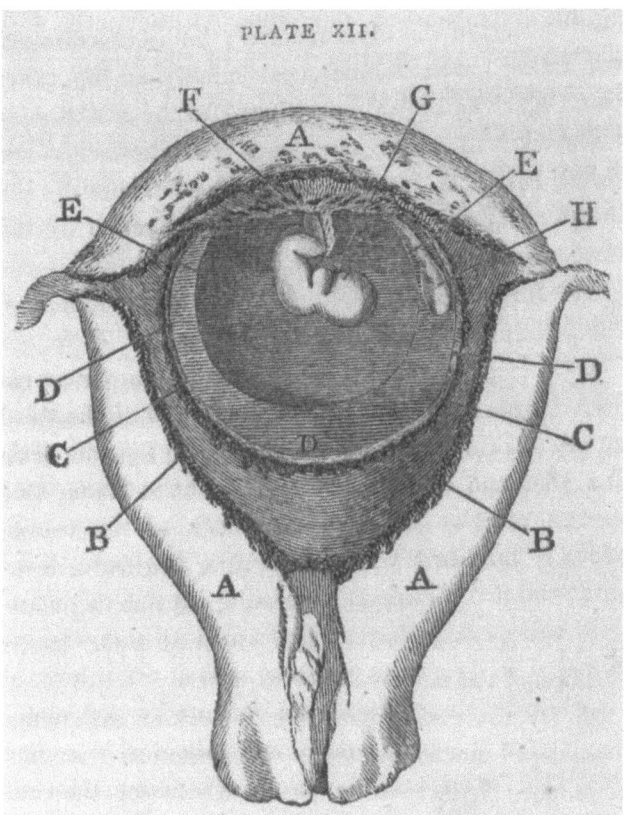

PLATE XII FROM BARD, 1807

David Daniel Davis

Elements of operative midwifery; comprising a description of certain new and improved powers for assisting difficult and dangerous labours... with cautionary strictures on the improve use of instruments. London, Hurst, Robinson & Co., 1825.

Davis (1777–1841), of Wales and graduate of Glasgow University (1801), after practicing in Sheffield, during which time he translated Philippe Pinel's (1745–1826) *Treatise on insanity* (Davis 1806), in 1813 he moved to London to specialize in obstetrics. In addition to teaching obstetrics privately, he served on the staff of the General Lying-In Hospital (Queen Charlotte's). In 1829, he was appointed Professor of Obstetric Medicine (or Midwifery) at the London University, which in 1836 became University College Hospital. He became physician-*accoucher* to Marie Luise Victoria of Saxe-Coburg-Saalfeld, the Duchess of Kent (1786–1861), presiding at the birth of Victoria Alexandrina (1819–1901), who became Queen of the United Kingdom of Great Britain and Ireland, and Empress of India.

In this volume Davis introduced modifications of a number of procedures and instruments. Many of these are illustrated in 20 double-page lithograph plates by Charles Joseph Hullmandel (1789–1850), and others. Davis worked to convince *accoucheurs* of the need to consider the mechanism of labor. Davis wrote on craniotomy, devising a pair of bone pliers termed "osteotomist" (Plate XVIII), and advocated decapitation in certain instances of obstructed labor (Plate XVI).

Davis later expanded this book to a two-volume work containing 70 lithograph plates, several of which are double page. As in the first [work] these depict both normal and abnormal delivery, and numerous instruments including forceps and instruments of destruction, many of which Davis invented (Davis 1836).

He was the first to associate *phegmasia alba dolens* with inflammation of the veins in cases of puerperal sepsis (Davis 1823).

References

Munk III, p. 117.

Davis, D.D. An essay on the proximate cause of the disease called phlegmasia dolens. *Med-Chir Trans* 12:419–460, 1823. (GM 6273).

Davis, D.D. *The principles and practice of obstetric medicine, in a series of systematic dissertations on midwifery, and on the diseases of women and children. 2 vols.* London, Printed for Taylor and Walton, 1836.

Davis, D.D. *Acute hydrocephalus, or, water in the head....* London, Taylor & Walton, 1840.

Munro Kerr, J.M., Johnstone, R.W. and M.H Phillips. *Historical Review of British Obstetrics and Gynaecology, 1800–1950,* Edinburgh and London, E. & S. Livingstone, 1954, pp. 35–36.

Pinel, P. *A treatise on insanity, in which are contained the principles of a new and more practical nosology of maniacal disorders than has yet been offered to the public.... Translated from the French, by D.D. Davis.* Sheffield, Printed by W. Todd, for Messrs. Cadell and Davies, Strand, London, 1806. (GM 4922 is original French edition)

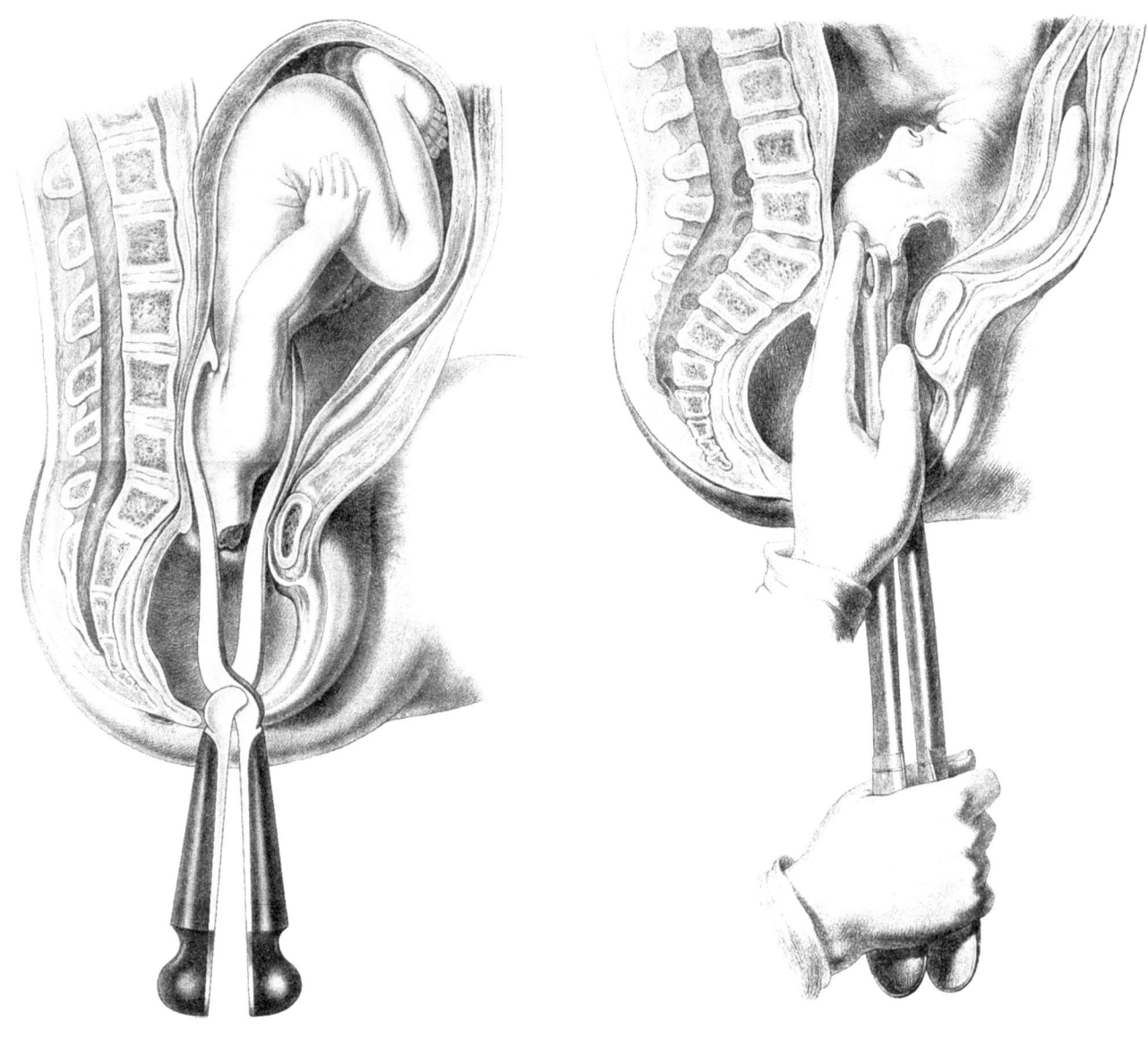

PLATES XVI AND XVIII FROM DAVIS, 1825
(COURTESY THE WELLCOME LIBRARY)

Matthias Mayor

Instructions sur L'art des Accouchemens a l'usage des sages-femmes du canton de Vaud. Lausanne, Imprimerie Hignou âiné, 1828.

The prominent Swiss physician, and 'membre du Grand conseil et du Conseil de Santé du Canton du Vaud', Matthias Mayor (1775–1847), authored this handbook for student midwives in the Canton of Vaud, Switzerland. The work provides a fascinating insight into the state of midwifery education during the early nineteenth century. As Mayor notes in his preface, by necessity midwifery manuals must differ from country to country, reflecting the educational systems in place, and more importantly, 'de l'intelligence du plus grand nombre des des élèves' [the intelligence of the largest number of students] (p. 1). In Vaud, only 3 months of practical schooling was provided. With the majority of midwifery students being young mothers from rural backgrounds who spoke a French dialect, Mayor noted that the work placed in their hands must be composed with care.

Thus, this volume was intended to provide a simple and easily accessible guide, and to introduce the basic principles of obstetrical theory, practice and care. In his opening address to the students, Mayor set forth a number of 'ground rules' which they were expected to attain. These included: to assid-uously attend all of the lectures; after their lessons the women were encouraged to discuss the topics amongst themselves; to help one another and avoid petty disagreements; to attend births and to help the poor and needy; and always when necessary to have the confidence to seek the help of more highly trained midwives.

The manual includes ten attractive lithograph illustrations depicting female anatomy and various fetal presentations. Plate 8 presents several views of the fetus and placenta. With the pelvic basin being an important aspect of obstetric knowledge, to help them with their study, Mayor instructed the midwives with practical assistance in the form of an artificial pelvis and mannikin. Upon returning home students could fabricate one using wire so that each pupil have such a model at their disposal.

The pioneer of a number of new surgical procedures (some of which now appear quite draconian!), Mayor did notable work with regards to the treatment of fractures and dislocations, however, he is known particularly for his simplification of the dressing of wounds.

References

Wellcome IV, p. 93; Hirsch IV, p. 184.

PLANCHE 8.

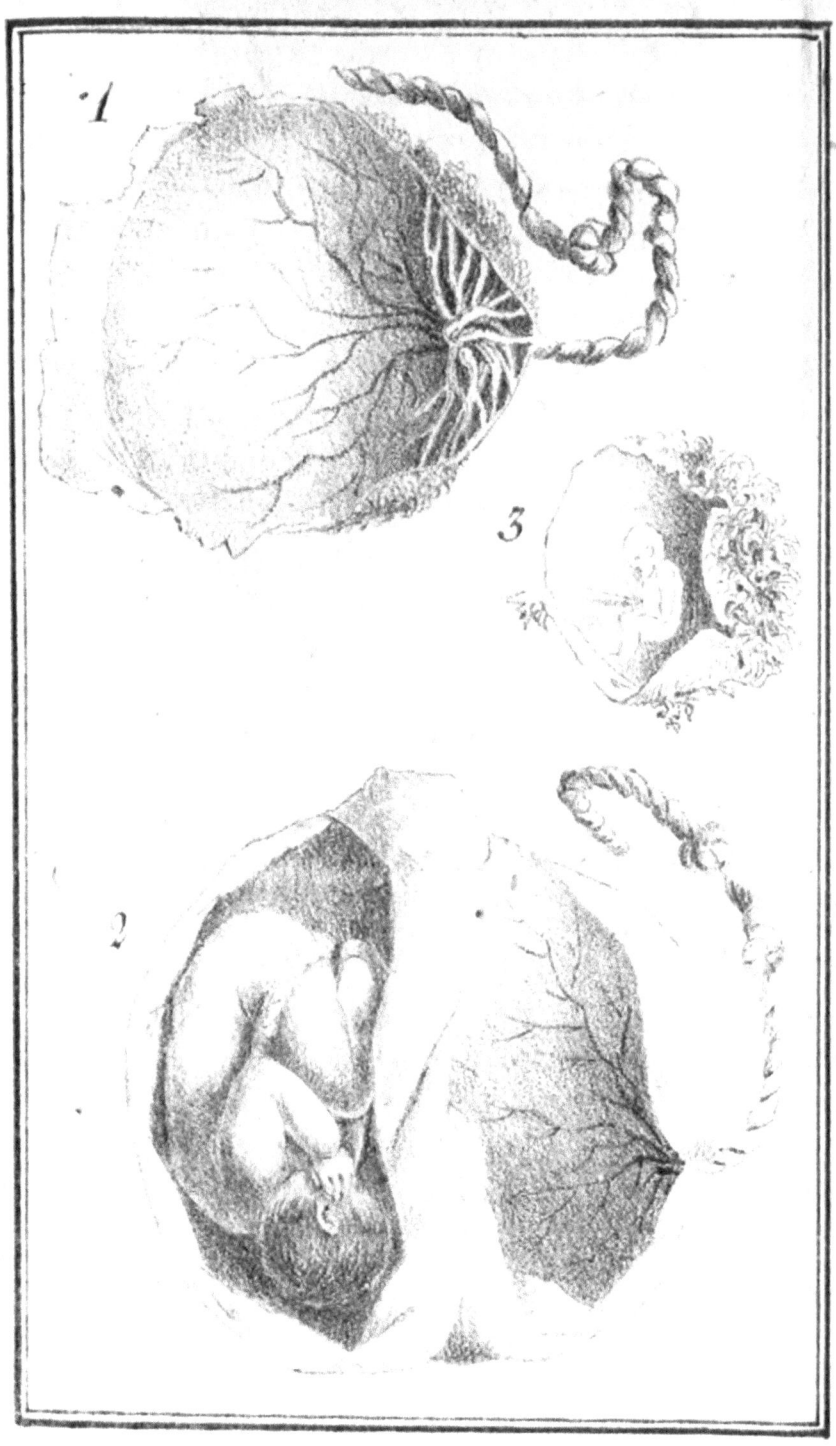

PLATE 8 FROM MAYOR, 1828

Evory Kennedy

Observations on obstetric auscultation, with an analysis of the evidences of pregnancy and an inquiry into the proofs of the life and death of the foetus in utero... with an appendix containing legal notes, by John Smith, Esq... Dublin, Hodges and Smith..., 1833.

Kennedy (1806–1886) of the Rotunda Hospital Dublin, was one of the first to recognize that abnormal fetal heart rate was a sign of compromised fetal well-being. Plate 3 illustrates the fetus in a normal cephalic (Figs. 1 and 2) and breech (Fig. 3) presentations. He described use of the stethoscope to aid in the diagnosis of "evidences of pregnancy". After noting its normal rate of "about 130 or 140 [beats] in the minute", Kennedy recorded the variation in fetal heart rate, "becoming suddenly more or less frequent, and then returning to its natural state without any apparent reason.... The external cause, which we shall find most frequently... is uterine action, particularly when long continued, as in labour" (pp. 90–91). He recorded an instance of fetal death following deep deterioration and disappearance of the fetal heart rate (pp. 92–93). Kennedy stressed the value of mediate auscultation in the diagnosis of twin pregnancy and their fetal positions (p. 129ff), location of the placental *soufflé* (p. 65ff), and as an aid to distinguish true from pseudopregnancy (p. 156ff). Kennedy concluded this tract by considering both the advantages and difficulties of auscultation, stating "We would merely beg, that those who have an opportunity, will give it a fair and impartial trial. As to the result we feel perfectly satisfied. And in conclusion, ...if its application be properly understood, it will afford us as satisfactory and unerring signs, as any diagnostic means relied on in medical practice" (p. 260). In an Appendix, Kennedy considered legal issues such as ascertaining the presence of pregnancy in the case of a woman "convicted of a capital felony", who pleads for a stay in execution because she "is quick with child" (p. 261ff).

Born in Carndonagh, Donegal, as with many Irish-Catholics, Kennedy went to Scotland for his medical education, receiving his degree from the University of Edinburgh in 1827. The following year, he was appointed lecturer in midwifery at Dublin's Richmond Hospital School, and became assistant to Robert Collins (1801–1868) master of the Rotunda Lying-in Hospital. Here, Kennedy used the stethoscope, recently invented by René Théophile Hyacinthe Laennec (1781–1826) of Paris for "mediate" auscultation (Laennec 1819), and brought to the attention of physicians in Great Britain by William Stokes (1804–1878) (Stokes 1825). Laennec, in the second edition of his work (1827) had, in fact, first referred to use of the stethoscope in auscultation of the fetal heart. (A decade later, Jean Marie Jacquemier (1806–1879) of Paris wrote a thorough thesis on auscultation of the fetal heart (Jacquemier 1837)). Also about this time, two of Kennedy's associates at the Rotunda published on the use of the stethoscope in obstetrics. In two patients, David C. Nagle employed it to diagnose twin pregnancy (Nagle 1830), while William O'Brien Adams (1833) used the stethoscope in cases of complicated labor to diagnose whether the fetus was dead prior to performing craniotomy. Adams noted, "by the aid of the stethoscope, we are enabled to determine with accuracy whether the time has arrived, which justifies the use of instruments", and that the "almost daily" use of the stethoscope presented "...ample evidence of its pre-eminent importance as a practical guide... entitling it to be ranked amongst the very greatest improvements made in practical midwifery during the past century" (Adams 1833, pp. 71–72). As in so often the case in medicine, the introduction of this new and valuable diagnostic modality met with skepticism and even some hostility among both laity and the medical profession.

In the year of publication of this work, at age 27 Kennedy succeeded Collins as master of the Rotunda for a 6 year term until 1840. One of Kennedy's additional, important contributions was to open a gynecological unit at the Rotunda, "for the humane and beneficial purpose of alleviating the sufferings of patients labouring under the diseases peculiar to women" (Kennedy,). He also was among the first to accept the idea of the contagious nature of puerperal sepsis, and fought for the establishment of cottages or *chalets*, to accommodate two patients and one nurse, rather than housing maternity patients in commodious hospital wards where the infection could be transmitted. In 1838, Kennedy founded the Obstetrical Society of Dublin, and served as Honorary President. He again held the office of President in 1849 and 1872. He also served as President of the King's and Queen's College of Physicians of Ireland (1853–1855), and the Irish Medical Association (1871). Following retirement from medical practice, Kennedy immersed himself in administrative and magisterial duties for County Dublin. In addition to working for sanitary reform, he was a supporter of the temperance movement, crusading against alcoholism and its evils.

References

Adams, O'B.W. Observations on mediate auscultation as a practical guide in difficult labours. *Dublin J Med Chem Sci* 3: 65–73, 1833.

Collins, R. *A practical treatise on midwifery, containing the result of sixteen thousand six hundred and fifty-four births, occurring in the Dublin Lying-in Hospital, during the period of seven years, commencing November, 1826.* London, Longman, 1835.

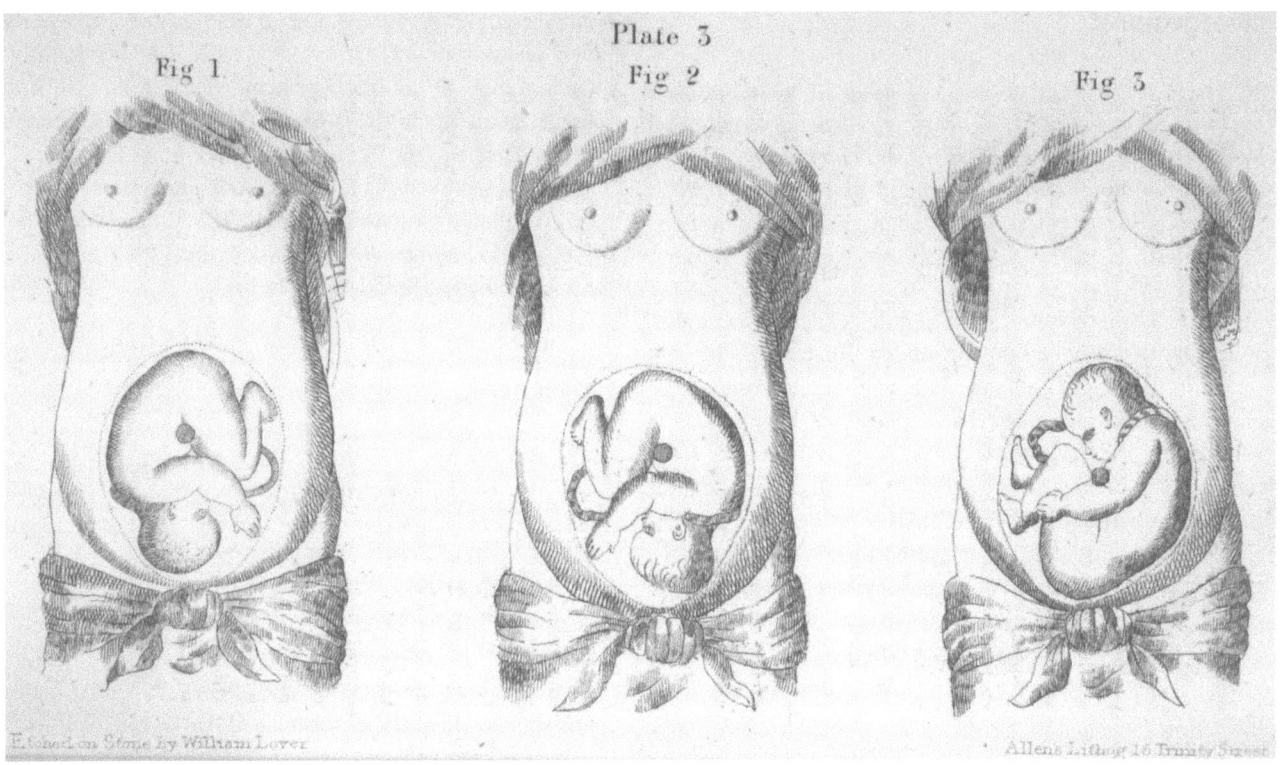

PLATE 3 FROM KENNEDY, 1833

Jacquemier, J.M. *De l'auscultation appliquée au système vasculaire des femmes enceintes, des nouvelles accouchées et du foetus, pendant la vie intra-utérine et immédiatement après la naissance....* Thèse..., Faculté de médecine de Paris, Paris, Rignaux et Ce, 1837.

Kennedy, E. *Observations on obstetric auscultation, with an analysis of the evidences of pregnancy, and an inquiry into the proofs of the life and death of the foetus in utero... with an appendix containing legal notes, by John Smith... with notes and additional illustrations by Isaac E. Taylor, M.D.* New York, J. & H.G. Langley, 1843.

Kennedy, [E]. *Statement, testimonials, and other documents, submitted in favour of Dr. Evory Kennedy, Master of the Dublin Lying-in Hospital, candidate for the Chair of Midwifery in the University of Edinburgh; with a descriptive catalogue of Dr. Kennedy's Museum [illustrative of his lectures on midwifery].* Edinburgh, Neill, 1840.

Kennedy, E. Zymotic diseases, as more especially illustrated by puerperal fever. *Dublin Quart J Med Sci* 47: 269–307, 1869.

Kennedy, E. An address delivered before the Dublin Obstetrical Society, 22nd November, 1873. *Dublin J Med Sci*, pp. 519–547, 1873.

Laennec, R.T.H. *De l'auscultation mediate, ou traité du diagnostic des maladies des poumons et du coeur, fonde principalement sur ce nouveau moyen d'exploration.* 2 vols. Paris, J.A. Brosson & J.S. Chaudé, 1819. (GM 2673).

Laennec, R.T.H. *Traité de l'auscultation mediate et des maladies des poumons et due coeur.* 2nd ed. Translated by John Forbes. London, T. and G. Underwood, 1827.

Nagle, D.C. On the use of the stethoscope for the detection of twins in utero, the presentation, &, &. *Lancet* 15: 232–234, 1830.

[Obituary notice] Evory Kennedy. *Brit Med J* 1: 911, 1886.

Pinkerton, J.H.M. John Creery Ferguson (1802–1865): Physician and fetologist. *Ulster Med J* 50: 10–20, 1981.

Pinkerton, J.H.M. Evory Kennedy, A master controversial. *Irish Med J* 77: 77–81, 1984.

Stokes, W. *An introduction to the use of the stethoscope.* Edinburgh, Maclachlan & Stewart, 1825. (GM 2674).

George Spratt

Obstetric tables: comprising coloured delineations on a peculiar plan, intended to illustrate elementary and other works on the practice of midwifery, elucidating particularly the application of the forceps, and other important points in obstetric science. London, For the author, by J. Churchill..., 1833.

Spratt (fl. 1830), about whose life little is known, was a self-styled surgeon-*accoucheur* in London. First issued by subscription, this original edition contained 12 complex engraved plates with movable flaps. These plates were innovative in adapting to obstetrical teaching the revived method of illustrating anatomical structures in depth by lifting two or three layers of superimposed folding flaps over the engraving. As each flap was lifted, in turn, by the reader, it exposed the explicit anatomical detail of pregnancy and childbirth. Most of the plates were prepared by Spratt with several members of his family, and are hand-colored. Table VI illustrates progressive stages in the progress of labor with cervical dilatation. Such use of superimposed plates to depict anatomical relations dates back to the late sixteenth and early seventeenth centuries (see Bartsich 1583, Remmelin 1619, Aselli 1627). Spratt dedicated this work to Sir Charles Mansfield Clarke (1782–1857) physician to Queen Adelaide, wife of King William IV (Clarke 1814–1821) and brother of John Clarke (1761–1815) a noted obstetrician and pediatrician (Clarke 1815). In his dedication Spratt noted "many improvements in these Tables" that were suggested by Clarke.

In his preface, Spratt considered the valuable contributions of illustrations to an understanding of anatomy and science. He stated, however, "... it does not appear that art has done as much towards the illustration of Obstetric Surgery as its great importance to all classes of the profession would appear to demand". He continued, "To the finished student ... this work will recall to his recollection many points of great interest; and to the experienced practitioner, it will form a supplementary volume to his obstetric library, which (it is hoped) will not unworthly fill a vacuum on his shelves". Spratt also included "a few practical remarks" to supplement the delineations (pages unnumbered). A second edition of 1835 includes seven additional plates (Spratt 1835). Many other editions followed, including in 1847 the first American edition from that of the fourth London (Spratt 1847). Some maintain that this American printing represents the illustrations in their finest form.

Spratt, a member of a family of book illustrators, also prepared the plates for the third edition of William Woodville's (1752–1805) *Medical Botany* (Woodville 1832), and edited the *Flora medica* (Spratt 1827–1830).

References

Cutter & Viets pp. 232–233.

Aselli, C. *De lactibus sive lacteis venis, quarto vasorum mesaraicorum genere novo invento.* Mediolani, Apud I.B. Bidellium, 1627. (GM 1094).

Bartsich, G. *Ophthalmodouleia; das ist, Augendienst....* Dresden, Matthes Stöckel, 1583. (GM 5817).

Clarke, Sir C.M. *Observations on those diseases of females which are attended by discharges.* 2 vol. London, Longman, Hurst, Rees, Orme, and Brown, 1814–1821.

Clarke, J. *Commentaries on some of the most important diseases of children. Part the first.* London, Longman, Hurst, Rees, Orme, and Brown, 1815. (GM 4825 & 6328).

Remmelin, J. *Catoptrum microcosmicum, suis aere incises visionibus splendens, cum historia, & pinace, de novo prodit....* Augustae Vindelicorum, Davidis Francki, 1619.

Spratt, G. *Flora medica containing coloured delineations of the various medicinal plants admitted into the London, Edinburgh, and Dublin pharmacopoeias....* London, Callow and Wilson, 1827–1830.

Spratt, G. *Obstetric tables: comprising graphic illustrations, with descriptions and practical remarks, exhibiting on dissected plates many important subjects in midwifery....* 2 vols. London, Published for the author by J. Churchill ... Machlachlan & Stewart, Edinburgh..., 1835.

Woodville, W. *Medical botany ... 3 vols & Supplement.* 3rd Ed., Edited by William Jackson Hooker. London, Phillips for the author, 1832. (First edition of 1790–1795 in GM 1831.1).

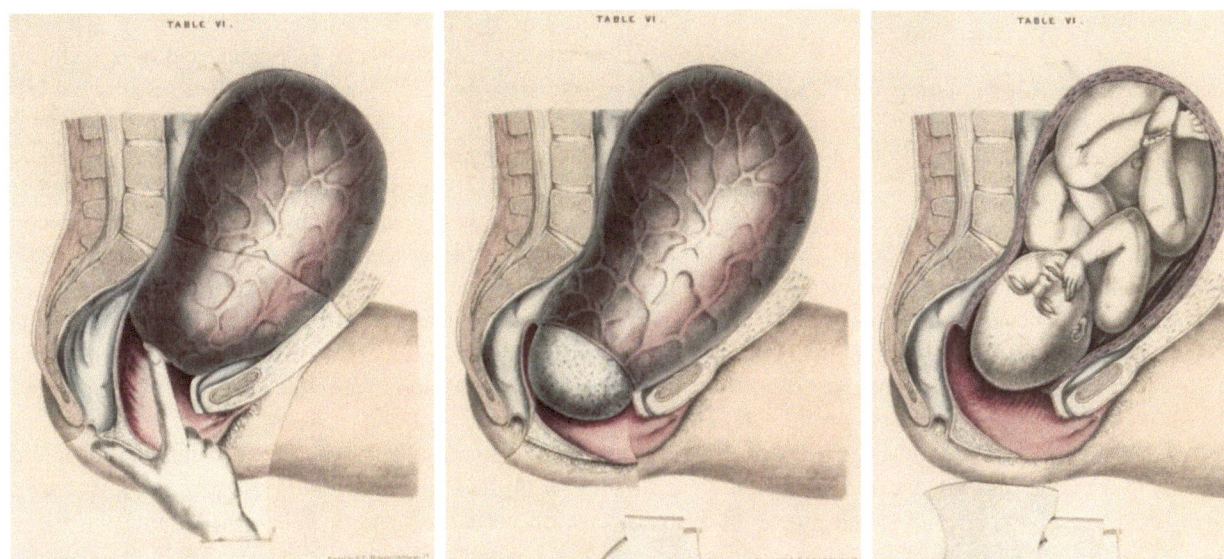

TAB. VI FROM SPRATT, 1833

Franz Carl Naegele

Das schräg verengte Becken nebst einem Anhange über die wichtigsten Fehler des weiblichen Beckens überhaupt. Mainz, Victor von Zabern, 1839.

In a short report of nine cases Naegele (1778–1851) first described the obliquely contracted pelvis (Naegele 1834). Several years later, in this more comprehensive study, he detailed 35 cases in women throughout Europe, including one which was found in an Egyptian mummy in Paris (Case No. 20). Two additional cases were in men. Naegele described the unusual characteristics of this deformity. "Complete ankylosis of the sacro-iliac symphysis or a complete fusion of the os sacrum with the os ilium on one side.... Arrest or imperfect development of the lateral part of the os sacrum and a narrower lumen of the foramena sacralia anteriora on the ankylosed side.... Narrowness of the os innominatum and of the sciatic notch on the same side.... The os sacrum appears pushed towards the ankylosed side to which its anterior surface is somewhat rotated. At the same time the symphysis pubis is forced towards the opposite side. The symphysis thus faces the promontory obliquely.... On the side of the ankylosis, the inner surface of the lateral wall and of the lateral wall and of the lateral part of the anterior wall of the pelvis is found less hollow and flatter than in the normal pelvis.... The other half of the pelvis, that on which the synchondrosis sacro-iliaca exists, likewise deviates from the normal.... This pelvis is obliquely contracted, i.e., in a diameter which crosses that other diameter extending from the ankylosis to the opposite acetabulum... Consequently, the pelvic inlet... and also another imaginary plane in the middle of the pelvic cavity... are both seen from the front as similar to an obliquely lying oval" (Naegele 1839, pp. 7–9).

The pelvis illustrated in Tafel 1 was from a 19-year-old primigravida patient. Following 2 days of labor, a 7 lb. stillborn infant was delivered by a difficult forceps extraction. On the fifth postpartum day the young woman died of puerperal sepsis. Naegele believed the deformity was a developmental anomaly, as it was not associated with rickets, traumatic injury, or infection. This view prevails today. In his time, the deformity never had been diagnosed in a living subject, as it caused little if any limp or other sign. Nonetheless, Naegele suggested several aids to its diagnosis in the living subject. Despite stressing the role of pelvic contraction as a cause of dystocia in labor, he had little to say regarding the role of cesarean section. In the preface to this work, Naegele wrote "... to obtain a correct conception of our deviations, it is very important to observe conscientiously and study carefully.... Nothing but the most profound study can supply the examiner with the necessary knowledge without which he would only discover the more evident deformities. The evil consequences to which such defective and false represent-

ations give rise may be easily inferred..." (p. iv). The artist and engraver are shown as F. Wagner and de Engelmann, respectively. In 1825, Naegele published his first work on pelvic anatomy and the obstetrical problems associated with pelvic and spinal deformities, "the female pelvis considered in relation to its position and the inclination of its curvature...." This work consists of two sections, the first of which is devoted to pelvic anatomy to establish "detailed knowledge of the form and conditions of all the points which constitute a well-made pelvis". The second part consists of a thorough review-history of the subject (Naegele 1825).

Born in Dusseldorf, Naegele studied at the Universities of Freiburg, Strassburg, and Bamberg. In 1807, he was called to the University of Heidelberg, and 3 years later was appointed director of the Lying-In Hospital. He also wrote on several other aspects of pregnancy including its duration, the mechanism of normal labor, and various aspects of obstetrical pathology. As one of Europe's most celebrated physicians, he was referred to by some as "the Euclid of obstetrics". In addition, he wrote a textbook for midwives (Naegele 1830), and stressed the importance of the female midwife (Naegele 1827). He also edited (1825–1847) the *Heidelberger Klinische Annalen*, later named *Medicinische Annalen*. Naegele's son, Hermann Franz Joseph (1810–1851) was also a prominent obstetrician, becoming professor at the University of Heidelberg, and giving the first clear account of auscultation of uterine sounds and the fetal heart (Naegele 1838). Over the century following publication of Naegele senior's *arbeit*, other pelvic abnormalities were described. In 1933, William Edgar Caldwell (1880–1943) and Howard Carman Moloy (1903–1953) published a classification of pelvic variations and their effect on labor (Caldwell and Moloy 1933) that remains valid to the present time.

References

Cutter and Viets p. 204; Garrison Morton 6257; Garrison p. 605.

Caldwell, W.E. and Moloy, H.C. Anatomical variations in the female pelvis and their effect in labor, with a suggested classification. *Am J Obstet Gynec* 26: 479–505, 1933. (GM 6266).

Naegele, F.C. *Erfahrungen und Abhandlungen aus dem Gebiethe der Krakheiten des Weiblichen Geschlechtes; nebst Grundzügen einer Methodenlehre der Geburtshülfe.* Loeffler, Mannheim, 1812, p. 281.

Naegele, F.C. Ueber den Mechanismus der Geburt. *Dtsch Arch Physiol* 5: 483–531, 1819.

Naegele, F.C. *Das weibliche Becken betrachtet in Beziehung auf seine Stellung und die Richtung seiner Höhle nebst Beyträgen zur Geschichte der Lehre von den Bechenaxen. Carlsruhe*, Müller, 1825.

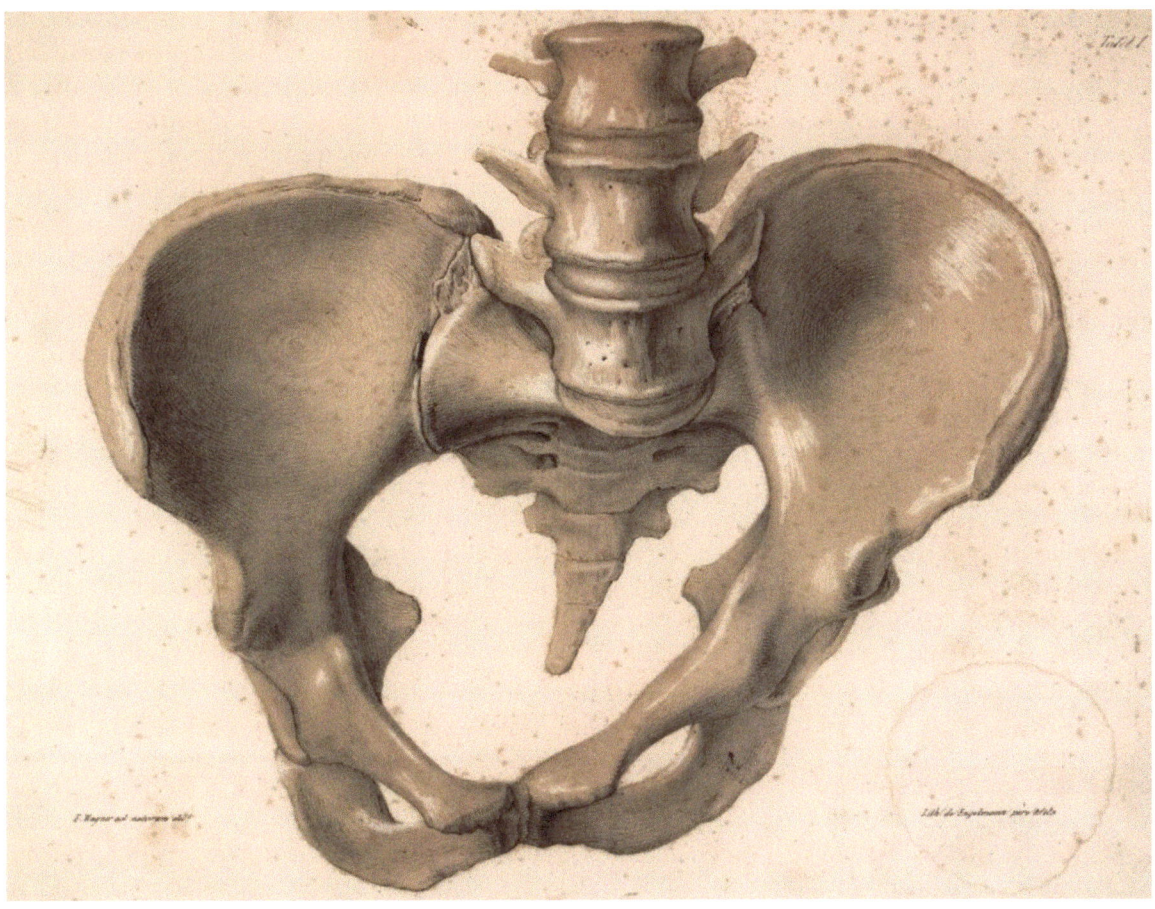

Taf. I from Naegele, 1839

Naegele, F.C. *Observations on the impropriety of men being employed in the business of midwifery.* London, Printed for Hunt and Clarke, 1827.

Naegele, F.C. *An essay on the mechanism of parturition.... Translated by Edward Rigby.* London, Callow and Wilson, 1829.

Naegele, F.C. *Lehrbuch der Gebürtshulfe für Hebammen.* Heidelberg, J.C.B. Mohr, 1830.

Naegele, F.C. Ueber eine besondere Art fehlerhaft gebildeter weiblichen Becken. *Heidelberg Klin Ann* 10: 449–472, 1834.

Naegele, F.C. *The obliquely contracted pelvis, containing also an appendix of the most important defects of the female pelvis.... Translated and with an introduction by Alfred M. Hellman and George Musa.* New York, Pynson Printers, 1939.

Naegele, F.C. *The obliquely contracted pelvis containing also an appendix of the most important defects of the female pelvis....* Birmingham, AL. The Classics in Obstetrics and Gynecology Library, Division of Gryphon Editions, 1991.

Naegele, H.F.J. *Die geburtshülfliche Auscultation....* Mainz, Victor von Zabern, 1838.

Speert, H. Franz Carl Naegele, Naegele's Rule, Naegele's Asynclitism, and the Naegele Pelvis.... In: *Obstetric and Gynecologic Milestones.* New York, Macmillan, 1958, pp. 159–179.

Julian Ledesma

Fenómeno raro de preñez, O historia de una hernia de la matriz, en estado de gestacion, terminada felizmente por la operacion cesárea, con la estraccion de una nina viva, sana y de todo tiempo sin ulterior lesion de la madre, con observaciones curiosas de auscultacion y percussion sobre el seno materno de la misma. Dos láminas finas representan el estado de embarazo y la operacion. Madrid, Imprenta de Boix, 1840.

In his native tongue, as well as with a sprinkling of Esperanto and Latin, Julian Ledesma, professor of surgery at the Hospital of Salamanca, Spain presented the remarkable case history of a "rare phenomenon of pregnancy." The patient, Elena Ramos, 42 years of age, suffered with a large right inguinal hernia that contained her pregnant uterus. Her infant girl was delivered by caesarean section, and the lives of both mother and child were preserved.

Ledesma first saw the patient near the end of January 1839 when she was about 5 months pregnant. In his assessment, beginning in April (at 6 months gestation), he described some "curious observations" on auscultation of the fetal heart, one of the earliest to do so, appearing only 2 years after the noted work on obstetric auscultation by Naegele, *Die geburtshülfliche Auscultation* (1838). Of interest, Ledesma quotes the first work on auscultation of the fetal heart by Jean Alexandre Kergaradec (1787–1877), *Mémoire sur l'auscultation appliqué a l'étude de la grossesse* (Kergaradec 1822), which appeared only several years following the invention of the stethoscope by René Hyacinthe Laennec (1781–1826) (Laennec 1819), and its subsequent modification by Pierre Adolphe Piorry (1794–1879) (Piorry 1828). Ledesma noted the "blowing noise" or "rumble" of the placental vasculature, as well as the fetal heart rate of "130–150" beats per minute, which was twice that of the mother. Originally thought by other consultants to be an extra-uterine pregnancy, upon percussing the fluid-filled amniotic sac surrounding the fetus, he diagnosed correctly that the pregnancy was intra-uterine, with the uterus within the hernia.

Near term on the morning of 7 June, Ms. Ramos went into labor. Her amniotic membranes ruptured releasing "… the water [which] was identical to that of any natural pregnancy, and the hernia sac became smaller" (p. 11). Ledesma attempted to deliver the infant through the vulva, however, because of the tight neck of the inguinal ring, this was not possible. He recorded that previously he had neither performed nor witnessed a caesarean section on a living patient, and quotes the work of both François Mauriceau (1637–1709) (Mauriceau 1668) and André Levret (1703–1780) (Levret 1747, 1753), as to the difficulties and hazards of performing the procedure, it being "… mortal, and if someone lived it was a miracle" (p. 14).

With an incision through the layers of the hernial sac and myometrium, Ledesma delivered the infant, extracting it by its feet and legs. Following delivery of the trunk, the uterus contracted strangulating the child's neck. He then had to extend the uterine incision to deliver the "… asphyxiated [baby which] … soon showed signals of life" following its "… exit into the light of the world" (p. 16).

Ledesma manually removed the adherent placenta, and the infant soon was baptized and given to the mother to nurse. Following manipulation, the uterus contracted, and the hernia incision, but not that of the uterus, was sutured.

Day by day, Ledesma records a rather stormy postoperative course with the mother experiencing some strong pains, signs of peritonitis, and foul discharge. Nonetheless, on 11 August, 5 weeks following the surgery, "… the patient with her daughter got up … in good health … the tumor [hernial sac] was small … [being] the size of a scrotum" (p. 22).

In his closing remarks, Ledesma apologized that "… my writing does not have the elegance that the public honors." He concluded, "… direct[ing] … attention to the important consequences derived from this case, for the glory of our homeland and the splendor of curative science" (pp. 22–23).

Of the two plates, the first illustrates the patient with her hernia at 6 months gestation. The second striking lithograph captures the dramatic delivery illustrating the opened uterus with its spurting blood vessels and the partially extracted fetal breech.

Of importance, at this time, caesarean delivery was associated with a maternal mortality rate of well over 50 %.

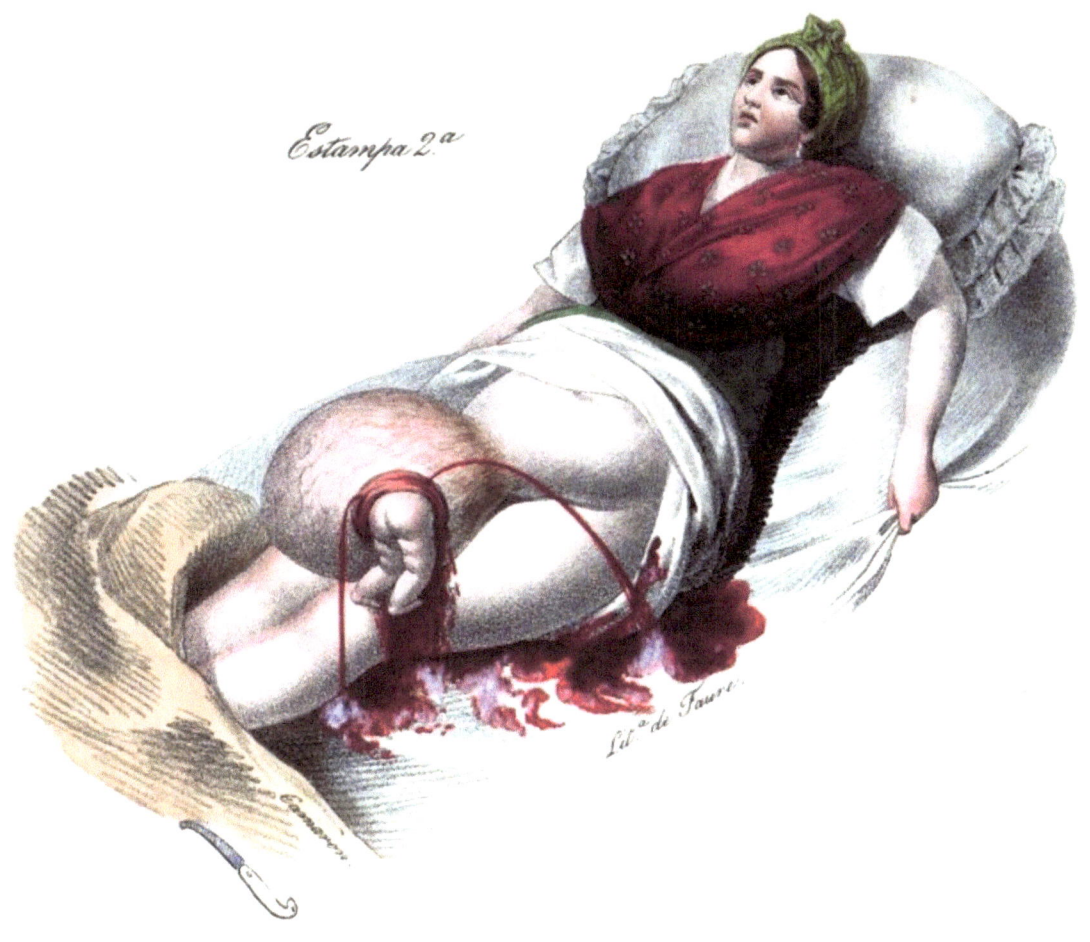

FIGURE FROM LEDESMA, 1840
Courtesy Hathi Trust

References

Kergaradec, J.A. *Mémoire sur l'auscultation appliquée à l'étude de la grossesse, ou recherches sur deux nouveaux signes propres à faire reconnaître plusieurs circonstances de l'état de gestation; lu à l'Académie royale de médecine, dans sa séance générale du 26 décembre 1821.* Paris, Méquignon-Marvis, 1822.

Laennec, R.T.H. *De l'auscultation médiate, ou Traité du diagnostic des maladies des poumons et du coeur, fondé principalement sur ce nouveau moyen d'exploration.* 2 vols. Paris, J.A. Brosson et J.S. Chaudé, 1819. (GM 2673, 3219).

Levret, A. *Observations sur les causes et les accidens de plusieurs accouchemens laborieux.* Paris, C. Osmont, 1747.

Levret, A. *L'art des accouchemens; démontré par des principes de physique et de mechanique; pour server de base & de fondement à des leçons particulières.* Paris, De l'imprimerie de Delaguette, 1753.

Mauriceau, F. *Des maladies des femmes grosses et accouchées.* Paris, I. Henault, I. d'Houry, R. de Ninville, I.B. Coignard, 1668.

Naegele, H.F. *Die geburtshülfliche Auscultation.* Mainz, Bei Victor von Zabern, 1838.

Piorry, P.A. *De la percussion médiate et des signes obtenus à l'aide de ce nouveau moyen d'exploration dans les maladies des organes thoraciques et abdominaux.* Paris, Chaudé, J.B. Bailliere, 1828. (GM 2675).

Joseph Hermann Schmidt

Lehrbuch der Geburtskunde fur die Hebammen in den Koniglichen Preussichen Staaten. Berlin, 1840.

In conjunction with the Prussian restoration in northern and central Germany (and which included the northern portion of what is now Poland) following the turmoil of the Napoleonic Wars (1799–1815), the Royal Health Ministry expressed the need for a definitive work on contemporary knowledge and practice for midwives. To reinvigorate the health of the nation, the subsection of Medical Affairs of the Prussian Ministry for Interior Affairs prepared a "white paper," published originally in 1815, which later was expanded into the present volume. A country physician and director of midwives at the Paderborn Hospital, Schmidt (1804–1852) is credited with preparing the final draft, which then was approved by a panel of overseers. The volume is representative of an effort to establish standards, and place midwifery on an academic basis, presenting both scientific and practical aspects of the craft. Another edition of this work, dated 1839, was co-authored with Frederich Carl Theodor Kanzow (b. 1820).

The two major sections include an extensive account of regulations and requirements, followed by fundamentals on reproductive anatomy and physiology, with details of the management of labor and delivery, as well as most conceivable complications. At this time, many medical professionals were not convinced of the value of vaccination against smallpox. As one interested in the obligation of the State to provide healthcare to everyone, Schmidt stressed the importance of midwives vaccinating patients, and providing other public health measures. Thus, this manual for midwives was embedded in the larger picture of public health administration. The 32 plates include illustrations on the female pelvis (Tafel 1 and 2), the external genitalia (Tafel 4), reproductive anatomy (Tafel 5 and 6), embryonic and fetal development (Tafel 7), methods of delivery of the normal (Tafel 14) and abnormal (Tafel 15, 18, 19, 20, 22, and others) presentations of the fetus (Tafel 21 and others). Illustrated in Tafel 16 are several aspects of breech presentation, and in Tafel 29 methods of delivery. Another illustration (Tafel 24) shows examples of transverse lie. Many of the steel etchings, which show considerable detail, are by E. Weber and E. Haas. The printing house for the plates is L. Sachse of Berlin.

References

Kanzow, F.C.T. and J.H. Schmidt. Lehrbuch der Geburtskunde für die Hebammen in den königlichen preussischen Staaten. Berlin, 1839.

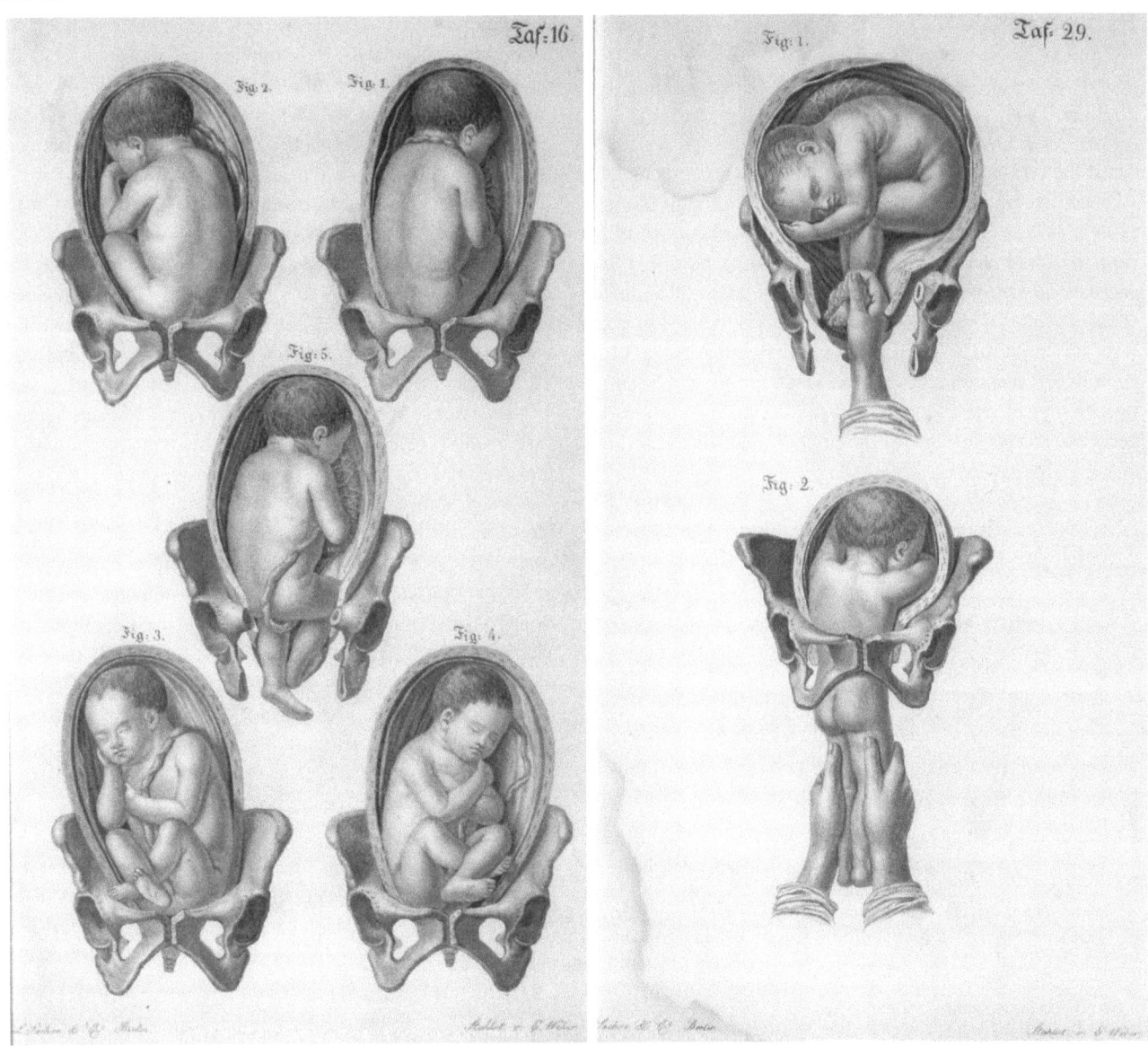

TAF. 16 (left) AND 29 (right) FROM SCHMIDT, 1840

Robert Lee

The anatomy of the nerves of the uterus. London, Hippolyte Ballière, 1841.

Until the mid-nineteenth century, there was no agreement as to whether the uterus possessed nerves. Although William Hunter (1718–1783), in his monograph of the anatomy of the human gravid uterus described such nerves, and noted their increase in size during pregnancy, many anatomists denied their existence. Lee (1793–1877) described the nerves in the non-gravid uterus, as well as their enlargement from 3 to 9 months of gestation, giving the first careful and accurate description of these nerves and the cervical ganglion. He recorded the background of the serendipitous discovery:

> On the 8th of April, 1838, while … dissecting a gravid uterus of 7 months, I observed the trunk of a large nerve proceeding upward from the cervix to the body of the uterus, with the right uterine vein, and sending off branches in its course to the posterior surface of the uterus, some of which accompanied the veins, and others appeared to be inserted into the peritoneum. A broad band, resembling a plexus of nerves was seen extending across the posterior surface of the uterus, and covering the nerve about midway from the fundus to the cervix. On the left side a large plexus of nerves was seen surrounding the uterine veins, where they were about to enter the hypogastric vein (p. 5).

Lee carefully described these nerves and their ramifications. Following a description of nerves in the non-gravid uterus, he concluded,

> These dissections prove that the human in unimpregnated uterus possesses a great system of nerves, which enlarges with the coats, blood-vessels and absorbents during pregnancy, and which returns after parturition to its original condition before conception takes place. It is chiefly by the influence of these nerves, that the uterus performs the varied functions of menstruation, conception, and parturition, and it is solely by their means, that the whole fabric of the nervous system sympathises [sic] with the different morbid affections of the uterus. If these nerves of the uterus did not exist, its physiology and pathology would be completely inexplicable (p. 8).

This illustration from the 1842 edition shows for the first time in color the left posterolateral aspect of the uterus at 9 months gestation, with associated ganglia and neuronal connections. Lee noted in the explanation of the plate: the fundus and body of the uterus (A); the vagina covered with nerves from the left hypogastric ganglion (B); the rectum (C); the left ovary and Fallopian tube (D); the great sympathetic nerve (K) passing along the anterior aspect of the aorta

(F) and dividing into two cords (M); the right (N) and left (O) hypogastric nerve: the left hypogastric ganglion (R); the nerves of the vaina (S); the left subperitoneal plexus covering the body of the uterus (X); and the left subperitoneal ganglion with numerous branches of nerves extending between it and the left hypogastric nerve and ganglion (Y).

These papers were also published separately as On the Ganglia and Other Nervous Structures of the Uterus (Lee 1842). Sadly, Lee's findings were misrepresented by his former student, Thomas Snow Beck (1814–1877), and only after a prolonged battle were Lee's claims recognized. Lee then extended his findings in his 1848 publication Engravings of the Ganglia and Nerves of the Uterus and Heart… (Lee 1848). Twenty-five years later Ferdinand Frankenhaeuser (1832–1894) of Zurich redescribed this innervation with its plexuses and ganglia in somewhat greater detail (1867).

Lee, a native of Scotland and Edinburgh graduate, was physician to the British Lying-In Hospital, and lecturer on midwifery at St. George's Hospital, London. Although he ardently opposed ovariotomy, Lee made numerous contributions to the specialty. He apparently first used the term "cervicitis". From dissections on cadavers, he demonstrated that leucorrhea originates from the uterus rather than from the tubes or vagina. Lee also ntoed the similarity of pathologic findings in cases of puerperal fever erysipelas.

References

Beck, T.S. On the nerves of the uterus. *Philos Trans R Soc Lond*, Part II, 136:213, 1846.

Frankenhauser, F. *Die Nerven der Gebaermutter und ihre Engidgung in den GlattenMuskelfasern.* Jena, F. Mauke, 1867.

Lee, R. *The anatomy of the nerves of the uterus.* London, Baillière, 1841.

Lee, R. On the nervous ganglia of the uterus. *Philos Trans R Soc Lond*, Part I, vol 269–275, 1841.

Lee, R. *An appendix to a paper on the nervous ganglia of the uterus, with a further account of the nervous structure of that organ.* London, Taylor, 1842.

Lee, R. *On the Ganglia and Other Nervous Structures of the Uterus.* London, Richard and John E. Taylor, 1842.

Lee, R. *Memoirs on the ganglia and nerves of the uterus.* London, John Churchill, 1849.

Lee, R. *Engravings of the Ganglia and Nerves of the Uterus and Heart.* London, Churchill, 1858.

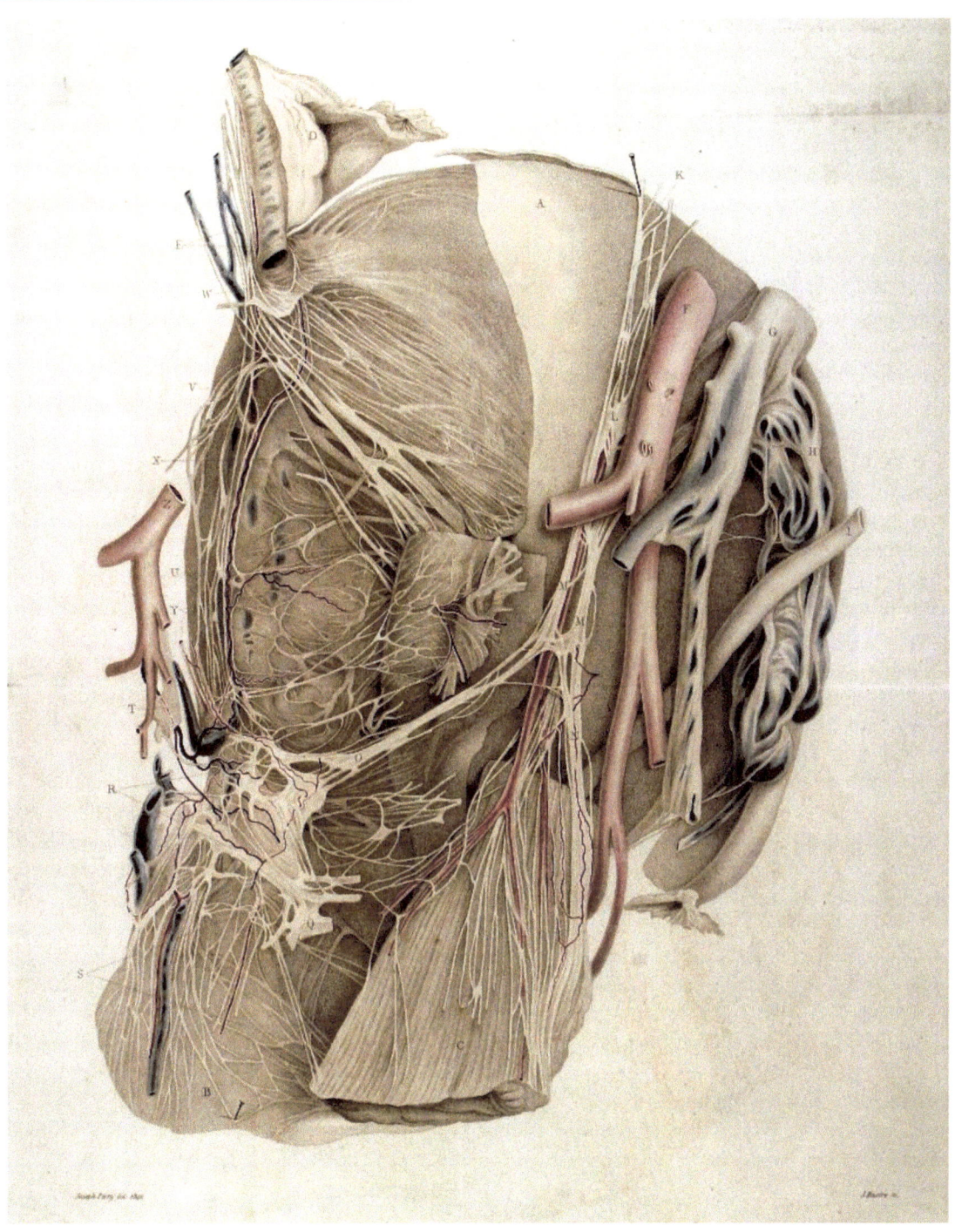

ILLUSTRATION FROM LEE, 1842

Wooster Beach

An improved system of midwifery: adapted to the reformed practice of medicine; illustrated by numerous paltes, to which is annexed, a compendium of the treatment of female and infantile diseases, with remarks on physiological and moral elevation. New York, By the Author, 1847.

In the early to mid nineteenth century the "heroic medicine" of Benjamin Rush (1745–1813) and his school (Rush 1789–1798) consisted, in large part, by treating the patient by purging with mercurials and other toxic compounds, and for surgery "blood-letting". Such therapy became quite popular with allopathic physicians. In reaction to these quasi-empirical practices, several schools of herbal medicine gained popularity in the United States. These included the "Botanics," Thompsonianism, Physio-medicalism, Eclecticism, and others. In opposition to the prevailing allopathic practices, Beach (1794–1868), the founder of "Eclecticism", rejected contemporary systems of medical practice, incorporating the treatment of disease with nature's vegetable remedies, various herbs and roots, into an allopathic frame. This botanico-medical movement flourished in the populist social and educational milieu of the Jacksonian era (Andrew Jackson [1767–1845], president 1829–1837), in which religious medical controversy warred against the more traditional theological and medical beliefs.

In the Preface to this volume, Beach stressed that he had "... endeavored to make it a practical work – a safe guide to the physician, midwife, and student ... A reformation ... in this branch, is imperiously required". He noted that obstetrics "... as taught in our colleges, and practiced by the generality of physicians ... is injudicious, injurious, and not infrequently fatal ... [consisting] principally of bleeding, mercury, untimely and improper interference, and the frequent and needless use of instruments...." Beach continued, *"Dame Nature is the best midwife in the world ... meddlesome midwifery is fraught with evil ... the less done, generally the better ... non-interference is the corner-stone of midwifery"* (pp. 5–6).

An itinerant healer, following an apprenticeship with a botanic physician in New Jersey, Beach first moved to New York where he is believed to have studied medicine for one term. Here, he founded the United States Medical Infirmary (1827). This evolved into a medical school, the Infirmary of the New York Reformed Medical Academy (1829), the name of which was changed to the Reformed Medical College of the City of New York (1830), and which operated without legal state charter. Initially practicing as an "Emperic," Beach taught that medical practice should be based on pathology, usually treating patients with small doses of a single "specific" remedy. Nonetheless, Eclecticism, so named because it drew its ideas from many sources, incorporated ideas from Homeopathy, Hydrotherapy, Botanical Medicine, and other naturalist schools. During the 1832 Asiatic cholera epidemic in New York, he cared for the poor with the disease, successfully treating nearly a thousand cases, and avoiding the use of calomel and other "heroic" measures.

In the early 1830s, Beach moved to Worthington, Ohio to found another medical school, the Medical Department of Worthington College; the first such chartered sectarian school in the United States. The author of at least a dozen medical works, Beach appreciated the importance of publication in spreading the "Eclectic" message. He founded a journal, the *Western Medical Reformer* (1836), which 3 years later continued as *The Eclectic Medical Journal* (1849). Beach also published *The Reformed Practice of Medicine* (1831), and the three volume *American Practice of Medicine* (1833). He condensed this latter work into his *The Family Physician; or the Reformed Practice of Medicine* (1842). Later, he moved to Cincinnati to found the Eclectic Medical Institute (1845), also known as the Reformed Medical School of Cincinnati. In 1852, Beach moved to Boston to establish the Reformed Medical College. A founder of the National Eclectic Medical Association, in 1855 he served as its president. A student of the Bible, Beach believed that many contemporary ideas and practices were opposed to biblical teaching; and for several years he edited the lay periodicals, *The Telescope* (1826–1828), and *The Ishmaelite*, and *The Battle-Axe*, each of which were devoted to religious and moral topics of contemporary interest.

References

Beach, W. *The reformed practice of medicine: as taught at the reformed medical colleges....* Boston, by the author, 1831.

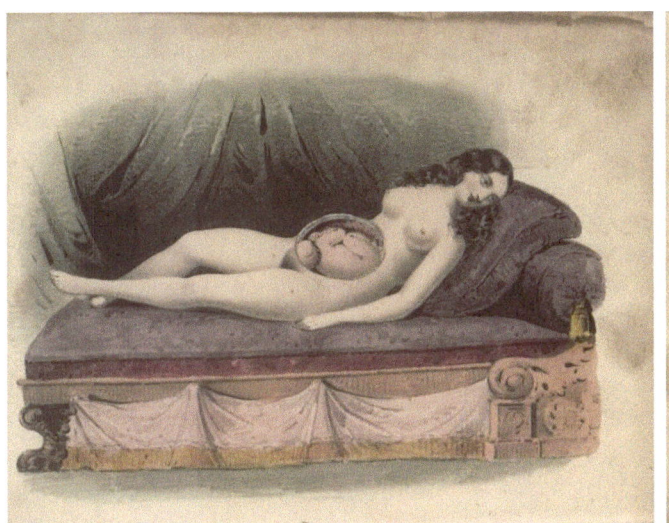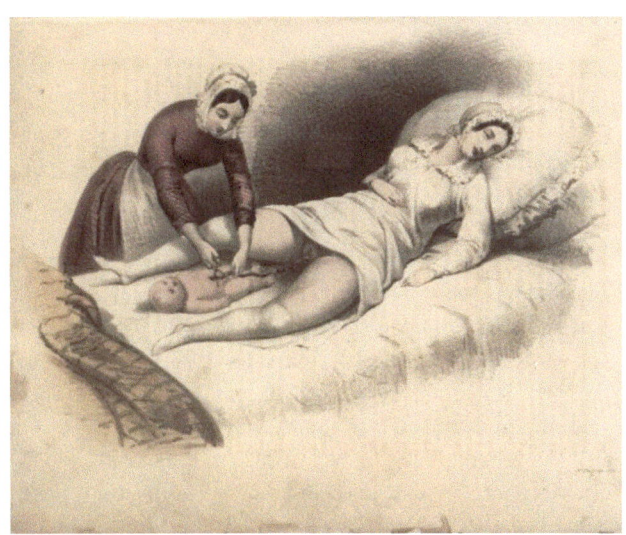

ILLUSTRATIONS FROM BEACH, 1847

Filippo Civinini

Lettera e Memoria Anatomica ... Intorno alla comunicazione diretta vascolare saguigna tra madre e feto. In risposta ad alcuni quesiti del Dottore B. Guglielmo Nob. De Seile ... Firenzi, Batelli e Compagni, 1859.

In this treatise, Civinini (1805–1844) confirmed the convincing demonstrations of the Hunter brothers and Mackenzie that there is no direct connection between the placental circulations of the mother and the fetus. The subject is presented in the form of autopsy reports, and accounts of experiments carried out on dissected human and animal cadavers. To establish the separate circulations of maternal and fetal blood in the placenta, Civinini injected coloured liquids into the placenta using an apparatus especially devised for the purpose, and which is depicted on the last folding plate. The fine, hand-coloured plates depict the gravid uterus at different stages, the placenta, and the vascular system, both human and animal. As noted earlier, almost a century earlier the Hunter brothers, William (1718–1783) and John (1728–1793), had established that the circulation of the fetal side of the placenta did not communicate with that of the mother (see Hunter 1774). Civinini published this work in a response to queries by Burkhard Wilhelm Seiler (1779–1843), professor of anatomy and surgery at Wittenberg University.

Perhaps surprisingly, as late as mid-nineteenth century, doubts remained concerning placental circulations. As an example, John Reid (1809–1849), anatomist of Edinburgh, wrote that in the half century since the Hunters' publications, the separateness of maternal and fetal circulations had been questioned by eminent investigators, "... but there cannot be a doubt that this opinion is erroneous, and ought now to be totally abandoned" (Reid 1841, p. 2). For instance, in his textbook of midwifery of 1833, William Campbell (1788–1848), of Edinburgh, described in "Peculiarities of the Foetus at Birth," the fetal circulation including that of the placenta. He observed, "It is still a disputed point whether the pulmonary arteries convey blood to the lungs" (Campbell 1833, p. 128). He continued presenting arguments for some admixture of blood from the fetal venous circulation with that arterialized blood returning from the placenta via the umbilical vein. Although writing some 60 years after the discovery of oxygen as a gas, it was 4 years later that Heinrich Gustav Magnus (1802–1870) first established by quantitative analysis that arterial blood contains more oxygen than that in veins (Magnus 1837). In this regard Campbell noted, "Though denied by the most celebrated physiologists of the day, that any difference exists between the blood in the arteries and that contained in the veins, yet I must still ... agree ... that in colour at least, they are dissimilar" (Campbell 1833, p. 129).

Civinini studied at the Medical and Surgical School of Pistoia and the University of Pisa, both in provinces of Tuscany. This region is of great historical interest, with central Italy being the center of Etruscan culture for many centuries B.C.E., the starting point of the Revival of Learning and the Renaissance in the late Middle Ages, and the center of reforms that resulted in the uniting of the Italian Kingdom in 1861. He became professor of surgical pathology and anatomy and director of the anatomical museum at the University of Pistoia (This college of medicine closed in 1839 with the Italian national reform in medical education). Civinini also prepared a catalogue of the collections of the Museum of Anatomy at Pisa. He was a defender of the law of recapitulation, i.e., the appearance of repeating the evolutionary stages of a species during the embryonic development of an individual organism. Civinini first reported on the pathologic abnormality, a neuroma, that causes a painful foot condition commonly referred to as (Thomas George) Morton's "metatarsalgia" (1835–1891) (Morton 1876). He also commenced work on *Commento medicofisico alla Divina Commedia* [a medico-physical commentary on the Divine Comedy] by Dante Alighieri (1265–1321); however, he did not live to complete this work.

References

Bargiacchi, L. *Storia degli Istituti di benedicenza d'istruzione ed educazione in Pistoia e suo circondario dalle respettive origini a tutto l'anno 1880.* Firenze, Tipografia della Pia casa di patronato pei minorenni, 1883–1884, pp. 268–273.

Campbell, W. *Introduction to the study and practice of midwifery, and the diseases of women and children.* Edinburgh, Adam and Charles Black, 1833.

Civinini, F. *Sulle conoscenze mediche de Dante Alighieri.* Manuscript, Pistoia, n.d.

Civinini, F. *Su d'un nervosa gangliare rigonfiamento alla planta del piede, lettera anatomica.* Pistoja, Dalla Tip Bracali, 1835.

Coturri, E. *Le scuole ospedaliere di chirurgia del Granducato di Toscana (secoli XVII-XIX).* Minerva Medica, 1958, pp. 3–18.

Hunter, W. *Anatomia uteri humani gravidi tabulis Illustrata... The Anatomy of the Human Gravid Uterus Explained by Figures.* Birmingham, John Baskerville, 1774.

Morton, T.G. A peculiar and painful affection of the fourth metatarso-phalangeal articulation. *Am J Med Sci* 71: 37–45, 1876. (GM 4341).

Pasero, G. & P. Marson. Filippo Civinini (1805–1844) e la scoperta del neurinoma plantare. *Reumatismo* 58: 319–322, 2006.

Reid, J. On the anatomical relations of the blood vessels of the mother to those of the foetus in the human species. *Edinburgh Med Surg J* 55:1–12, 1841.

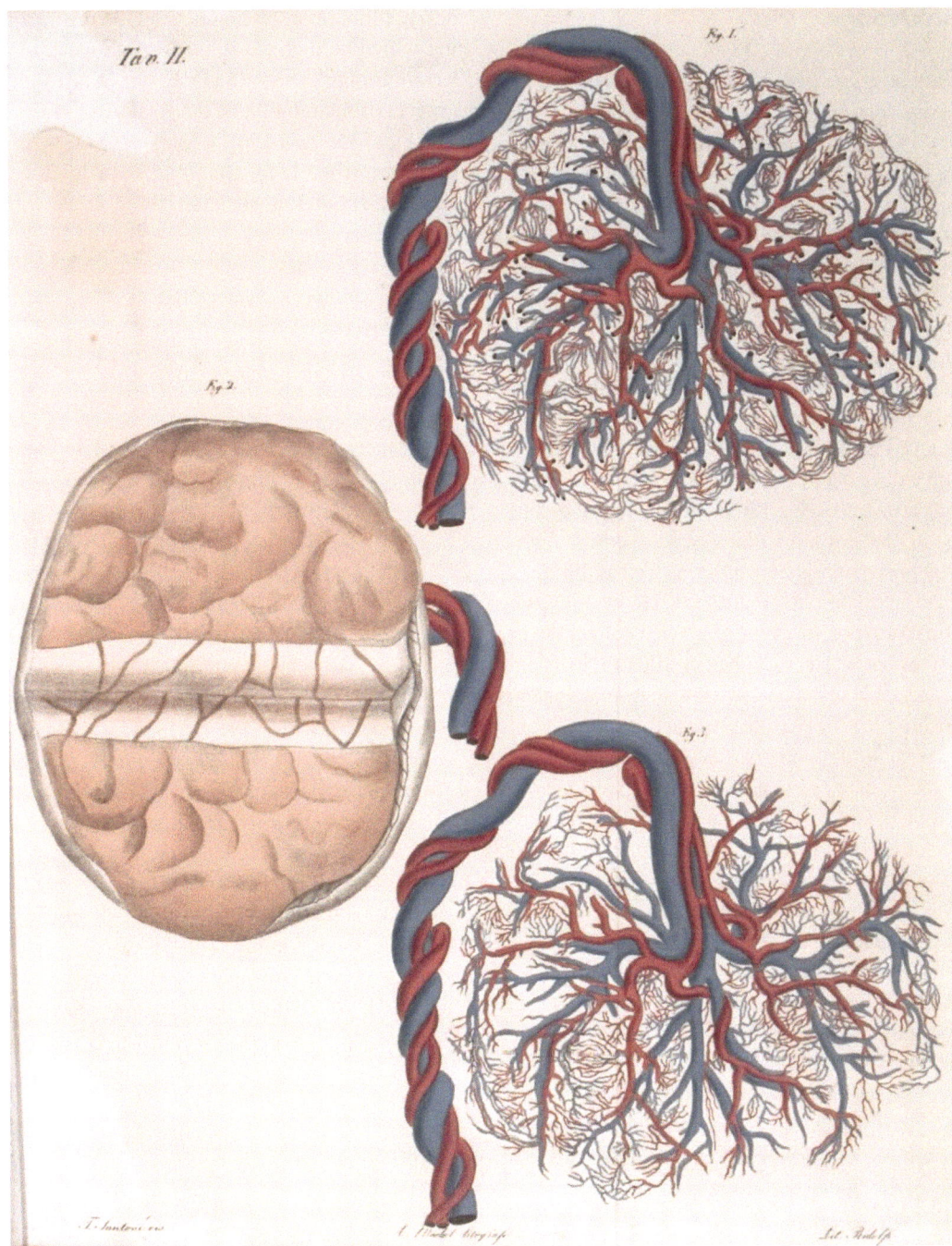

TAV. 2 FROM CIVININI, 1859

Hugh Lenox Hodge

The principles and practice of obstetrics. Illustrated with one hundred and fifty-nine lithographic figures from original photographs, and with numerous woodcuts. Philadelphia, Blanchard & Lea, 1864.

A highly regarded teacher of obstetrics and gynecology, and accurate observer of great experience, Hodge (1796–1873), Professor of Obstetrics at the University of Pennsylvania, described "parallel planes" at the various levels of the pelvis and stressed the inadequacy of external pelvimetry in assessing pelvic capacity for delivery. This was in conjunction with his analysis of the mechanism of labor in the several pelvic types. He invented the Hodge pessary to correct uterine retroversion, and the placental forceps to aid in uterine evacuation following miscarriage (Hodge 1860). Hodge opposed the use of anesthesia in obstetrics, the use of accouchement force in cases of severe preeclampsia-eclampsia, as well as the concept of the contagiousness of puerperal sepsis. He also inveighed against criminal abortion. In cases of pelvic contraction, and/or when previous pregnancies resulted in stillbirth, in an effort to avoid the hazards of cesarean section with its maternal mortality approaching 90 %, Hodge advocated the induction of labor prior to term. In addition to modifying the Hodge obstetrical forceps, he invented several useful instruments including modifications of Baudelocque's cephalotribe and craniotomy scissors. In mid-century America, this was the authoritative work on this subject.

Hodge was a graduate of the University of Pennsylvania (1818), and subsequently taught there, first in anatomy and then in surgery. In 1826 he was a founding editor of the *North American Meddical and Surgical Journal*. Because of failing eyesight, had to present his lectures from memory and also switched from surgery to obstetrics. In 1832 he became physician-in-charge of the lying-in department of Pennsylvania Hospital, after which he was appointed Professor of Obstetrics and the Disease of Women (1834–1863). As his eyesight had deteriorated to the point of being nearly blind, in 1863 he resigned his professorship, following which he dictated this 550 page work to his son Hugh Lenox Hodge (1836–1881). Plate XXXII illustrates several aspects of the relation of the fetal head to the maternal pelvis (Figs. 155 and 156), a transverse lie with shoulder presentation (Fig. 157), the skull of a hydrocephalic infant (Fig. 158), and a skeleton of a fetus with two heads (Fig. 159).

References

Cutter & Viets, pp. 157–159; Garrison Morton 6185; Heirs of Hippocrates 1556.

Goodell, W. *Biographical memoir of Hugh L. Hodge.* Philadelphia, Collins, 1874.

Hodge, H.L. *On diseases peculiar to women, including displacements of the uterus.* Philadelphia, Blanchard & Lea, 1860. (GM 6043.1).

Hodge, H.L. *Foeticide, or criminal abortion....* Philadelphia, Lindsay and Blakiston, 1869.

Kaufman, M., S. Galishoff, and T.L. Savitt (Eds). *Dictionary of American medical biography.* Vol. 1. Westport, Conn., Greenwood Press, 1984, p. 354.

Kelly, H.A. *A Cyclopedia of American Medical Biography: Comprising the Lives of Eminent Deceased Physicians and Surgeons from 1610 to 1910.* Vol. 1. New York, W.B. Saunders Company. 1920, p. 535.

Penn Biographies. Hugh L. (Hugh Lenox) Hodge (1796–1873). http://www.archives.upenn.edu/people/1700s/hodge_hugh_l.html, 1995–2013, University of Pennsylvania University Archives and Records Center.

Speert, H. *Obstetric and gynecologic milestones; essays in eponymy.* New York, Macmillan, 1958.

Thoms, H. Hugh Lenox Hodge. A master mind in obstetrical science. *Am J Obstet Gynec* 33:886–892, 1937.

Thoms, H. *Chapters in American obstetrics.* Second Edition. Springfield, Ill., Charles C. Thomas, 1961, pp. 74–82; 548–549.

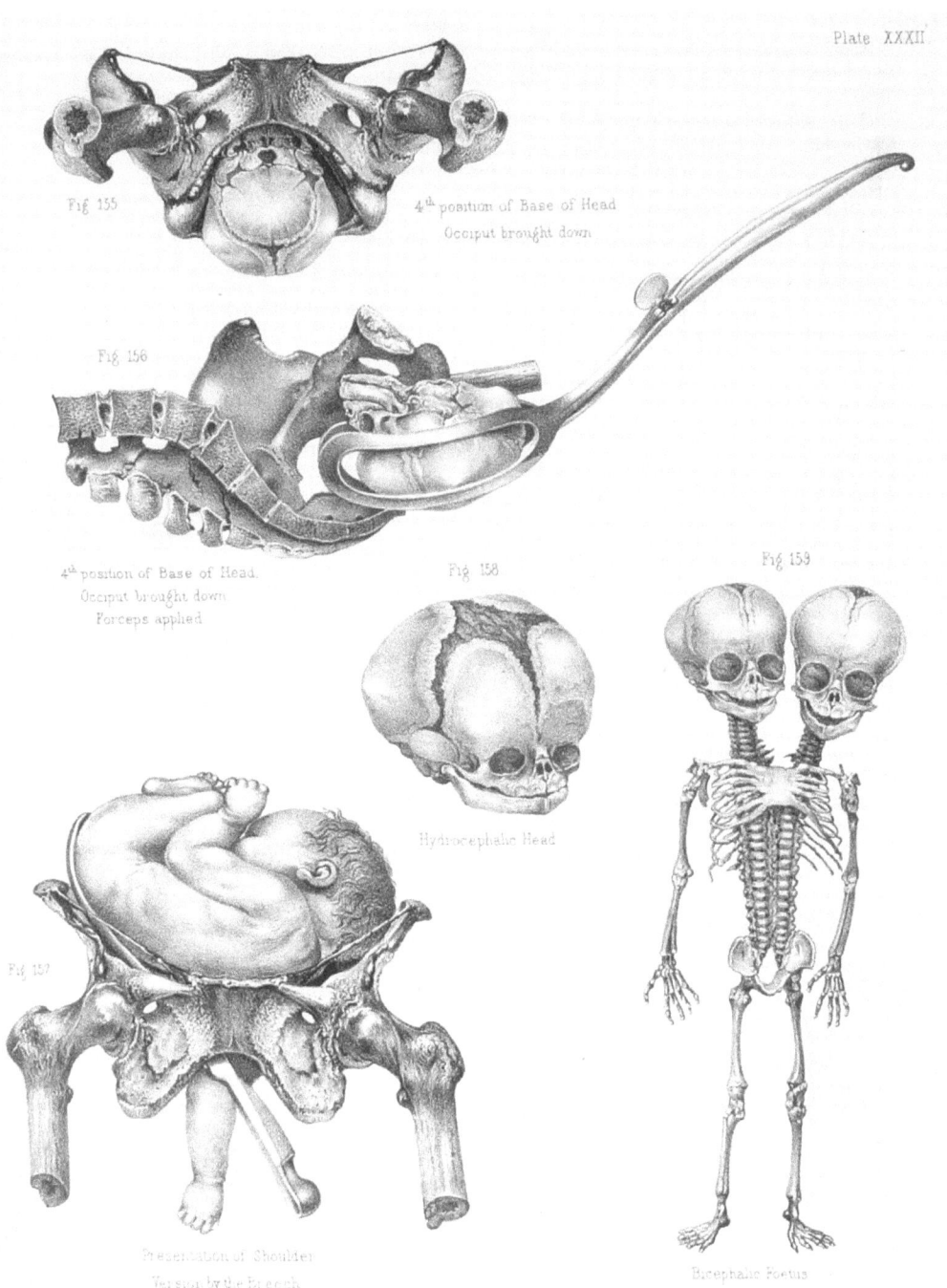

Plate XXXII

Fig 155

4ᵗʰ position of Base of Head
Occiput brought down

Fig 156

4ᵗʰ position of Base of Head.
Occiput brought down
Forceps applied

Fig 158

Fig 159

Hydrocephalic Head

Fig 157

Presentation of Shoulder
Version by the Breech

Bicephalic Foetus

PLATE XXXII FROM HODGE, 1864

Adolphe Lenoir, Marc Sée and Stéphane Tarnier

Atlas complémentaire de tous les traité d'accouchements....
2 volumes. Paris, Victor Masson…, 1865.

Lenoir (1802–1860) of l'hôpital Necker and the University of Paris died before the text of this work could be completed. The volume was finished and seen through the press, however, by Sée (b. 1827) and Tarnier (1828–1897), fellow members of the Paris faculty, and the latter of which was one of the great *accoucheurs* of that city. The 105 delicate, coloured plates of the atlas were drawn and lithographed by Emile Beau (b. 1810). These illustrate in detail normal and abnormal anatomy of the pelvis, embryology, the mechanism of parturition for various presentations, and obstetrical instruments. The first 15 pages of the atlas contain descriptions of each plate. Plates 45 and 46 depict modifications of the ovum within the oviduct, with segmentation of the cell mass into the blastocyst (Fig. 5, Plate 46), and early embryo (Fig. 7, Plate 46). Plate 72 shows methods of delivery with fetal malpresentation of the trunk. Plate 99 shows the method of craniotomy of a fetal head at the pelvic brim (Fig. 4), and several instruments for this purpose (Figs. 1–3).

Additional plates dealt with varioius aspects of pregnancy. In Plate 11 he illustrates the changes in the female pelvis from infancy (center), a mature woman (upper), and old age (lower). Plate 76 shows twin pregnancy with both fetuses in cephalic-presentation (Fig. 1), and one cephalic with the other breech (Fig. 2). Plate 79 illustrates the twin placentas, separate (Fig. 1), combined with two umbilical cords within a single amniotic sac, and a true knot in the umbilical cord to the left (Fig. 2), and a single placenta with bifurcating umbilical cord (Fig. 3).

Lenoir was also a Chevalier of the French Légion f'honneur.

References

Waller 5716 (Atlas only).

Tarnier, E.S. *Description de deux nouveaux forceps.* Paris, Lauwereyns, 1877. (GM 6192).

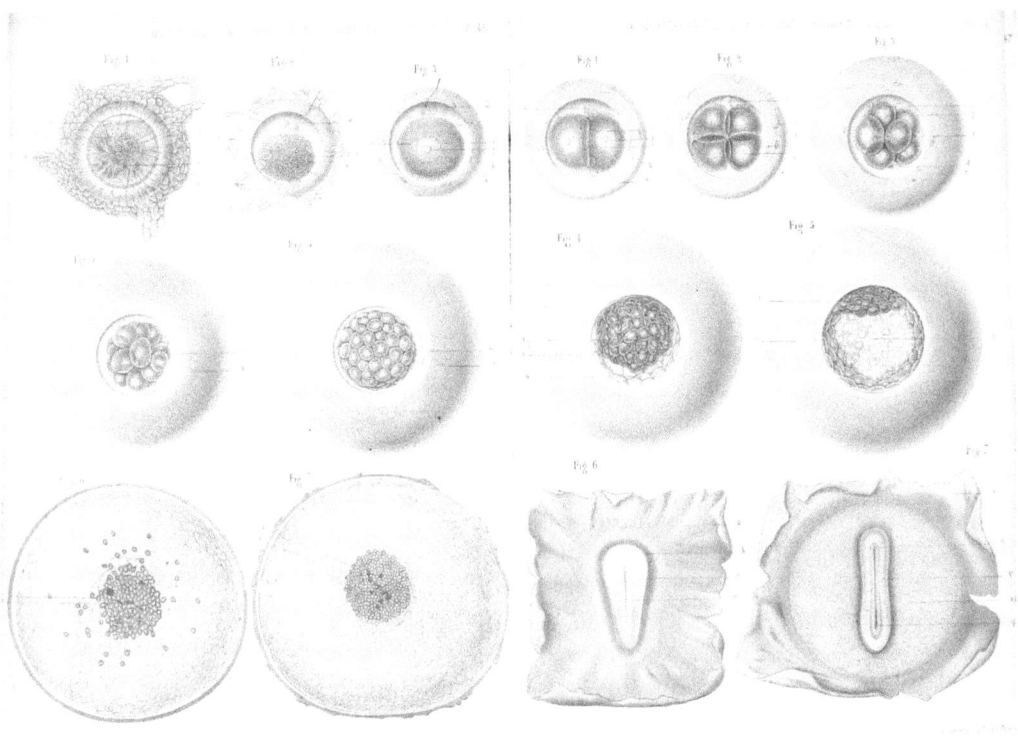

PLATES 45 AND 46 FROM LENOIR ET AL., 1865

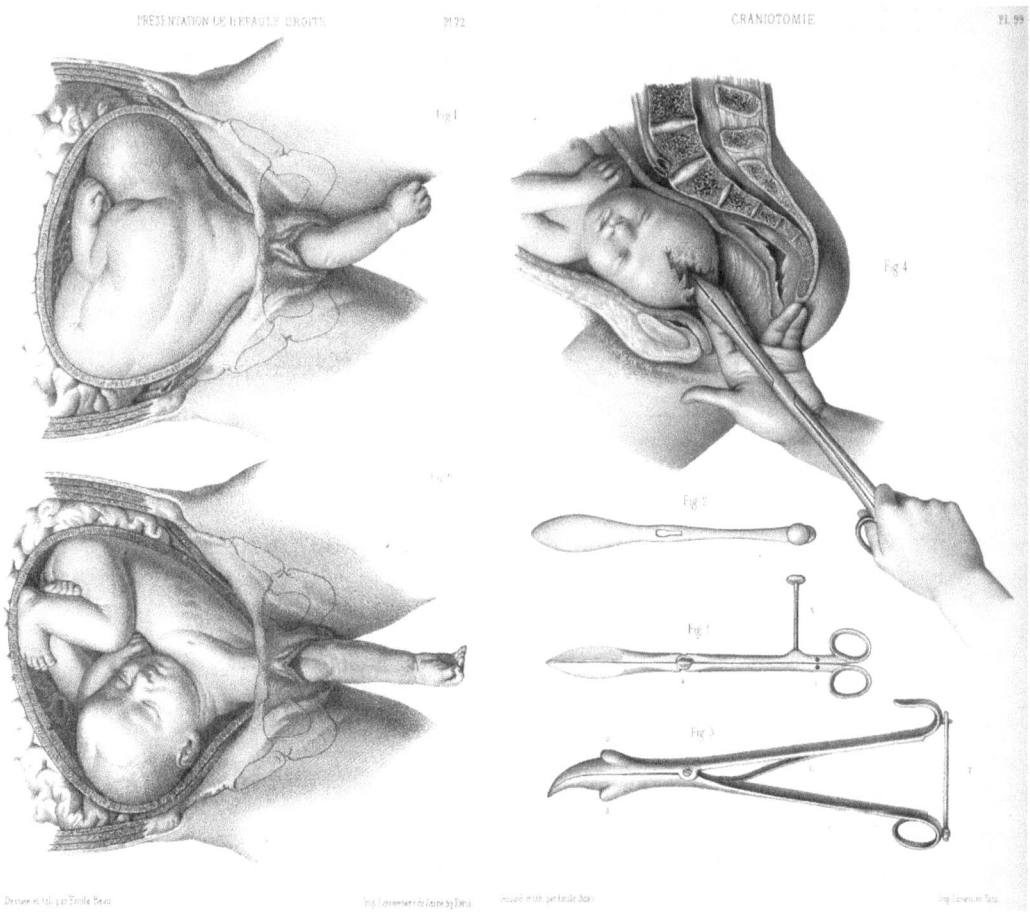

PLATES 72 AND 99 FROM LENOIR ET AL., 1865

Etienne Stéphane Tarnier

Description de deux nouveaux forceps. Paris, Lauwereyns, 1877.

Following the introduction and use of obstetrical forceps to aid delivery in the early to mid-eighteenth century, literally hundreds of modifications were introduced by various practitioners. In part, this was a consequence of the recognition that the fetus plays a passive role in the process of birth. Thus, consideration of the pelvic dimensions, shapes, planes, and axes played an increasing role in understanding the mechanism of childbirth. In this report, Tarnier (1828–1897), one of the great *accoucheurs* of Paris and chief of obstetrics at *L'Hôpital Maternite* described his axis-traction forceps for obstetrical delivery. Tarnier, who devoted a considerable portion of his life to study of the mechanical principles of forceps assisted delivery, maintained that with axis traction, the force required to deliver the infant was much less than without its use, with less possible injury to fetus or mother. He noted that the fetal head would follow the axis of the pelvic cavity whatever its orientation. This "arrangement allows the head the liberty of always following the curvature of the birth canal. Thanks to the mobility of the articulation of the traction rods on the grasping handles, the head can still easily and spontaneously execute its rotating movement around the axis of the pelvis."

Figure 5, from Tarnier 1877 (*Annales de Gynécologique* (Paris) 7: 201–215) shows the forceps, with the blades for encasing the fetal head (S), the shank for traction (I), handle for pulling (P), appendages for depression (O), and pressure screw (V). The axis traction principle became widely used, as by guiding the fetal head in its descent through the pelvis, it minimized trauma to either "passenger or passage". In an earlier report Tarnier had discussed the use of a pointer (*auguilles*) to aid in forceps delivery (Tarnier 1877, p. 161). The same year, Tarnier published a report in which he discussed axis traction *in extenso* and defended its use against the objections of others (Tarnier 1877). As recorded by John Whitridge Williams (1866–1931), over the two decades from the introduction of these forceps until his death, Tarnier continued to modify and improve the instrument (Williams 1896).

Tarnier was born in a village near Dijon, and qualified in medicine in Paris. His doctoral thesis addressed the contagiousness and transmission of puerperal fever, which at that time claimed the lives of one in every six women who delivered in hospital. With this death rate being over tenfold greater than that of women who delivered at home, Tarnier was convinced that the disease was contagious (Tarnier 1857). He later expanded upon this thesis with a major work on infection in pregnancy, with emphasis on antisepsis and asepsis (Tarnier 1894). Tarnier was one of the first in France to adopt the antiseptic principles of Joseph Lister (1827–1912) and, under his direction, over a period of a few years the mortality rate from puerperal sepsis fell to near zero.

Following delivery, Tarnier also advocated waiting to ligate the umbilical cord until it had ceased pulsating, thus fostering optimal blood being transfused from the placenta to the newborn infant. He also gave special attention to the care of the newborn, and played a key role in developing incubators for prematurely born and other ill infants. In 1891, he was elected president of the *Academié de Medicine* in Paris.

References

Cutter Viets pp. 66, 194; Garrison Morton 6192.

Lister, J. *The Collected Papers of Joseph Baron Lister.* 2 vol. Oxford, Clarendon Press, 1909. (GM 85).

Pajot, C. La seconde sur le forceps a aiguille. *Ann de Gynécol* (Paris) 7: 321–362, 1877.

Speert, H. Étienne Stéphane Tarnier and his axis traction forceps. In: *Obstetric and Gynecologic Milestones.* New York, Macmillan, 1958, pp. 481–491.

Tarnier, E.S. *Recherches sur l'état puerperal et sur les maladies des femmes en couches.* Thesis, Paris, Rignoux, 1857.

Tarnier, E.S. *Des cans dans lesquels l'extraction du foetus est nécessaire et des procédés opératoires relatifs à cette extraction.* Paris, Baillière, 1860.

Tarnier, E.S. Travaux originaux. Examen du forceps a aguilles. *Ann Gynéc* (Paris) 7: 161–174, 1877.

Tarnier, E.S. Du nouveau forceps de M. Tarnier. Discussion relative au nouveaux de M. Tarnier. A Monsieur le Professor Pajot. *Annales de Gynécologique* (Paris) 7: 201–215 and 241–264, 1877.

Tarnier, S., G. Chantreuil, and P. Budin. *Traité de l'art des accouchements....* 2 vol. Paris, G. Steinheil, 1886–1888.

Tarnier, E.S. *De l'asepsie et de l'antisepsie en obstétrique... Recueillies et Rédigées par le Docteur J. Potocki.* Paris, G. Steinheil, 1894. (GM 5639).

Thoms, H. Etiene Tarnier 1828–1897. In: *Classical contributions to obstetrics and gynecology.* Springfield, IL, C.C. Thomas, 1935, pp. 153–157.

Williams, J.W. Demonstration of instruments illustrating the history and development of the axis traction forceps. *Bull Johns Hopkins Hosp* 7: 24–25, 1896.

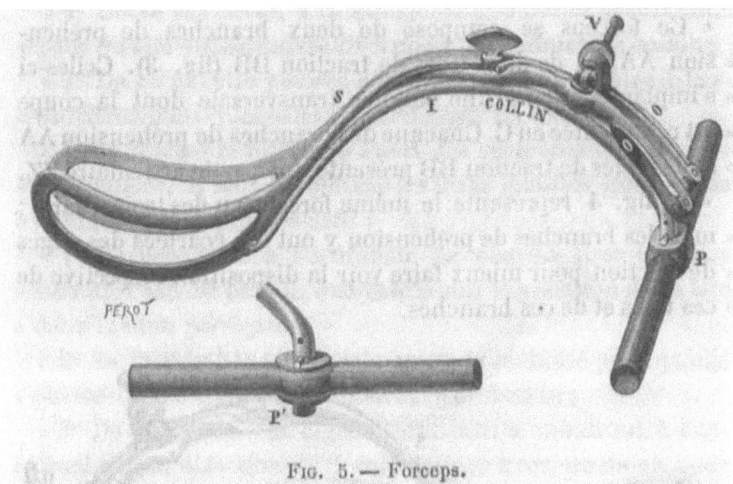

FIG. 5. — Forceps.

S, branches de préhension. — O, oreilles pouvant s'élever ou s'abaisser. —
P, poignée dans laquelle s'implantent les tiges de traction. — P', même poignée
vue à part.

FIG. 5 FROM TARNIER, 1877

Theodor Langhans

Untersuchungen uber die menschliche Placenta. Archiv für Anatomie und Entwickelungsgeschichte 188–267, 1877.

Until the mid-nineteenth century, there existed considerable disagreement regarding the exact nature of the outermost cells of the placental villi in the human. Langhans (1839–1915) first clearly described the cells of these villi, the *chorion frondosum* [leafy membrane] and *chorion laeve* [disappearing membrane], which formed a continuous layer from the early stages of development, and were of fetal origin (Langhans 1870). In his essay "knowledge/information of the human placenta," Langhans described the maternal portion of the placenta, the manner in which the anchoring villi attach to the decidua, intervillous anastomosis, and syncytial knots (Langhans 1870). In this report, contrary to what is commonly stated, he did not describe the cytotrophoblast. Several years later, in the present essay "inquiry/examination of the human placenta" he distinguished between the inner *Zellschicht* [cell layer], the large, individual, pale-staining, polyhedral epithelial cells with large nuclei and individual cell boundaries/membranes, and the outer *Chorionepithel* [chorionic epithelium]. We now know these as the cytotrophoblast, or "Langhans" cells and the syncytium or syncytiotrophoblast, respectively. Langhans noted that the double layered trophoblast covering the outer surface of the villi decreases in thickness with advancing age, and that lines representing what might be cell boundaries are visible among the nuclei only in younger ova, and very rarely even there (Langhans 1877, pp. 201–206). In a later report, he described in greater detail the cell layers of the chorionic membranes (Langhans 1882).

In plate VIII, Fig. 15B depicts the *chorion leave* (smooth, or non-villous, chorion) from a placenta at 14 weeks gestation: *a*, chorion membrane, *b*, cell layer, and *c*, the epithelial layer or syncytium. Figure 16A depicts the chorion with an atrophic villus. In this report Langhans also elucidated features including the intervillous circulation, fibrin deposition, and calcification. Langhans also reported that the trophoblast thickness decreases with advancing gestational age, noting, "… in the mature placenta the cell layer can no longer be demonstrated with certainty".

In 1880, he correctly surmised that these trophoblastic cells were of fetal origin (Langhans 1880). Langhans did not use the term *trophoblast* [nutrition + germ]; this was introduced by Ambrosius Arnold Willem Hubrecht (1853–1915) of Utrecht, to indicate that part of the blastocyst not contributing to formation of the embryo *per se*, but, rather, forming the placental villi for its nourishment. Hubrecht appreciated the double cellular layer, with the outer layer of nuclei in "nests" not demonstrating mitosis, and the innermost layer of cylindrical cells. Although in HUbrecht's original descrip-

tion, the term trophoblast had limited morphologic significance, soon, however, it came to apply to the epithelial derivatives of the outer layer of the blastocyst, the two cell layers described by Langhans. Later it became recognized that the outermost layer was a syncytium of cells, the syncytiotrophoblast [cells together + trophoblast], and innermost Langhans layer, the cytotrophoblast [cellular + trophoblast] (Bonnet 1903, Boyd and Hamilton 1966, Hubrecht 1888, 1889). Several decades later it was established that the cytotrophoblast cells give origin to the syncytiotrophoblast [cells together + trophoblast] (Bonnet 1907).

A further technical and conceptual contribution to understanding placental morphology, was that of Charles Sedgwick Minot (1852–1914), who in 1886 invented the automatic rotary microtome for cutting ultra-thin tissue sections. Minot gave a definitive account of the microscopic structure of the human placenta, in one instance describing a section through the uterus and placenta in situ at 7 months gestation, with amnion, chorion, villus trunk, sections of villi in the substance of the placenta, decidua, muscularis, uterine blood-vessel opening into the placenta, and so forth (Minot 1889). In a later review Minot noted, "the chorion is separated by a dense forest of villi from the decidua, … the ends of some of the villi touch and are imbedded in the decidual tissue … the decidua is plainly divided into two strata … the section passes through a wide tube, … which opens directly into the interior of the placenta and contains blood … this opening is … a vein …" (Minot 1900–1904, p.). These contributions did much to help clarify the nature of the fetal-maternal (e.g., placental) barrier.

Langhans trained under several of the great pathologists including Rudolf Ludwig Karl Virchow (1821–1902) in Berlin, Friedrich Gustav Jakob Henle (1809–1885) in Göttingen, and Friedrich Daniel von Recklinghausen (1833–1910) in Würzberg. Following several years at Marburg, he became chairman of pathology at the University of Berne (1872), where he had a wide influence. In addition to his contributions on the microscopic structure of the placenta, Langhans also observed the multinucleated giant cells of tuberculosis granuloma (1867) and a variety of other granulomatous disorders such as sarcoidosis (Sakula 1987), and the giant cells in the lymph nodes of Hodgkin's disease (1872). He also made fundamental contributions to the pathologic changes in the thyroid gland, kidneys, and other tissues. In teaching, his "… encouragement to exact observation" is stated to have given his students "… a priceless preparation for their careers in after-life" (Anonymous 1916). One of Langhans' students, Raissa Nitabuch (_____), in her inaugural dissertation, described a further placental layer, the eosinophilic honeycombed fibrinoid deposition that develops between the invading trophoblast and its contact with the boundary zone of maternal decidua (Nitabuch 1887).

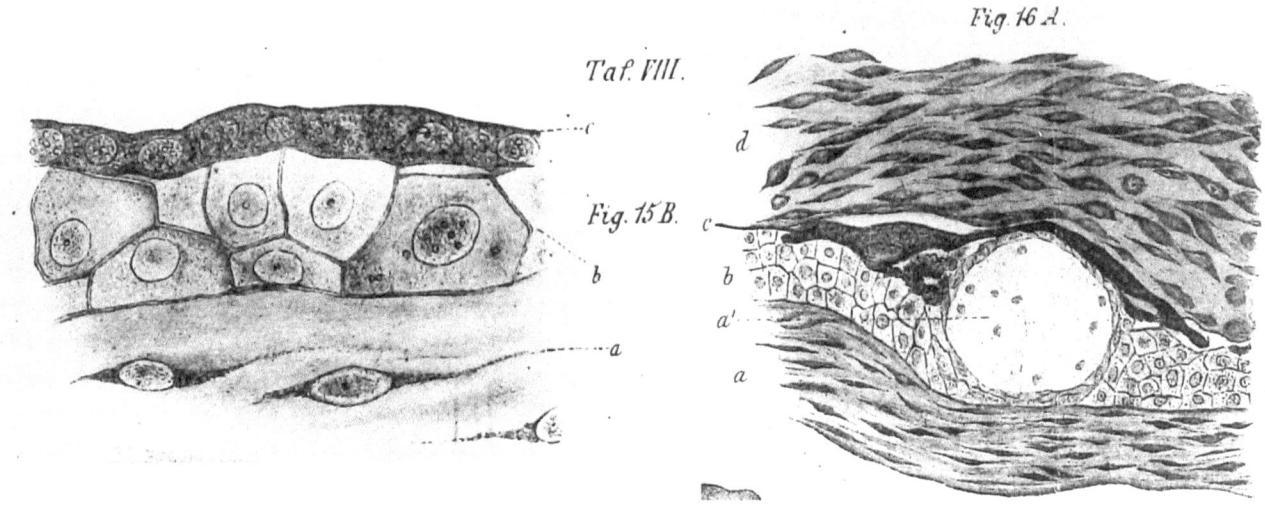

FIGURES 15B AND 16A FROM TAF. VIII OF LANGHANS, 1877

References

Anonymous. Obituary. Dr. Theodor Langhans. *Lancet* 1:161, 1916.

Bonnet, R. Über Syncytien, Plasmodien und symplasma in der Placenta der Säugetiere und des Menschen. *Mschr Geburtsh Gynäk* 18:1–51, 1903.

Boyd, J.D. and W.J. Hamilton. *The human placenta.* Cambridge, Heffer, 1970.

Holubar, K. Looking Back: Theodor Langhans, Paul Langerhans and Carol Touton – contemporary histopathologists. *Am J Dermatopath* 12:534–535, 1990.

Hubrecht, A.A.W. Keimblätterbildung und Placentation des Igels. *Anat Anzeiger* 3:510–515, 1888.

Hubrecht, A.A.W. Studies in mammalian embryology. 1. The placentation of Erinaceus europaeus with remarks on the phylogeny of the placenta. *Quart J Microsc Sci* 30:283–404, 1889.

Langhans, T. *Die Ubertragbarkeit der Tuberkulose auf Kaninchen Habilitationsschrift.* Marburg, Druck von J.A. Koch, 1867.

Langhans, T. Zur Kenntnis der menschlichen Placenta. *Archive für Gynäkologie* 1:317–334, 1870. (GM 491)

Langhans, T. Das maligne Lymphosarkom (Pseudoleukämie). *Virchows Arch Path Anat* 54: 509–537, 1872. (GM 3767)

Langhans, T. Ueber die Zellschicht des menschlichen Chorion. *Beitr Anat* (Vol #):69–79, 1882.

Minot, C.S. 1900–1904.

Nitabuch, R. *Beiträge zur Kenntniss der menschlichen Placenta. Inaugural-Dissertation....* Berne, Stämpflische Buchdruckerei, 1887.

Sakula, A. In memoriam. Theodor Langhans (1839–1915). *Sarcoidosis* 4:77–78, 1987.

Adolphe Pinard

Traité du palper abdominal au point de vue obstétrical et de la version par manoeuvres externs. Paris, H. Lauwereyns, 1878.

In his monograph on the external version of the fetus, Pinard (1844–1934) one of the leading *accoucheurs* in Paris, professor at the University of Paris and director of the obstetrical clinic, stressed the importance of abdominal palpation in obstetric diagnosis to determine the orientation of the fetus. Previously, obstetric examination had been based chiefly on the internal examination, and physicians had not appreciated the information that could be derived from suprapubic palpation of the abdomen. Aware of the dangers of too frequent or indiscriminate vaginal examination, and the hazard of puerperal infection, and with the recent understanding of the bacterial basis for infection, Pinard worked to determine the fetal presentation and orientation by external palpation. Because of the relative hazards of breech delivery, Pinard emphasized the value of external cephalic version in cases of breech presentation. Regarding the method of examination of the abdomen of the parturient, Pinard stressed that the physician's hands should be warm, and the patient's bladder and bowels be empty. He wrote, "…it is necessary that the *accoucheur* should stand at about the height of the umbilicus. One should then examine the thickness of the abdominal wall, for the sensations perceived will be more or less distinct and superficial, just in proportion as the abdominal wall is more or less thick … For these reasons I have endeavored to simplify the method, to render it rational, and make it rest entirely on the exact knowledge of the various attitudes the foetus may assume during the last month of gestation; i.e., the accomodation". Pinard continued, "It is necessary to find the pubis and its horizontal rami, i.e., the superior opening to the excavation or the anterior part of the superior strait. It is absolutely indispensable to recognize this point, as it is only after this, that it will be possible to appreciate the degree of engagement of the foetal part, which will be more or less marked, according as the presentation is found above or below this point". He also noted, "The examiner should then interrogate the excavation. For that purpose, placing the hands about five or six centimetres to the right and left of the median line, the extremities of the fingers being in relation with the anterior curve of the pelvis, he depresses the abdominal wall from above downwards and from before backwards, just grazing over the horizontal rami of the pubes (See Fig. 1). When properly palpating, only two sensations may be perceived, viz: the fingers experiencing a sensation of resistance, resulting from contact with a hard, round, and voluminous body which fills the excavation, can not penetrate deeper; or, on the contrary, they only meet with the resistance offered by the soft parts, and can therefore sink more or less deeply into the excavation" (Pinard, 1878, p. 110 ff).

A decade and a half later, Christian Gerhard Leopold (1846–1912) elaborated on Pinard's method, to define four maneuvers by external palpation to determine the position, presentation, and degree of engagement of the fetus (Leopold and Spörlin 1894).

As noted above, Pinard emphasized the need to determine the fetal orientation in the case of performing external cephalic version to correct breech presentation. He wrote, "… apply one hand over the fetal head, and the other over the breech, and by gentle and sustained pressure exerted inversely over one and the other extremity, turn the two poles of the fetus… the pressure made over the breech is more efficient than that made over the head, in as much as it is more directly transmitted to the trunk" (Fig. 2). At a time when the rachitic flat pelvis was a common cause of brim disproportion, Pinard also advocated using the fetal head to assess engagement in the pelvic brim in order to evaluate the cephalopelvic relationship. As was said later, "The fetal head is the best pelvimeter".

Pinard was a medical student at the time of the Franco-Prussian War of 1870–1871. He enlisted in the army, and was awarded *le chevalier de la Légion d'honneur* [the Cross of the Legion of Honor] for his service. Upon returning to Paris, his medical graduate thesis "The defects of Conformation of the Pelvis, studied from the stand-point of the Conformation and the anterior-posterior Diameter", included 100 plates based on his analysis of as many pelves in the Anthropological gallery of the museum of the Paris maternity Hospital, *Hôpitaux de la Maternité* (Pinard 1874). He soon became an assistant to the master *accoucheur* Professor Stéphane Tarnier (1828–1897). It was Tarnier, in fact, who suggested that Pinard contribute to the neglected field of abdominal palpation.

In cases of obstructed labor, Pinard helped to popularize symphysiotomy (separation of the pubic/symphysis) to enlarge the pelvic cavity several centimeters, and thus allow delivery of the infant. This was an important development in avoiding the cruel dilemma of on one hand crushing the living infant's head to effect delivery, or performing a cesarean section (which while in many instances saved the baby, resulted in 90–95 % death of the mothers).

Importantly, Pinard helped to develop the idea of *la puériculture intrauterine* (intrauterine pediatric care). In this regard, Pinard's commitment to the health and welfare of women was demonstrated in a number of ways. He worked for the establishment of prenatal care, shelters for expectant mothers, nurseries, and a School for Pediatric Care. With the help of a French midwife, he opened a hostel, the *Refuge de L'Avenue du Maine,* and later the *Michelet Asylum,* for pregnant women who were without means. Deliveries were then conducted in the *Hôpital Baudelocque.* By 1895, he was able to show that the mothers who had rested and been looked after in the *Refuge* were more likely to deliver larger and

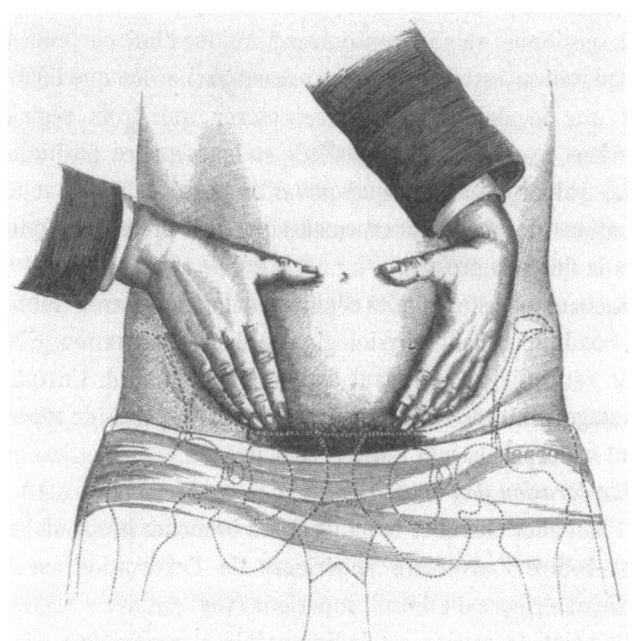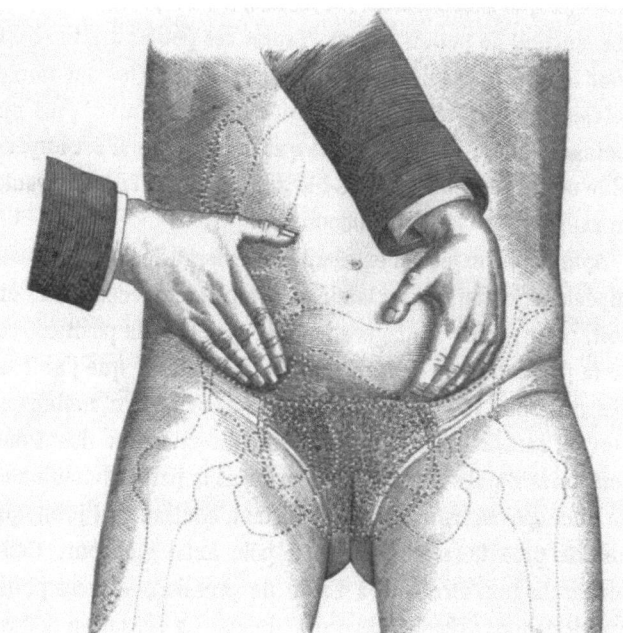

Fig. 1. — Position des mains au début de l'exploration de l'excavation.

Fig. 2. — Mains explorant l'excavation. Main droite arrêtée par le front à droite.

FIGURES 1 AND 2 FROM PINARD, 1878

more healthy babies at term. This new antenatal approach also provided the opportunity for checking mothers for pelvic disproportion, correcting malpresentations, detecting renal and other problems, in addition to attempting to ensure the normal development of the fetus. This in turn was followed by *extrauterine puericulture*, ensuring maternal and infant hygiene in care of the newly delivered mother to avoid puerperal sepsis, and insisting on non-separation of mother and baby and on breast feeding. Gradually, other hostels opened, to be followed by antenatal departments and wards in maternity hospitals.

In addition to his many contributions to medicine, Pinard served as mayor of his city of origin, *Méry-sur-Seine*, and was a member of Parliament.

References

Garrison Morton 6193.

Auvray, M. Adolphe Pinard (1844–1934). *Bull Soc Obstét Gynecol de Paris* 23: 335–342, 1934.

Bar, P. Adolphe Pinard. *Gynéc et Obstét* 29: 497–512, 1934.

Dumont, M. Adolphe Pinard. *La Press Médicale* 13: 1658–1660, 1984.

Dumont, M. Le cent-cinquantenaire de la naissance d'Adolphe Pinard (1844–1934). *J Gynecol Obstet Biol Reprod* 23: 351–357, 1994.

Dunn, P.M. Adolphe Pinard (1844–1934) of Paris and intrauterine paediatric care. *Arch Dis Child Fetal Neonatal Ed* 91: F231-F232, 2006.

Leopold, G . & Spörlin. Die Leitung der regelmässigen Geburten nur durch äussere Untersuchung. *Arch für Gynäk* 45: 337–368, 1894.

Pinard, A. *Les vices de conformation du bassin étudiés au point de vue de la forme et des diamètres antéropostérieurs....* Paris, Libraire J.B. Baillière et Fils, 1874.

Pinard, A. *De l'action comparée du chloroforme, du chloral, de l'opium et de la morphine chez la femme en travail.* Paris, Octave Dione, Libraire-Editeur..., 1878.

Pinard, A. *A treatise on abdominal palpation, as applied to obstetrics, and version by external manipulations.... Translated by L.E. Neale.* New York, J.H. Vail & Co., 1885.

Pinard, A. *La puériculture du premier âge....* Paris, Librairie Armand Colin, 1904.

Pinard, A. *A treatise on abdominal palpation, as applied to obstetrics, and version by external manipulations... Translated by L.E. Neale, edited by L.D. Longo.* New York, N.Y., The Classics of Obstetrics and Gynecology Library, Division of Gryphon Editions, 1995.

Thoms, H. Adolph Pinard, 1844–1934. In: *Classical contributions to obstetrics and gynecology.* Springfield, IL, Charles C. Thomas, 1935, pp. 43–46.

Witkowski, G.J. *Accoucheurs et sages-femmes célèbres.* Paris, G. Steinheil, 1890, p. 256.

David Berry Hart

The structural anatomy of the female pelvic floor. Edinburgh, Maclachlan and Stewart, 1880.

Following his graduation from the University of Edinburgh (1877), Hart (1851–1920) served as assistant to Alexander Russell Simpson (1835–1916), nephew of Sir James Young Simpson (1811–1870), and who succeeded him in the chair at Edinburgh. This volume is his graduation thesis, for which he received a gold medal. Later Hart became an extramural teacher in Edinburgh. For his anatomical studies, he introduced the technique of studying large frozen sections to investigate complications in obstetrics. Hart defined the term "pelvic floor" for the muscles crossing the lower pelvis. He demonstrated that this muscular sheet consisted of two distinct portions, an anterior or pubic segment, and a posterior or sacral segment, that were separated by the vagina, and which with the peritoneum and connective tissue helped to provide support of the uterus. In Plate II of a multiparous woman, Hart displays the right half of the vagina (A), as well as a vertical (coronal) section of the pelvis (B). The sectioned uterus is retroflexed and contains an embryo of 2 months gestation.

In his study of internal rotation of the fetal head in its descent through the pelvis, Hart noted, "… the part which first persistently strikes a lateral half of the sacral segment will be rotated to the front; internal rotation is never a direct movement to the back" (Hart 1912, p. 177). By use of his frozen section technique, Hart demonstrated an ectopic pregnancy in the broad ligament (Hart 1883), and he was the first in Scotland to perform successfully on ruptured ectopic pregnancy.

References

Anonymous. David Berry Hart, M.D., F.R.C.P. Edin. *Br Med J* 1:852–853, 1920.

Hart, D.B. *Atlas of female pelvic anatomy.* Edinburgh, W. & A.K. Johnston, 1884.

Hart, D.B. *Contributions to the topographical and sectional anatomy of the female pelvis.* Edinburgh, W. & A.K. Johnston, 1885.

Hart, D.B. *Guide to Midwifery.* London, Rebman, 1912, p. 177.

Hart, D.B. and A.H. Barbour. *Manual of gynecology.* Edinburgh, W. & A.K. Johnston, 1883.

Hart, D.B. and J.T. Carter. A contribution to the sectional anatomy of advanced extrauterine gestation. *Edinburgh Med J* 33:332–343, 1887.

Munro Kerr, J.M., R.W. Johnstone and M.H. Phillips. *Historical Review of British Obstetrics and Gynaecology, 1800–1950.* Edinburgh and London, E. & S. Livingstone, 1954, pp. 50–55.

Sanders, W.R. *Lectures in Pathology.* Taken by David Berry Hart, Edinburgh, 1875–1876.

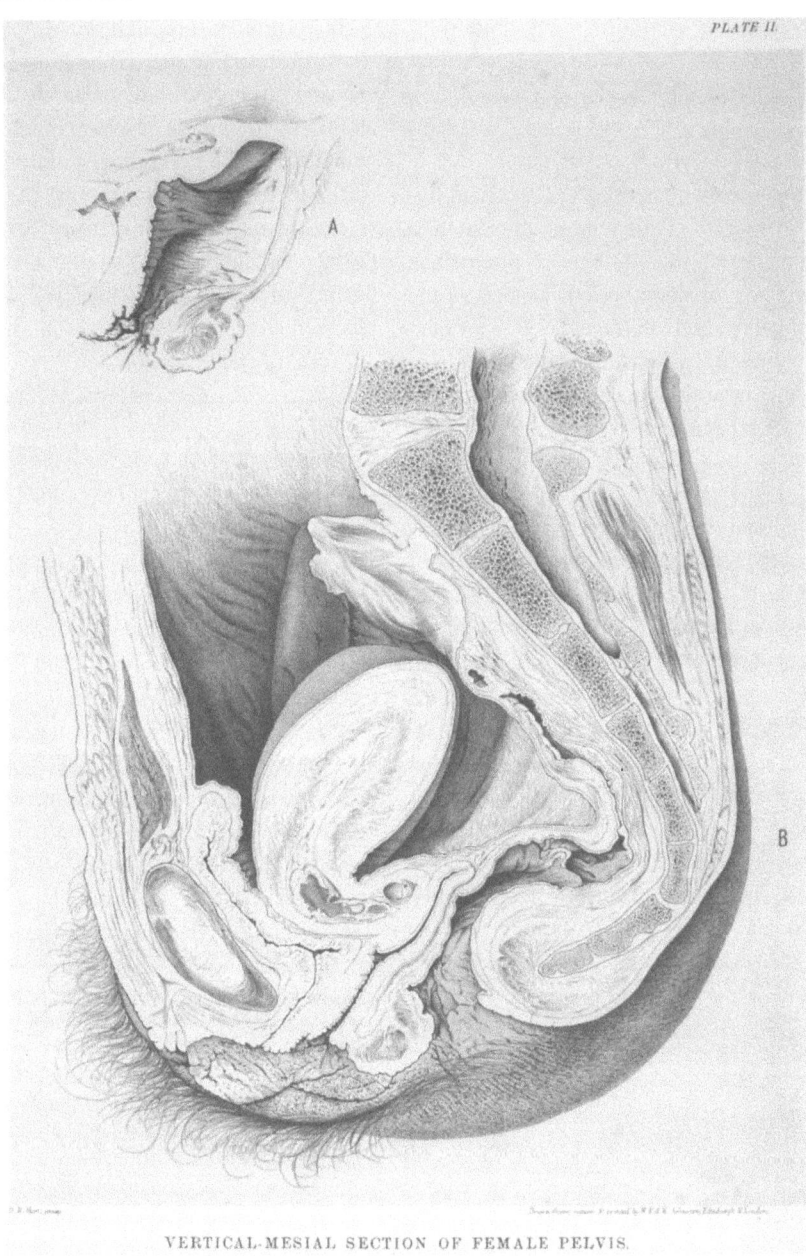

PLATE II.

VERTICAL-MESIAL SECTION OF FEMALE PELVIS.

PLATE II FROM HART, 1880

Frank Horace Getchell

An illustrated encyclopedia of the science and practice of obstetrics. Philadelphia, Gebbie & Co, 1885.

With 84 full page steel engraved plates and 133 woodcut figures, Getchell (1836–1907) of Philadelphia presented a series of anatomically correct illustrations of the pelvis, uterus, and fetus. The plates illustrate various aspects of the normal and malformed pelvis, and the female organs of generation. Following a series of plates that depict different views of the fetus during natural labor are about 30 figures that illustrate abnormal fetal positions and the step-by-step management of their resolution. For instance, Plate XXXV illustrates successive stages of delivery of the fetal head when in the "first" or normal occiput anterior position. This is followed by plates of the fetus delivering in the "third" or mentum anterior position, the placenta and its delivery, other figures of the mechanism of labor, version and extraction by breech delivery, presentation of the shoulder and arm, and delivery by forceps.

The accompanying text by a number of authorities expands upon the several topics to edify the reader in the most up to date medical views on various complications of labor and delivery. For the most part, these were taken directly or edited from textbooks on obstetrics by Paulin Cazeaux (1808–1862), William Leishman (1834–1894), William Smoult Playfair (1836–1903), Francis Henry Ramsbotham (1800–1868), and others; however, Getchell did not provide the citations for these essays.

In his preface, Getchell stated, "Nothing is more important to the obstetrician than to be prepared to meet the emergencies and difficulties that constantly arise in the lying-in room, and for the treatment of which there is then no time to consult authorities". He continued, pointing out the value of "correctly drawn" illustrations to aid the *accoucheur* in his work.

Following graduation from Dartmouth Medical School (1862), Getchell served as assistant surgeon with the Third Maine Volunteers during the Civil War (1861–1865). He also received a degree from Jefferson Medical College (1871).

References

Frank Horace Getchell. *JAMA* 49:162, 1907.

Cazeaux, P. *A theoretical and practical treatise on midwifery, including the diseases of pregnancy and parturition.... Translated from the second French edition, with occasional notes and a copious index by Robert P. Thomas....* Philadelphia, Lindsay & Blackiston, 1850.

Getchell, F.H. *The maternal management of infancy.* Philadelphia, J.B. Lippincott & Co., 1868.

Leishman, W. *A system of midwifery, including the diseases of pregnancy and the puerperal state.* Philadelphia, Henry C. Lea, 1873.

Ramsbotham, J. *Practical observations in midwifery, with a selection of cases.* London, Highley, 1832.

Ramsbotham, F.H. *The principles and practice of obstetric medicine & surgery, in reference to the process of parturition.* London, Churchill, 1841.

PLATE XXXV FROM GETCHELL, 1885
(COURTESY MEDACCESSUSA.COM [OBTAINED ON PINTEREST.COM])

Alexander Hugh Freeland Barbour

The anatomy of labour, including that of full-time pregnancy and the first days of the puerperium exhibited in frozen sections reproduced ad naturam. Edinburgh, W. & A.K. Johnston, 1889.

Inspired by William Hunter's (1718–1783) *Gravid Uterus*, Barbour (1856–1927), lecturer in midwifery and the diseases of women at the University of Edinburgh, presented 11 superb colored and 15 uncolored engraved plates, several of which were double page and folding. These plates, reproduced life-size, and as nearly as possible *ad naturam* from both frozen and fixed sections, illustrated the gravid uterus and its contents before the onset of labor, as well as during the first, second, and third stages of labor, and immediately following delivery.

Barbour's studies contributed to an understanding of the musculature of the gravid uterus, the dilatation and effacement of the cervix during labor and the anatomy of the pelvic musculature. What Barbour referred to as the "Ice Age" in obstetrical anatomy, the study of frozen sections of the bodies of women who had died during pregnancy, labor, or the puerpium, began with the work of Pieter de Reimer (1760–1831) who first used human frozen sections for anatomic illustration (1818). This innovation was followed by Nikolai Ivanovich Pirogov (1810–1881), a Russian surgeon, who introduced the use of frozen sections in teaching applied topographical anatomy for surgeons (1852–1859).

References

Barbour, A.H.F. *Atlas of the anatomy of labour.* Edinburgh, 1889.

Munro Kerr, J.M., R.W. Johnstone and M.H. Phillips. *Historical Review of British Obstetrics and Gynaecology, 1800–1950.* Edinburgh and London, E.& S. Livingstone, 1954, pp. 50–55.

Pirogov, N.I. *Anatome topographica sectionibus per corpus humanum congelatum triplici directione ductis illustrate.* 8 pts. Petropoli, J. Trey, 1852–1859. (GM 416)

Riemer, P.D. *Afbeeldingen van de juiste plaatsing der inwendige deelen van het menschelijk ligchaam.* 's Gravenhage, J. Allart, 1818. (GM 408)

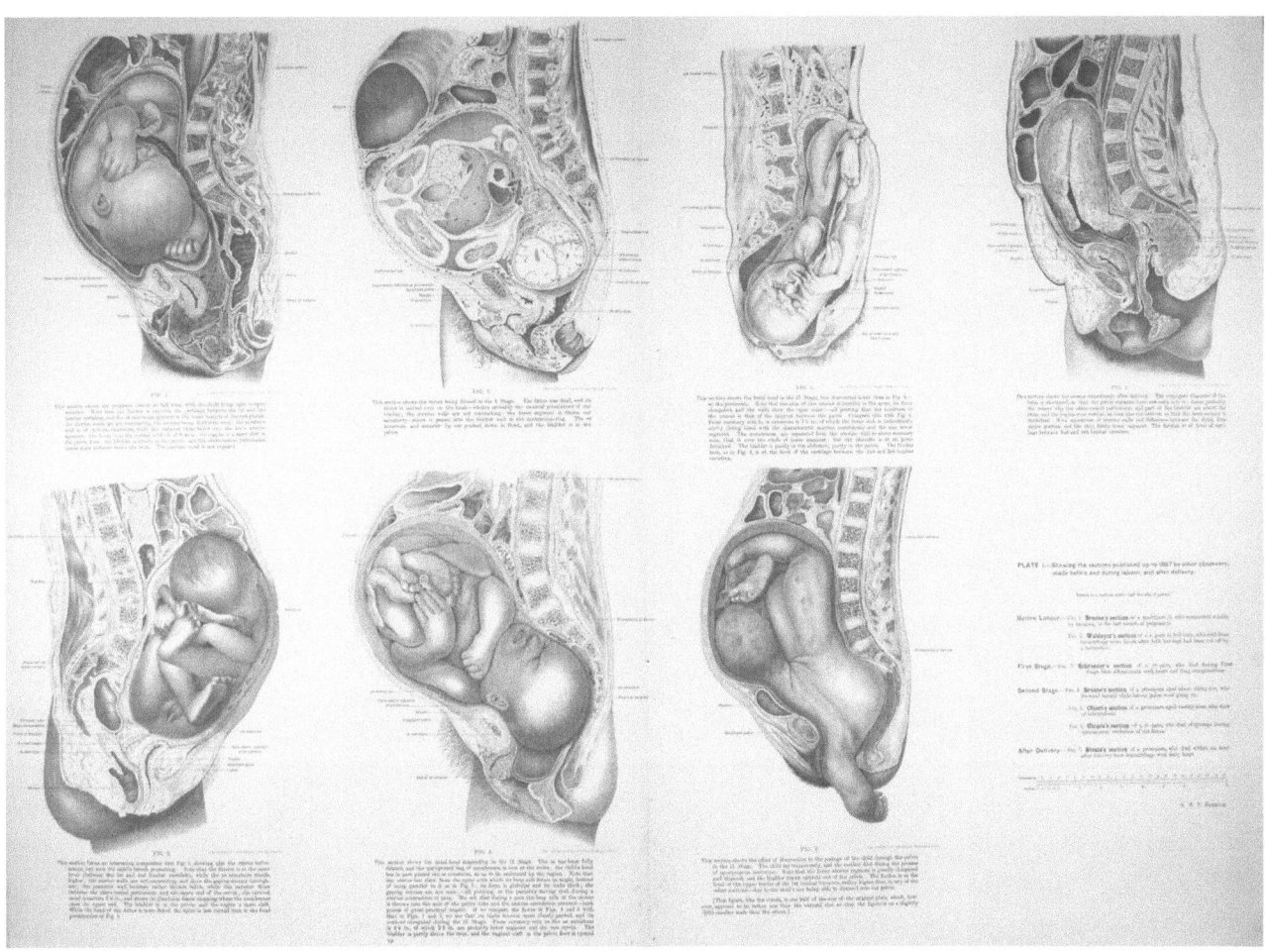

PLATE I FROM BARBOUR, 1889

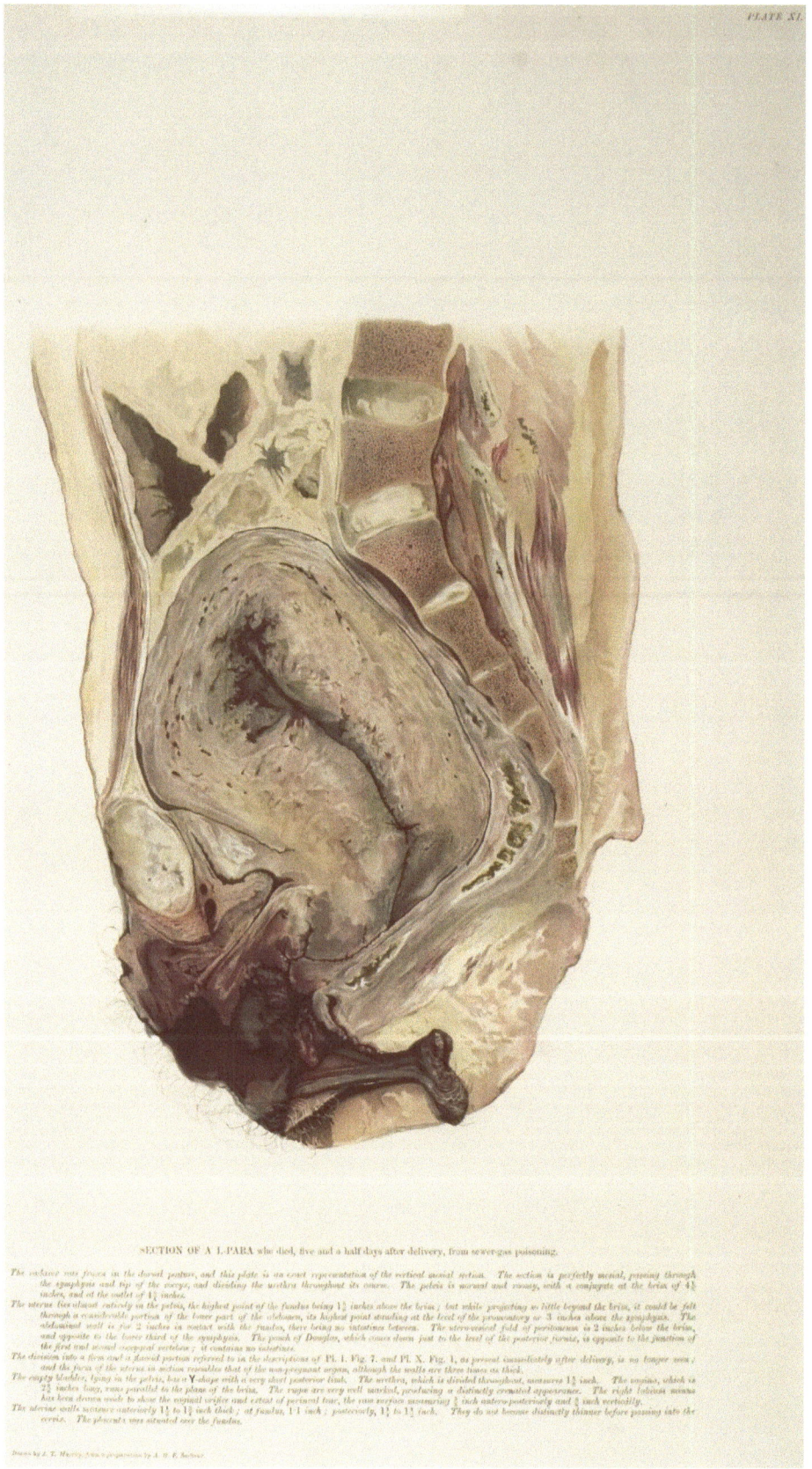

PLATE XI FROM BARBOUR, 1889

Additional Authors of Significance to the Nineteenth Century

James Read Chadwick

The value of the bluish coloration of the vaginal entrance as a sign of pregnancy. Transactions of the American Gynecological Society for the year 1886. Vol 11, pp. 339–418, 1887.

Chadwick (1844–1905), a Boston gynecologist, presented a thorough account of the bluish discoloration of the vagina as early evidence of pregnancy, Chadwick's sign, based on the examination of about 6,000 patients, 281 of whom were pregnant, over a period of 10 years. In early pregnancy, the vulvo vaginal mucosa undergoes a change in coloration from pink or rose to bluish-purple due to the increased vascularity of the uterus, vagina, and associated structures. Because of restraints placed on their examination, recognition of this phenomenon escaped the notice of many early workers. Until the mid-nineteenth century, patient modesty and the mores of the time required physicians to examine the pelvic organs of their female patients under drapes. In fact, many deliveries were conducted in this manner.

In his report, Chadwick emphasized that the violet coloration was present at the vaginal entrance or introitus, and did not require a speculum to examine the entire vaginal walls. He wrote, "the color begins as a pale violet in the early months, becomes more bluish as pregnancy advances, until it often assumes finally a dusky, almost black, tint; this last is familiar to every obstetrician. It is not due to pigmentation, but to an hypertrophy of the venous plexuses in the mucous membrane of the vagina (or a dilatation of the minute veins), induced by the afflux of blood to the uterus under the stimulus of pregnancy. The predominance of the veins in this location could alone account for the bluish color; moreover, when, toward the end of pregnancy, the color is most intense, varicose veins are plainly visible in the labia and in the legs" (p. 406).

Chadwick added, "in scrutinizing the color of this part in a large number of women I early discovered that, while in the majority the bluish tinge appeared over the whole vaginal entrance, there was a fair proportion in which the violet tint was confined to the anterior wall of the vagina, just below the urinary meatus, whence it shaded off into the normal pink color laterally. This, when distinctly perceptible, I soon found to be, in my practice, an absolutely sure sign of pregnancy ... The recognition of this peculiar localization of the blue tint on the anterior wall as a sure sign of pregnancy I feel is the most important new point in this communication" (p. 407).

Chadwick emphasized, "*... that its absence is not to be accepted as evidence that pregnancy does not exist, especially in the first three months, when satisfactory evidence is most needed*" (p. 409) and "*... that from (and including) the second month, this color is generally present, and often of such character as to be diagnostic*" (p. 410). He followed with an analysis of the months of pregnancy at which this sign was diagnostic, as opposed to being only suggestive; and concluded by listing several complications such as uterine retroversion and extrauterine pregnancy in which presence of this bluish-purple coloration was of particular value.

Chadwick, a graduate of Harvard Medical School (1871), studied in Europe for 2 years before returning to work at the Boston City Hospital and later establish his own gynecology dispensary. One of the founders and guiding lights of the American Gynecological Society (1876), and later its president (1897), he adopted Johann Wolfgang von Goethe's (1749–1832) *motto Ohne Hast aber onhe Rast* [without haste but without rest/repose] as his own, as well as that of the Society. The author, poet, and professor of anatomy at Harvard, Oliver Wendell Holmes (1809–1894), is quoted as speaking of Chadwick as "the untiring, imperturbable, tenacious, irrepressible, all-subduing agitator, who gave no sleep to his eyes, no slumber to his eyelids, until he had gained his ends, who neither rested nor let rest until the success of his project was assured" (Burrage 1906, p. 439). A serious bibliophile, Chadwick also played a key role in the founding and development of the Boston Medical Library (Burrage 1906, Farlow 1906). He was an advocate of cremation and served in a leadership role in the New England Cremation Society (Burrage 1906).

One of the earliest descriptions of the purplish discoloration of the mucous membrane lining the vagina as being diagnostic of pregnancy was, in fact, by Jean Marie Jacquemier (1806–1879) in his "anatomical and physiological researches of the vascular system of the human uterus during gestation and the blood vessels of the placenta" (Jacquemier's sign), five decades earlier (Jacquemier 1838, 1846). In turn, in his textbook of obstetrics, Jacquemier credited his contemporary Etienne Joseph Jacquemin (1796–1872), chief physician to the prison at Mazas and *medecin des prisonières des madelonnettes et de la Prison de la Force* (physician to the repentant prisoners in the Prison of Strength [or Authority]) with originally describing this sign in the prostitutes for whom he cared (Jacquemier 1846). Another contemporary Alexandre Jean Baptiste Parent-Duchâtelet (1790–1836) also credited

Jacquemin with this observation, stating that the violet coloration resembled "wine dregs" (Parent-Duchâtelet 1837). An additional contemporary report described this sign being evident in pregnanct syphilitic patients in the clinic of Professor Klug in Berlin (Sommer 1835, 1837). Unfortunately, neither Jacquemier or Jacquemin illustrated this sign in their publications.

References

Beer, E., G. Mangiante and D. Pecorari. *Distocia delle spalle. Storia ed attualità*. Roma, CIC Edizioni Internazionali, 2006.

Burrage, W.L. James Read Chadwick, M.D. (1844-1905). *Trans Am Gynecol Soc* 31:437–445, 1906.

Farlow, J.W. A tribute to the memory of Dr. James Read Chadwick. Excerpted from the Thirtieth Annual Report of the Boston Medical Library. *Med Library Hist J* 4:114–116, 1906.

Gleichert, J.E. Étienne Joseph Jacquemin, Discoverer of 'Chadwick's Sign'. *J Hist Med* 26:75–80, 1971.

Jacquemier, J.M. Recherches d'anatomie et de physiologie sur le système vasculaire sanguine de l'utérus humain pendant la gestation, et plus spécialement sur les vaisseaux utéro-placentaires. *Arch gén Med* 3 sér. 3:165–194, 1838. (GM 6174).

Jacquemier, J.M. *Manuel des accouchements et des maladies des femmes grosses et accouchées, contenant les soins à donner aux nouveaux-nés*. Paris, J.B. Baillière, 1846, Vol 1, p. 215.

Parent-Duchâtelet, A.J.B. De la prostitution dans la ville de Paris…. 2nd Ed. Paris, J.B. Baillière, 1837, Vol.1, pp. 217–218. (GM 1607).

Sommer, J.G. [Letter] *Medicin og Chirurgie* 1835.

Sommer, J.G. [Letter] *Berliner Medicinische Centr-Ztg* 6:34–38, 1837.

Speert, H. James Read Chadwick and his pregnancy sign. In: *Essays in Eponymy Obstetric and Gynecologic Milestones*. New York, Macmillan, 1958, pp. 229–235.

Midwives and Midwifery

With their role in the care of women, it is assumed that midwives have practiced their craft since earliest times. Of importance to the context of this volume is the question of the extent to which midwives authored books on midwifery, contributed to depictions of "the gravid uterus and its contents," and/or improved understanding of antenatal development. Attested by papyrus records and Bas reliefs, midwifery was a recognized practice by women in ancient Egypt. The Hebrew Torah contains several references to midwives, their characteristics and work. In medieval times, in addition to their care of the parturient, midwives baptized some newborn infants. Unfortunately, little is known of the early history of the midwife [middle English *mid-wif*, Latin *cummater*, with woman]. In part, this is a consequence of childbirth being regarded as an inferior task, or even forbidden for the physician. Also because few midwives had the benefit of formal education, for the most part their work went unrecorded and they learned their craft by serving as an apprentice. Because about 95 % of deliveries present few or no complications, the system worked quite well. Also, with it being considered highly inappropriate for a male to view, much less touch, a woman's pelvic region, "women's work" in serving and caring for other women was the norm.

In considering the rise of the science of *tocology* [Greek, knowledge of childbirth], the practice of obstetrics or midwifery, one might ask about the role of midwives in advancing this science. Several sixteenth century works by physicians had been written for women as well as for men-midwives. As noted earlier, these included the first printed book devoted to obstetrics *Der schwangern frauwen und hebammen roszgarten* (1513) by Eucharius Rösslin (d. 1526) and *Ein schön lustig Trostbüchle...*, of 1554 by Jacob Rueff, and others. In 1609, the midwife, Louise [Louyse] Bourgeois [dite Boursier] (1563–1636), *accoucheuse* to the French court, published the first book by a midwife on the subject, *Observations diverses sur la stérilité, perte de fruits, fécon-* *dité, accouchements....* With many editions her writings had great influence, not only in France, but were translated into German, Dutch, and English and used throughout Europe. Unfortunately, her work included no illustrations, and thus is not included in this volume. In 1671 Jane Sharp (fl. 1670), of London, wrote *The midwives book*, the first such work by an English midwife. In Prussia, in 1690, the midwife to the court of the Elector of Brandenberg, Justine Siegemundin (1650–1705) wrote the first book for midwives of the German speaking lands, *Die Chur-Brandenburgische Hoff-Wehe-Mutter....* Formulated as a catechism with questions and answers, Siegemundin included a number of illustrations of value, the plates of which she had drawn and engraved at her own expense. Her work went through a number of editions, and its illustrations were copied widely.

In the eighteenth century the English midwife Elizabeth Nihell (b. 1723) authored *A treatise on the art of midwifery...* (1760), which again included no illustrations. After spending several years as an apprentice midwife in Paris at the *Hôtel Dieu*, she returned to London where she became embroiled in controversy with William Smellie (1697–1763) and other man-midwives over the care for women at the time of delivery. A major factor in the attraction of male *accoucheurs* was the early eighteenth century revelation to the public of the obstetrical forceps, following invention of the "secret" instrument by the Chamberlens a century earlier. In contemporary medicine, midwives continue to play an important role in most countries of the world. In mid- to late-nineteenth century Britain, a physician leader, a champion for the education and certification of midwives, was James Hobson Aveling (1828–1892), in part, by his active campaigning in the medical and lay press. As noted, a number of volumes included in this assemblage were written by physicians for midwives, and then adapted for other physicians. The descriptions that follow are limited to those works written by midwives.

© Springer International Publishing Switzerland 2016
L.D. Longo, L.P. Reynolds, *Wombs with a View*, DOI 10.1007/978-3-319-23567-7_3

Jane Sharp [Sharpe]

The midwives book, or the whole art of midwifery discovered; directing child-bearing women how to behave themselves in their conception, breeding, and nursing of children … London, Simon Miller, 1671.

Midwifery has been practiced since earliest times. Mentioned in the Bible, writings of the Greeks, Romans, and others, the profession has an honorable history, and was occupied in many instances by a *femme-sage*, virtuous and wise woman. Jane Sharp (fl. 1650) was the first English midwife to write a book on the subject; unfortunately, no copies of her original edition are known. (The following is from the third edition of 1724). "A practitioner in the art of midwifery above 30 years" (p. 182), in addition to medical knowledge of that era, in six "books" she included considerable personal anecdote. Although she was against formal training, she recommended that midwives study anatomy, and emphasized her view that the field of midwifery should be reserved for women. In her preface, addressed to her "sisters," the "celebrated midwives of Great Britain and Ireland," Sharp confessed, "I have often sat down sad in the consideration of the many Miseries Women endure in the Hands of unskilful Midwives; many professing the Art (without any skill in Anatomy which is the Principal part effectually necessary for a Midwife) merely for Lucres-sake" (Preface).

Sharp considered at length the relationships between female and male-midwives, an issue which, with the secret of the obstetrical forceps revealed, became a source of controversy. In the introduction, Sharp stated, "This Art is doubtless one of the most useful and necessary of all Arts, for the Being and Well-being of Mankind, and therefore it is extremely requisite that a Midwife be both Fearing God, Faithful, and exceeding well Experienced in that Profession." She continued, "As for their Knowledge, it must be Two-fold, *Speculative* and *Practical*, she that wants the Knowledge of Speculation, is like to one that is Blind or wants her Sight: She that wants the Practice, is like one that is Lame and wants her Legs; the Lame may See, but they cannot Walk, the Blind may Walk, but they cannot See. Such is the Condition of those Midwives that are not well versed in THESE. Some perhaps may think, that then it is not proper for Women to be of this Profession, because they cannot attain so rarely to the Knowledge of Things as Men may, who are bred up in Universities, Schools of Learning, or serve their Apprenticeships for that End and Purpose, where Anatomy Lectures being frequently read, the Situation of the Parts both of Men and Women … are often made plain to them" (pp. x–xi). She maintained, nevertheless, that it was not necessary that midwives be given university educations nor be instructed in foreign languages in order that they might read the treatises of learned authorities. Staunchly maintaining that midwifery was the legitimate "business" of women, she emphasized, "It is not hard Words that perform the Work, as if none understood the Art that cannot understand Greek. Words are but the Shell … It is commendable for Men to employ their spare Time in some Things, of deeper Speculation than is required of the female sex; but the art of MIDWIFERY chiefly concerns us" (pp. xi–xii).

The frontispiece woodcut print illustrates a birthroom scene (upper panel) with the parturient reclining in bed while one attendant serves her a bowl of broth, and two others [attend] the newborn infant. The middle panel illustrates a group of 11 men and women. The lower panel shows another group sharing a repast around a table. A plate from Book II displays a pregnant woman "Being a Dissection of the WOMB, with the usual Manner how the CHILD lies therein, near the time of its Birth," showing the inner parts of the chorion "branched out" (B; D), the extended Amnion (C), "the fleshy substance, called the *cake* or *placenta*, which nourishes the infant; it is full of vessels" (E), "the Navel-string, carrying nourishment from the *Placenta* to the Navel" (G), and the infant (H) (pp. 96–97). A believer in astrology, Sharp noted that none of the published tables for calculating the influence of the planets upon parturition had any truth; thus, she prepared one upon which her patients could rely.

Little is known of Sharp's life. Some have pointed out that although this work expresses considerable good sense, it does not display the originality of Louise Bourgeois [dite Boursier] (1563–1636) *Observations...* written a half a century earlier (see Additional Author(s) of Significance to Midwives and Midwifery; Boursier 1609; Donnison 1977). During the eighteenth century, several other English midwives wrote manuals of substance to help educate their sister practitioners. Notable among these were the works of Sarah Stone (fl. 1730) of Taunton, Bristol, and London (Stone 1737), and Elizabeth Nihell (b. 1723) of London (Nihell 1760).

Frontispiece from Sharp, 1671

References

Blake, p. 415 lists later edition

Aveling, J.H. *English midwives: their history and prospects*. London, J.A. Churchill, 1872, pp. 47–54.

Aveling, J.H. *English midwives: their history and prospects*. London, J.A. Churchill, 1872, pp. 47–54.

Bourgeois, L. *Observations diverses sur la sterilite, perte de fruict, foecondite, accouchements, et maladies des femmes, et enfants nouveaux naiz*. Paris, A. Saugrain, 1609. (GM 6145).

Culpeper, N. *A directory for midwives; or, a guide for women, in their conception, bearing, and suckling their children*. London, Peter Cole, 1651. (Krivatsy, 2957).

Donnison, J. *Midwives and medical men. A history of inter-professional rivalries and women's rights*. London, Heinemann, 1977, pp. 15–17.

Nihell, E. *A treatise on the art of midwifery: setting forth various abuses therein, especially as to the practice with instruments....* London, A. Morley, 1760. (Blake, p. 325).

Sharp, J. *The complete midwife's companion: or, the art of midwifry improv'd: directing child-bearing women how to order themselves in their conception, breeding, bearing, and nursing of children....* 4th Ed. London, John Marshall, 1725. (Blake, p. 415).

Stone, S. *A complete practice of midwifery. Consisting of upwards of forty cases or observations in that valuable art....* London, T. Cooper, 1737. (Blake, p. 435).

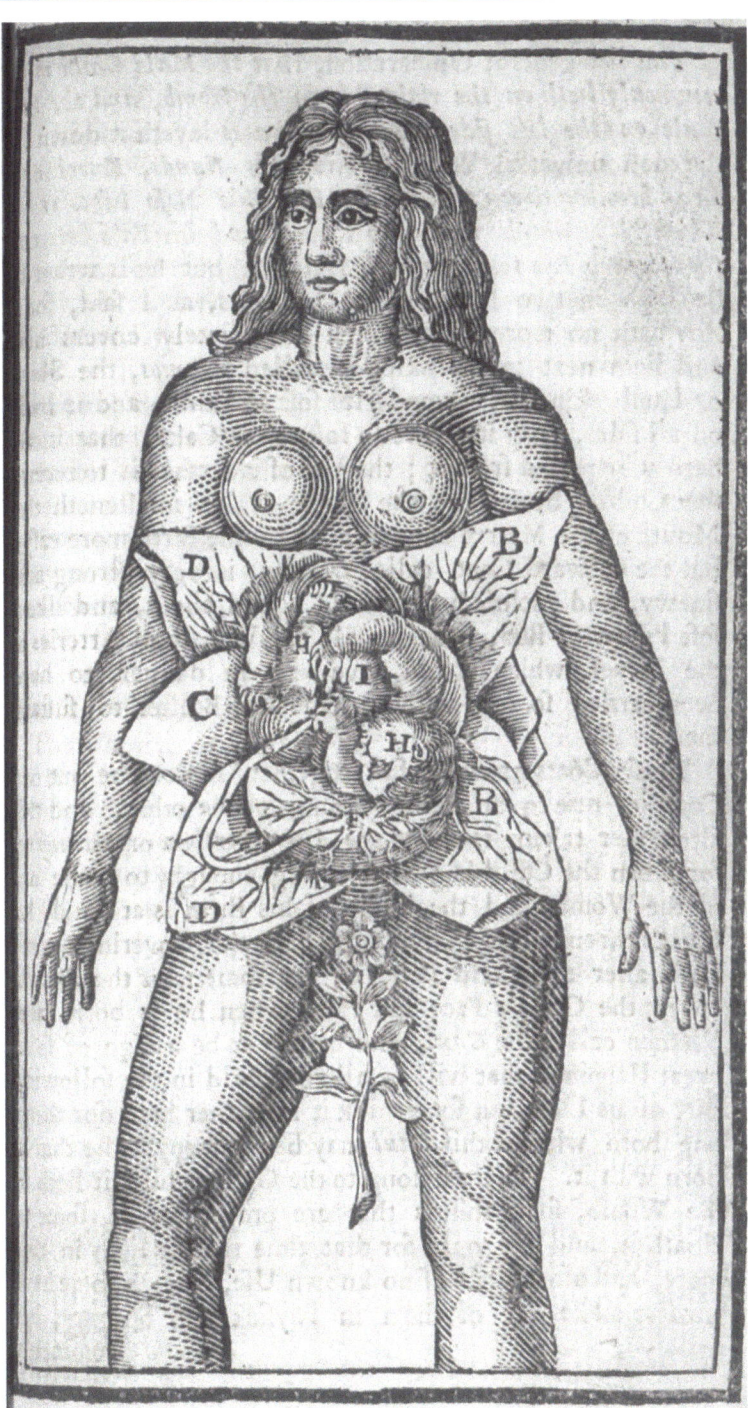

PLATE FROM SHARP, 1671

Justine Siegemundin [Nee Dittrichin; Dittrich]

Die Chur-Brandenburgische Hoff-Wehe-Mutter, das ist: Ein hochst-nothiger Unterricht, von schweren und unrecht-stehenden Geburten... Colln an der Spree, ...Ulrich Liebperten, 1690.

From the midst of the Germanic world of the late seventeenth century, which consisted of several hundred individual states with territorial sovereignty, and with complex cultural, political, and religious divisions, came one of the most renowned and celebrated midwives of that era. Basically self-taught, Siegemundin (1636–1705) developed skills such that in 1683 she was appointed the Court Midwife by Frederick William (1657–1713), Elector of Brandenburg (1688–1713). When he later was crowned Frederick I, first King in Prussia (1701–1713), she became court midwife of Prussia.

Siegemundin's treatise the *Court Midwife...* is based upon her own experience and case records of common complications of labor, including malpresentation and placenta previa. In the initial portion of this work, Siegemundin discusses difficult labors and their management. The second section consists of a dialogue or catechism of questions and answers in a conversation between midwife, Justine, presumably the author's *alter ego*, and her pupil Christina. Initially, the pupil asks the questions, while later these roles are reversed, so that Justine may test the extent to which the pupil has grasped a proper understanding of her teachings.

The lovely frontispiece includes the statement "An Gottes hilff und Seegen, Geschickten hand bewegen, Ist all mein Tuhn gelegen" [On gracious God relying, My skillful hand applying, Devoted deeds allying]. Following the title page, she quotes the Torah "Therefore God dealt well with the midwives: and the people multiplied... And it came to pass, because the midwives feared God, that he made them houses" (Exodus 1:20–21 KJV). Despite being childless, a condition anathema for midwives, Siegemundin attests to considerable expertise, having had many parturients under her hands. In the text, she offers words of encouragement that should be spoken to the mother in labor:

My dear child, fear not the pains and be not frightened. Remain as firm and reassured as you possibly can and let not your courage and great expectations flag. I assure you, with God's help it will go better than you think! Just hang on with both hands tightly so that you tremble not. It will soon pass. You will see that Our Dear God will soon help you. How quickly a pain passes: who would want to let her courage flag so soon, for God's help will soon be here (p. 136).

She includes her ideas regarding the character traits that should characterize a midwife: having nimble hands, understanding the proper dilatation of the uterine cervix and its examination, the ability to distinguish the true pains of labor, and the ability to perform a version with breech extractions when required. She also maintains that for their own deference, midwives should not dispense pharmaceuticals.

Siegemundin published this work at her own expense and, to establish authenticity with authorities, included attestation from several chaplains of the Brandenberg court and testimonials from her patients. The volume includes 43 fine copper plate engravings, which she had prepared in Holland. These are the first original obstetrical observations not copied from Eucharius Rösslin (ca. 1470–1526) (1513) or Jacob Rueff (1500–1558) (1554). The initial engraving is of a normal infant at full-term, lying within the uterus, with identification of anatomical parts: A, the placenta; D, uterine wall; G, amnion; H, umbilical cord; L, knee; M, hand; N, chest; P, foot. This plate was taken from Plate 56 of Govert Bidloo's (1649–1713) great treatise on anatomy (Bidloo 1685). Plates 17 and 18 illustrate methods used by the midwife to aid in version with breech extractions. As is apparent in these figures, the uterine cavity is depicted as being much more commodious than in real life.

With François Mauriceau (1637–1709) (1668), Siegemundin introduced the practice of puncture of the amniotic membranes to arrest hemorrhage in cases of marginal placental previa. She also recognized that, contrary to contemporary opinion, the pubic bones did not separate to aid parturition. During her career Siegemundin delivered over 6000 infants, and despite her Royal patronage, she says essentially nothing about her life at court, or the birth of royalty. This work represents a remarkable record of the most practical and enlightened thought at that time, and was republished in seven subsequent editions.

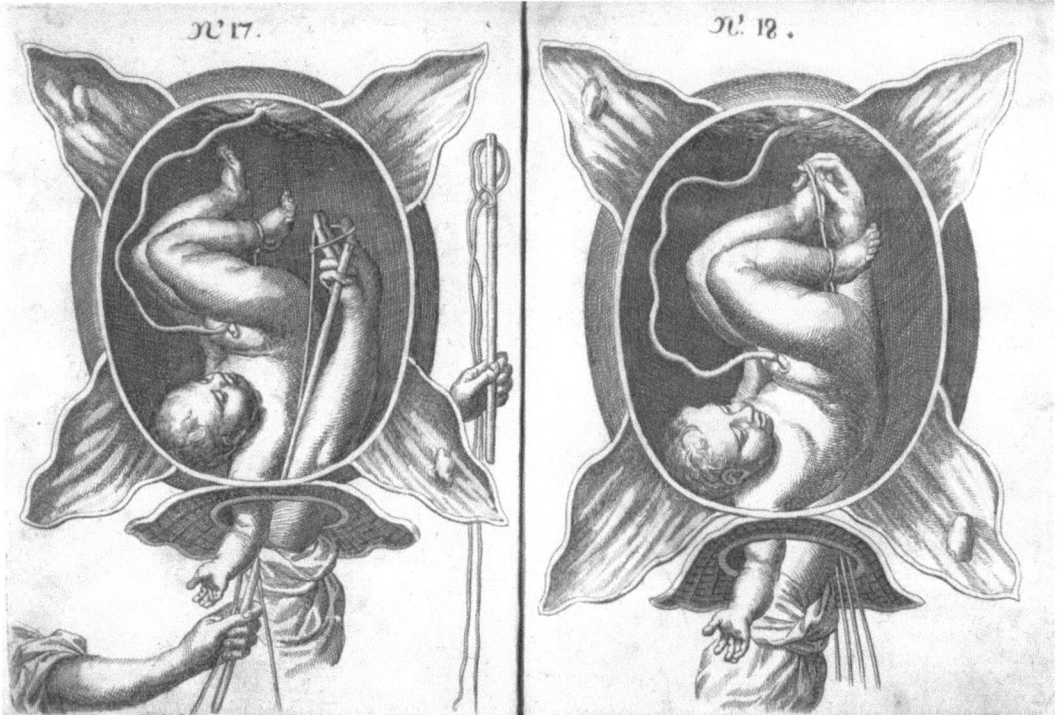

FRONTISPIECE AND PLATES FROM SIEGEMUNDIN, 1690

References

Garrison Morton 6149; Krivatsy 11085.

Bidloo, G. *Anatomia humani corporis, centum et quinque tabulis...ad vivum delineates.* Amstelodami, Joannes a Sumpt Someren, etc., 1685. (GM 385; Krivatsy 1238).

Bourgeois, L. *Observations diverses sur la sterilité, perte de fruict, foecondité, accouchements, et maladies des femmes, et enfants nouveaux naiz.* Paris, A. Saugrain, 1609. (GM 6145, Krivatsy 1625).

Mauriceau, F. *Des maladies des femmes grosses et accouchées. Aved la bonne et veritable Méthode de les bien alder en leurs accouchemens naturels,....* Paris, Chez Iean Henavlt *et al...*, Imprimeries de Charles Coignard, 1668. (GM 6147).

Rösslin, E. *Der schwangern Frauwen und Hebammen Roszgarten.* Hagenau, H. Gran (also Strassburg, Martin Flach, Junior), 1513. (GM 6138).

Rueff, J. *De conceptu et generatione hominis et iis quae circa lec potissimum confyderantur.* Tiguri, Apud Frosch, 1554. (GM 6141).

Siegemund, J. *The court midwife. Edited and translated by Lynne Tatlock.* Chicago, University of Chicago Press, 2005.

Tatlock, L. Speculum Feminarum: gendered perspectives on obstetrics and gynecology in early modern Germany. *Signs* 17:725–760, 1992.

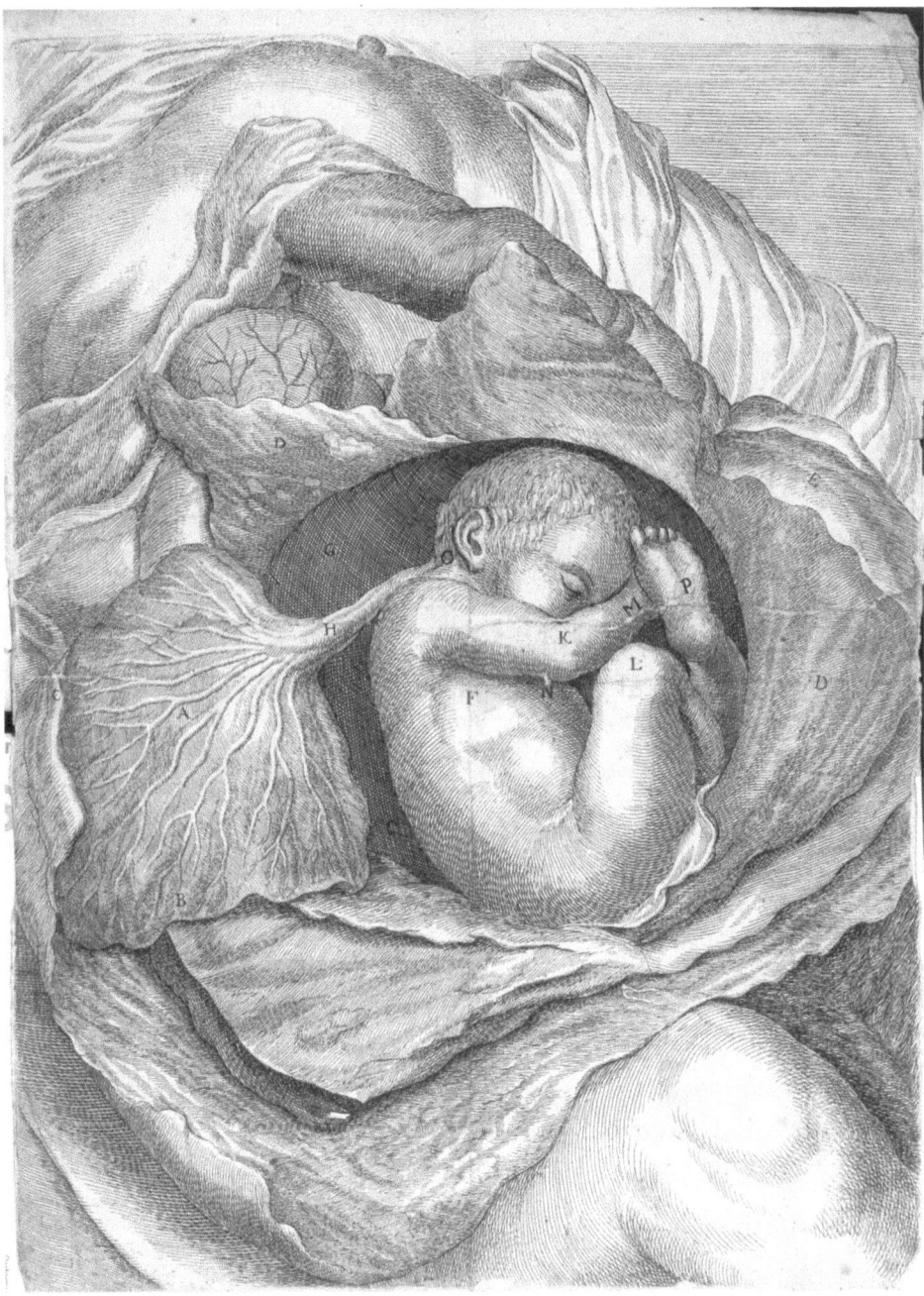

PLATE FROM SIEGEMUNDIN, 1690

Barbara Widenmann

Kurzte, jedoch hinlängliche und gründliche anweisung christlicher hebammen, wie sie so wohl bey ordentlichen, als allen ausserordentlichen schwehren Geburten denen kreissenden Frauen Hülffe leisten, den Handgriff gewiß und sicher verrichten... Nebst einem Anhang, Wie eine zu diesem Beruff sich angehende Hebamme Obrigkeitlich zu examiniren. Augsburg, druckts und verlegts Johann Jacob Lotter, 1735.

In her book of instructions for midwives, the leading Augsburg midwife Barbara Widenmann (1695–ca. 1760) advised against the use of labour-inducing prescriptions, and emphasized that the causes of difficult births are often found in the midwife's ignorance, rather than in the parturient woman or the baby. She favored the use of the birth-stool, and although recommending improvements, she noted that it was not an absolutely necessary implement for a successful delivery. She also was highly skeptical about the ability to predict the sex of the child, stating that it is something that only God can know. Some have suggested that Widenmann was stimulated to write this work by the 1690 midwifery manual written by the midwife to the Court of the Elector of Brandenburg, Justine Siegemundin (1636–1705) (Siegemundin 1690). This volume contains an engraved allegorical frontispiece, four plates each of which depict various expedients for delivery of four abnormal and difficult presentations of the fetus, with one plate illustrating assorted obstetrical instruments, and one of a birth-chair. The plates illustrating fetal malpositions and manipulations clearly were based on those of Siegemundin, although they cannot compare in artistic merit with those figures. For instance, Tab. 2 illustrated maneuvers by

the midwife to effect version with breech extraction, with Fig. 6 being reversed, but otherwise similar to No. 17 in Siegemundin. Many of the other illustrations, which are about one-ninth the size of those of Siegemundin, are derived from that text (e.g., Widenmann No. 9 is similar to Siegemundin No. 1, No. 14 to that of 10, No. 13 to that of 21).

Beginning as a country practitioner, in the spring of 1729 Widenmann was permitted to follow her profession in Augsburg. Herself a mother of 15 children, Widenmann records that in her 19 years of practice she assisted more than 1800 lying-in women. As she relates in the preface, she received basic instruction from her husband, the Augsburg physician and surgeon Frantz Widenmann (fl. 1730), who is believed to have written part or all of this volume. One of the early eighteenth century references to an operating table is that depicted in Frantz Widenmann's *Neuer curieuser und ausführlicher Bericht*, 1719.

References

Blake p. 489.

Ricci, J.V. *The development of gynecological surgery and instruments....* Philadelphia, Blakiston, 1949. (GM 6310).

Siegemundin, J. *Die Chur-Brandenburgische Hoff-Wehe-Mutter....* Cölln an der Sree, U. Liebperten, 1690. (GM 6149; Krivatsy 11085).

Tatlock, L. Speculum Feminarum: gendered perspectives on obstetrics and gynecology in early modern Germany. *Signs* 17:725–760, 1992.

Widenmann, F. *Neuer curieuser und ausführlicher Bericht, Stein und Brüche, so wohl mit als ohne Castrierung, zuschneiden, wie auch Staaren zustechen.* Augspurg, 1719. (Blake, p. 489).

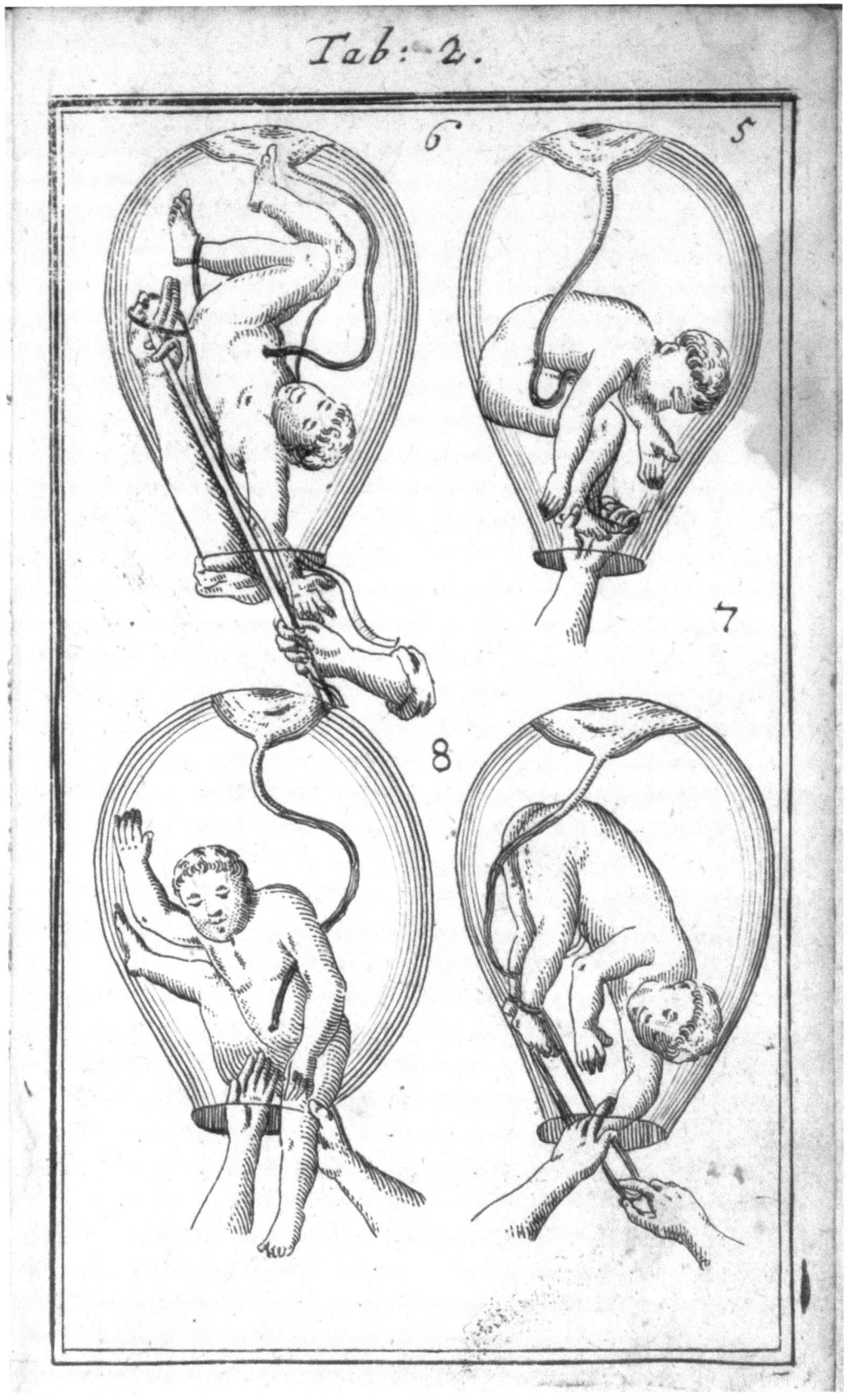

TAB. 2 FROM WIDENMANN, 1735

Angélique Marguerite le Boursier du Coudray

Abbrégé de l'art des accouchemens, dans lequel on donne les préceptes nécessaires... Nouvelle Edition, enriche de figures en taille-douché enluminees. Saintes, Chez Pierre Toussaints, libraire, imprimeur..., 1769.

Following the excesses of King Louis XIV (1638–1715; King 1643–1715), with almost continuous warfare for two and a half decades, and wars during the reign of Louis XV (1710–1774; King of France from 1715 until his death) with the war of Austrian Succession (1740–1748), war on land and sea with the British (1744–1748), and the Seven Years War (1756–1763) the manpower of France (as well as its Treasury) were severely depleted. In an attempt by the monarchy to counteract the decline in births and population, Louis XV and his court advisors developed a plan to increase midwives' knowledge of how to manage normal pregnancy and its complications, and to reduce maternal and infant mortality.

At the behest of Louis XVI (1754–1793; Monarch from 1774 until 1792), beginning in 1760 and for over two decades, Madame le Boursier du Coudray (1714/5–1794), distinguished midwife for the French Royal Court, was given a *brevet* [certificate or warrant] and pension to travel to over forty provincial cities, educating midwives in the practice of childbirth. du Coudray, a pioneer in scientific midwifery, first published her *abrégé* [abbreviated or synopsis of the art of obstetrics], a practical manual for midwives, in 1759. This original edition did not include illustrations.

In the second edition, however, to aid in her instruction du Coudray included twenty-six delicately engraved plates that depict the female pelvis, uterus, and various fetal presentations. This was the first work in obstetrics to use multichrome color plates, and it has been suggested that du Coudray used a considerable portion of her own funds to produce these. Jean Robert, a Paris engraver of the latter half of the eighteenth century, used the technique developed by Jacob Christoph Le Blon (1667–1741), that of mezzotinting with separate imprints, for the several colors. In these illustrations the pelvis is pale cream, the fetus and hands of the delivering midwife are in red. In some plates darker red as well as black are added. In addition to details of the mode of delivery of normal and complicated cases, du Coudray wrote on many aspects of childbirth including the characteristics required of the midwife, the anatomy of the bony pelvis, and the dangers of convulsions in pregnancy (eclampsia) with the benefit of delivery as rapidly as possible.

In her *Avant-Propos* [Forward or Preface], du Coudray presented her rationale for writing this volume, and subjecting herself to a peripatetic life-style for her parturient sisters and the sake of France. She pledged to educate midwives to preserve life and thus subjects for France. "I have put together these lessons and am venturing to publish them ... to reduce infant mortality, less out of presumption, which twenty years of experience might have inspired in me, than out of the desire to make myself useful to my *Patrie* [native land]" (p. viii). du Coudray stressed that it was her compassion for childbearing women that compelled her to become a *femme auteur* [female author], and that her zeal fueled her determination to freely give these lessons. The first chapter considers "Qualities required of women intending to practice the art of childbirth," the last reviews "Qualities required of a good wet-nurse." Six editions of this work appeared during her lifetime.

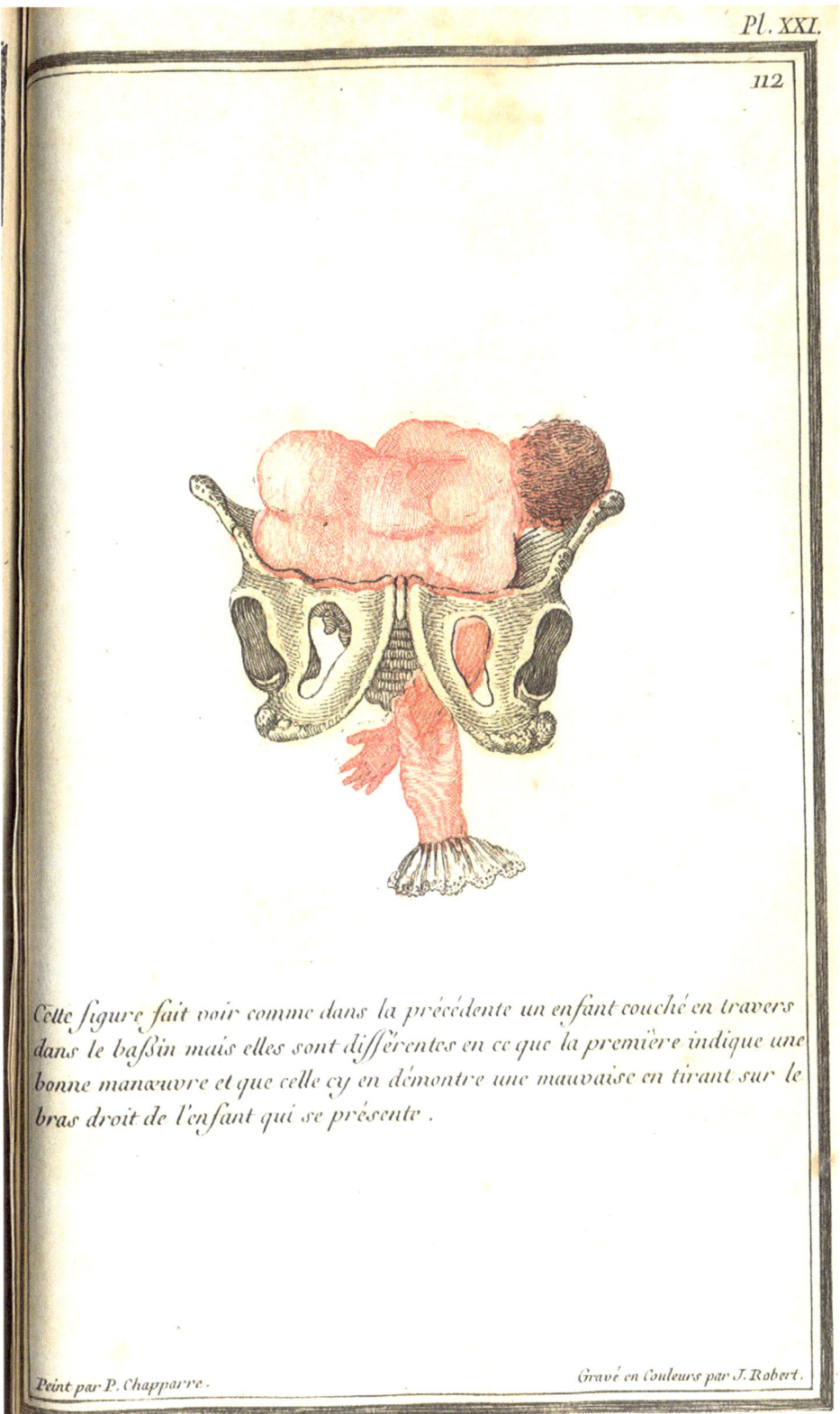

Pl. XXI.

112

Cette figure fait voir comme dans la précédente un enfant couché en travers dans le bassin mais elles sont différentes en ce que la première indique une bonne manœuvre et que celle cy en démontre une mauvaise en tirant sur le bras droit de l'enfant qui se présente.

Peint par P. Chapparre.

Gravé en Couleurs par J. Robert.

PLATE XXI FROM DU COUDRAY, 1769

In a series of courses each of which lasted two or three months, it is estimated that over a quarter of a century this royal ambassadress, this national midwife of France, helped to train about 10,000 young women in the art and science of midwifery, a remarkable accomplishment. This effort to spread obstetrical enlightenment and the *bien de l'humanite* [good of humanity] constituted the first such national effort in Europe, and is believed to have reversed a decline in population which some had regarded as a demographic crisis.

Importantly, Madame du Coudray devised a life-size "machine" (phantom pelvis or mannikin), a "monument to humanity" as she called it, with a fetal doll to teach obstetrical maneuvers to midwives as well as to surgeons. Both André Levret (1703–1780), inventor of the long forceps with pelvic cure, and Royal *accoucheur*, and the surgeon-*accoucheur* Jean-Joseph Süe (1710–1792) were strong supporters of du Coudray. Among her male midwife pupils was Jean Bernard Jacobs (1734–1790), who taught with her "machine," and wrote an important obstetrical text in Flemish (Jacobs 1784). Among her teachings was that the three most important things in childbirth were patience, patience, and more patience. She also was a staunch advocate of breast feeding. On the frontispiece of this volume, her motto emblazoned over her portrait, *ad operam* [to work], characterized her life. Beneath the portrait are the words, "pensioned and sent by the King to teach the practice of midwifery throughout the Relm." Hers was a life well lived. du Coudray died a natural death during *le regné de la terreur* [the Reign of Terror]. The first book on obstetrics to be written by a midwife was that of Louise Bourgeois [dite Boursier] (1563–1636) (Bourgeois 1609), who at 36 years of age was called to attend Marie de'Medici (1573–1642), wife of King Henry IV (1553–1610; King 1589–1610) as *accoucheuse* to the French court. After the birth of the future Louis XIII in 1601 at *Fontainebleau* she wrote a *Récit véritable de la naissance de meisseigneurs et dames les enfans de France…* in which she described events and individuals attending the birth (Bourgeois 1601). Bourgeois also served the court of Louis XIII (1601–1643; King 1610–1643). Unfortunately, after the death in 1627 by puerperal sepsis of one of the members of the Royal family, her service at the court was terminated (Goodell 1876).

References

Blake p. 260; Cushing L-107 (under Le Boursier…); Waller 5656.

Bourgeois, L. *Observations diverses sur la sterilité, perte de fruict, foecondité, accouchements, et maladies des femmes, et enfants noiveaux naiz.* Paris, A. Saugrain, 1609. (GM 6145).

Du Coudray, A.M.L.B. *Abrégé de l'art des accouchemens…*. Paris, Chez la veuve Delaguette…, 1759.

Gelbart, N.R. Midwife to a nation: Mme du Coudray serves France. In: *The Art of Midwifery. Early modern midwives in Europe.* Hilary Marland (Ed). London, Routledge, 1993, pp. 131–151.

Gelbart, N.R. Books and the birthing business: The midwife manuals of Madame du Coudray. In: *Going Public. Women and publishing in early modern France.* E.C. Goldsmith and D. Goodman (Eds). Ithaca, New York, Cornell University Press, 1995, pp. 79–96.

Gelbart, N.R. Delivering the goods: patriotism, property and the midwife mission of Madame du Coudray. In: *Early modern conceptions of property.* J. Brewer and S. Staves (Eds). London; New York, Routledge, 1995, pp. 467–480.

Gelbart, N.R. The monarchy's midwife who left no memoirs. *French Hist Studies* 19:997–1023, 1996.

Gelbart, N.R. *The King's midwife. A history and mystery of Madame Du Coudray.* Berkeley, University California Press, 1998.

Gélis, J. *History of childbirth: fertility, pregnancy, and birth in early modern Europe.* Boston, Northeastern University Press, 1991.

Goodell, W. *A sketch of the life and writings of Louyse Bourgeois, midwife to Marie de' Medici, the Queen of Henry IV of France.* Philadelphia, Collins, 1876.

Jacobs, J.B. *Vroedkundige oeffenschool.* Gend, 1784. (Blake, p. 232).

Levret, A. *L'art des accouchemens.* Paris, Delaguette, 1753. (GM 6153).

Siegemundin, J. *Die Chur-Brandenburgische Hoff-Wehe-Mutter….* Cölln an der Spree, U. Liebperten, 1690. (GM 6149).

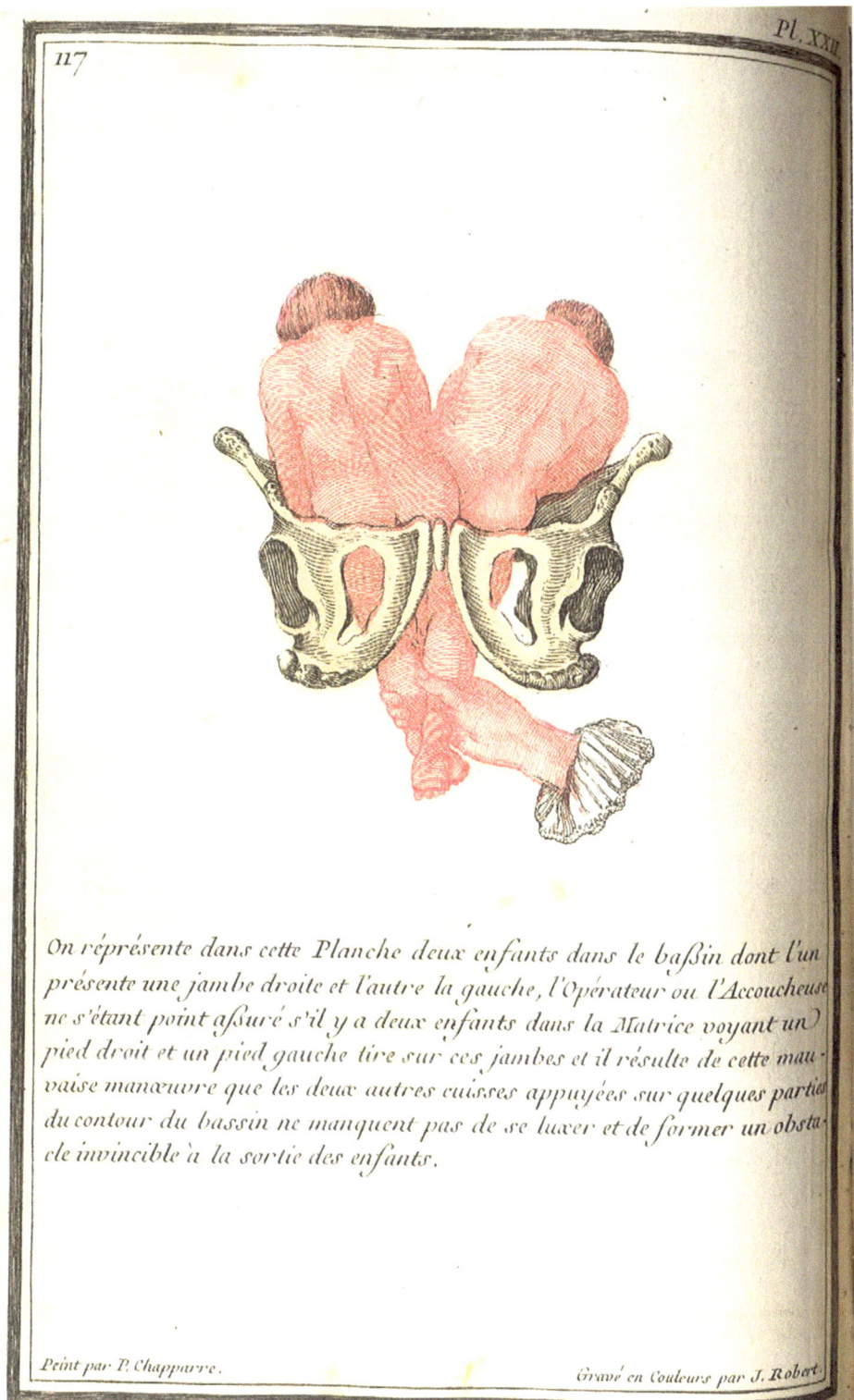

PLATE XXII FROM DU COUDRAY, 1769

Additional Authors of Significance to Midwives and Midwifery

Louise Bourgeois [dite Boursier]

Observations diverses sur la stérilité, perte de fruits, fécondité, accouchements, et maladies des femmes et enfants nouveau-nés : suivi de Instructions à ma fille. A. Saugrain, 1609.

Louise Bourgeois (1563–1636), accoucheuse to the French court, was the first midwife to publish a book on obstetrics. She induced premature labor in patients with pelvic contraction.

Reference

Garrison Morton 6145.

Marie-Louise La Chapelle, née Dugès

Pratique des accouchemens, ou mémoires, et observations choisies, sur les points les plus importans de l'art; Par Mme. Lachapelle... Publiés par Antoine Dugès, son neveu... 3 vols. Paris, J.B. Baillière, 1821–1825.

In her noteworthy series of obstetrical memoires, the midwife, La Chapelle (1769–1821) presented her considerable experience based on supervising nearly 4000 deliveries in 1811. She reduced the more than 90 theoretical presentations of the fetus proposed by her medical colleague Jean-Louis August Baudelocque (1746–1810) to 22. Her deductions forming the basis of present teaching in this respect. La Chapelle advocated that the obstetrical forceps and other instruments be used as seldom as possible, and never with the goal of shortening labor. She also opposed the practice of forced dilation of the cervix, which was not uncommonly practiced to shorten labor. At the close of Volume I she classified the characteristics of labor in 15,481 patients, with 99 % (15,313) singletons, 165 twins, and 3 triplets. Of a total of 15,652 infants, 4.4 % (689) died. In Table II, she enumerated the fetal positions/presentations (vertex, breech, footling, knee, face, shoulder, *et cetera*).

Table III presents the type of delivery, whether spontaneous 99 % (15,380), and less that 1 % each of forceps (93), version and extraction (155), and other (24) for a total of only 1.7 % (272) with interference. Cesarean section was performed on only one patient who died, but the infant survived. Volume II of this work was on extraordinary cases and unusual presentations (transverse lie and shoulder, placenta previa, *et cetera*). Volume III considered puerperal fever, eclampsia, premature labor, the use of forceps, craniotomy, and symphysiotomy, this latter operation which she believed was overrated, and she never performed.

Mme La Chapelle received her fundamental training from her mother, who was a midwife at the Hôtel Dieu. For several years she was married to the surgeon Lachapelle who worked at the *Hôpital Saint-Louis*, before dying at a young age. Following a period of further experience in Heidelberg, she was asked to help establish a proper school for midwifery training in Paris, the *Hospice de la maternité* (later renamed the Maison d'accouchement), and was made its director. There she taught together with Baudelocque. The school gained a great reputation in Europe, and from it many distinguished midwives went to practice and to teach. At her death, she left numerous notes with her original observations and important views in manuscript form. Her nephew, Antoine Louis Dugès (1797–1838), prosecutor of the medical faculty in Paris and from 1824 professor of obstetrics at Montpelier, collated these to form the present work. Dugès also authored the *Manuel d'obstétrique* (1826). Volume II of La Chapelle's work contains a review of her productive life and contributions to improved obstetrical care, presented by professor Francois Chaussier (1746–1828) of the *École de Sante*.

References

Cutter & Viets pp. 93–95 and 199; Garrison Morton 6170; Waller 5478; Wellcome II, p. 495; Hirsch III, pp. 580–581; Garrison, Hist of Med, p. 605; Roy Coll Obst & Gyn, London, p. 43.

Baudelocque, J.L. *L'art des accouchemens.* 2 vols. Paris, Mequignon, 1781.

Duges, A.L. *Manuel d'obstétrique, ou, précis de la science et de l'art des accouchemens....* Paris, Gabon, 1826.

Robb, H. Mme. Lachapelle, midwife. *Bull Johns Hopkins Hosp* 2:163–164, 1891.

Fertilization and Embryology

One of the wonders of life is the manner in which the single-cell fertilized ovum develops into a sentient human being, with several trillion cells of over 200 individual types. This is not the place to consider the details of embryonic development; however, a brief overview may be appropriate. Following ovulation and fertilization of the ovum, the latter of which normally occurs in the fallopian tube within minutes of ovulation, the zygote (the cell that results from fertilization) divides sequentially to form the morula (a solid ball with 16 cells), then the blastocyst in which a fluid filled cavity forms. At the end of the first week post-conception upon reaching the uterus, the blastocyst implants into the endometrium/decidua that lines the uterine cavity. The impetus for implantation is derived from the blastocyst and its metabolic products, which result in invasion of the uterine decidua and small blood vessels. By week 3 post-conception, cell division continues with the inner cell mass giving rise to the embryo. The blastocyst outer cell mass develops into trophoblast cells, the early placenta. Soon the blastocyst develops three layers of cells, the germ layers from which body organs and tissues arise. These include the innermost layer, endoderm (or entoderm), that gives rise to the epithelial cells of the lungs, digestive organs and other intra-abdominal viscera; the middle layer, mesoderm, from which arise the skeleton, muscles, cardiovascular system, reproductive system, and connective tissues; and the outer cell layer, ectoderm, which gives rise to the brain, nervous system, and integument including the skin. Recognition of pregnancy by the maternal organism includes a number of processes, including prolongation of the life-span of the ovarian corpus luteum to ensure secretion of progesterone, and tolerance by the maternal decidua of the semiallogenic graft of the placenta and fetus. The Embryonic period extends until the end of the seventh week of gestation, at which time the major organ systems have commenced their development. From the eighth week onward, the developing conceptus is referred to as a fetus.

That obstetrical research had its origins in embryology may have stemmed, in part, from the development (as early as 3000 BCE) of a system for artificial incubation of bird eggs by the Egyptians, and probably also by the Chinese. This original "biotechnology" provided not only an abundant supply of poultry for the table, but also a ready source of embryological material for investigation. In addition, as Joseph Needham (1900–1995) so clearly pointed out (Needham 1934, p. 1), "… even at the most remote times children were being born, and, though the practitioners of ancient folk-medicine might confine their ideas for the most part to simple obstetrics, they yet could hardly avoid some slight speculation on the growth and formations of the embryo." This speculation included the placenta, which was "an easily observed biological phenomenon," and "was regarded as of great importance, since it was thought to be the especial seat of the external soul" (Needham 1934, p. 4).

Embryology, the science that treats of the formation of the embryo, has been of interest since earliest times (Montague 1949). Perhaps based in part on the study of bird embryos, the ancient investigators recognized the correct function of the placenta and umbilical cord. For example, Aristotle (384–322 BCE), the first observational biologist and one time tutor to Alexander the Great (356–323 BCE), in his great embryological treatises *De generatione animalium* [On the generation of animals] and *De animalibus* [of animals] wrote at length on generation and classified animals on the basis of their embryological characteristics. Included in this first work is the report of his examination of the developing chick on successive days of embryogenesis. Aristotle stated that "The [umbilical] vessels join on the uterus like the roots of plants and through them the embryo receives its nourishment." The ancient investigators also recognized the similarities of early development of the chick and mammals. Along with the development of practical methods to incubate fertilized chicken eggs, this revelation led to the use of the developing chick as the most important vertebrate model organism for embryological research from that time forward.

It is of interest to speculate on obstetric research in ancient times. For example, Rabbinic legend holds that Cleopatra VI (69–30 BCE) "… investigated the process of foetal development by the dissection of slaves at known intervals of time

© Springer International Publishing Switzerland 2016
L.D. Longo, L.P. Reynolds, *Wombs with a View*, DOI 10.1007/978-3-319-23567-7_4

from conception, following the precepts of Hippocrates with regard to hen's eggs" (Needham 1934, p. 47). Although this story may be apocryphal, Cleopatra's Alexandria experiments were often cited by the Ancient Investigators as evidence that sexual differentiation of the male fetus was "complete" about 8 weeks before that of the female fetus. It seems difficult to explain such detailed knowledge of human sexual development unless investigations such as those attributed to Cleopatra had actually been performed. Ancient Indian texts (ca. the sixth century BCE) also demonstrate a relatively detailed knowledge of human embryology, and knowledge in this area was "… likely to have passed in one direction to the other [i.e., between India and the Mediterranean]" (Needham 1934, p. 2).

During the Renaissance, the artist, engineer, and polymath, Leonardo da Vinci (1452–1519) contributed much to the embryology of mammals and birds, although his early sixteenth century drawings did not come to public knowledge until two and a half centuries after his death. Giacomo Beregario da Carpi (ca. 1460–ca. 1530), professor of Surgery at the University of Bologna, advocated the study of fetal development as the tissues are simpler and less well developed than in the adult, and in some instances only vestigial in the adult. Andreas Vesalius, the founder of modern anatomy, in the second edition of the *Fabrica…* (1555) differentiated the discoidal placenta of man from the annular, or zonary, placenta in the dog, and the cotyledonous placenta of ruminants. Gabriele Falloppio (1523–1562) in his *Observationes anatomica*, [anatomical observations] (1561), noted that the human fetus has a single umbilical vein, in contrast to two in ruminants. Giolio Cesare Aranzi [Arantius] (1530–1589) in his *De humano foetu* (1564) was the first to maintain the

separation of maternal and fetal circulations in the placenta, and that organ's role in "purifying" fetal blood. Volcher Coiter (1534–1576) in his *Externarum et internarum* [of that that is outside and inside]… (1572) described development of the chick embryo on 20 successive days; unfortunately, this account was not illustrated.

It was in the seventeenth century, in the spirit of the new age in science, that embryology became an experimental discipline. At this time, unable to explain how such unique organs as the brain, heart, and uterus could have arisen from a single cell during development, the idea of "preformation" held sway. According to this hypothesis, development occurred as a consequence of unfolding and increase in size of organisms that preexisted within germ cells. An illustration of this view is the 1694 figure by Nicolas Hartsoeker (1656–1725) of a miniature human being within a sperm, a so called *homunculus* [little man]. As the field of embryological research matured, the concept of "epigenesis" (which originally had been advanced by Aristotle), with gradual and progressive development from the fertilized egg into specific organ structures came to the fore.

References

Montagu, M. F. Ashley. *Early History of Embryology: Embryological Beliefs of Primitive Peoples; Embryology from Antiquity to The End of the Eighteenth Century.* January–February 1949, CIBA Symposia; Volume 10, Number 4, Caspari-Rosen, Beate (editor).

Needham, J. *A history of embryology*. Cambridge, The University Press, 1934.

Girolamo Fabrizio [Hieronymi Fabrici, Fabricius ab Aquapendente]

De formatione ovi, et pulli tractatus accuratissimus. Patavii [Padua], ex officina Aloysii Bencii, 1621.

It was Fabricius (ca. 1533–1619) who first applied the Vesalian method of direct observation to the study of the embryo. With seven full page plates, he presented the earliest printed figures of the development of the chick. He was the first to establish with any degree of accuracy the role played by the ovary and oviduct in the formation of the hen's egg. He also was the first person to describe the germinal disc distinctly. *De formatione ovi et pulli*...[The formation of the egg and chick] is divided into two parts. The first, in three chapters, deals with formation of the egg. The first chapter discusses the three bases of animal generation given by Aristotle, the egg, the seed, and spontaneous generation from decomposing material. In the second chapter, Fabrici describes two functions of the "uterus": the formation of the egg and its subsequent nutrition. The third chapter presents in more detail these functions. The second part of the treatise, also in three chapters, deals with the generation of the chick within the egg, and begins with a description of the eggs of various species. The second chapter deals with the three basic functions of the egg: the formation, growth, and nutrition of the chick. He concludes his discussion with the trophic functions of both yolk and albumen. Fabricius then speculates on the various possible causes and conditions on generation, including a discussion of the order in which various parts of the embryo are formed during its development. The last chapter of the treatise returns to teleology to consider the utility of both the egg and the semen of the rooster. In this work, which was published posthumously, Fabricius made several erroneous assumptions, including that for impregnation the sperm did not enter the ovum, but rather stimulated conception from a distance.

In this second of the plates, Fabricius presents in over a dozen illustrations the development of the chick embryo up to day 13, with emergence of the wing buds. Some considerations of the life and work of Fabricius are given in Fabrizio 1604. According to Joseph Needham (1900–1995), Fabricius does not deserve the elevated status accorded him as an embryologist. Needham emphasizes his "scholasticism" and "argumentativeness", and the erroneous ideas that he promulgated. For instance, he presented a complex and confused view of the origin of the hen's egg, whether it was formed in the oviduct by transudation through blood vessels, and whether the yolk is more "earthy" than the white. He also states that the human placenta is cotyledonous, that the fetal heart and other organs have no function, and that the liquors and humors around the fetus consist of sweat and urine. Nonetheless, Needham credits Fabricius for his illustrations, which "were far better than anything before and for a long time afterwards" (Needham 1934, pp. 87–90).

References

Garrison Morton 466; Osler, 2559. Krivatsy 3826.

Adelmann, H.B. *The embryological treatises of Hieronymus Fabricius of Aquapendente. The formation of the egg and of the chick* [*De Formatione Ovi et Pulli*]. *The formed fetus* [*de Formato Foetu*]. *A facsimile edition.* Ithaca, NY, Cornell Univ Press, 1942.

Fabrizio, G. *De venarum ostiolis.* Patavii, ex typ. L. Pasquatin, 1603. (GM 757).

Fabrizio, G. *De formatione ovi et pulli....* Patavii, ex off. A. Bencÿ, 1621. (GM 466).

Falloppio, G. *Observationes anatomicae.* Venetiis, Apud M.A. Ulmum, 1561 (GM 1208, GM 1537).

Needham, J. *A history of embryology.* Cambridge, at the University Press, 1934.

Zanobio, B. Girolamo Fabrici (or Fabricius ab Aquapendente, Geronimo Fabrizio).... In: *Dictionary of Scientific Biography.* Vol IV. Charles Coulston Gillispie (Ed.). New York, Charles Scribner's Sons, 1971, pp. 507–512.

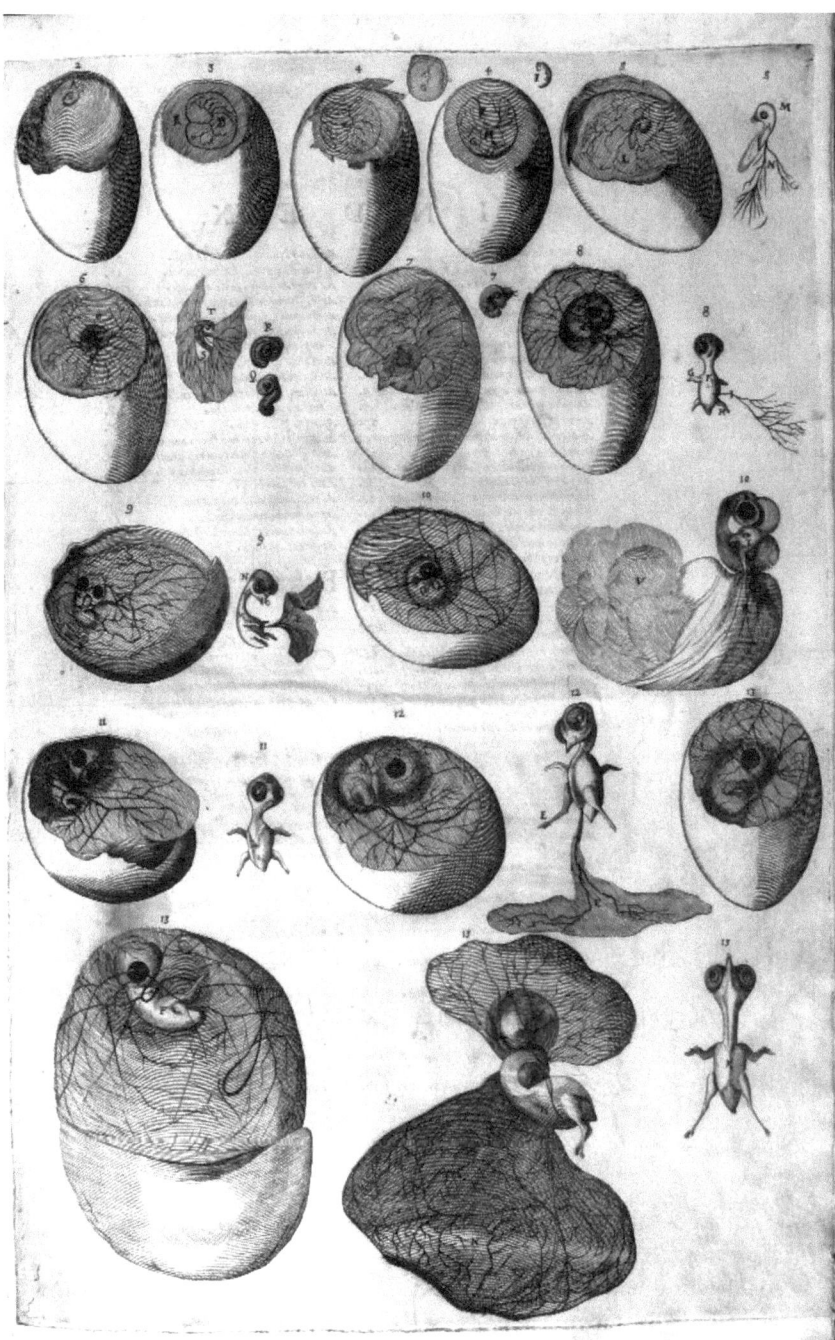

PLATE FROM FABRIZIO, 1621

Ulisse Aldrovandi [Aldrovandus]

Monstrorum historia, cum paralipomenis historiae omnium animalium. Bartholomaeus Ambrosinus (ed)... volumen composuit. Marcus Antonius Bernia in lucem edidit ... Bononiae, Typis Nicolai Tebaldini, 1642.

Aldrovandi's (1522–1605) treatise on monsters and prodigies, illustrated with woodcuts of human and animal malformations, as well as anomalous plants, forms a portion of his encyclopedic work on natural history. Although comprising 13 volumes, only the first three on birds and a fourth on insects were published during his lifetime. In this work, published almost four decades after his death, Aldrovandi presented both observed and mythical descriptions of developmental anomalies, including chimeras and unicorns. The title page presents an illustration of Aldrovandi with curling mustache, and *puttos* that display globes with figures and the captions *sapientiae symbolum* [wise symbols], *fertilitatis indicium* [fruitful information/evidence], *firmitudo bene consuitorum* [strengthen good customs], *aevi perennitas* [lasting forever], and so forth. Among the over 450 woodcuts are 18 that illustrate the fetus *in utero* in various positions, and the placenta, not unlike those of Jacob Rueff (1500–1558) (for instance Rueff 1554, pp. 34 and 37). Page 64 illustrates an aborted fetus, shown as an adult, with placenta and the three-vessel umbilical cord.

Following the lead of Aristotle (384–322 BCE), in his embryologic study of the chick he opened and examined the egg during its incubation period day by day for 20 days, to describe in detail embryonic development. By this technique, Aldrovandi established that the heart is formed in the vitelline sac, rather than in the albumin, as was believed at that time. He also made important observations on the development of teratologic anomalies. Although not profound contributions to the progress of embryology, his studies pointed future investigators to the need for personal direct observation and confirmation, rather than relying on the descriptions of Aristotle, Galen, or others. This laid the groundwork for the subsequent studies of Volcher Coiter (1534–1576; 1572; 1574), Fabricius ab Aquapendente (1533–1619; 1604; 1621), William Harvey (1578–1657; 1651), and Marcello Malpighi (1628—1694; 1673).

Aldrovandi, called the Pliny of his age and the "Bolognese Aristotle," was a classical Renaissance scholar. Born of noble parentage in Bologna, and following an adventurous tour of Spain, he returned to Bologna to graduate in medicine and philosophy in 1553. Several years earlier, in 1549, he had been accused of heresy for espousing anti-trinitarian beliefs. He remained in custody in Rome for almost a year, before being absolved. During this time, he became acquainted with the physician Guillaume Rondelet [Rondeletius] (1507–1566) who was in the process of preparing a major work on fishes of the Mediterranean (Rondelet 1554–1555). This experience stimulated Aldrovandi to study natural history. As many other naturalists, to amass a complete collection, he became gripped with a passion to gather and possess specimens of every species known. Holding that "nothing is sweeter than to know all things", he commenced gathering specimens for his own "cabinet", which grew into a major museum with over 18,000 specimens. As a student of many aspects of natural history, he traveled widely to collect the unusual and obscure. Returning to Bologna in his academic post, Aldrovandi stimulated many students, and in 1561 was appointed full professor of logic and philosophy, as well as of medicine. He wrote on many aspects of natural history, and worked to establish the botanical gardens of Bologna, the *Orto Botanico dell'Università di Bologna*, one of the first in Europe. During this time he prepared an *Antidotario* [Book of Antidotes] an official pharmacopeia (Aldrovandi 1574). This work on monsters was published posthumously by Bartolommeo Ambrosini (1588–1657) who added a number of cases he had observed, and who had succeeded Aldrovandi as director of the Botanical Garden in Bologna. Based upon the specimens and drawings in his collection, many of the illustrations for the copper plate engravings were prepared by the prolific artist to the Medici court, Jacopo Ligozzi (1547–1627) (Aldrovandi 1559–1667).

It was during this period of the Renaissance that modern descriptive and experimental science was being born through the work of investigators such as Aldrovandi. Both the Swedish classifier Carl von Linné [Linnaeus] (1707–1778) and the French naturalist Georges Louis Leclerc *Comte* de Buffon (1707–1788) held him to be the founder of studies in natural history. Aldrovandi assembled a spectacular collection of plants, both living for his botanical garden, and dried for his herbarium. He willed his vast collection of botanical and zoological specimens to the Senate of Bologna. These were conserved in the *Palazzo Pubblica*

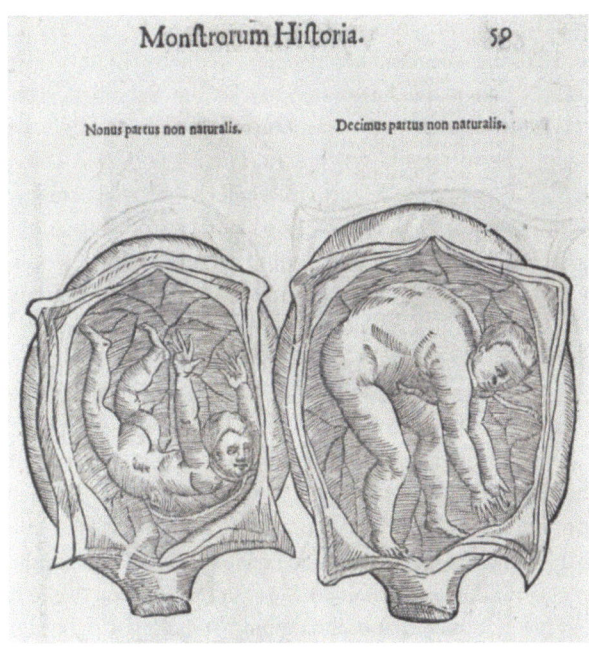

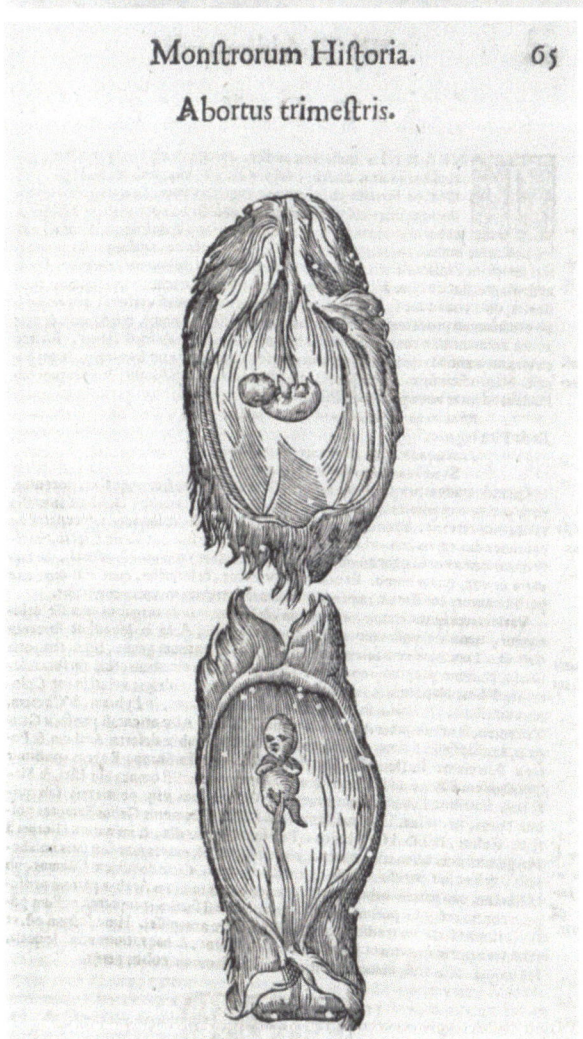

ILLUSTRATIONS FROM ALDROVANDI, 1642

[Public palace or large building], and then later in the *Palazzo Poggi* [the palace of the Poggi brothers, Alessandro and Cardinal Giovanni]. In the early twentieth century a number of Aldrovandi's specimens were brought together at the *Palazzo Poggi*. Aldrovandi also wrote a survey of the statuary in Rome (1556), and several works in ornithology in which he described their zoological and physiological characteristics. The *Dorsa Aldrovandi* on the moon was named in his honor.

References

Garrison Morton 534.53; Krivatsy 187.

Aldrovandi, U. *Le antichita de la citta di Roma*. In Venetia, Appresso Giordano Ziletti…, 1556.

Aldrovandi, U. *Opera omnia. 13 vols*. Bononiae, J.G. Bellagamba…, 1559–1667. (GM 290).

Aldrovandi, U. *Antidotarii Bononiensis, siue De vsitata ratione componendorum, miscendorumq[ue] medicamentorum, epitome….* Bononiae, Apud Ioannem Rossium, 1574.

Aldrovandi, U. *Ornithologiae, hoc est de avibus historiae, libri XII….* Bononiae, apud Franciscum de Franciscis Senensem, 1599–1603.

Aldrovandi, U. *De animalibus insectis libri septem, cum singulorum inconibus ad viuuum expressis*. Bonon, Apud Ioan. Bapt. Bellagambam, cum consensu superiorum, 1602.

Aldrovandi, U. *Ornithologiae hoc est de avibus historiae libri XII… tomus tertius, ac postremus*. Bononiae, Apud Ioannem Baptistam Bellagambam, MDCIII, 1603.

Coiter, V. *Externarum et internarum principalium humani corporis partium tabulae, atque anatomicae exercitationes observationesque variae, novis, diversis, ac artificiosissimis figuris illustratae, philosophis, medicis, in primis autem anatomico studio addictis summè utiles*. Noribergae, in officina Theodorici Gerlatzeni, 1572. (GM 464.1).

Coiter, V. *Lectiones Gabrielis Fallopii de partibus similaribus humani corporis, ex diversis exemplaribus a Volchero Coiter… collectae; his accessere diversorum animalium sceletorum explicationes iconibus… illustratae… autore eodem Volchero Coiter*. Noribergae, in officina T. Gerlachii, 1575. (GM 284).

Fabricius ab Aquapendente, G. *De formato foetu. Venetiis. per Franciscum Bolzettam 1600*. Colophon, Laurentius Pasquatus, 1604.

Fabricius ab Aquapendente, G. *De formatione ovi, et pulli tractatus accuratissimus*. Patavii [Padua], ex officina Aloysii, Bencii, 1621.

Harvey, W. *Exercitationes de generatione animalium. Quibus accedunt quaedam de partu: de membranis ac humoribus uteri: & de conceptione*. Londoni, Octavian Pulleyn, 1651.

Malpighi, M. *Dissertatio epistolica de formatione pulli in ovo*. London, John Martyn, 1673.

Rondelet, G. *Libri de piscibus marinis, in quibus verae piscium effigies expressae sunt… (Universae aquatilium historiae pars altera, cum veris ipsorum imaginibus…). 2 vols*. Lugduni, apud Matthiam Bonhomme, 1554–1555. (GM 282).

Rueff, J. *Ein schön lustig Trostbüchle von dem Empfengknussen und Geburten der Menschen*. Tiguri, Apud Frosch [overum], 1554. (GM 6141).

Ruggieri, M. and A. Polizzi. *From Aldrovandi's "Homuncio" (1592) to Buffon's girl (1749) and the "Wart Man" of Tilesius (1793): antique illustrations of mosaicism in neurofibromatosis? J Med Genet* 40:227–232, 2003.

William Harvey

Exercitationes de generatione animalium. Quibus accedunt quaedam de partu: *de membranis ac humoribus uteri*: & *de conceptione* ... Londoni, Octavian Pulleyn, 1651.

In Harvey's (1578–1657) fundamental treatise on the generation of animals and embryology, "the most important book on the subject to appear during the seventeenth century" (Garrison and Morton), he rejected the prevailing doctrine of the preformation of the fetus, and advanced the theory, radical for its time, of epigenesis (*per epigenesin*), that all living beings derive from the ovum "by the gradual building up and aggregation of it's parts." Regarding Harvey's theory of epigenesis, Thomas Henry Huxley (1825–1895) claimed this should give him an even greater claim to the veneration of posterity than his better known discovery of the circulation of the blood (Keynes 1953). Harvey reported a wealth of observations on many aspects of reproduction in a wide variety of species. As representatives of vivipara, his attention chiefly was devoted to the deer, while that for ovipara was the domestic foul. For Harvey, all life develops from the egg. This is expressed on the frontispiece which depicts the supreme Roman god Jupiter [Jove] opening a large egg, inscribed with the fundamental dictum of embryology, *ex* (upper half of egg shell) *ovo omnia* (lower half of egg shell), which translates as, "from the egg everything," and from which the liberated animals and insects fly. In addition to a small human figure, these include a bird, stag, fish, lizard, snake, grasshopper, butterfly, and spider. Although the phrase *omne vivum ex ovo* [all life originates from the egg] is often attributed to Harvey, he does not state this explicitly in the text.

An opponent of the theory of spontaneous generation, Harvey speculated that humans and other mammals must reproduce through the joining of an egg and sperm. No other theory was credible. By positing and demonstrating for viviparous animals the same mechanism of reproduction as that observed in oviparous animals, he thus initiated the search for the mammalian ovum, which was not discovered until 1827 by Carl Ernst von Baer (1792–1876). Harvey maintained that "... *Jovis omnia plena* [All things are full of Deity], so that in the chicken and all its functions and actions the *digitus Dei* [the Finger of God] or the God of nature, reveals himself" (1651, p. 170).

In addition to its pioneering embryology, this work of 72 chapters or "Exercises" (two exercises are numbered 4, so that the numbered total is 71) with eight appendices, includes an epilogue on *De Partu* [of the birth]. This latter is important for being the first original work on obstetrics by an Englishman (Spencer 1921, 1927). The first English translation appeared in 1653. The first edition to contain illustra-

tions of the embryological development of the chick was that in 1674 by the Dutch physician Wilhelm Langly (1616–1668), under the editorship of Justin Schrader (Harvey 1674). Rather than a complete edition of Harvey's work, Langly performed his own experiments, and used Harvey's text as a template. Figure III depicts the chick embryo at 72 h, with the *puncta salientio* [salient or cicatricula], the point (a), great artery as a small cloud leaving the central body (b), the vena cavae carrying the blood to the heart (c), a circular vein (d), vitelline umbilical vessel (e), chorionic umbilical vessel (f), albumen (g), and fractured shell of the egg (h). Figure V illustrates the embryo on day 5, with blood vessels entering and leaving the centrally placed heart.

Harvey, celebrated for his discovery of the circulation of the blood (1628), was the first to adopt the scientific method for the solution of a biological problem. In fact, it was Harvey's work on the function of the heart and the circulation in general that led him to consider the circulation of the fetus and its relation to that of the mother. To help grasp the significance of Harvey's monumental achievement, one must consider the traditional Galenic concept of the heart, blood vessels, and their contents at this time. Briefly, this view held that blood originated in the liver, which was the dynamic organ that provided vascular pulsations. The heart, in contrast, was believed to be a fibrous sac that dilated or collapsed passively as a consequence of the motion of the blood. In the liver, venous blood was mixed with an imaginary essence, the "natural spirits", and distributed to the several body organs. Venous blood within the right side of the heart was believed to seep through "pores" in the septum dividing the two sides of the heart, to mix on the left side with air to produce arterial blood. Also at this time, most workers believed that the heart consisted of only two chambers, the ventricles, and that the atria were but extensions of the veins joining the heart. In left ventricle, blood was charged with a second essence, "vital spirits", before being distributed to various organs. In the brain, "animal spirits" were added to pass through the nerves. Interaction of the "vital" and "animal" spirits was believed to provide movement, such as muscular activity.

Regarding the placental circulation, Harvey stated, "the Extremities of the *Umbilical* vessels, are no way conjoined to the *Uterine* vessels by an *Anastomosis*; nor do extract blood from them..." (1653, p. 439). By logic based on his knowledge of the circulation, he held the maternal and fetal circulations to be separate, each following in an opposite direction to the placenta by way of the arteries and returning by the veins. He also noted that the "*Embryo* is no other manner sustained in the *Uterus*, than the *chicken* is in the *Egge*" (1653, p. 440). Based on knowledge that *in utero* the fetus can grow and mature in the absence of air, however, at delivery it dies if the umbilical cord is compressed and it does not breathe, Harvey

FRONTISPIECE FROM HARVEY, 1651

postulated that substances absorbed by the umbilical cord stimulated organ development, while the fetus received its main nourishment from the amniotic fluid. Following Aranzi, he also referred to the placenta as the *hepar uterinum* [uterine liver] and *mammas uterinas* [uterine breasts]. It was not until a decade later that Marcello Malpighi (1628–1694) first described the capillary bed connecting arteries and veins (Malpighi 1661), and made possible an understanding of the anatomical basis of regional circulation. Lacking this knowledge, Harvey could not understand completely certain details of the circulatory system in either the adult or the fetus.

In this work, Harvey raised an additional question of fundamental importance to developmental biology and life in utero; that is, the extent to which the placenta serves as a lung for the fetus.

> ... I shall propose this *Probleme* to the Learned; namely, How the *Embryo* doth subsist after the *seventh moneth* in his *Mothers* womb when yet in case he were borne, he would instantly breath: nay he could not continue one small hour without it? And yet remaining in the *womb*, though he pass the *ninth moneth*, he lives, and is safe without the help of *Respiration* ... How commeth it to pass, that the *Foetus* being now borne ... if he have but once attracted the Aire unto his *Lungs*, he cannot afterwards live a minute without it, but dyeth instantly?

(Harvey 1653, pp. 482–483)

When it was raised, 'Harvey's, question', as it came to be known, aroused the interest of both philosophers and experimentalists. However, with no clear understanding of respiration or metabolism, and without knowing of the existence of oxygen (the discovery of which did not occur until the following century), Harvey could only speculate on this question. Some details of the path of this discovery two centuries later by a young Swiss obstetrician, Paul Zweifel (1848–1928), who showed that the fetus consumes oxygen *in utero* (Zweifel 1876), have been given by Donald Henry Barron (1905–1993). Barron concluded that this discovery "... marks the beginning of the modern era of research on foetal physiology" (Barron 1976).

Harvey studied at Gonville and Caius College, Cambridge University, and in 1602 graduated from the University of Padua. At this latter institution, he studied under Girolamo Fabrizio [Fabricius ab Aquapendente] (1533–1619), whose work in veins and their valves stimulated his thinking about the flow of blood in veins and arteries. Physician at St. Bartholomew's Hospital, London, Harvey served as personal physician to the court of King James I (1566–1625) and to his successor Charles I (1600–1649). It was these royal patrons who supported Harvey's experiments and, in fact, provided deer and other animals from the Royal preserve for his studies. Of this patronage, Harvey wrote, that "he had daily opportunity of dissecting and studying the reproductive tract and genital tract. He also credits his Royal patrons with taking a great interest in his work, for instance, "...my Royal master, whose Physitian I was, was himself delighted in this kinde of curiosity, being many times pleased to be an eyewitness, and to assert my new inventions" (p. 397).

References

Garrison Morton 467; Krivatsy 5342; see Norman, 27; Osler 710; Russell 375.

Baer, C.E. von, *Ueber Entwicklungs geschichte der Thiere.* 2 vols in 3. Königsberg, Bornträger, 1828–1888.

Barron, D.H. *Paul Zweifel, pioneer fetal physiologist.* A centenary tribute. Arch. Gynäk. 221: 1–4, 1976.

Bylebyl, J.J. William Harvey. In: *Dictionary of Scientific Biography. Vol VI. Charles Coulston Gillipsie (Ed).* New York, Charles Scribner's Sons, 1972, pp. 150–162.

Harvey, W. *Anatomical exercitations, concerning the generation of living creatures: To which are added particular discourses of births, and of conception, &c.* London, printed by James Young..., 1653.

Harvey, W. *Exercitatio anatomica de mortu cordis et sanguinis in animalibus.* Francofurti, sumpt Guilielmi Fitzeri, 1628. (GM 759).

[Harvey, W] *Observationes et historiae omnes singulae è Guiljelmi Harvei libello De Generatione Animalium... item Wilhelmi Langly... studio Justi Schraderi....* Amstelodami, Abrahami Wolfgang, 1674.

Keynes, Sir G.L. *A bibliography of the writings of Dr. William Harvey.* Cambridge, University Press, 1953.

La Courveé, J.C. *De nutritione foetüs in ütero paradoxa.* Dantisci [Danzig], Georgii Försten, 1655. (GM 6146.1).

Malpighi, M. *De pulmonibus observations anatomicae.* Bononiae, B. Ferronius, 1761. (GM 760, 915).

Meyer, A.W. *An analysis of the De Generatione Animalium of William Harvey.* London, Oxford University Press, 1936.

Needham, J. *A history of embryology.* Cambridge, at the University Press, 1934, pp. 112–133.

Spencer, H.R. The Harveian Oration on William Harvey, obstetric physician and gynaecologist. *Br Med J* 2:621–626.1, 1921.

Spencer, H.R. *The history of British Midwifery from 1650–1800. The Fitz-Patrick Lectures for 1927 delivered before the Royal College of Physicians of London.* London, John Bale, Sons & Danielsson, 1927, p 1ff.

Zweifel, P. Die Respiration des Fötus. Arch. Gynäk. 9: 291–305, 1876.

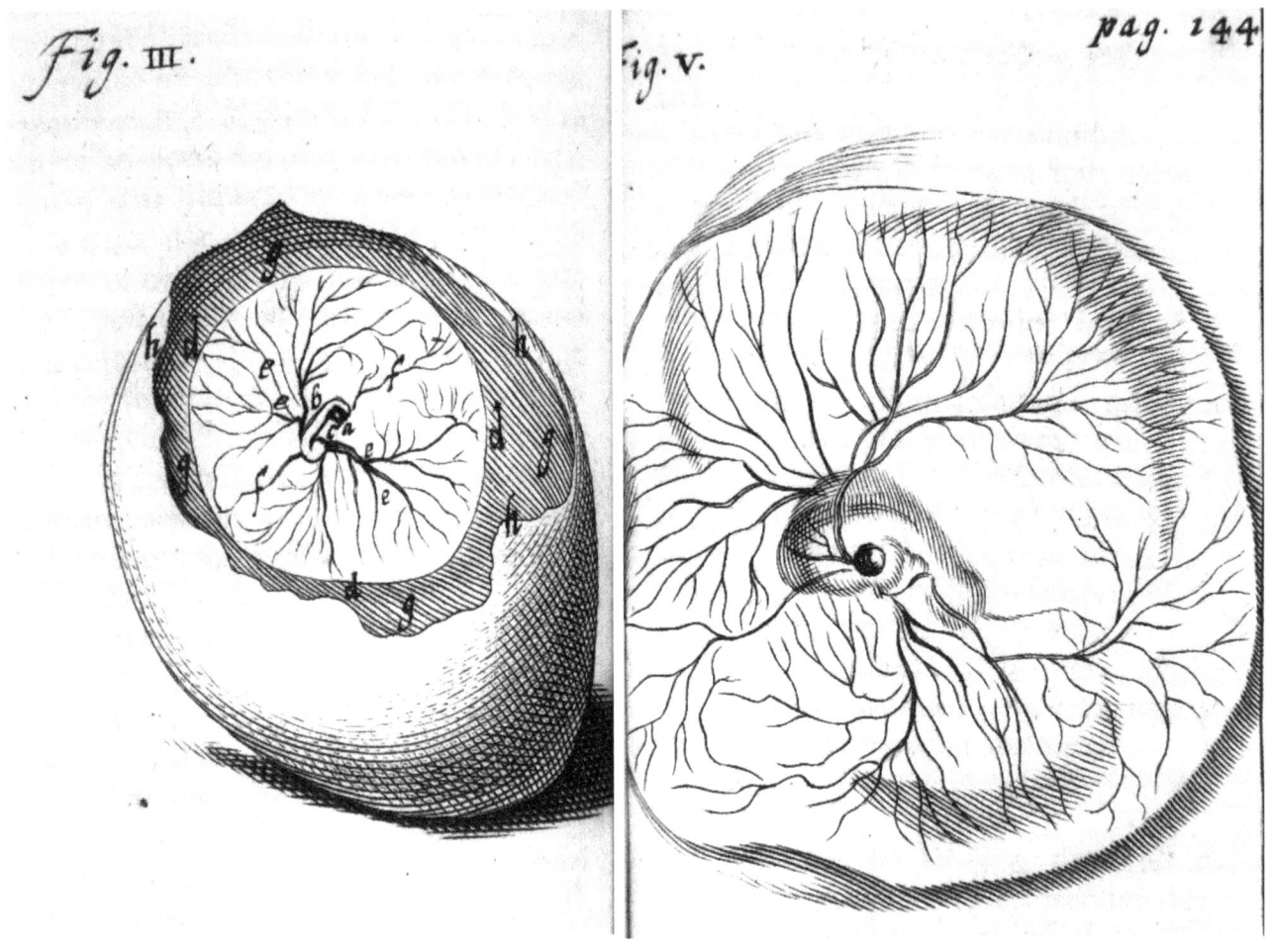

FIGURES III AND V FROM HARVEY, 1651

Walter Needham

Disquisitio anatomica de formato foetu. Londini, Typis Gulielmi Godbid... Radulphum Needham, 1667.

Needham (ca. 1631–1691), a Cambridge medical graduate who studied anatomy at Oxford, was the first to report chemical experiments on the developing mammalian embryo. This volume is divided into seven chapters. In the first, Needham maintained that the uterine arteries must supply nutrients to the uterus and developing fetus. He also refuted the idea that "uterine milk" is identical with lymph. In Chap. 2, Needham considers the anatomy of the placenta, its comparative anatomy among species, and its function. By careful dissection, he demonstrated the *chorion frondosum* [leafy membrane] and *chorion leave* [disappearing membrane], and confirmed that the placenta consists of separate maternal and fetal portions. Importantly, he supported the doctrine of William Harvey (1578–1657) regarding the separate maternal and fetal placental circulations. He also upheld Harvey's concept that the umbilical veins convey "nourishing juice" to the fetus for its nutrition. Chapter 3, "the membranes and humours" presents a comparative description of the amniotic and allantoic membranes and their fluid contents. On the basis of his studies, he stated, "these liquors... proceed from the blood and seem similar to its serum but they are different from it. For when fire is applied to them in an evaporating basin... they do not coagulate, as the blood-serum always does.... In the same way humors differ from themselves before and after digestion, filtration, and the other operations... of nature. All, when distilled, give over a soft and clear water... very like distilled milk" (pp. 49ff). Needham also described small solid bodies in the allantoic and amniotic fluid. Chapter 4 is devoted to the vessels of the umbilical cord. In Chap. 5, Needham discusses the *foramen ovale* and the fetal circulation. In Chap. 6, he argues (incorrectly) against the idea of the lungs as an organ of respiration, stating rather that its function is to "comminute the blood and so render it fit for a due circulation." The seventh chapter presents instructions for dissection of the embryo and fetus of various species.

The figures illustrate several fetuses with their membranes. The upper figure in Plate 1 is of a fetal cow with placental cotyledons and separated membranes: umbilical cord (A); vessels (B) to cotyledons (C); Allantoic membranes (D); Amniotic membranes (F); and urachus (G). The lower figure, Plate 2, shows a fetal horse (A); umbilical cord (B); Amnion (C); and Chorionic membranes (D).

In the late seventeenth century there were few tools for accurate chemical analysis. Thus, Needham's experiments were limited to distilling and coagulating the amniotic, allantoic, and other fluids, observing the salt remaining after evaporation, and so forth. It would remain for a century before further advances were made in the field of chemical embryology. As a physician, Needham practiced in Shropshire for much of his professional life.

References

Garrison Morton 467.2; Krivatsy 8283; Wing p. 411.

Harvey, W. *Exercitationes de generatione animalium. Quibus accedunt quaedam de partu: de membranis ac humoribus uteri: & de conceptione....* Londoni, Octavian Pulleyn, 1651.

Needham, J. *A history of embryology.* Cambridge, at the University Press, 1934, pp. 137–141.

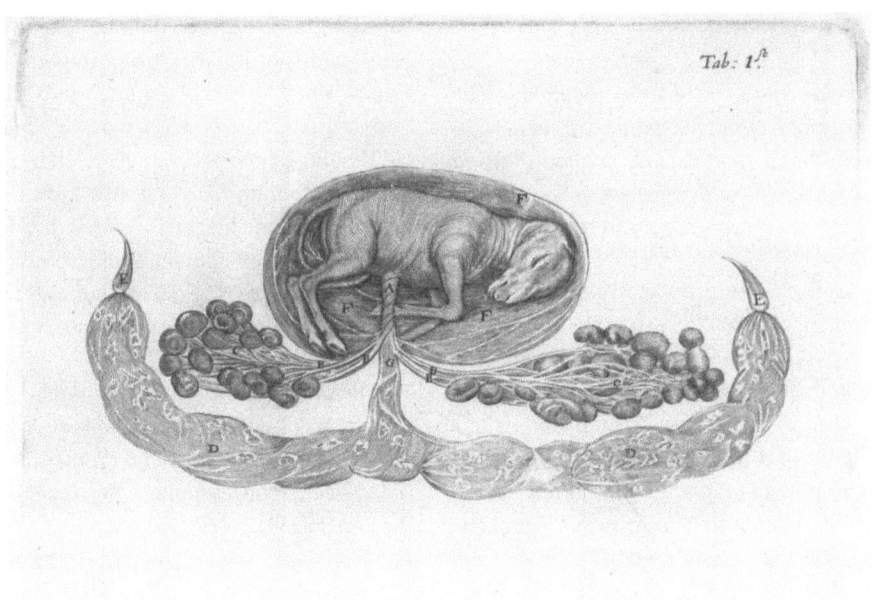

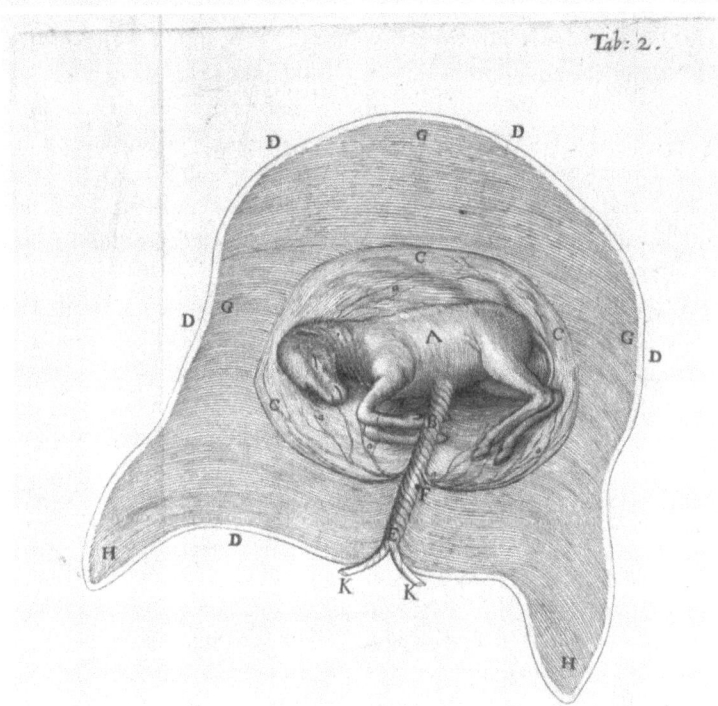

TAB. 1 AND 2 FROM NEEDHAM, 1667

Thomas Theodor Kerckring

Spicilegium anatomicum, continens observationum anatomicarium rariorum centuriam unam: nec non osteogeniam foetuum... (with) Osteogeniam foetuum, in qua quid cuique ossiculo singulis accedat mensibus .... Amstelodami, Sumpt Andreas Frisius, 1670.

In his "A gleaning or anthology of anatomical observations with development of the bones of the fetus" published with another work on fetal osteology, Kerckring (1640–1693) presented an *omnium gatherum* [a collection of everything] of clinical observations, curiosities and autopsy findings, many of which were rare. In this compendium he presented several original ideas on fetal development, particularly in regards to osteogenesis. Kerckring correctly noted that by the second month only a small amount of the skeleton had formed, and that the bones develop by a transformation of a membranous structure into cartilage into bone. He stated in error, however, that the younger fetal skeleton is merely a miniature of that of the older fetus; basing this assertion on "observations" of very early specimens.

Tabula II (Observation II) illustrates two fetal skeletons in a case of superfetation, e.g., the fertilization and subsequent development of an ovum when a fetus is already present in the uterus. This occurs as a consequence of fertilization of ova during different ovulatory cycles, and yielding fetuses of different ages. The larger fetus appears to be at about 28 weeks gestation, the younger at about 14 weeks.

Several other plates show such things as a polydactylous infant (Tabulae VIII [Observation XXII]) that had been found in an Amsterdam river. This probably had been disposed of as a fearsome monster with a number of anomalies, not rare when taken singly, but rather unique when combined, including short arms and legs, as compared with the trunk, indicating that the infant was an achondroplasic (Greek, negative cartilage formation) dwarf. This is a hereditary, congenital condition, in which incomplete bone formation results in an individual with dwarfism, short limbs, and a normal trunk. A perhaps more remarkable feature of this infant is the polydactyly, with the right hand possessing seven digits; the left, six, with an accessory. The right foot had eight digits, four of them being partly

fused in pairs; the left foot had nine digits, the first, second, third, and fourth partly fused in pairs. There were also certain anomalies in the ossification of the bones of the skull and thorax.

Tabula IX (Observation XXIII) illustrated a monster more resembling a demon than a human being, which was a case of craniorachischisis (Greek, head spine fissure), a congenital fissure of the skull and vertebral column. The head was prognathous, sessile, and set well forward, suggesting an ape, with the front part of the skull represented by a thin, lid-like structure surmounted by bony bosses. The posterior part was lacking and the spinal column had failed to close posteriorly, to a point just above the lower angles of the scapulae.

The last plate depicts, in a rather fanciful manner, embryonic and fetal development. Fig. I, eggs of different size from the "testicles" (ovaries) of a woman; Fig. II, an egg opened 4 days (A, chorionic membrane; B, the early embryo); Fig. III, a larger egg, opened 14 days after conception; Figures IV to VI, the skeletons of fetuses at 3 weeks, 4 weeks, and 6 weeks after conception. The copperplate engravings are by Bloteling, a student of Vissher.

Kerckring, a graduate of the University of Leiden, worked as a physician in Amsterdam and was a colleague of the anatomist Frederik Ruysch (1638–1731). He described the large ossicle sometimes present at the posterior margin of the *foramen magnum* in the skull, at about the sixteenth week of gestation, which serves as a center of ossification, uniting with other bony parts at birth (Kerckring's center). He also described the circular folds in the small intestine (the *valvulae conniventes*; or *plicae circulares* or folds or valves of Kerckring). Speculating on generation, Kerckring correctly concluded that the ovaries of both oviparous and vivaparous animals contain ova. He also stated that expulsion of the ovum is induced by the onset of menstruation or by sexual irritation; and that at autopsy he had found ova the size of peas (illustrated) and that the ova contained fluid of disagreeable taste and odor (*sapore ingrato*). In addition, he held that the male semen reach the ovary via the Fallopian tubes, and that the fertilized ovum enters the uterus *per vas ejaculatorium* [by vessels of ejaculation] (Lindeboom 1984, p. 1031). As with Ruysch and other Dutch anatomists, Kerckring assembled a highly regarded "Cabinet" of anatomical specimens that later were incorporated into the museum of the University of Leiden.

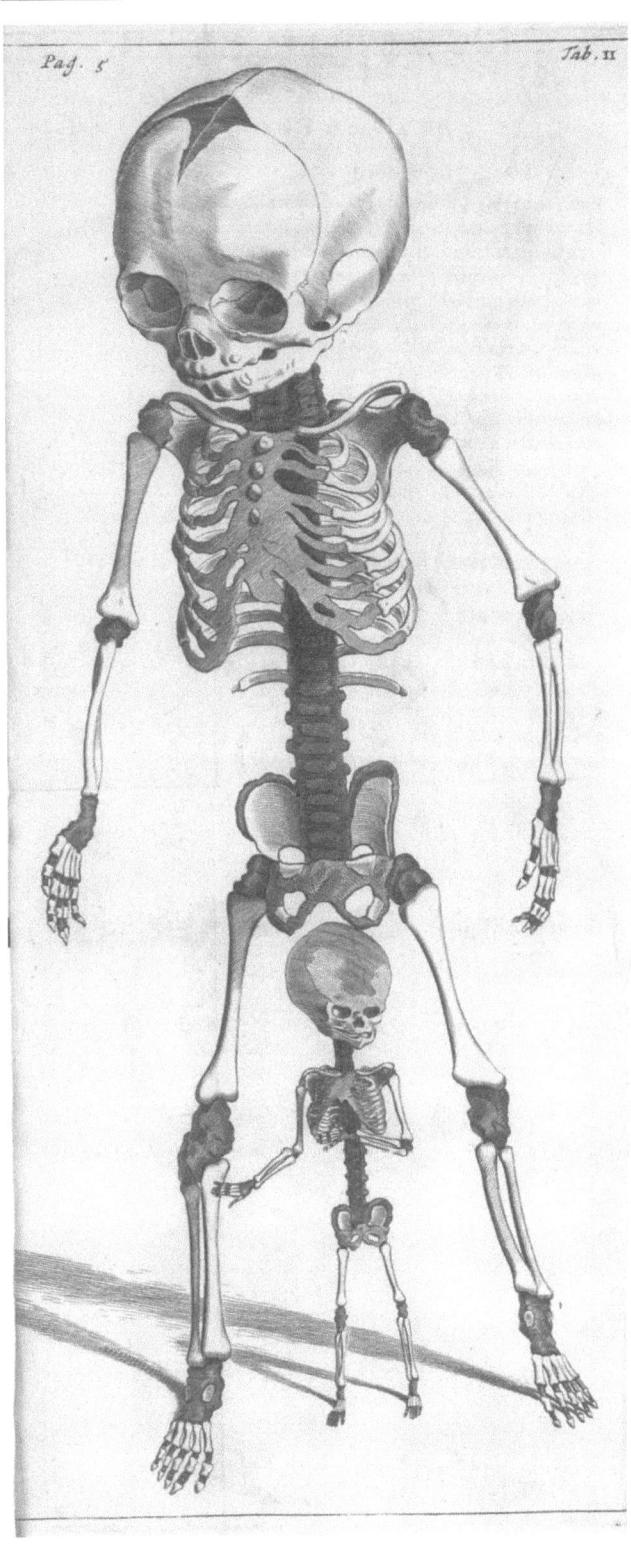

Pag. 5   Tab. II

TAB. II FROM KERCKRING, 1670

References

Cushing K-52; Garrison Morton 383; Heirs of Hippocrates 632; Krivatsy 6346; Norman 1209, 1210; Osler 3115; Ricci, p. 284.

Kerckring, T.T. *Anthropogenieae ichnographia*; *sive*, *Conformatio foetus ab ovo usque ad ossificationis principia*.... Amstelodami, Sumptibus Andreae Frisii, 1671. (Krivatsy 6345).

Kerckring, T.T. An account of what hath been of late observed by Dr. Kerkringius concerning eggs to be found in all sorts of females. *Phil Trans Royal Soc* 7: 4018–4026, 1672.

Kerckring, T.T. *Opera omnia anatomica. Continentia spicilegium* [*sic*] *anatomicum, osteogeniam foetuum....* Lugduni Batavorum, Apud Vid. & Fil. Corn. Boutesteyn, 1717.

Lindeboom, G.A. *Dutch medical biography: a biographical dictionary of Dutch physicians and surgeons, 1475–1975.* Amsterdam, Rodopi, 1984.

Nicholls, A.G. Theodore Kerckring and his "spicilegium anatomicum". *Canad Med Assn J* 42: 480–483, 1940.

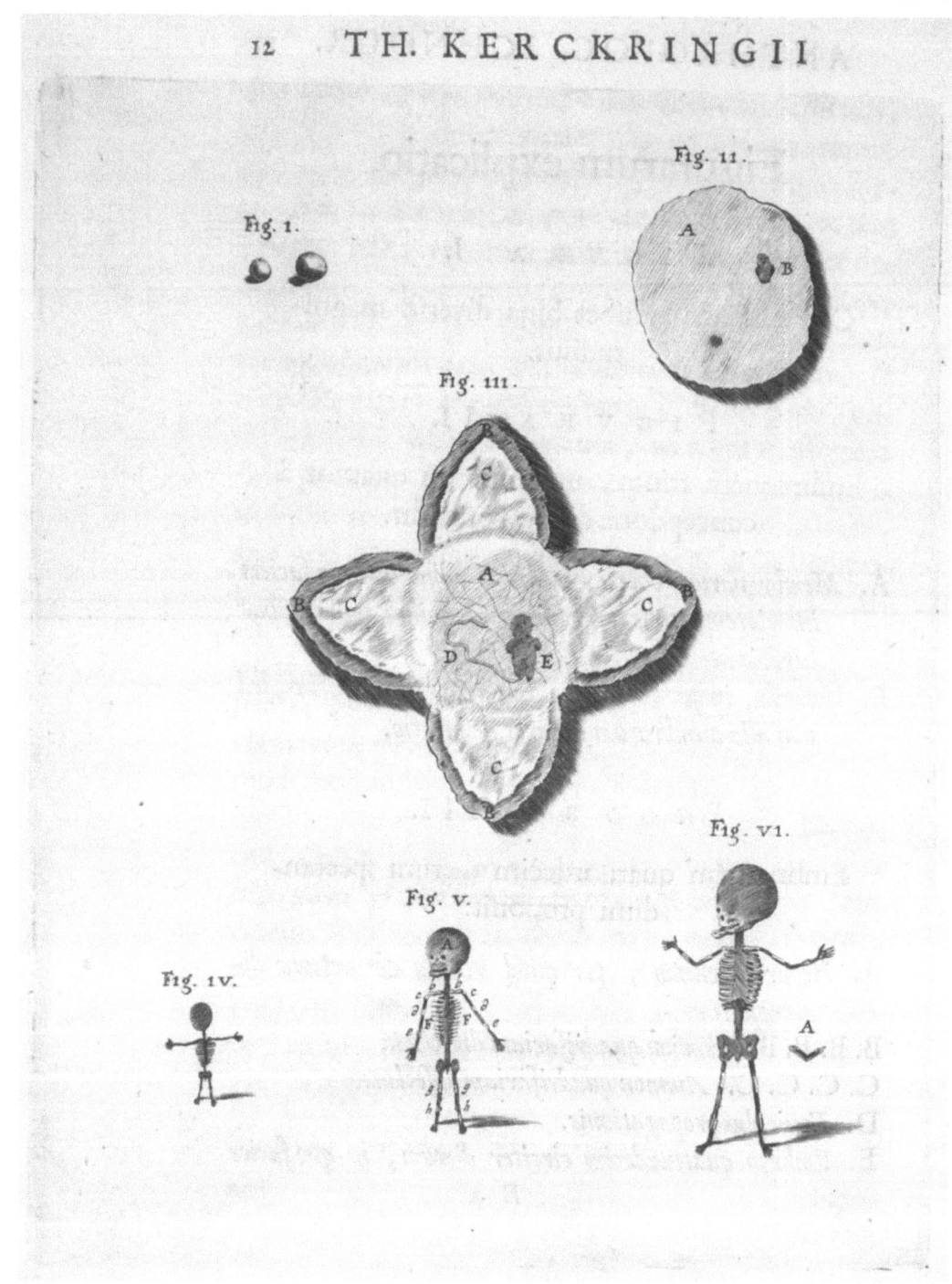

FIGURES FROM KERCKRING, 1670

Regnier [Reinier] De Graaf

De mulierum organis generationi inservientibus tractatus novus. Demonstrans tam homines & animalia caetera omnia, quae vivpara dicuntur, haud minus quam ovipara ab ovo originem ducere. Lugduni Batavorum. [Leyden], ex officina Hackiana, 1672.

During the seventeenth century, many still believed that living organisms, including animals, could arise from spontaneous generation. An alternate theory was the Aristotelian doctrine of the egg being formed in the uterus as a result of activation of the menstrual blood by the male semen, with subsequent differentiation into the embryo and fetus (for a detailed review, see Needham 1934). It was at this time that the idea gained credence that the female testes, as the ovary of birds, was the site of egg formation. De Graaf (1641–1673), a practicing physician in Delft, in his *De mulierum organis generationi* [concerning the generative organs of women] ... first thoroughly described the mammalian female gonad and established that this organ produces the ovum. On the basis of a comparative study of the ovaries of mammals and birds, De Graaf concluded that the cell-like protuberances, which had been observed by Andreas Vesalius (1514–1564) (1543) and Gabriele Fallopio (1523–1562) (1561) in the ovary of mammals correspond to the egg of the bird's ovary, and that the process of fertilization is similar in every animal. Just as the bird's fertilized egg acquires albumin and a shell, the egg of a mammal becomes fertilized in the Fallopian tube, traverses to the uterus, and there develops into the embryo and fetus.

De Graaf assumed incorrectly that the entire follicle was the ovum, an understandable error in this era when the microscope was used only in a limited fashion. Besides describing the follicle, which several authors had noted previously, he described for the first time the corpus luteum. He thus rejected the Aristotelean assertion that the embryo originates solely in the male semen. De Graaf concluded, "Thus, the general function of the female testicles is to generate the ova, to nourish them, and to bring them to maturity, so that they serve the same purpose in the woman as the ovaries of birds. Hence, they should rather be called ovaries than testes..., many have considered these bodies useless, but this is incorrect, because they are indispensable for reproduction." De Graaf credited Johannes Van Horne (1621–1670) of Leiden who, with Jan Swammerdam (1637–1680) and Niels Stensen (1638–1686) of Copenhagen, independently developed this hypothesis. The work of Swammerdam was published later, however.

The frontispiece (lower left) shows two cherubs observing with the aid of a microscope an egg taken from a rabbit, and a woman holding a diagram of the female reproductive system. Although he was a friend of the microscopist Antoni van Leeuwenhoek (1632–1723), there is no evidence that De Graaf used a microscope in his studies. The true eggs of mammals were first described a century and a half later by Karl Ernst von Baer (1792–1876) (1827). Plate XIII shows the female reproductive system (although somewhat different from that seen in the frontispiece), with the uterine fundus (A), cervix (B), vagina (C), Fallopian tubes (E, G), ovaries ("testes", D, F), enlarged uterine (N, O) and ovarian (H, I) blood vessels of pregnancy, and external genitalia (a to m). Plate XIV illustrates the ovaries of cows (Fig. 1 and II), showing cross-sections of ovarian follicles (Fig 1, D), as well as ovaries of ewes (Figs. III–V) with a cross-section of a corpus luteum (Fig. 5, B), which De Graaf first described. Plate XVI shows a bisected human ovary with follicles (B).

De Graaf also was a comparative anatomist and apparently was the first to document the considerable enlargement of the pelvic blood vessels in pregnancy. Plate XXVI shows the rabbit eggs at various stages after coitus (days 3–7 in Figs. 1–5, respectively) as well as embryos in the opened uterus on days 8, 9, 10, 12 and 14 after coitus (Figs. 6, 7, 8, 9, and 10, respectively). Plate XXVII shows a rabbit fetus in its amnion on day 29 after mating (Fig. I) as well as the maternal (or uterine; Fig. II) and fetal (Fig. III) faces of the placenta. Plate XXII illustrates a human fetus (A) with umbilical cord (B) and attached placenta (H to M) that aborted at 3 months gestation (although the infant shown is obviously much older than that). The extensive branching ramifications of the chorioallantoic (fetal) placental vessels (L) are particularly noteworthy. Concerning this drawing, he comments "... it may be that nutritive fluid is received through the mouths of the [umbilical] veins and delivered to the fetus and that part is then released through the arteries into the amnion as though into the fetus's treasury" (Jocelyn and Setchell 1972; p. 158).

De Graaf first studied at Leyden, then obtained his medical degree at Angers, France, with a dissertation on pancreatic secretion (1664). He also published the first detailed account of the male reproductive system (1668). On one hand, De Graaf stands at the summit of the achievement of Renaissance anatomists, whose work he summarized, corrected, and extended. On the other hand, he stands at the beginning of modern microscopical anatomy—in the year of his death at the age of 32, he reported Leeuwenhoeck's first discoveries to the Royal Society of London.

FRONTISPIECE FROM DE GRAAF, 1672

References

Garrison Morton 1209; Krivatsy 4908

Baer, K.E.V. *De ovi mammalium et hominis genesi*. Lipsiae, L. Vossius, 1827. (GM 477).

Corner, G.W. The discovery of the mammalian ovum. In: *Lectures on the history of medicine: A series of lectures at the Mayo Foundation, 1926–1932*. Philadelphia, W.B. Saunders Co., 1933, pp. 401–426.

Corner, G.W. On the female testes or ovaries by Regner de Graaf. In: *Essays in Biology, in honor of Herbert M. Evans*. Berkeley, CA, University California Press, 1943, pp. 121–137.

De Graaf, R. *De succi pancreatici natura ex usu exercitation anatomico-medica*. Lugduni Batavorum, ext off Hackiana, 1664.

De Graaf, R. *De virorum organis generationi inservientibus, declysteribus et de usu siphonis anatomia*. Lugduni Batavoru, ex off. Hackiana, 1668.

De Graaf, R. *De mulierum organis generationi inservientibus 1672. – Facsimile with an introduction by J.A. van Dongen*. (Dutch Classics on History of Science, XIII). Utrecht, Hes & De Graaf, 1965.

Falloppio, G. *Observationes anatomicae*. Venetiis, apud M.A. Ulmum, 1561.

Jocelyn, H.D., Setchell B.P. *Regnier de Graaf on the human reproductive organs. An annotated translation of "Tractatus de virorum organis generationi inservientibus"* (*1668*) *and "De mulierum organis generationi inservientibus tractatus novus"* (*1672*). Oxford, Blackwell Scientific, 1972 (*J Reprod Fertil Suppl* 17:1–222, 1972).

Needham, J. *A history of embryology*. Cambridge, at the University Press, 1934. (GM 533).

Speert, H. Reinier de Graaf and the Graafian Follicles. In: *Essays in Eponymy. Obstetric and gynecologic milestones*. New York, Macmillian Co., 1958, pp. 9–17.

Swammerdam, J. *Miraculum naturae; sive,uteri muliebris fabrica….* Lugduni Batavorum, apud Severinum Mathaei, 1672. (GM 1211).

Van Horne, J. *Mikrotechne; seu, Methodica ad chirurgicam introductio….* Lugd. Batav., Apud Gaasbekios, 1668.

Vesalius, A. *De humani corporis fabrica libri septum*. Basileae, ex off. Ioannis Oporini, 1543.

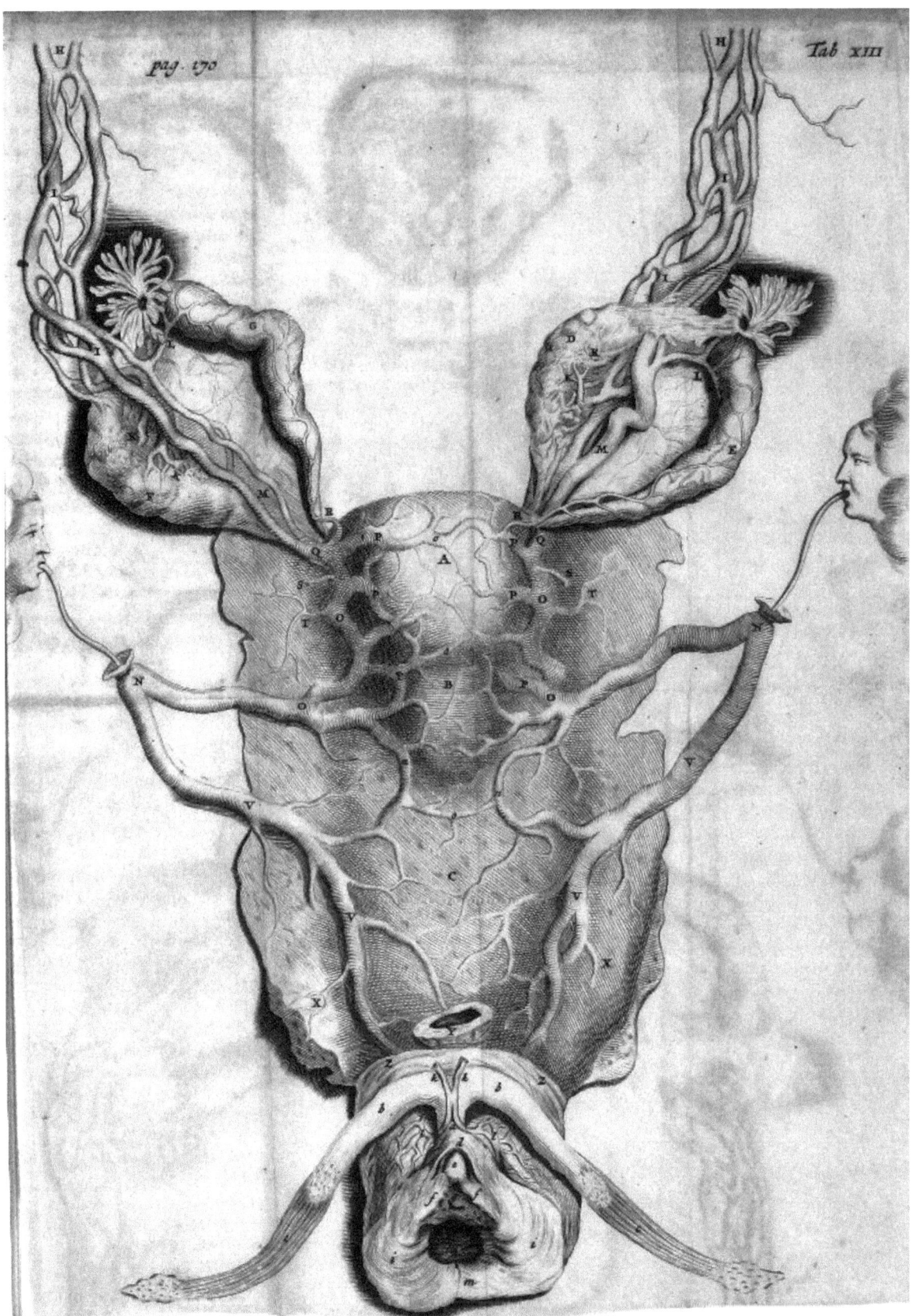

TAB XIII FROM DE GRAAF, 1672

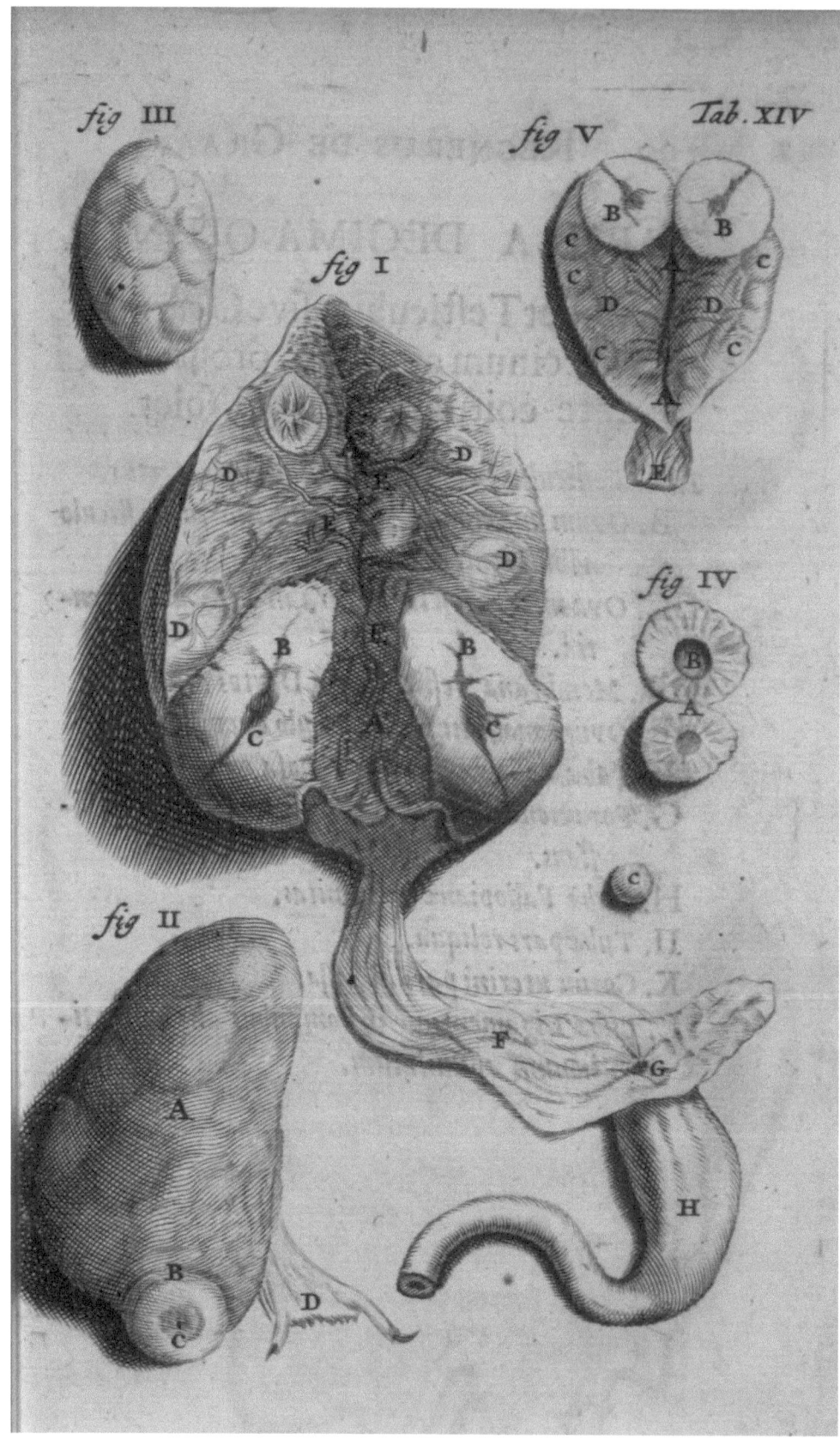

Tab XIV from De Graaf, 1672

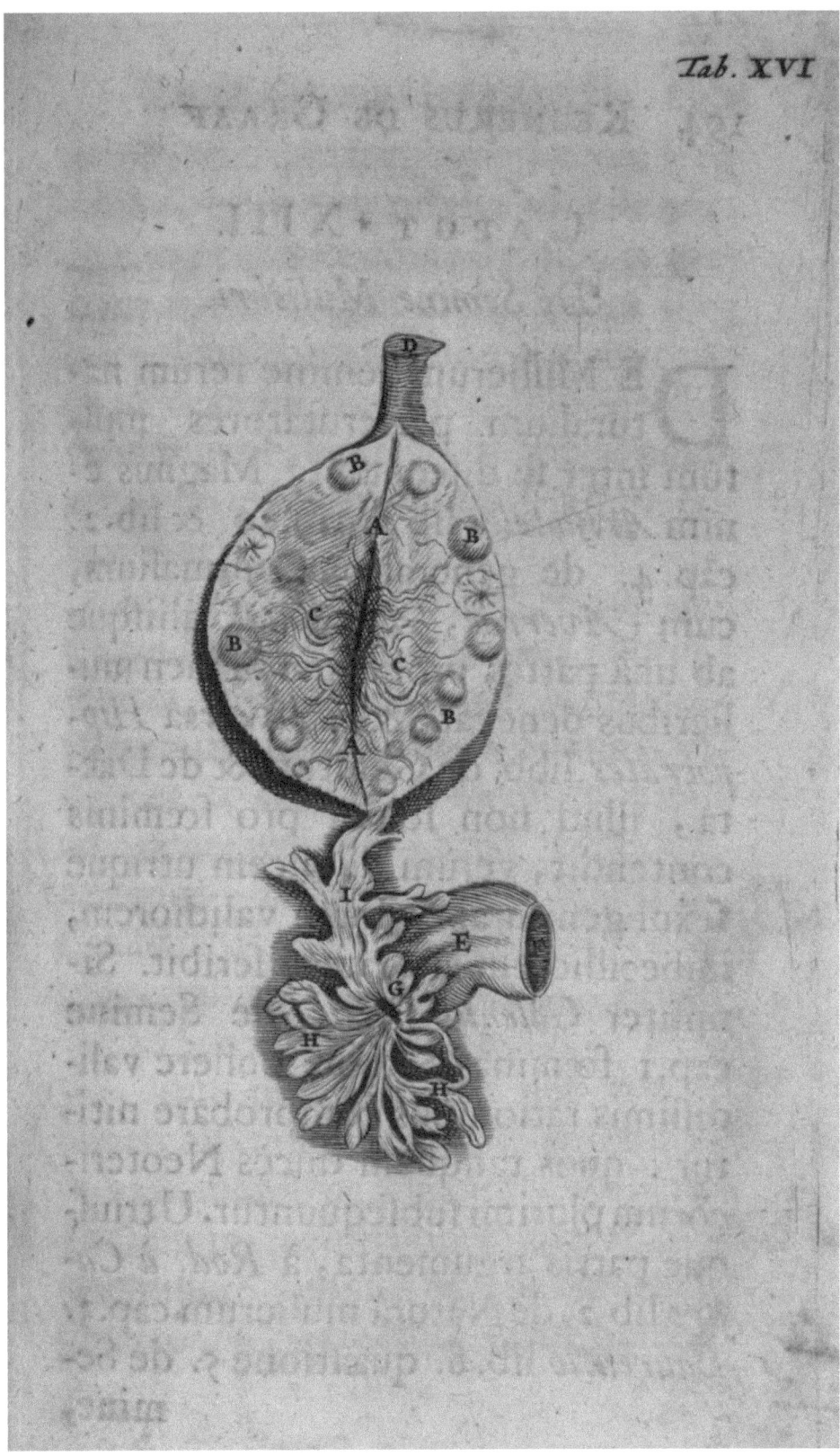

Tab. XVI

TAB XVI FROM DE GRAAF, 1672

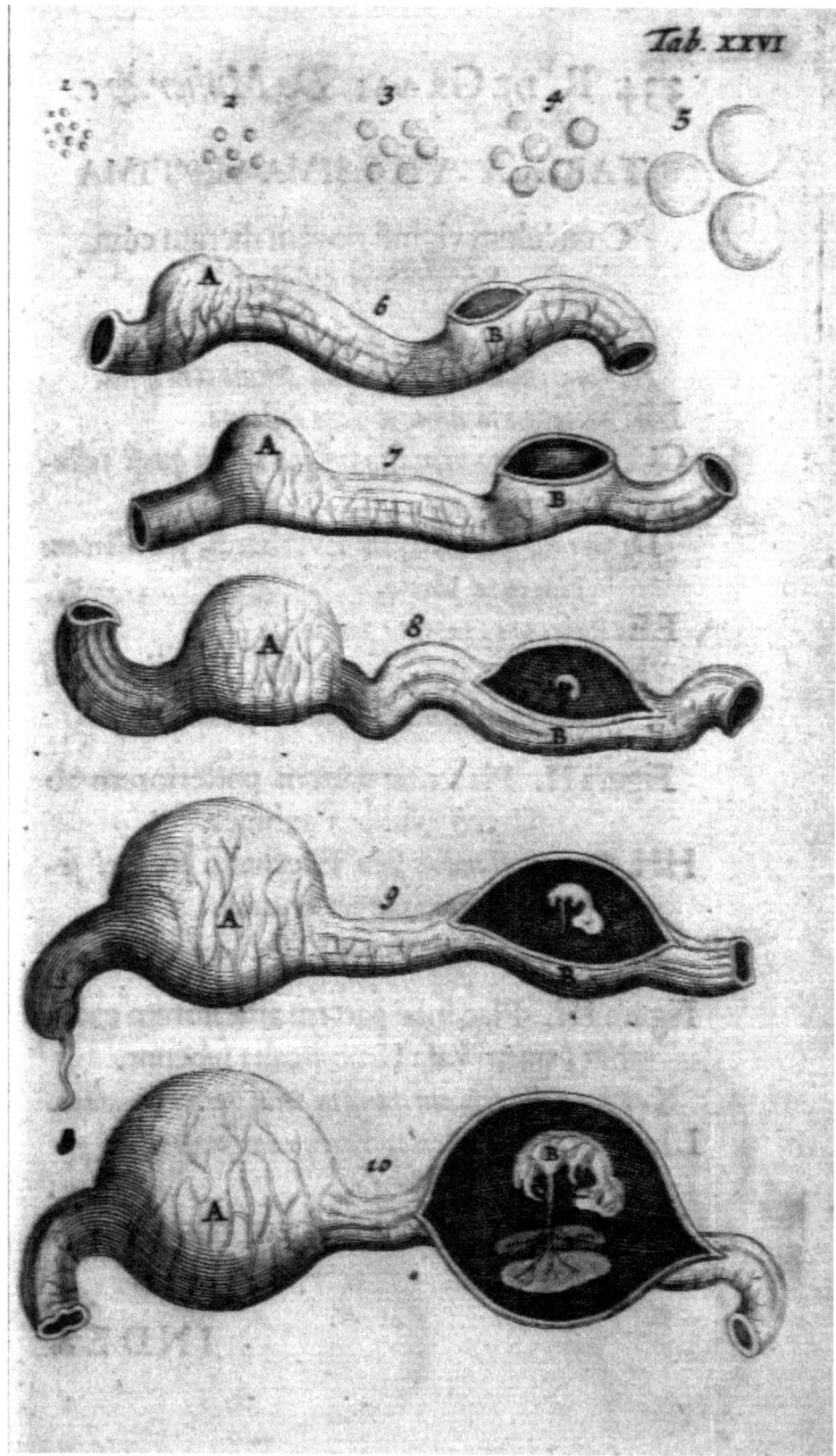

Tab XXVI from De Graaf, 1672

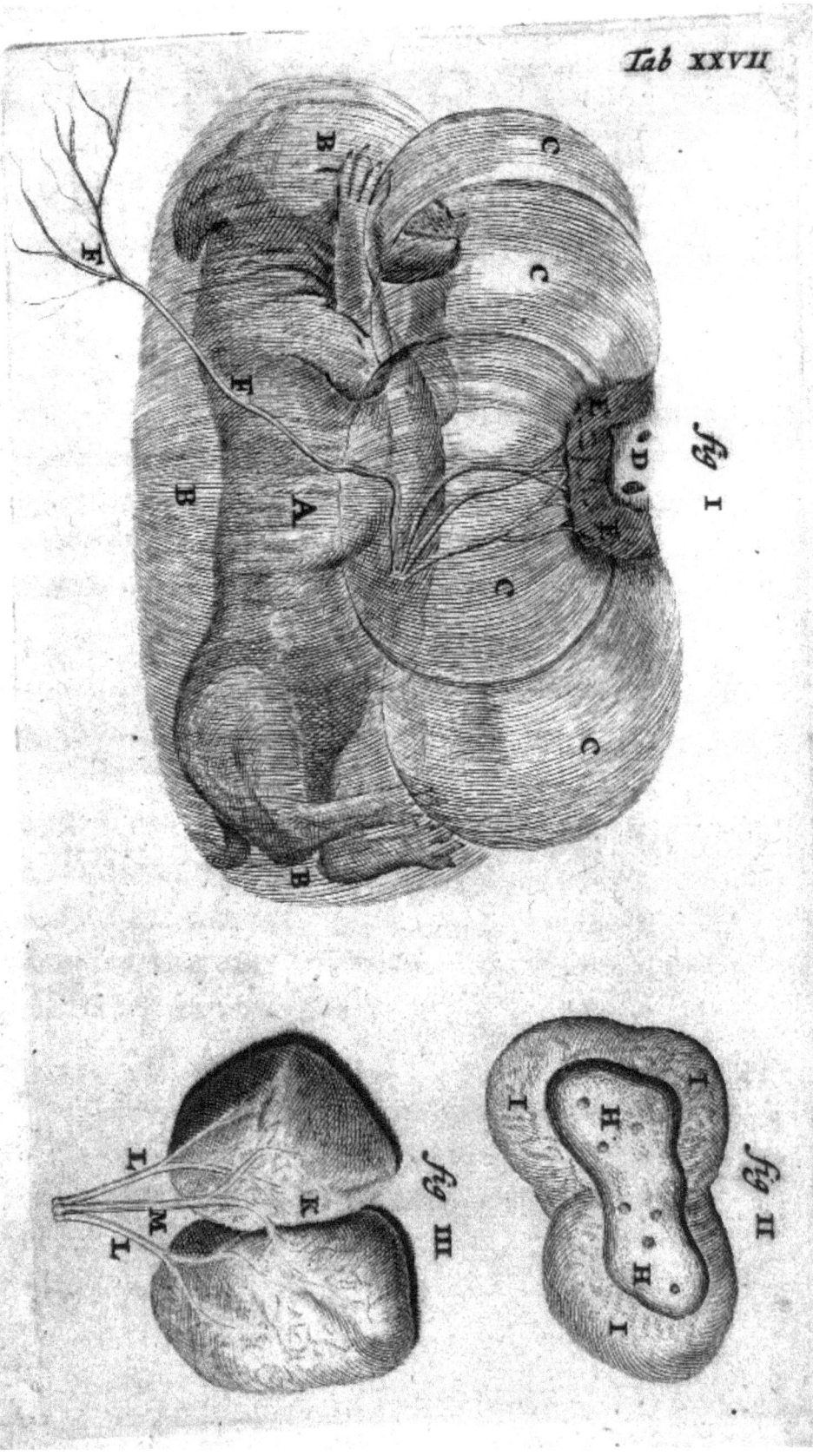

TAB XVI FROM DE GRAAF, 1672

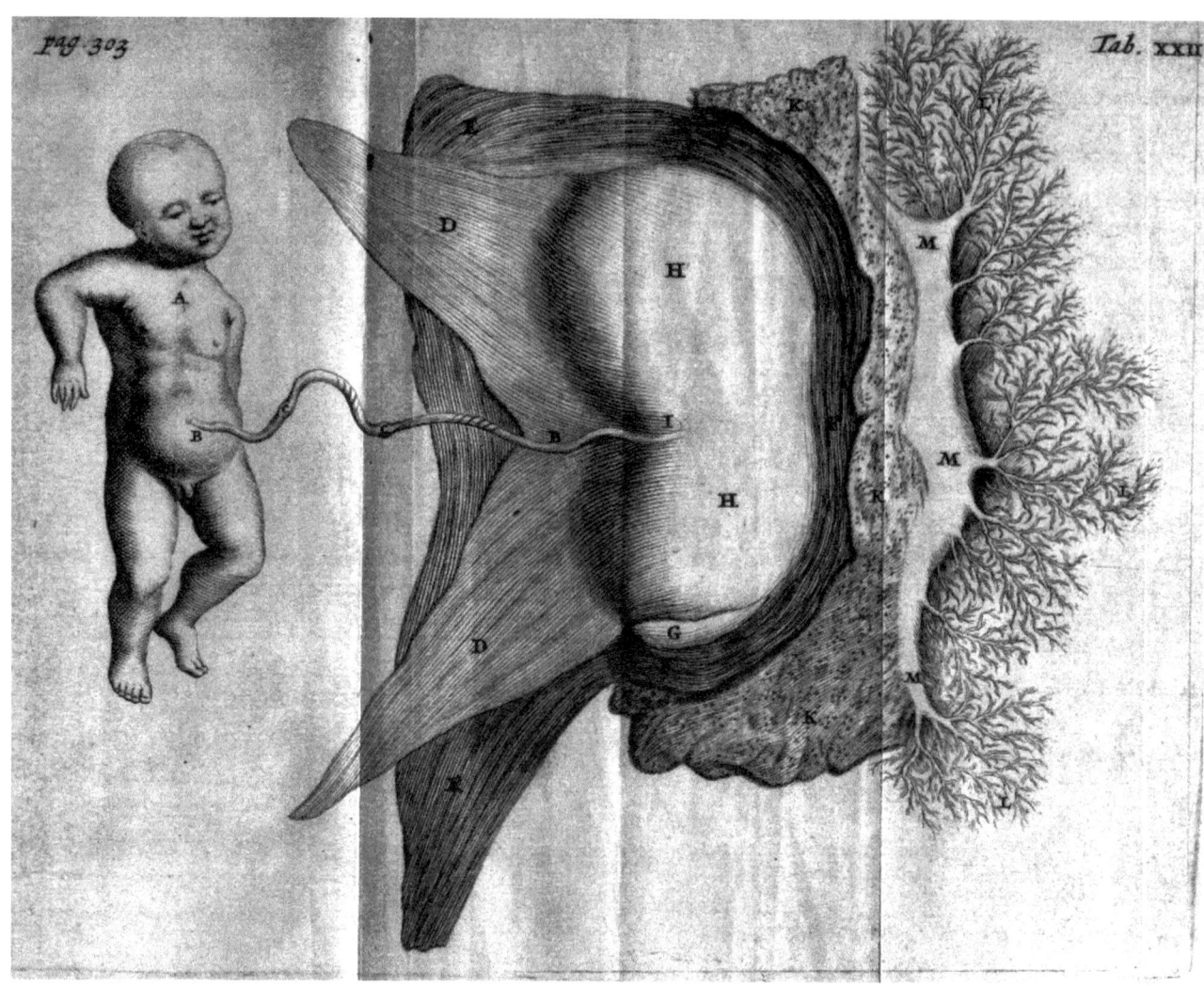

TAB XXII FROM **DE GRAAF**, 1672

Marcello Malpighi

Dissertatio epistolica de formatione pulli in ovo. Londini, Apud Joannen Martyn, 1673.

With his microscopic studies, Malpighi (1628–1694) helped to establish the descriptive study of embryology and histology on a sound basis. This work describing the developmental stages of the chicken in the egg contain four plates accompanying the text, which far surpass previous work in this area, and show the development of such fine structures as the neural groove, cerebral vesicles, optic vesicles, aortic arches, and the earliest blood vessels. As noted by Joseph Needham (1900–1995), Malpighi's studies could not have been done without the stimulus of William Harvey (1578–1657) to identify the *cicatricula* [scar or mark of origin] as the site where embryonic development began, and therefore as Malpighi reasoned, the locus where microscopic examination would be most profitable (Needham 1934, pp. 144–145). On Tabula III, Malpighi illustrates a number of features of early development in the chick. Figure XVI on the left shows the *area vasculosa* [vascular region] with the embryo at about 85 h incubation; (B) is the anterior vitelline veins; (F) and (G) are the posterior vitelline vessels, and (H) the marginal vein. In (D) are depicted somites (paired mesodermal segments) of which 20 are shown. Figures XVI (right), XVII, and XVIII illustrate the formation of the heart. Figure XVIII illustrates the developing vena cava (K), the auricle (I), the ventricle (L), and the bulbus (M). Three aortic arches connecting the heart to the aorta are shown by (N). The developmental significance of these structures was not appreciated for another century and a half. Because of misinterpretation of his findings, Malpighi held to the preformation theory, as opposed to epigenesis. Also included with the treatise are seven letters between Malpighi and Henry Oldenburg (1619–1677), secretary of the Royal Society of London, under whose auspices the work was published. This with his *De ovo incubato observationes* (Malpighi 1673) placed the study of embryology on a firm scientific basis. An English translation was published by Howard Bernardt Adelmann (1898–1988) in 1966.

Malpighi graduated in medicine from the University of Bologna (1653). Shortly thereafter, he spent 3 years at the University of Pisa where he worked with the mathematician/physiologist Giovanni Alfonso Borelli (1608–1679). In 1659, he returned to Bologna, where, except for several years, he remained as professor of medicine. Here, Malpighi used the microscope to make fundamental contributions in biology and physiology. In his *De pulmonibus...* [on the lungs] of 1661, he described the fine structure of the pulmonary alveoli (air sacs) and their circulation in the frog, observing for the first time the pulmonary capillaries. These observations confirmed Harvey's theory of the circulation of the blood (Harvey 1628), serving greatly to assure its acceptance; however, because of limitations in the resolution of the microscopes he had available, he was not able to distinguish capillaries in mammals. Malpighi also contributed to understanding the structure of the papillae or taste buds of the tongue, the structure of neurons in the brain, and the structure of secretory glands. Elected to Fellowship in the Royal Society of London (1668), in 1691 he moved to the Vatican to become physician to Pope Innocent XII (1615–1700), where he died 3 years later.

References

Garrison Morton 469; *Heirs of Hippocrates* 369; Krivatsy 7335.

Adelmann, H.B. *Marcello Malpighi and the evolution of embryology. 5 Vols.* Ithaca, N.Y., Cornell University Press, 1966. (GM 534.1).

Belloni, L. Marcello Malpighi. In: *Dictionary of Scientific Biography. Charles Coulston Gillispie (Ed).* New York, Charles Scribner's Sons, 1974, pp. 62–66.

Harvey, W. *Exercitatio anatomica de motu cordis et sanguinis in animalibus.* Francofurti, sumpt Guilielmi Fitzeri, 1628. (GM 759).

Malpighi, M. *De pulmonibus observationes anatomicae.* Bononiae, B. Ferronius, 1661. (GM 760).

Malpighi, M. *Dissertatio epistolica de bombyce.* London, J. Martyn & J. Allestry, 1669. (GM 293).

Malpighi, M. *De ovo incubato observationes.* London, J. Martyn, 1673. (GM 468).

Malpighi, M. *Anatome plantarum....* London, Impensis Johannis Martyn, 1675–1679.

Malpighi, M. *Opere scelte. L. Belloni (Ed).* Turino, Unione tipografico-editrice torinese, 1967.

Needham, J. *A history of embryology.* Cambridge, at the University Press, 1934, pp. 144–149.

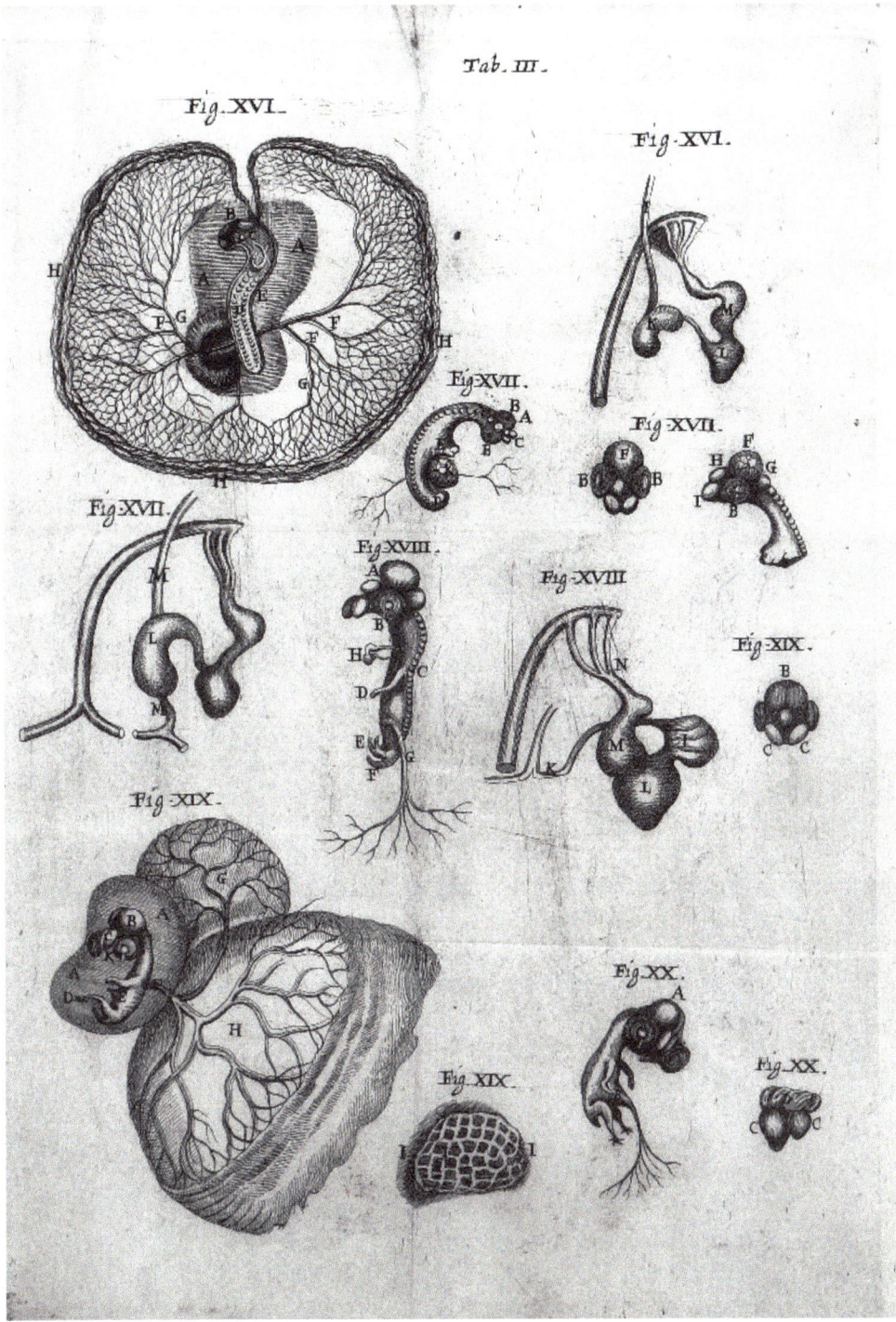

TAB. III FROM MALPIGHI, 1673

Francesco Emmanuele Cangiamila

Embryologie sacra, sive, de officio sacerdotum, medicorum, et aliorum circa aeternam parvulorum in utero existentium salutem. Libri quatuor. [Palermo], Panormi, 1758.

In his "Sacred embryology, or office of the priesthood..." Cangiamila (1702–1763), Cannon of Palermo, described some aspects of embryonic development, reproductive function, and aspects of motherhood. Because of the belief that all who were unbaptized were condemned to damnation and eternal torment in hell, Cangiamila became an advocate of the liberal use of Cesarean section in recently dead pregnant women and in those in whom death was imminent. He held that, in the absence of a physician, this should be performed by a priest. Although such infants saved by baptism were believed to be subject to original sin, but innocent of other sins, they could be transported to heaven. The plate illustrates in 17 figures intrauterine development from early embryo to relatively late in gestation.

Cangiamila, born in Palermo, became Archpriest at Girgenti, and then vicar general and provincial inquisitor-general. He held the distinction of baptizing the first infant delivered by Cesarean section in Sicily. Cangiamila's ideas that priests must reform many customs that relate to reproduction, and work to prevent miscarriage, abortion, and infanticide received the endorsement of Pope Benedict XIV (original name Prospero Lambertini; 1675–1758), who in several *Breves* [letters], had these doctrines disseminated throughout Europe. Cangiamila's alleged personal encounters with the devil were recorded by Father Luis Crema of Palermo (Crema 1764). The *Embryologica sacra...* became a popular work,

and was even translated into modern Greek for the benefit of Orthodox Catholics. As observed by Neeham (1934, pp. 182–183) "theological embryology" appeared to reach its climax about this time with the invention of intra-uterine baptism by means of a syringe (Deventer 1734). In his *Traité sur divers accouchemens laborieux ...* (1782), George Herbiniaux of Brussels illustrated an instrument with which to perform baptism *in utero* (Herbiniaux 1782). In addition to this and other works on embryology, Cangiamila wrote a number of works in theology, jurisprudence, and literature (Vitello 1955).

References

Blake p. 76 lists other editions; Osler 2225.

Boldrini, B. *Riv di Storia d Sci Med e Nat* 18: 1,1927.

Cangiamila, F.E. *Embriologia sacra, overo dell'uffizio de'sacerdoti, medici, e superiori, circa l'eterna salute de'bambini racchiusi nell'utero.* Palermo, Francesco Valenza, 1745.

Cangiamila, F.E. *Embryologia sacra, sive, de officio sacerdotum, medicorum, & aliorum circa aeternam parvulorum in utero existentium salutem. Libri quatuor.* Venetiis, Typis Sebastiani Coleti, 1763.

Crema, L. *Elogio historico de don Francisco Manuel Cangiamila....* Palermo, 1764.

Deventer, H.V. & J.J Bruhier. *Observations importantes sur le manuel des accouchemens ....* Paris, Guillaume Cavelier, 1734.

Herbiniaux, G. *Traité sur divers accouchemens laborieux, et sur les polypes de la matrice; ouvrage dans lequel on trouve la déscription d'un nouveau levier, imité de celui de Roonhuysen, & mis en parallel avec le forceps: ainsi que d'un nouvel instrument, proper à la ligature des polypes, approuvé pa l'Académie Royale de Chirurgie de Paris.* Brussels, chez J.L. De Boubers, 1782.

Needham, J. *A history of embryology.* Cambridge, at the University Press, 1934.

Vitello, A. Francesco Emanuele Cangiamila e la sua opera ostetrica. *Rass Clin Ter* 54: 110–127 and 165–177, 1955.

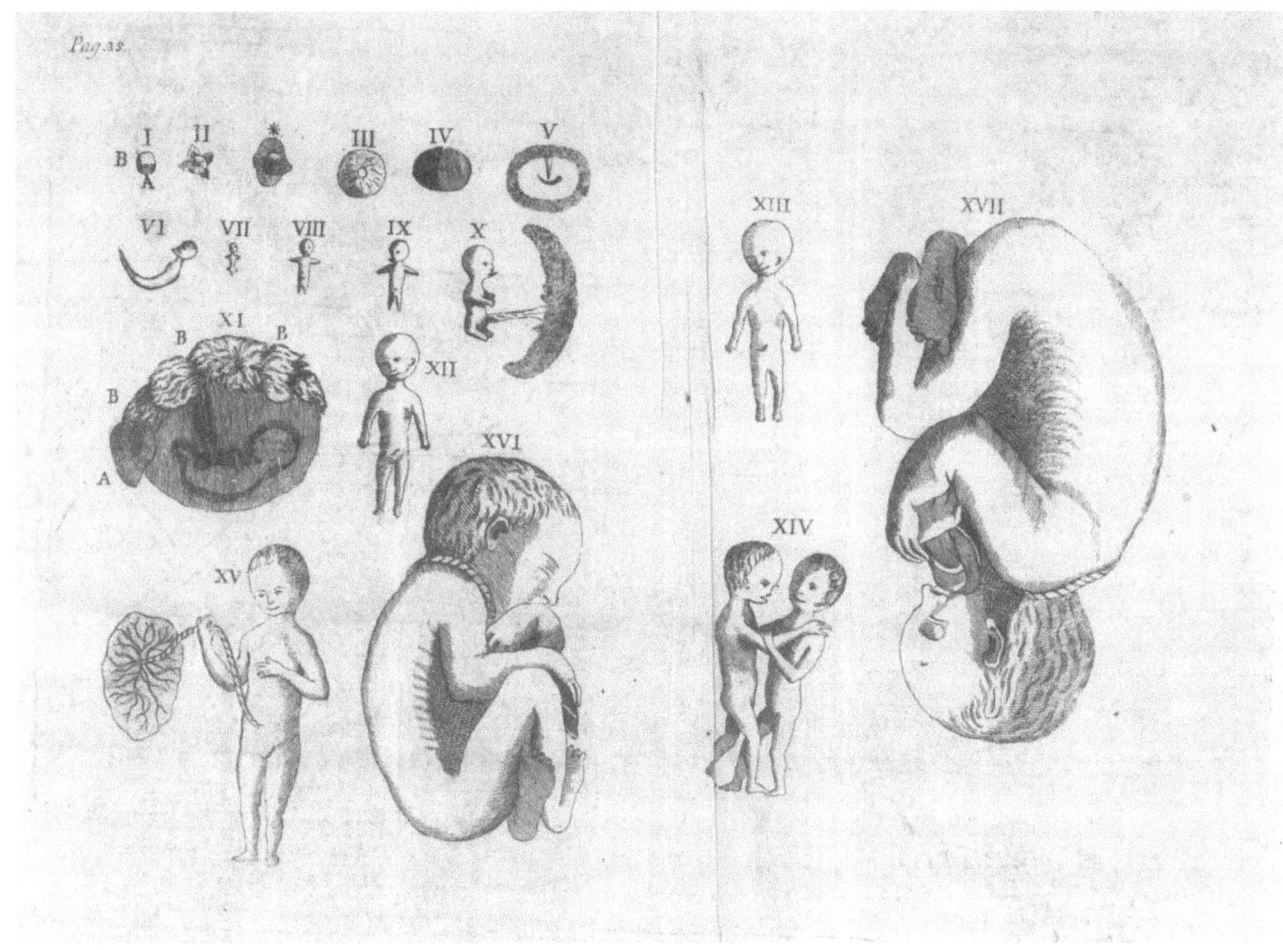

FIGURES FROM CANGIAMILA, 1758

Robert Wallace Johnson

A new system of midwifery, in four parts; founded on practical observations; the whole illustrated with copper plates. London, printed for the author, and sold by D. Wilson & G. Nical, 1769.

In addition to his excellent representation of the female pelvis, Johnson, a pupil of William Smellie (1697–1763) and friend of William Hunter (1718–1783), to whom he dedicated this work, devised a method to determine accurately the internal pelvic measurements. He designed an obstetrical forceps with a pelvic curve of the blades, to obtain "axis traction" and to assist with delivery of the infant when it is in other than the normal occiput anterior position. Nonetheless, regarding forceps Johnson warned, "But as much mischief may be done by using it improperly, I shall beg leave to be as explicit as I can on this subject; though, perhaps, some of it may seem very dry and tedious to the reader" (p. 268). He also introduced the lower segment incision for cesarean section, although this did not come into common use until a century later. In his introductory "Advertisement," Johnson stated his modest claim for this work, that "It contains, if… [I am] not mistaken, some things that are new and important."

In Plate 4, based on the observations of Marcello Malpighi (1628–1694) (1673), is shown stages of development of the chicken egg at several hours of incubation: 12 h (Fig. 1), 30 h (Fig. 2), 40 h (Fig. 3), 3 days (Figs. 4 and 5). Figures 6–8, based on the work of Regnier de Graaf (1641–1673) (1672), illustrate the development of the embryo in the rabbit from day 8–12.

Plate V depicts an early developing human embryo expelled as a spontaneous abortion (Figs. 1 and 2). Figures 3–8 illustrate various aspects of another human embryo that aborted at about 10 weeks gestation, while Figures 9–13 show another embryo at 6 or 7 weeks. From another patient, Fig. 14 illustrates a male fetus at about 7 weeks, while Fig. 15 shows another at 12–14 weeks gestation. Johnson accompanied these illustrations with expanded and detailed descriptions of embryonic and fetal development (pp. 55–98).

References

Blake 236.

Spencer, H.R. *The history of British midwifery from 1650 to 1800. The Fitz-Patrick Lectures for 1927 delivered before the Royal College of Physicians of London.* London, John Bale, Sons and Danielson, Ltd, 1927, pp. 87–90.

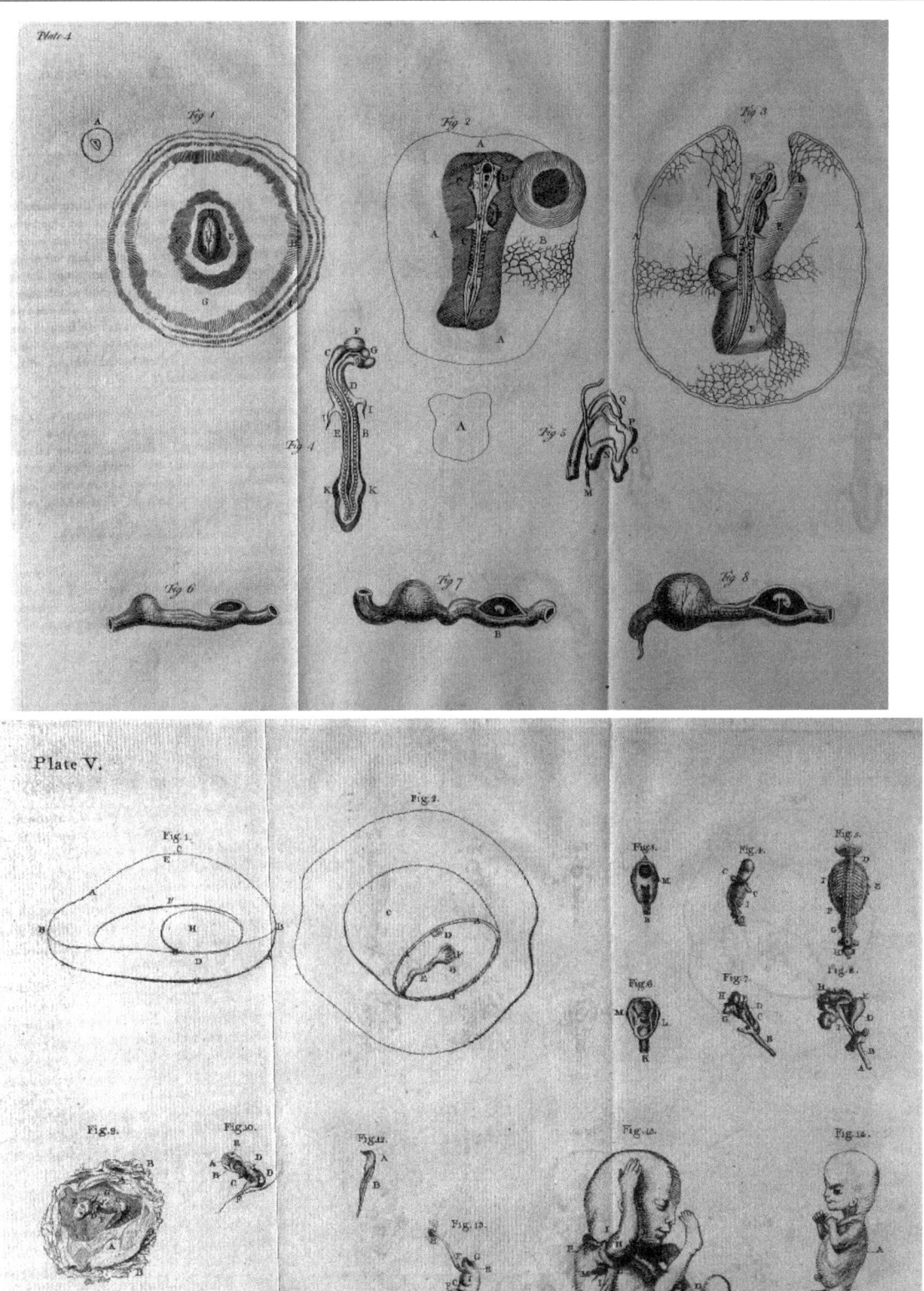

PLATES IV AND V FROM JOHNSON, 1769

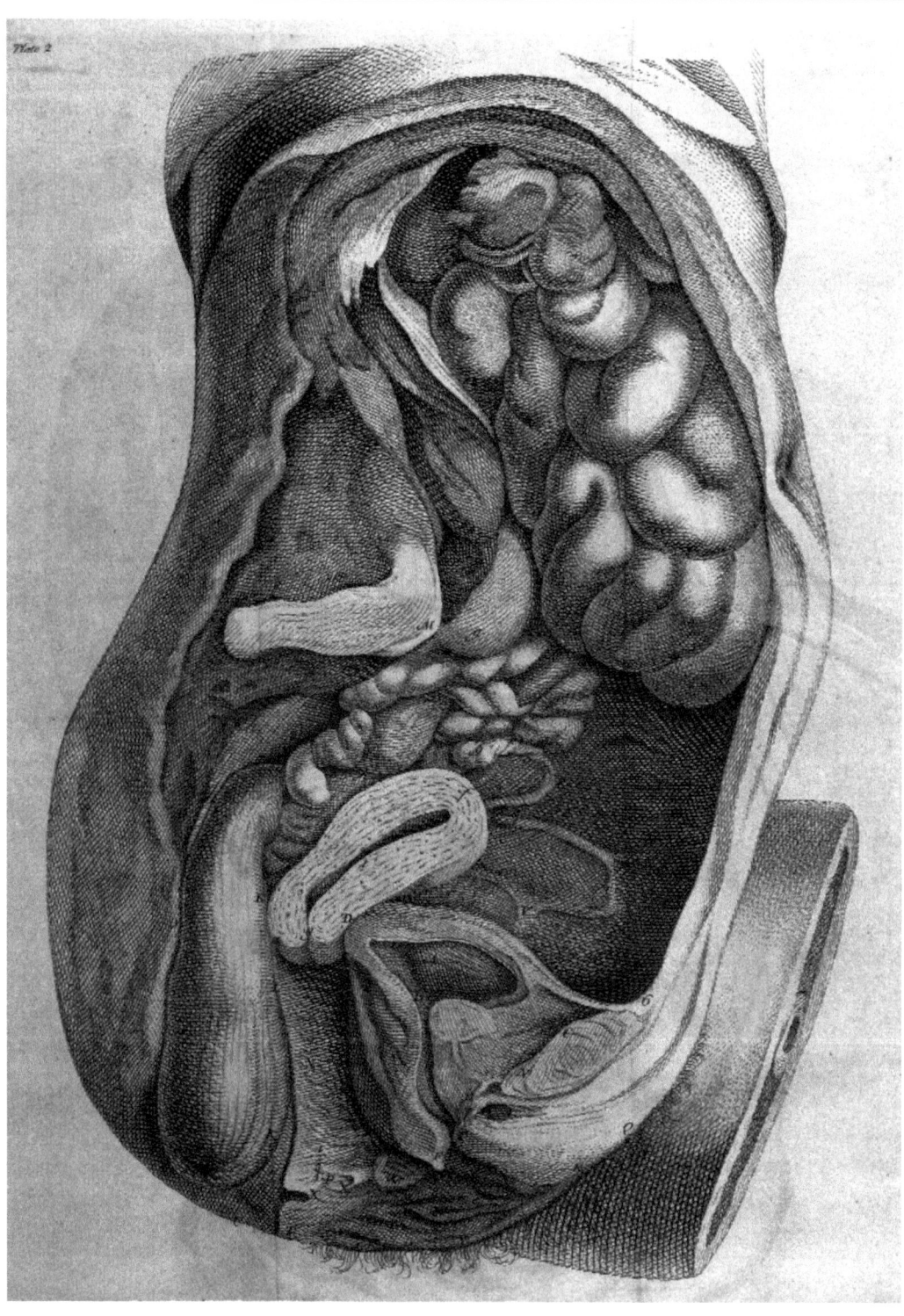

PLATE 2 FROM JOHNSON, 1769

Ebenezer Sibly

The medical mirror; or, treatise on the impregnation of the human female showing the origin of diseases, and the principles of life and death. London, printed for the Author, [1794].

In this curious work on generation, Sibly (1751–1800), a physician, astrologer, and writer on the occult, considered sexual intercourse, human conception, and nourishment and growth of the fetus. In his Preface he noted that, "In this MIRROR, every Patient may behold, not only the true picture of his own disorder, whether hereditary or accidental—chronical or acute—but may also perceive the direct and obvious road to an immediate cure … particularly in relaxed and debilitated constitutions; in lowness of spirits, and weakness of nerves …" (p. iii). Writing at the time of the French Revolution (1789–1799), Sibly referred to "the thunderbolts of war" stating, "In this Mirror such a balm is discovered which, if applied in time to gun-shots, stabs, and wounds, may be the means of preserving to their relatives and friends, some thousands of valuable members of society" (p. iv).

In introducing the present work, Sibly stated that in explaining the nature of human generation he might, "… thence deduce the origin of hereditary diseases, and point out with more facility those which are accidental …. I shall endeavor to furnish my readers with such obvious directions for *eschewing the evil and choosing the good*, that, if resolutely followed, will not fail to preserve health and long life, and prove of no small benefit to future generations" (p. 6). Following an exegesis of Holy Scripture and the legacy of Adam, Sibly described the organs of generation. In rather florid terms, he detailed copulation and impregnation, and early embryonic development. In Plate I the embryo and fetus are illustrated from conception to the fourth month. Plate II continues this monthly sequence from the fifth to ninth months. Plate III illustrates the female body with opened thorax and abdominal cavity with mulberry-like ovaries, and a fetus of 4 or 5 months gestation at the time of quickening (similar to that seen in Plate II). Sibly incorrectly described the placenta as allowing vascular communication between the mother and fetus with the uterine arteries discharging "their blood into the branches of the umbilical vein. So that, after quickening, the blood of the mother is constantly passing in at one side of the placenta, and out again at the other, for the nourishment of the child" (p. 116). This despite the definitive demonstration several decades earlier by William (1718–1783) and John (1728–1793) Hunter that the circulations of maternal and fetal blood in the placenta were completely separate (see Hunter 1774). Regarding "the *action of quickening*," Sibly described as "instantaneous, yet undescribable motion of the vital principle, which, the instant the foetus has acquired a sufficient degree of animal heat … rushes like an electric shock, or flash of lightning, conducted by the sanguineous and nervous fluids, from the heart and brain of the mother, to the heart and brain of the child. At this moment the entire circulation begins; the infant fabric is completely set in motion; and the child becomes a living *soul*" (p. 116).

Sibly's volume includes testimonials from his patients extolling his new Tinctures. For women he claimed that his LUNAR TINCTURE, "… has been calculated to act upon the menstrual and vegetative fluids." An Aristotelian, Sibly stated that the invention of his *Tinctures* "… hath been the result of long and laborious application to the study of unveiled nature—of the properties of fire, air, earth, and water, in the propagation of animal and vegetable life" (p. 67). Sibly's objective was to treat multiple female "complaints," including menstrual disorders and diseases that relate to pregnancy. He maintained "That the vegetative or procreative faculties of women are universally governed by the lunations [phases] of the moon" (p. 75). He considered topics such as "chlorosis, or … Love Fever" (p. 80), "Fluor Albus" [white flow, used to describe leucorrhea or vaginal discharge] (p. 86), "Infertility (p. 91), "Indispositions attendant on Pregnancy" (p. 94), and the "State of Women at the Turn of Life" (p. 102). Sibly also reviewed "Solar Diseases" of the male (p. 106ff), including rheumatic gout and onanism. He presented a number of case reports illustrating the various conditions and their treatment. The plates were engraved by J. Pass after Dodd.

Plates I and II from Sibly, 1794

A student of scientific and medical history, Sibly held that beyond natural science one must consider the sciences of alchemy, astrology, and mysticism to understand the harmony of the universe. In 1792, Sibly became a fellow of the Harmonic Philosophical Society, which was founded in Paris to promote the doctrines of Frans Anton Mesmer (1734–1815) regarding animal magnetism and hypnotism (Mesmer 1779). A follower of the theories of Theophrastus Bombastus von Hohenheim Philippus Aureolus [pseud Paracelsus] (1493–1541) and the Hermeticists (followers of Hermes Trismegistus, the Greek name for the Egyptian god Thoth, author of works in alchemy and magic), Sibly attempted to incorporate the "truths" of mystical philosophy, mesmerism, and Freemasonry with the natural philosophy and scientific concepts of his time. He held that medicine should be based on "universal sympathy," and that with vegetable and herbal remedies we obtain the "purest minerals" for therapeutic use. He was an advocate for exercise, active work, with temperance and moderation in all aspects of life.

Sibly studied surgery in London, and received his M.D. from the King's College, Aberdeen. Below his portrait on the frontispiece of the present volume, he lists himself as a member of the "Royal College of Physicians of Aberdeen." (The society was, in fact, The College of Physicians of Aberdeen). Sibly's *Medical mirror* was a reasonably popular work, reaching six editions by 1814. He authored other works, including a key to physic and the occult sciences (Sibly 1794), and edited several editions of Nicholas Culpeper's (1616–1654) *The English Physician* (1794). Accused of being a "quack", Sibly may be considered a forerunner of the Botanical movement (see Beach 1847) of the nineteenth century, and "alternative medicine" of the present age.

References

Blake p. 417

Beach, W. *An improved system of midwifery: adapted to the reformed practice of medicine; illustrated by numerous plates, to which is annexed, a compendium of the treatment of female and infantile diseases, with remarks on physiological and moral elevation.* New York, By the Author, 1847.

Culpeper, N. *The English physician, containing, admirable and approved remedies, for several of the most usual diseases….* Boston, Re-printed for Nicholas Boone, 1708. (GM 1828).

Culpeper, N. *Culpeper's English Physician, and Complete Herbal: to which are now first added upwards of one hundred additional herbs….* London, Printed for the Author, and sold at the British Directory-Office… and by Champante and Whitrow [1794].

Debus, A.G. Scientific truth and occult tradition: The medical world of Ebenezer Sibly (1751–1799). *Med Hist* 26: 259–278, 1982.

Hunter, W. *Anatomia uteri humani gravidi tabulis Illustrata… The Anatomy of the Human Gravid Uterus Explained by Figures.* Birmingham, John Baskerville, 1774.

Mesmer, F.A. *Mémoire sur la découverte du magnétisme animal.* Genève, Paris, P.F. Didot le jeune, 1779. (GM 4992.1).

Sibly, E. *A new and complete illustration of the occult sciences, or, The art of foretelling future events and contingencies by the aspects, positions and influences of the heavenly bodies….* London, Printed for the proprietor and sold by C. Stalker, 1784–1791.

Sibly, E. *A new and complete illustration of the celestial science of astrology, or, The art of foretelling future events and contingencies by the aspects, positions, and influences of the heavenly bodies… In four parts….* London, Printed for the proprietor, and sold by W. Nicoll [etc.], 1784–1788.

Sibly, E. *A key to physic, and the occult sciences. Opening to mental view, the system and order of the interior and exterior heavens….* London, Printed for the author, and sold by Champante and Whitrow, [1794].

Sibly, E. M*agazine of natural history comprehending the whole science of animals, plants, and minerals; divided into distinct parts, the characters separately described, and systematically arranged.* London, Printed for the proprietor, and sold by Champante & Whitrow, and at the British Directory-Office, Ave Maria Lane, St. Pauls, 1794–1808.

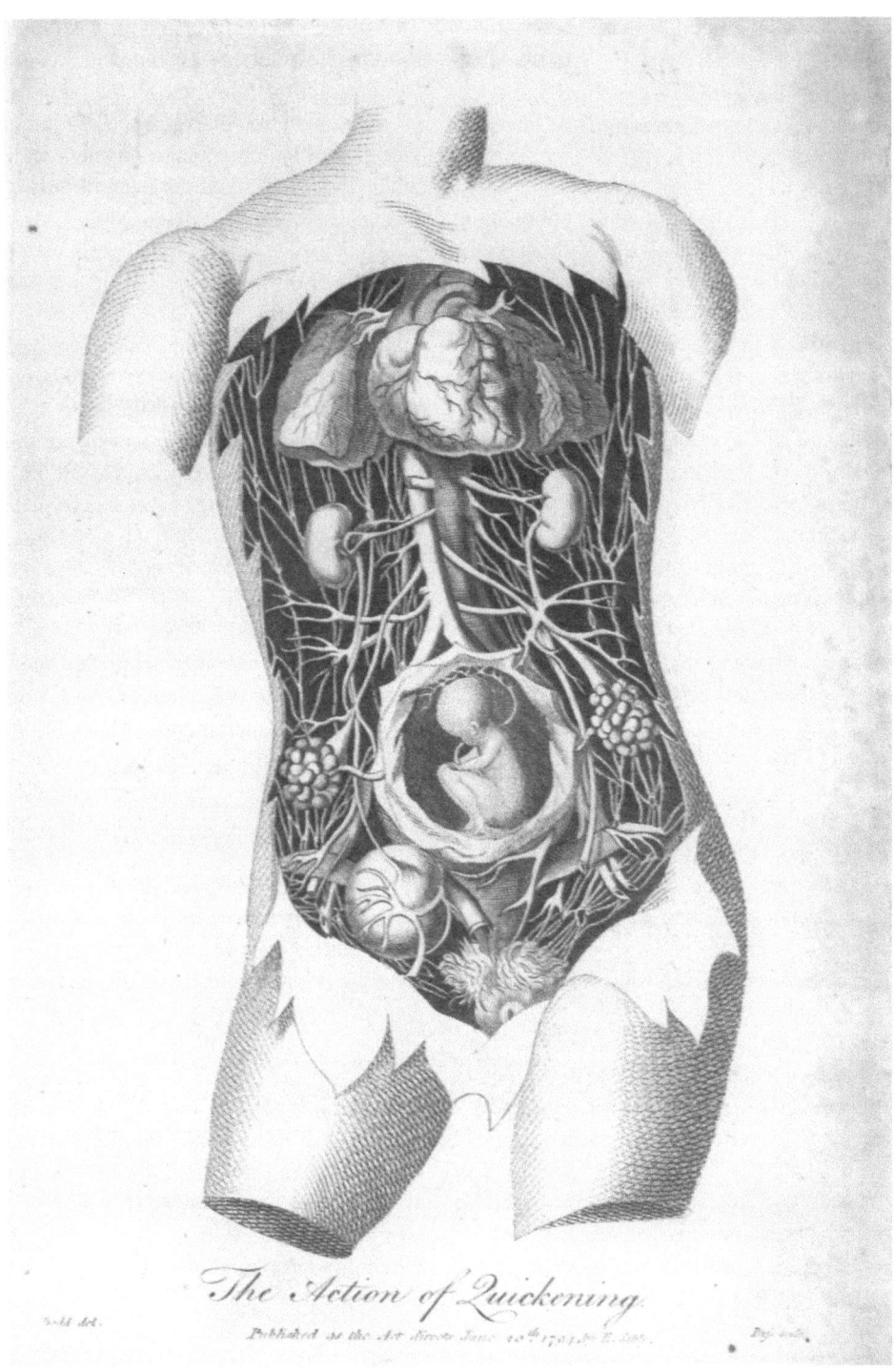

PLATE III FROM SIBLY, 1794

John Burns

The anatomy of the gravid uterus. With practical inferences relative to pregnancy and labour. Glasgow, The University Press, 1799.

In the preface, Burns (1774–1850) justified the rationale for this work. "Dr. Hunter's posthumous work is, without doubt, truly valuable and useful; but it is not so explicit, on some points, as could be wished; and it is entirely deficient in those practical inferences and conclusions which are so essential to the student" (p v). This was in reference to William Hunter's (1718–1783) *Anatomia uteri humani gravidi* (1774), which served as a stimulus for this expository essay. Burns then described the problem for the student of determining from an anatomical atlas that which is of most importance, and the need to assist the student in learning not only the fundamentals of anatomical knowledge, but how to use that information when operations are required. He continued, "For many ages, the art of midwifery was founded on false and mistaken doctrines. Even at present, there are too many who attempt to practice it without any fixed and certain principles, proceeding upon a confused jumble of directions, unconnected with each other, and arising from no sure and evident source" (p. viii). Burns argued that, as with other aspects of medicine, midwifery is founded on both basic anatomic structural realities and physiologic principles. Burns continued, "The study, then, of any of the departments of the healing art, will require our greatest attention, exerted for the longest lifetime. Whoever aspires at eminence and respectability, must, by unremitting application, and diligent study, purchase that honor which he is solicitous to obtain. I know, that it is an opinion with many, that success in the medical world depends more upon interest than abilities. But I shall venture to affirm, that he who trusts to this maxim, and neglects the means of improvement, shall find himself most miserably mistaken. No man will trust his own life, or the safety of those whom he holds dear, to any man, however powerful his recommendations may be, if he once detects him to be a blockhead …. Sooner or later, difficult and important cases must occur; his treatment of these will not pass without observation; and his real character must be made known. If possessed of many friends, he may, for a time, procure concealment or palliation of his faults; but blunders, frequently repeated, must at last become notorious" (pp. x–xi).

To the aspiring student, Burns counseled, "Let him then rouse himself, and, by diligence, steadiness, and a thorough knowledge of his profession, prove, that he is not inferior to those with whom he is to compete. Let him early lay down a fixed resolution to become learned: Let him attend diligently wherever attention is requisite: Let him mark out the road of industry, which others have pursued; let him follow steadily in that path, and, sooner or later, in spite of every opposition, he must succeed" (p. xiii).

After listing many of the complication which can befall the woman during the course of pregnancy and childbirth, Burns cautioned, "Should he stop to deliberate, if the reasoning of such a man can be called deliberation, may not the woman die before his eyes, and without assistance? Can he, without uneasiness, attend the more lingering illness, produced by the fruitless efforts of the uterus, to push the child through an ill-formed pelvis? Will he dare, in any one instance, to determine, upon his own authority, when the head should be opened and the crotchet employed? Must the child be wantonly sacrificed, because he, in his ignorance, believes it to be requisite? or, must the woman perish, because he foolishly hopes that assistance is still unnecessary? Must both parent and child become victims to his awkwardness? It is a very poor excuse for these crimes, to say, that he had no malice in his heart. The laws of his country will indeed acquit him; but his own conscience must tell him that he is a murderer. It is only a small alleviation of his guilt, to say, that he did the best he could. It was unwarrantable and criminal to undertake the practice of a profession for which he was not qualified" (pp. xix–xx).

Burns concluded the Preface noting that originally he had intended to have included a number of plates to illustrate the points made; however, in view of Hunter's "very elegant and accurate… engravings," he had abandoned that idea (p. xxi). Nonetheless, he included two plates, one depicting the changes that occur in the uterine cervix, and a second illustrating "a plan of the membranes in the different stages of gestation." Figure I shows the uterine interior with Fallopian tubes (B), layers of the uterine decidua (C), embryo (D), and chorion (E). Figure II shows a more advanced stage, the chorion covered with decidua reflexa (A), amnion (B), and placenta (C). Figure III depicts gestation at about 7 weeks, while Fig. IV illustrates the fetus at about 16 weeks, with the chorion in contact with the decidua as the developing fetus and membranes fill the uterus.

After qualifying in medicine from the University of Glasgow, in 1792 at the opening of the Royal Infirmary of Glasgow, Burns was appointed surgeon's clerk. He also

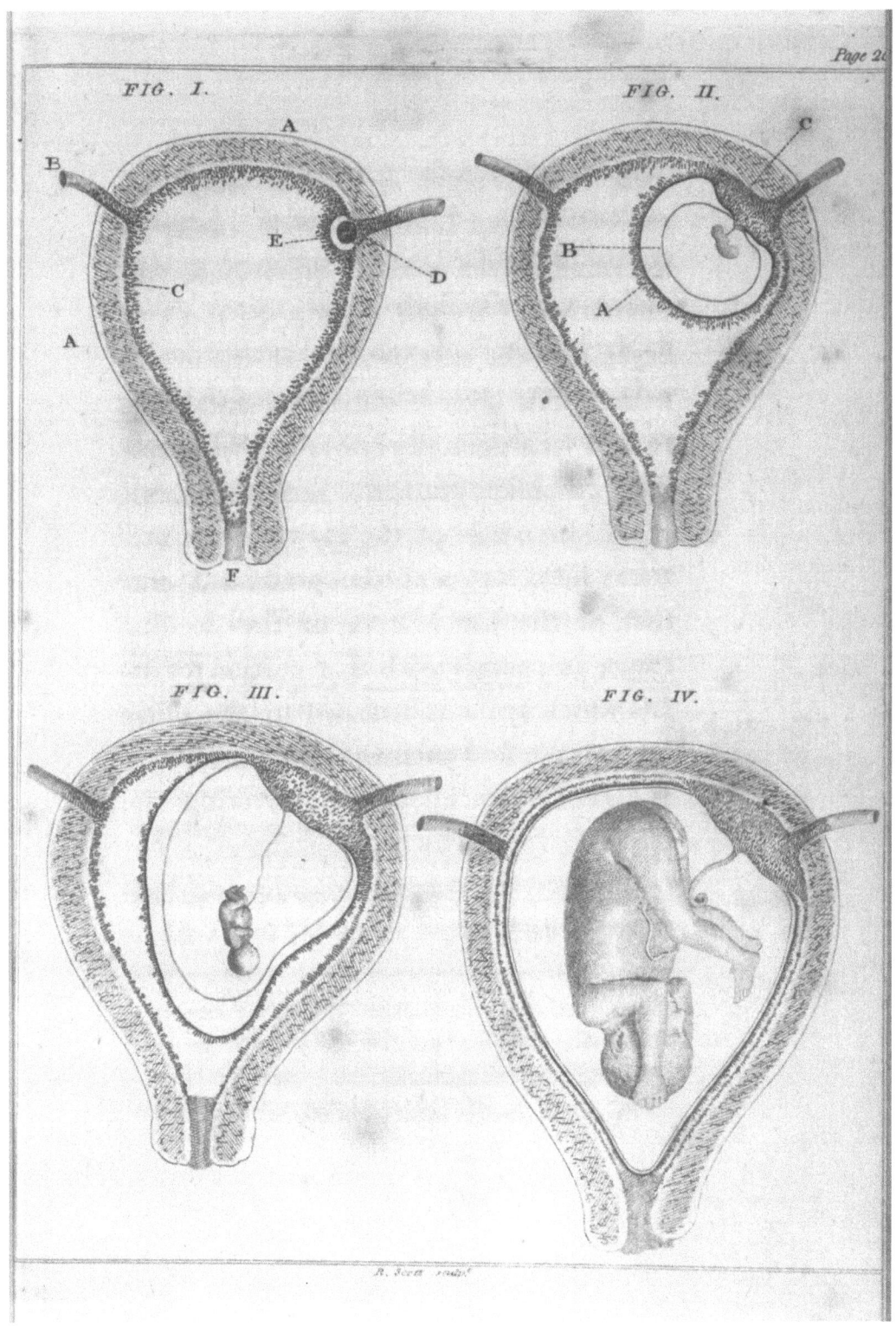

Page 2

FIG. I.   FIG. II.

FIG. III.   FIG. IV.

Plate from Burns, 1799

offered a course of extramural lectures to students in anatomy. Although a popular lecturer, he was forced to abandon that role when it was discovered that some of the cadavers he used in dissection had been obtained by illegitimate means. In 1794, Burns became House Surgeon to the Royal Infirmary and shortly thereafter established a general practice. Although twice serving as Surgeon to the Royal Infirmary (from 1797 to 1799 and 1809 to 1810), for the most part his patients were from his clinical practice. In 1800, he accepted a post teaching anatomy, midwifery, and surgery at Anderson College, Glasgow. Upon establishment of the Chair of surgery at the University of Glasgow in 1815, Burns was appointed the first professor. In addition to his writings in midwifery and surgery, Burns had great interest in the diseases of the infants and children, distinguishing between obstructive and non obstructive jaundice in the newborn infant (Dunn 1989). He also was involved in Scottish parliamentary affairs that related to regulation of the medical profession. When the steamer *Orion*, upon which he was returning to Glasgow from London, was wrecked near Portpatrick, Burns died by drowning. He was a Fellow of the Royal Society of London.

References

Blake p. 72; Cutter and Viets pp. 210–22; Russell 116.

Burns, J. *Dissertations on inflammation*. 2 vol. Glasgow, Printed for James Mundell, for John Murdoch; 1800.

Burns, J. *Observations on abortion. Containing an account of the manner in which it is accomplished, the causes which produced it, and the method of preventing or treating it*. London, Longman, Hurst, Rees, and Orme, 1806.

Burns, J. *Practical observations on the uterine hemorrhage; with remarks on the management of the placenta*. London, Longman, Hurst, Rees, and Orme, 1807.

Burns, J. *The anatomy of the gravid uterus. With practical inferences relative to pregnancy and labour*. Salem, Mass, Cushing & Appleton and Joshua Cushing, 1808.

Burns, J. *The principles of midwifery; including the diseases of women and children*. London, Longman, Hurst, Rees, and Orme, 1809.

Burns, J. *Popular directions for the treatment of the diseases of women and children*. London, Longman, 1811.

Coutts, J. *A history of the University of Glasgow, from its foundation in 1451 to 1909*. Glasgow, James Maclehose and Sons, 1909.

Dunn, P.M. Dr John Burns (1774–1850) and neonatal jaundice. *Arch Dis Child* 64: 1416–1417, 1989.

Hunter, W. *Anatomia uteri humani gravidi tabulis Illustrata…The Anatomy of the Human Gravid Uterus Explained by Figures*. Birmingham, John Baskerville, 1774. (GM 6157).

Samuel Thomas Sömmerring

Icones embryonum humanorum. Frankfurt am Main, Varrentrapp & Wenner, 1799.

In 1778, on a visit to London, Sömmerring (1755–1830) met and was inspired by the anatomist-obstetrician William Hunter (1718–1783). After seeing the latter's classic work *Anatomia uteri humani gravidi…* (1774), which illustrated the gravid uterus and fetus during the latter half of pregnancy, Sömmerring decided to prepare a supplementary volume with illustrations depicting the embryo and fetus during the first half of gestation. In this work based on embryos, most of which were spontaneous abortions, he attempted to convey a vision of true development in addition to that of growth *per se* (Hopwood 2000). In Plate I, Sömmerring illustrated the external forms of the developing embryo and fetus at a succession of stages from several weeks (I–III) to mid-gestation (XVII). Plate II illustrates a fetus with and without its placental membranes. In conjunction with another of his works, Sömmerring noted that his goal was "to select only that which proved to be the most excellent or most perfect specimen among many, in other words, the anatomic norm…" (Choulant, p. 302). Typical of those who prepared such atlases, he sought to depict "types," rather than individuals. In the Preface, Sömmerring reviewed the major works in embryology from that of Fabricius (Girolamo Fabrizio; 1533–1619) of 1604 and 1621, to his own time. The illustrations were by Christian Köck (1758–1818), whom Sömmerring trained specifically for this atlas and later works.

A graduate of the University of Gottingen (1778), Sömmerring worked under the professor of anatomy Heinrich August Wrisberg (1739–1808), and the professor of medicine Johann Friedrich Blumenbach (1752–1840), who was a noted anthropologist and supporter of the theory of epigenesist (as opposed to "preformation"). He also worked briefly with Pieter Camper (1722–1789). Sommerring served as professor of anatomy at the Universities of Kassel and Mainz, and Geheimrat [privy councillor] at the Bavarian Academy of Sciences in Munich. Much of his work was in the field of neuroanatomy: he established the origin of the 12 cranial nerves, the crossing of the optic nerves, and that rather than being a "great nerve," the spinal cord is part of the central nervous system. He also made important contributions to paleontology, and he was considered a first-rate astronomer. Sommerring published a number of studies in anatomy including his *Abbildungen und Beschreibungen einiger Missgeburten…* [Illustrations depicting or describing monsters] (1791) on anencephalic and bicephalic monsters. He was awarded many honors, including a Knight of the Order of the Civil Service of the Bavarian Crown (1808). Eponyms associated with Sommerring include: foramen, ganglion, ligament, muscle, nerve, and others.

References

Garrison Morton 473; see also Garrison Morton 6157; Heirs of Hippocrates 704.

Anonymous. Biography of Sömmerring. *Lancet*, London; Vol II, pp. 248–249, 1830.

Bast, T.H. The life and work of Samuel Thomas von Sömmerring. *Ann Med Hist* 6:369–386, 1924.

Hintzsche, E. Samuel Thomas Soemmerring. In: *Dictionary of Scientific Biography*. Vol XII. Charles Coulston Gillispie (Ed). New York, Charles Scribner's Sons, 1975, pp. 509–511.

Hopwood, N. Producing development: the anatomy of human embryos and the norms of Wilhelm His. *Bull Hist Med* 74:29–79, 2000.

Sömmerring, S.T. *De basi encephali et originibus nervorum cranio egredientium*. Goettingae, 1778.

Sömmerring, S.T. *Abbildungen und Beschreibungen einiger Missgeburten, die sich ehemals anf dem anatomischen Theater zu Cassel befanden*. Mainz, Universitätsbuchhandlung, 1791.

Sömmerring, S.T. *Vom Baue des menschlichen Körpers*. 5 vols. Frankfurt am Main, Varrentrapp und Wenner, 1791–1796.

Wagner, R. & S.T. Sömmerring. *Samuel Thomas von Sömmerring's Leben und Verkehr mit seinen Zeitgenossen*. Leipzig, L. Voss, 1844.

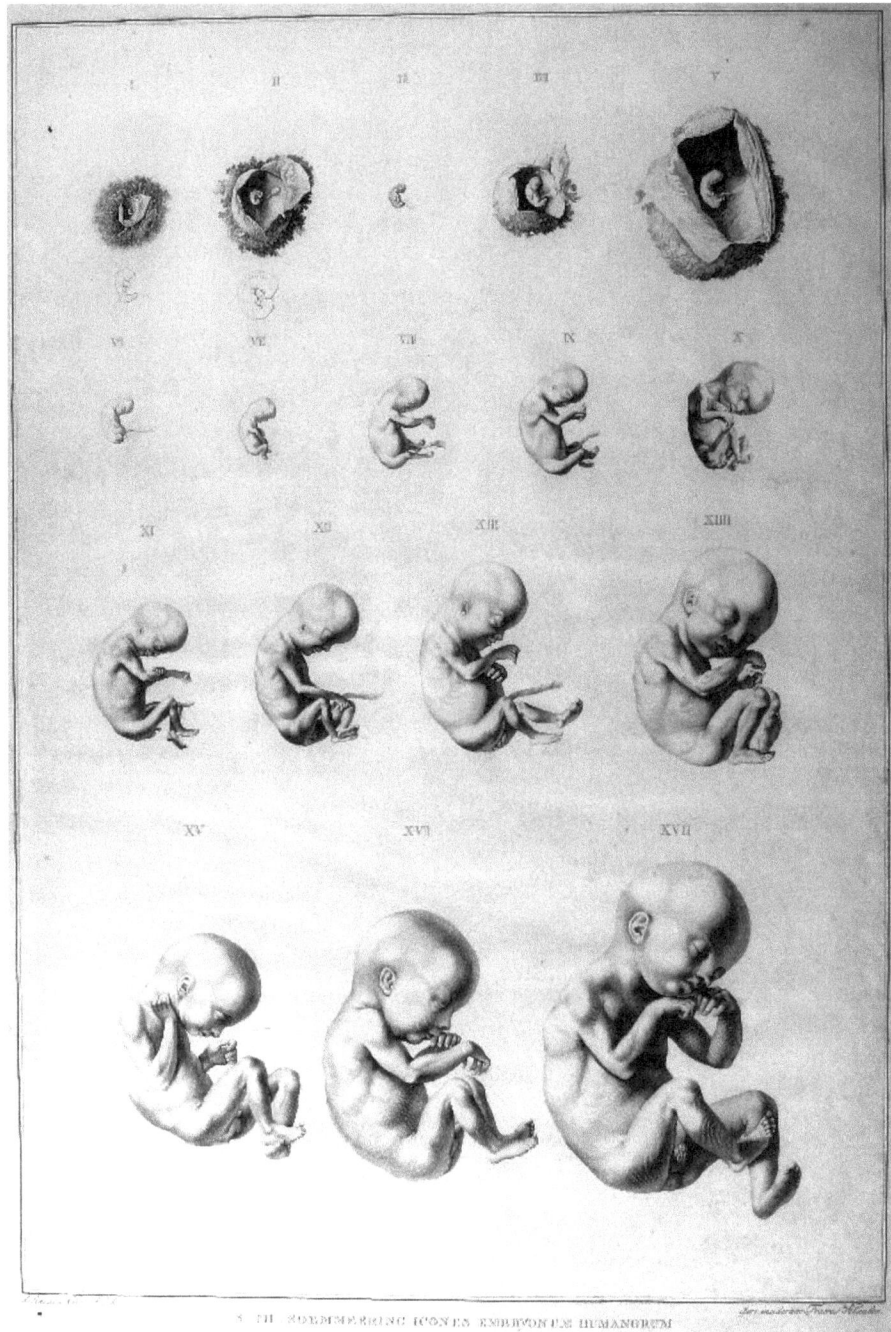

PLATE I FROM SÖMMERRING, 1799

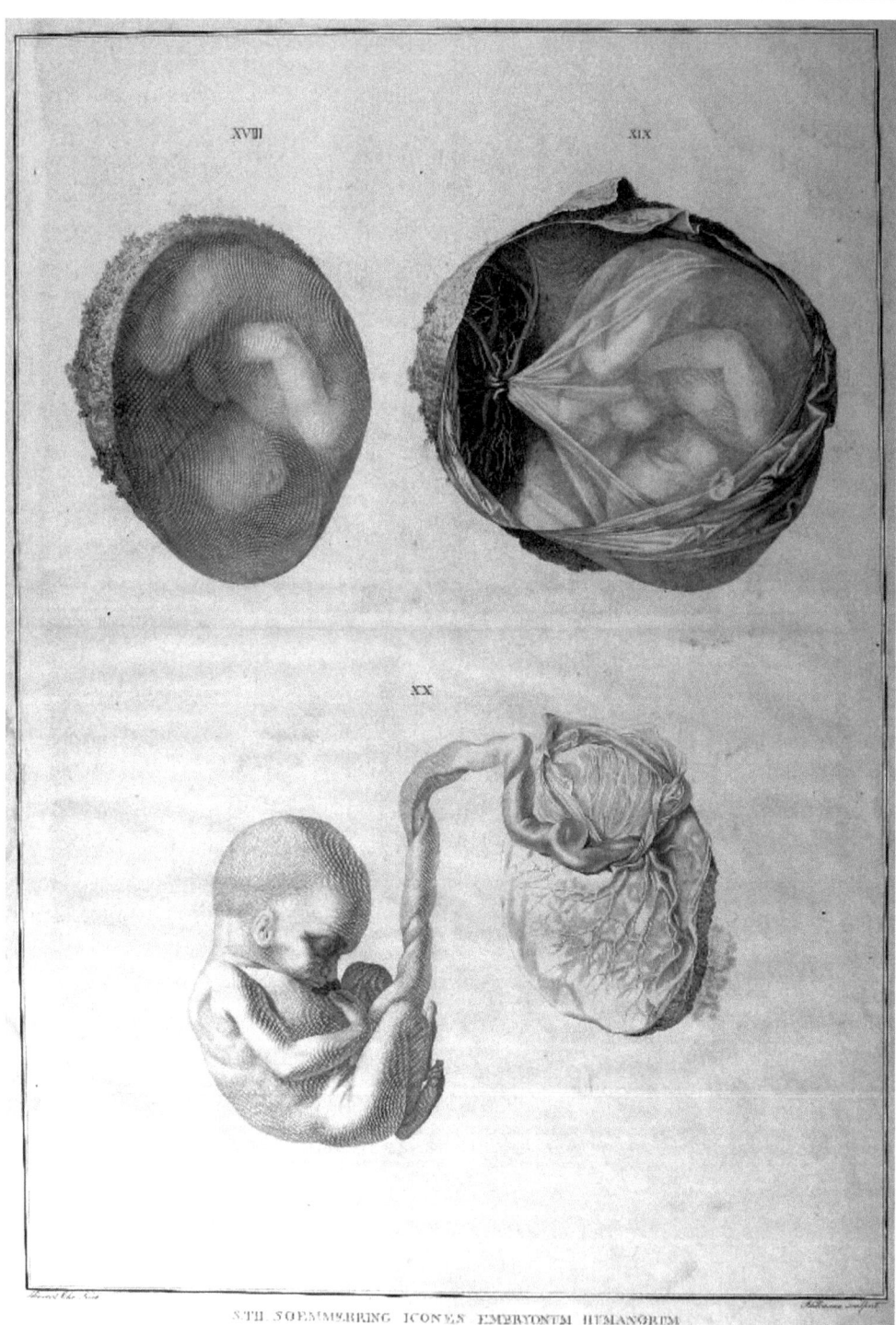

PLATE II FROM SÖMMERRING, 1799

Hermann Friedrich Kilian

Geburtshulflicher Atlas in 48 Tafeln und erklarendem Texte, Dusseldorf, Arnz, 1835–1838.

In a series of 48 beautiful lithograph plates, Kilian (1800–1863) portrayed pelvic skeletal structure, and embryonic, fetal and placental development *in utero* (Tab. XXIV and Tab. XXVIII), the pregnant uterus and its vascular development (Tab. XXVII), and childbirth. Tabulae XXXIV (shown) to XXXVI illustrate over 200 obstetrical instruments. A volume of text was issued separately.

Kilian also authored two classics on pelvic anomalies (Kilian 1854a, b).

References

Choulant, p. 307.

Kilian, H.F. *Schilderungen neuer Beckenformen und ihres Verhaltens im Leben; der Praxis entnommen.* Mannheim, Bassermann & Mathy, 1854. (GM 6261).

Kilian, H.F. *De spondylolisthesi gravissimae pelvangustiae caussa nuper detecta commentatio anatomico-obstetricia.* Bonnae, Litteris Caroli Georgii, [1854]. (GM 6261).

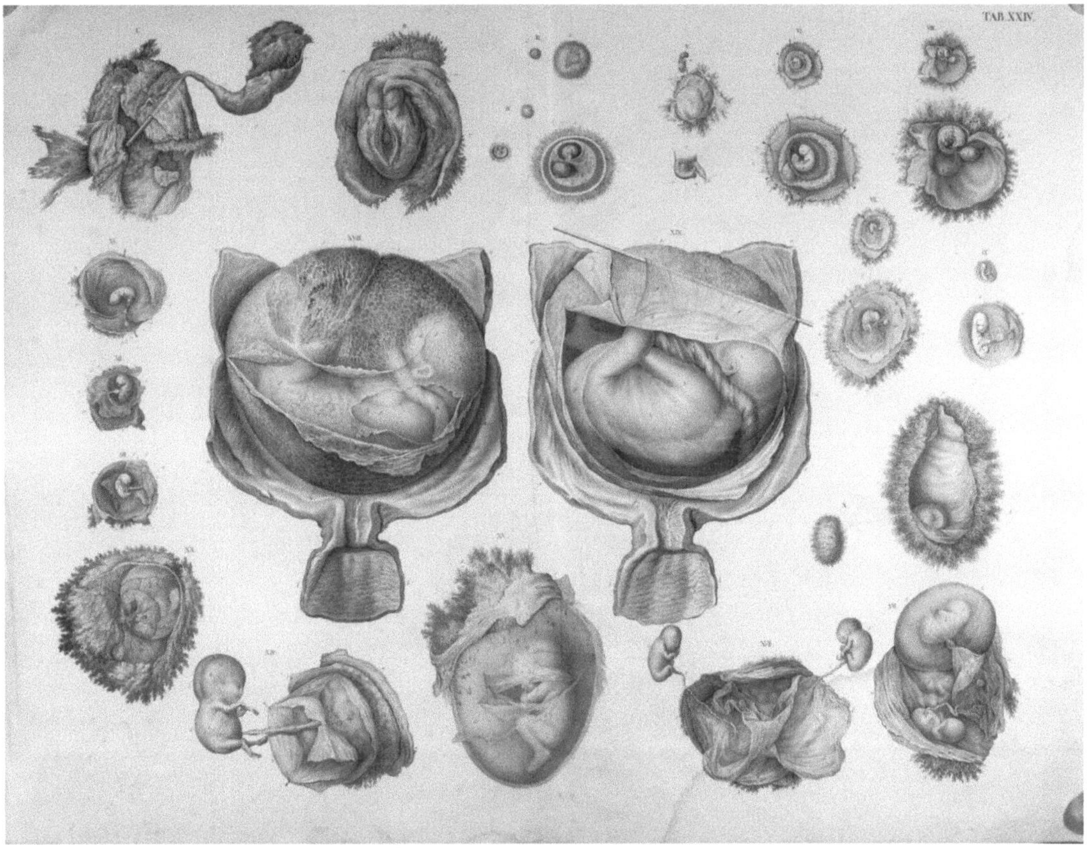

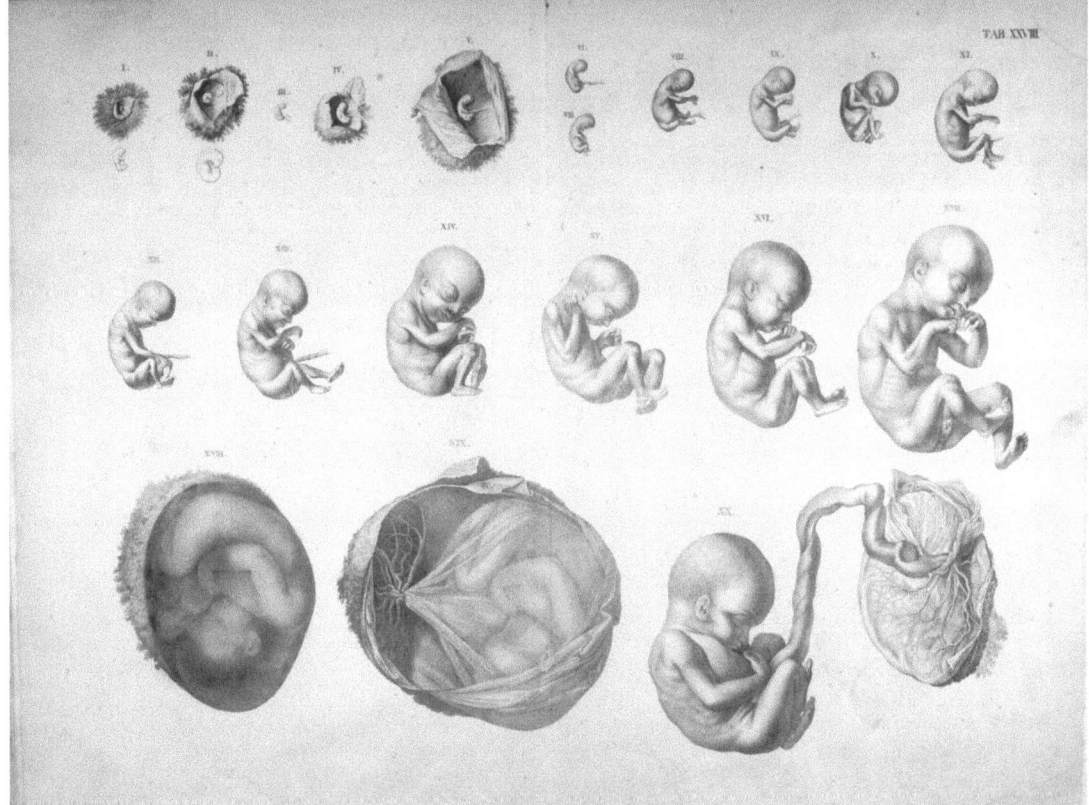

TAB. XXVII AND XXXIV FROM KILIAN 1835–1838

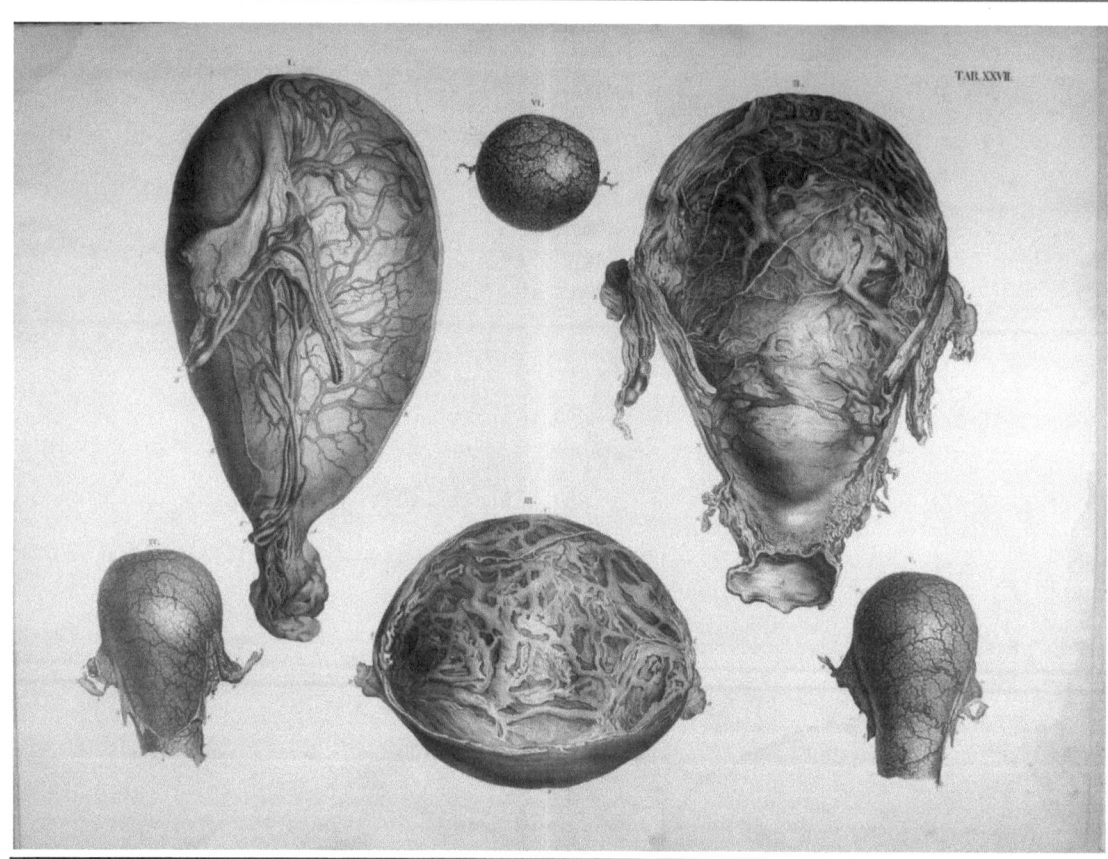

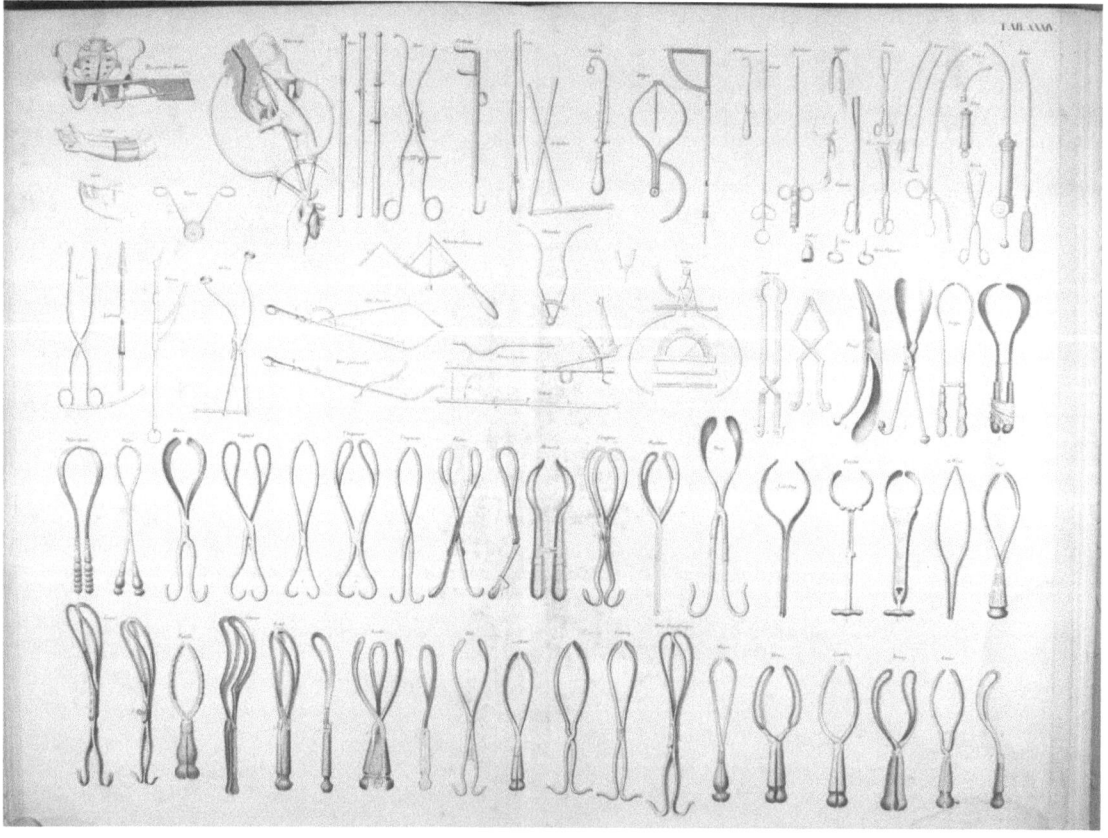

Tab. XXIV and **XXVIII** from **Kilian** 1835–1838

Rudolph Albert von Kölliker

Entwicklungsgeschichte des Menschen und der höheren Thiere, Leipzig, W. Engelmann, 1861.

In his series of *Vorlesungs* [lectures], "History of the development of humans and higher animals," Kölliker (1817–1905), professor of anatomy and physiology at the University of Würzburg, presented the first definitive work on comparative embryology. Kölliker studied the development of many species including cephalopods (e.g. octopus, squid), amphibians, aves (birds including chickens), and a number of mammals. In this regard, he carefully reviewed the embryonic development of the several systems, including that of bone and muscle, and in addition the cardiovascular, brain and nervous, gastrointestinal, urinary, and reproductive systems. Of particular importance, he interpreted the structure of these tissues in terms of their cellular elements.

Only two decades earlier, the cell had been established as the elementary structure of tissues by Matthias Jakob Schleiden (1804–1881) for plants and Theodor Schwann (1810–1882) for animals These studies established that the animal ovum is a single cell consisting essentially of a mass of protoplasm (cytoplasm) containing a nucleus and other structures. Several years previously the pathologist-anthropologist Rudolph Ludwig Karl Virchow (1821–1902) of Berlin had published his monumental *Die Cellularpathologie ...*, which laid the foundation for the cellular basis of normal bodily function and its abnormal function in disease (Virchow 1858). Figure 78 in Kölliker's twenty-first lecture illustrates the earliest stages of embryonic development in the human, based on development in other mammals (as shown previously in Plate 47). Shown is the developing embryo with the placental membranes, the chorion and amnion.

Kölliker, a native of Zurich, was educated at the Universities of Zurich, Bonn, and Berlin, receiving his Ph.D. at Zurich (1841) and his M.D. at Heidelberg (1842). Following a brief return to Zurich, in 1847 he accepted the chair of Physiology and of Comparative Anatomy at the University of Würzburg, where he remained until his death. With his colleague and friend, the comparative anatomist Carl Theodor Ernst von Siebold (1804–1885), in 1848 he founded and edited the *Zeitschrift far Wissenschaftliche Zoologie* (Journal of Scientific Zoology). Among his many scholarly contributions to embryology and histology, he did much to improve the science and art of microscopy: fixation of tissues, sectioning, and staining to help elucidate the fine details of cellular structure.

Kölliker first demonstrated the cellular origin of spermatozoa, establishing that they were, indeed, living cells (Kölliker 1841a, b). He also was the first to isolate and recognize smooth muscle cells found in arteries and various bodily organs (Kölliker 1852). With Heinrich Müller (1820–1864), Kölliker first measured action currents in cardiac muscle (Kölliker and Müller 1856). Kölliker also was the first to investigate the effects of various poisons on muscular contraction (Kölliker 1856). With Wilhelm Roux (1850–1924), he recognized that hereditary characteristics were transmitted by the cell nucleus (Kölliker 1885; Roux 1883).

Eponyms associated with his name include Kölliker's column (the sarcostyle or myofibril bundle, the myofilaments in striated muscle), Kölliker's granule (various sized small particles seen in the sarcoplasm of muscle fibers), and Kölliker's membrane (the reticular membrane of the ductus cochlearis). Because of his many scholarly contributions, the title *von* was added to his name. A member of many learned societies in Europe, he was a Fellow of the Royal Society (1830), which in 1867 awarded him with its highest token of esteem, the Copley Medal.

References

Aristotle. *Quinque de animalium generatione libri*. Venetiis, Per Joannem Antonium & Stephanum ac fratres de Sabio, 1526. (GM 17,18).

Daintith, J. (ed.). *Biographical Encyclopedia of Scientist, 3rd edition*. Taylor and Francis, 2012.

Kölliker, R.A. von. *Beiträge zur Kenntniss der Geschlechtsverhältnisse und der Samenflüssigkeit wirbellaser Theire nebst einem Versuch über das Wesen und die Bedeutung der sogenannten Sarnenthiere*. Berlin, W. Logier, 1841. (GM 1219).

Kölliker, R.A. von. Ueber das Wesen der sogenannten Saamenthiere. *N Notiz a.d. Geb d. Natur und Heilk* (*Weimar*) 19: 4–8, 1841. (GM 1220).

Kölliker, R.A. von. *Handbuch der Gewebelehre des Menschen für Aerzte und Studirende*. Leipzig, Wilhelm Engelmann, 1852. (GM 546).

Kölliker, R.A. von. *Manual of human histology. Translated and Edited by George Busk and Thomas Huxley. 2 vols*. London, Sydenham Society, 1853–1854.

Kölliker, R.A. von. Physiologische Untersuchungen über die Wirkung einiger Gifte. *Virchows Arch Path Anat* 10: 3–77, 235–296, 1856. (GM 2078).

Kölliker, R.A. von. *Uber die Darwin' sche Schöpfunstheorie*. Leipzig, W. Engelmann, 1864.

Kölliker, R.A. von. Die Bedeutung der Zellenkerne für die Vorgänge der Vererbung. *Z Wiss Zool* 42: 1–46, 1885. (GM231).

Kölliker, R.A. von. *Erinnerungen aus meinem Leben*. Leipzig, W. Engelmann, 1899.

Kölliker, R.A. von & H. Müller. Nachweis der negativen Schwankung des Muskelstroms am natürlich sich contrahirenden Muskel. *Verh Phys-med Ges Wüürzburg* 6:528–533, 1856. (GM 618).

Pierce C.S. *Writings of Charles S. Pierce: A Chronological Edition, Volume 8:1890–1892*. Bloomington, Indiana University Press, 2009.

Roux, W. *Uber die Bedeutung der Kerntheilungsfiguren*. Leipzig, F. Engelmann, 1883. (GM 229).

Schleiden, M.J. *Beiträge zur Phytogenesis*. ä137–176, 1838 (GM 112).

Schwann, T. *Mikroskopische Untersuchungen über die Uebereinstimmung in der Struktur und dem Wachsthum der Thiere und Pflanzen*. Berlin, Sander, 1839. (GM 113).

Virchow, R.L.K. Die *Cellularpathologie in ihrer Begründung auf physiologische und pathologische Gewebelehre*. Berlin, A. Hirschwald, 1858. (GM 2299).

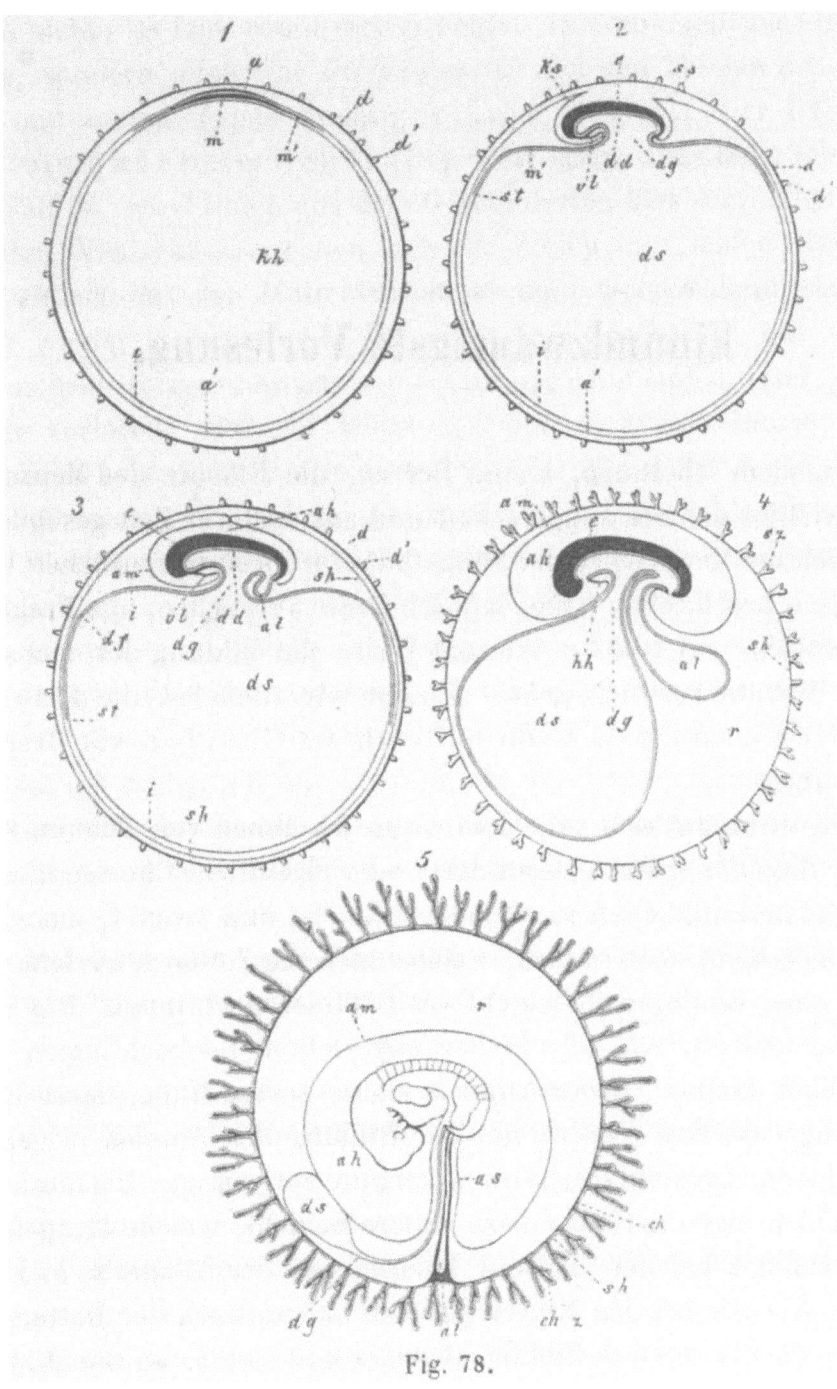

Fig. 78.

FIG 78 FROM KOLLIKER 1861

Willhelm His, Sr.

Anatomie Menschlicher Embryonen, 3 vols: vol. 1, Embryonen des ersten Monats (1880); vol. 2, Gestalt-und Grossenentwicklung bis zum Schlus des 2 Monats (1882); vol. 3, Zur Geschichte der Organe (1885). Leipzig, F.C.W. Vogel, 1861.

In his "Anatomy of the Human Embryo in three volumes (vol 1, Embryos of the first months; vol 2, The form and important development until conclusion of the second month; vol 3, Along with the history of the organs)," William His, senior (1831–1904), one of the great embryologists of the nineteenth century and professor of anatomy at the University of Leipzig, by scrupulous attention to detail and careful selection, presented for the first time a highly accurate series of illustrations of the developing human embryo in sequence from the second week to the second month of gestation. His approached embryology in an analytic fashion, with numbered stages of development, although later he abandoned this approach for representative "norms." Although usually bound together, the three installments of text and figures are paginated independently. The first and third volumes were accompanied by folio atlases, and two series of eight wax models and a set of glass photographs that were sold separately. In conjunction with his many contributions to developmental embryology, His invented the microtome, an instrument to slice tissue into a series of thin sections for microscopic examination. This allowed him to section the embryos and then perform careful three-dimensional reconstructions. For this treatise, Honikel photographed the specimens. Many of the drawings were composed by His himself, while the lithography was done by "a very careful artist," C. Pausch. His, with his colleagues, used a specially devised "embryograph," a prismatic drawing apparatus that allowed him to project the images onto paper to ensure more accurate and reliable drawings of the microscopic sections. With his assistant F. Steger, they developed the His-Steger wax models of embryonic development that are part of the collections of many museums. His collection and models of human embryos inspired Franklin P. Mall to establish the human embryo collection at the Carnegie Institution of Washington.

The plate shown, Normentafel (plate of norms) from volume 3, illustrates in 25 successive images the first 10 weeks of normal embryonic development. These expand upon the images for the same duration of gestation depicted by Samuel

Thomas Sömmering (1755–1830, pg. 407), almost a century earlier (Sömmerring 1799). The additional figures shown illustrate development of human embryos at the 4–5 mm (Figs. 7–13) and 7–8 mm (Fig. 1–6) stages. His helped to establish a framework for dating and otherwise assessing new specimens, and did much to establish modern-day descriptive embryology. Although he was one of the first to seek a causal-mechanical explanation of embryonic development, His relied on morphological rather than experimental studies. In addition, and importantly to his detailed description of embryonic development, His presented an important classification of different types of tissues in development, tissue histogenesis, the study of embryonic origins (His, 1865a, b). Overall, it was the work of His that helped to establish embryology as a modern day science.

Born in Basel Switzerland, His (1831–1904), His studied medicine at several universities including Basel, Bern, Berlin, Prague, and Vienna. At Würzburg, he studied anatomy under Rudolf Virchow (1821–1902), and embryology under Rudolph Albert van Kölliker (1817–1905), who wrote the first book on comparative embryology (1861). His also founded and edited several important scientific journals and served as vice-chancellor at both the Universities of Basel and Leipzig. He inspired many investigators with his leadership (e.g., he was co-founder and President of the German Anatomical Society). His studies included contributions to the microscopic anatomy of the cornea, presenting a dissertation of normal and pathological histology of the cornea in 1855 while still a student at Würzburg, as well as the lymphatic system, vascular endothelium (which he named), and mesodermal bodily cavities. Importantly, he was one of the originators of the "neuron doctrine," that individual nerve cells constitute the brain tissue. This was opposed to the "reticulist" view of a complex network, in which all nerves of the brain interconnect to all other cells in a vast reticular network. The neuron theory was established finally and definitely at the end of the nineteenth century by Santigo Ramón y Cajal (1852–1934) (Ramón y Cajal 1892, 1899–1904).

His son, Wilhelm His, junior (1863–1934), with whom he is sometimes confused, was a cardiologist, and also made important contributions to anatomy. Among these is his description of the atrioventricular bundle of the heart, the "bundle of His," that conducts electrical impulses from the atria to the ventricles (His 1893). The younger His also first described "heart block" when these impulses are interrupted.

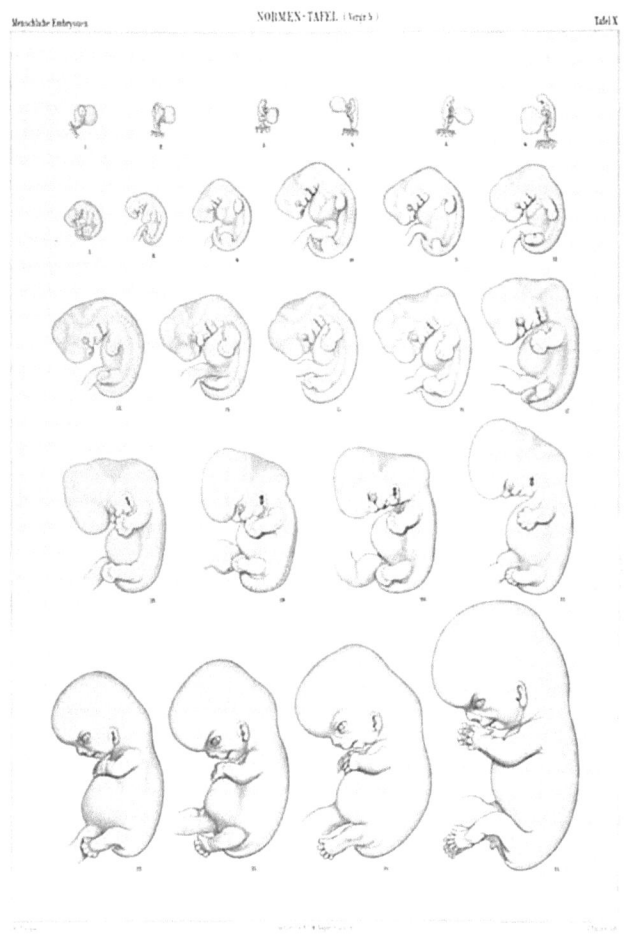

NORMENTAFEL FROM HIS 1880–1885.
COURTESY OF THE SYNDICS OF CAMBRIDGE UNIVERSITY LIBRARY

References

GM 489, 490, 494, 501.

Buettner, Kimberly A., *Wilhelm His, Sr.* Embryo Project Encyclopedia (2007–11–01). ISSN: 1940–5030 http://embryo.asu.edu/handle/10776/1705.—See more at: http://embryo.asu.edu/pages/wilhelm-his-sr#sthash.0RSEmwQb.dpuf.

Garrison, F.H. *An introduction to the history of medicine* .... 4th ed., Philadelphia, W.B. Saunders, 1929, pp. 526–527. (GM 6408 is 1913 ed).

His, W. Beobachtungen überden Bau des Säugethier-Eierstockes. *Arch Mikr Anat* 1:151–202, 1865a. (GM 489).

His, W. *Die Häute und Höhlen des Körpers.* Basel, Schwighauser, 1865b. (GM 490).

His, W. *Unsere Körperform und dos physiologische Problem ihrer Entstenhung.* Leipzig, F.C.W. Vogel, 1874.

His, W. Jnr. Die Thätigkeit des embryonalen Herzens und deren Bedeutung für die Lehre von der Herzbewegung beim Erwachsenen. *Arb Med Klin Leipzig* 14–50, 1893 (English translation in F.A. Willius and T. E. Keys, *Cardiac classics. A collection of clas-sic works on the Heart and Circulation with comprehensive Biographic accounts of the authors.* St. Louis, C.V. Mosby, 1941, p. 695). (GM 836).

His, W. Jnr. Ueber eine neue periodische Fiebererkrankung (Febris Wolhynica). *Berl Klin Wschr* 53:322–323, 1916. (GM 5387).

Hopwood, N. Producing development: the anatomy of human embryos and the norms of Wilhelm His. *Bull Hist Med* 74:29–79, 2000.

Kolliker, R.A. von. *Entwicklungsgeschichte des Menschen und der höheren Thiere.* Leipzig, W. Engelmann, 1861. (GM 487).

Louise, E.D., Stapf, C. Unraveling the neuron jungle: the 1879–1886 publications by Wilhelm His on the embryological development of the human brain. *Arch Neurol* 58:1932–1935, 2001.

Querner, H. Wilhelm His. In: *Dictionary of Scientific Biography.* Vol. VI. Charles Coulston Gillispie (Ed.). New York, Charles Scribner's Sons, 1972, pp. 434–436.

Ramón y Cajal, S. Nuevo concepto de la histologia de los centros nervi-osos. *Rev Cienc med Barcelone* 18:457–476, 1892.

Ramón y Cajal, S. *Textura del sistema nervioso de hombre y de los vertebrados* 2 vols in 3. Madrid, Moya, 1899–1904.

Sommerring, S.T. *Icones embryonum humanarium.* Frankfurt am Main, Varrentrapp & Wenner, 1799.

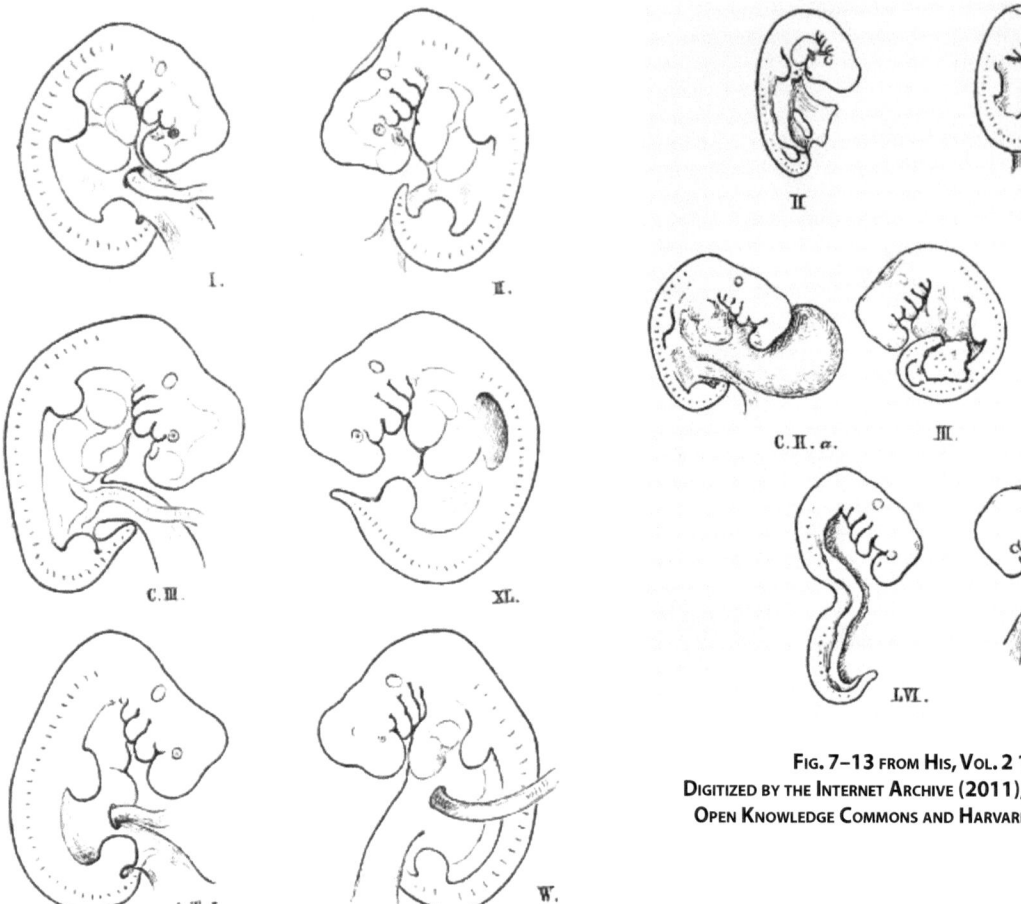

Fig. 1–6 from His, Vol. 2 1882.
Digitized by the Internet Archive (2011), with funding from
Open Knowledge Commons and Harvard Medical School

Fig. 7–13 from His, Vol. 2 1882.
Digitized by the Internet Archive (2011), with funding from
Open Knowledge Commons and Harvard Medical School

Charles Sedgwick Minot

Uterus and Embryo. I. Rabbit; II. Man. Boston, Harvard Medical School, Ginn and Company. 1889. (Reprinted in The Journal of Morphology, 2: 341–462, 1889).

Charles Sedgwick Minot (1852–1914) was born in Roxbury, MA (now a part of Boston) and as a child exhibited an intense interest in the natural sciences (Lewis 1920). In 1869, at the age of 16, he presented a paper to the Boston Society of Natural Sciences describing the male of a small butterfly species, *Hesperia metea* (only the female had previously been described). His interest in anatomy, and especially at the microscopic level, continued throughout his life. After obtaining a B.S. from the Massachusetts Institute of Technology in 1872, he studied the effects of anesthetics on vasomotor centers with Henry P. Bowditch, then Assistant Professor of Physiology at the Harvard Medical School. He also visited the physiological institute in Leipzig and studied production of carbonic acid in resting and active muscle with Carl Ludwig. While in Leipzig, Minot learned the latest microscopic methods during investigations of Turbellarian flatworms in the laboratory of Rudolf Leuckart, famous for his studies of Trichina. After returning to America her continued his research at Harvard, and received the Doctor of Science degree in 1878 for his work on the physiology of muscle contraction. He was appointed Lecturer in Embryology at Harvard Medical School in 1880, rising to the rank of Professor by 1892, and later James Stillman Professor of Comparative Anatomy.

Minot made important contributions to the understanding of embryonic development. He amassed a superb collection of embryological material, and in 1886 invented the automatic rotary microtome for cutting ultra-thin tissue sections. He cut embryos into serial sections and arranged them in steel cabinets, which served as an unrivaled resource for teaching and research (Councilman 1918). His *Human Embryology*, published in 1892, presented a comprehensive summary of human development and introduced several novel theories. In his *The Problem of Age, Growth, and Death* (1908) he introduced the term "cytomorphosis" to describe the structural alterations that cells undergo during the course of their development. This work included studies of growth and senescence from the embryonic period through old age, with observations in humans, dogs, cats, rabbits, guinea pigs, chickens, snakes, frogs, salamanders, dogfish, snails, and woodlice.

Minot also contributed to our understanding of placental morphology. Minot suggested the term trophoderm to describe the mature placental cells; however, this never became widely accepted. Nonetheless, he gave a definitive account of the microscopic structure of the human placenta. In Fig. 213 he describes a section through the uterus and placenta *in situ* at 7 months gestation, with amnion (Am), chorion (Cho), villous trunk (Vi), sections of villi in the substance of the placenta (vi), decidua (D), muscularis (Mc), compact layer of the decidua (D'), uterine blood vessels opening into the placenta (Ve), and so forth (Minot 1889, 1891). In Fig. 211A he describes the placenta at full term, again depicting the decidua (D), chorion (Cho), a vascular sinus (Si), placental villi (Vi), etc., and in Fig. 211B a more magnified portion of that in 211A showing a blood vessel (v), a decidual cell with one nucleus (d) and a decidual cell with several nuclei (d'). In a later review Minot noted, "the chorion is separated by a dense forest of villi from the decidua, … the ends of some of the villi touch and are imbedded in the decidual tissue … the decidua is plainly divided into two strata … the section passes through a wide tube, … which opens directly into the interior of the placenta and contains blood … this opening is … a vein …" (Minot 1903). He also was the first to use the term "subplacenta," (in this case for the rabbit placenta [Minot 1889]) to refer to the zone of the chorion between the placental disc and the decidua basalis (Olio et al. 2014). These observations did much to explain the nature of the fetal–maternal (i.e., placental) barrier.

References

Councilman, W.T. Charles Sedgwick Minot (1852–1914). *Proc Am Acad Arts Sci* 53: 840–847.

Lewis, F.T. In: *American Medical Biographies*. Kelly H.A. and Burrage W.L., eds. Baltimore, The Norman Remington Company. 1920. pp. 797–799.

Longo, L.D. *The Rise of Fetal and Neonatal Physiology. Basic Science to Clinical Care*. American Physiological Society. New York, Springer, 2013.

Longo, L.D., Reynolds L.P. Some historical aspects of understanding placental development, structure and function. *Int J Develop Biol* 54: 237–255, 2010.

Minot C.S. *Human Embryology*. New York, William Wood and Company. 1892. (2nd edition, Macmillan Company, 1897).

Minot, C.S. A theory of the structure of the placenta. *Anat Anz* 6: 125–131. 1891.

Minot, C.S. *A laboratory text-book of embryology*. P. Blakiston's Son & Co., Philadelphia. 1903.

Minot C.S. *The Problem of Age, Growth, and Death. A Study of Cytomorphosis. Based on Lectures at the Lowell Institute, March, 1907*. London, John Murray (also New York, Putnam). 1908.

Olio R.L., Lobo L.M., Pereira, M.A., Santos A.C., Viana, D.C., Favaron P.O., Miglino M.A. Accessory placental structures—A review. *Open J Anim Sci* 4: 305–312. 2014.

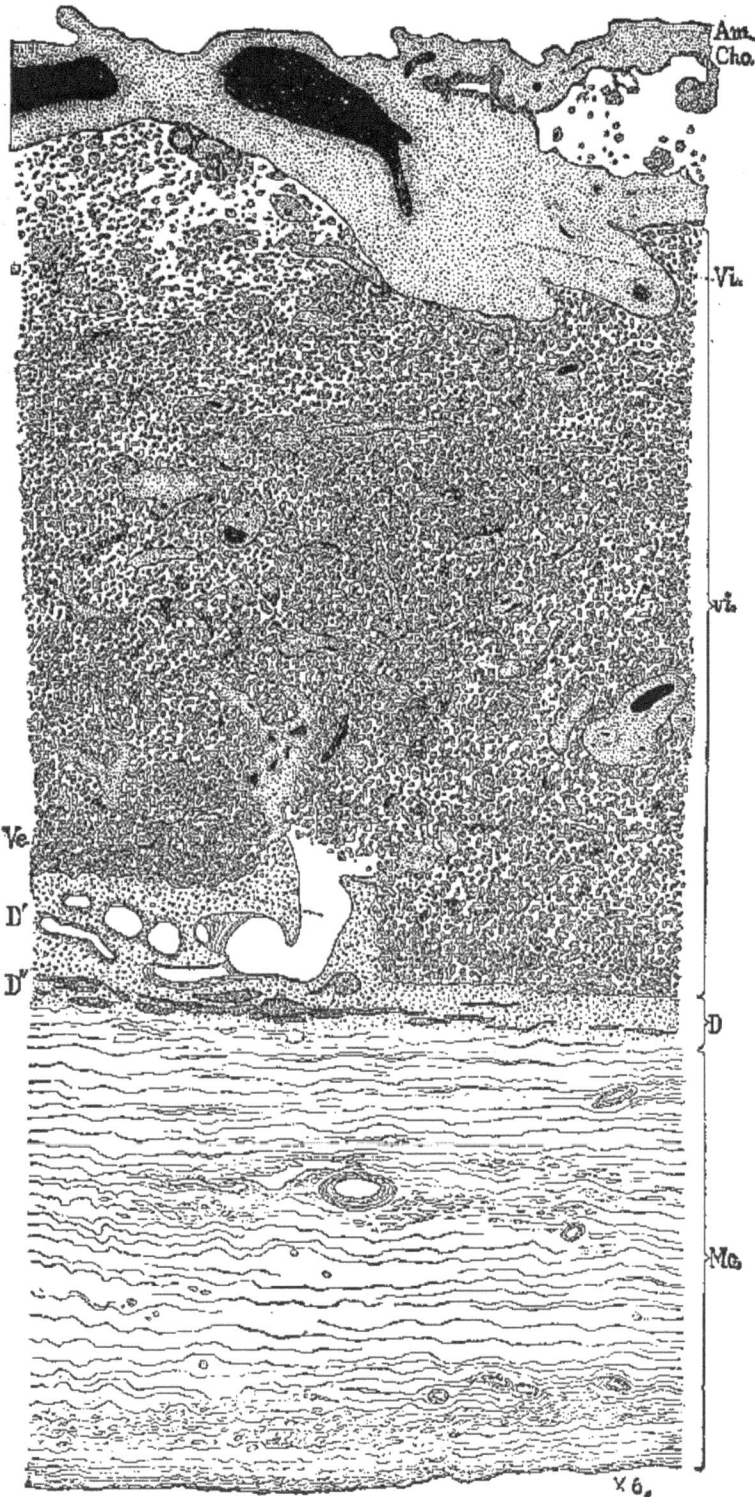

Fig. 213 from Minot 1889

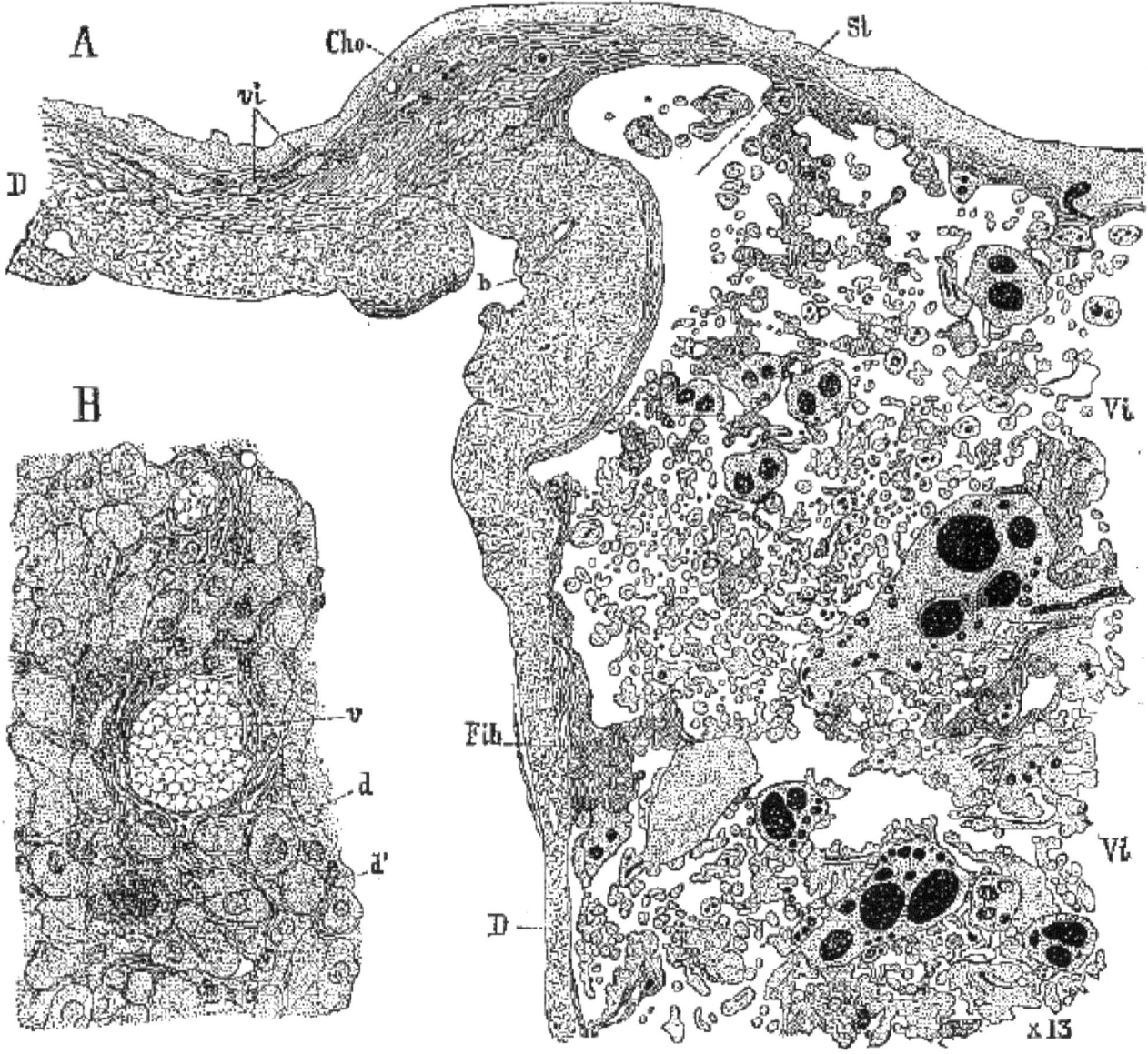

FIG. 211A AND 211B FROM MINOT 1889

Edmund Beecher Wilson

*An atlas of the fertilization and karyokinesis of the ovum...
With the cooperation of Edward Leaming.* New York,
Published for the Colombia University Press by Macmillan
and Co., 1895.

It was during the latter part of the nineteenth and early twentieth centuries that biology was transformed from a descriptive accounting of natural history, to a science dominated by
rigorous, quantitative experimental analysis. Up to this time,
fertilization was poorly understood, and the phenomenon of
chromosomes, mitosis, and other features of reproduction
and embryology had not been discovered. Jean Louis Prévost
(1790–1850) and Jean Baptiste Andre Dumas (1800–1884)
had first demonstrated in the frog that fertilization occurs by
union of a spermatozoa and an ovum (Prévost and Dumas
1824), while Wilhelm August Oscar Hertwig (1849–1922)
established that fertilization requires union of the nuclei of
the male and female sex cells (Hertwig 1876).

Wilson (1856–1939) stressed the importance of cytological structure and progressive cell development in the
study of experimental embryology and heredity, and is
credited as being one of the founders of the fields of cell
biology and cytogenesis. He was consumed by the problem
of how an entire individual may lie implicit in a single cell.
Thus he dedicated considerable effort to exploring factors
that result in cell segregation into tissues and organs, with
their functional specialization. In his study of the cell-lineage of the marine worm *Neresis*, Wilson traced egg cleavage step-by-step through the formation of germ layers and
the principal organs of the embryo (Wilson 1892). His findings helped to establish the idea that the mosaic-like character of ontogeny emerges at early stages and different
periods in the several species (Wilson 1894). With fate
maps, he followed embryonic cell-by-cell development to
establish the normal morphogenic processes from fertilization
onward (Wilson 1904).

The present work by Wilson was the first successful
attempt to produce clear photographic illustrations of the
early history of the ovum. The atlas of 10 phototype plates
contain 40 photomicrographs that depict fertilization and
mitosis of the ovum with sections of eggs of the sea-urchin
Toxopneustes variegatus enlarged 950–1000 diameters. Plate
X illustrates the two cell stage (Fig. 37), preparation for the
second cleavage (Fig. 38), the fourth cleavage showing
micromere spindles (Fig. 39), and the 16 cell blastula, a hollow sphere surrounding the blastocoel (Fig. 40). In his preface, Wilson stressed the need to understand the minutest

details of the early development of the fertilized ovum. He
noted that no drawing, however exact, can convey these delicate and complicated subcellular structures. He also noted
the challenges of preparing a photograph in this regard, the
difficulty in making the preparations without distortion, and
the extreme care that had gone into illustrating the fine structural detail. Wilson's associate, Edward Leaming (1861–
1916), photographed the specimens.

A graduate of the Johns Hopkins University, Wilson
spent much of his professional life at Columbia University
and was among the most creative and prolific biologists during this period. He obtained recognition for his *The cell in
development and inheritance* (1896), which, with critical
observations and analysis, laid the foundation for understanding cellular structure and functions. Other of his
reviews that became classics include Experimental studies
on germinal localization (1904) and *The cell in development
and heredity* (1925).

Independently in 1900, Carl Franz Joseph Erich Correns
(1864–1933) (Correns 1900), Hugo Marie de Vries (1848–
1935) (de Vries 1900), and Erich Tschermak von Seysenegg
(1871–1962) (Tschermak 1900), rediscovered Gregor
Johann Mendel's (1822–1884) paper on the laws of genetic
inheritance (Mendel 1866), with the laws of segregation and
that of independent assortment or "ratios" in inheritance. In
his other lasting contributions to embryology and genetics,
Wilson promoted the idea that the nucleus contains the basis
of inheritance, with chromatin as a key element of information, and that chromosomes contain the essential "factors" or
"genes." He also discovered the X–Y chromosome system of
sex determination (Wilson 1905; 1906; 1910; 1914), detailing aspects of gene activation which are still under investigation (Lyon 1961). With his studies in *Drosophila
melanogaster*, Wilson's colleague at Columbia University,
Thomas Hunt Morgan (1866–1945), also played an important role in integrating the tenets of Mendelian genetics with
the chromosome theory of inheritance (Morgan et al. 1915).
To a great extent, modern day genetics is founded upon the
work and ideas of Wilson and Morgan.

A highly regarded scientist, Wilson received honorary
degrees from a number of universities, and also was an
accomplished cellist. His *The cell in development and heredity* was awarded the Daniel Giraud Elliot Medal of the
National Academy of Sciences (1928) and the gold medal of
the Linnean Society, London. He also received the National
Academy of Science's John J. Carty Medal and Award for
the Advancement of Science (1936), and other honors. He
served as president of the American Association for the
Advancement of Science in 1913.

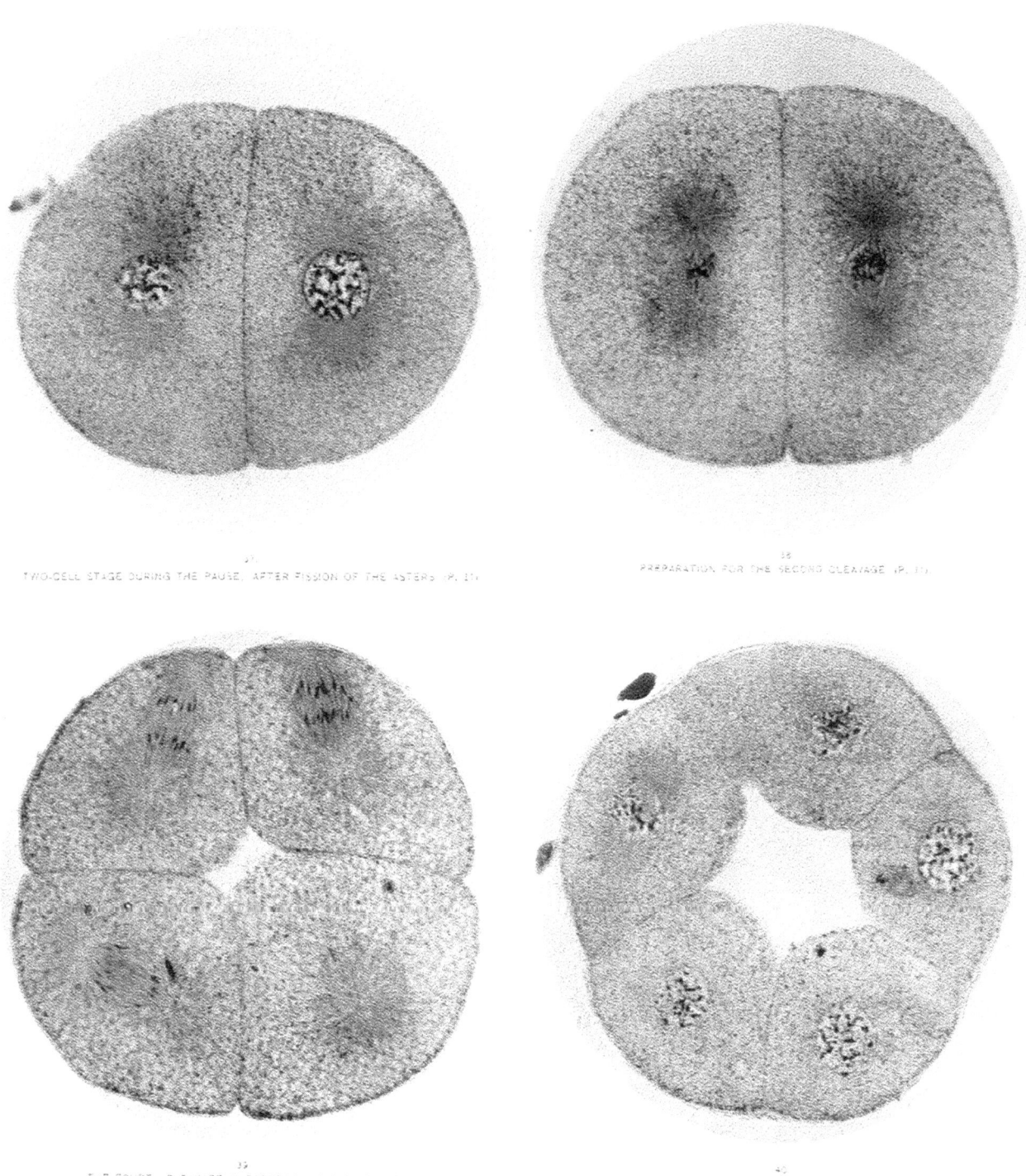

37.
TWO-CELL STAGE DURING THE PAUSE, AFTER FISSION OF THE ASTERS. (P. 31)

38.
PREPARATION FOR THE SECOND CLEAVAGE. (P. 31)

39.
THE FOURTH CLEAVAGE IN PROGRESS, SHOWING THE
MICROMERE-SPINDLES (P. 32)

40.
BLASTULA, SIXTEEN-CELL STAGE IN SECTION. (P. 33)

PLATE X FROM WILSON, 1895

References

Allen, G.E. Edmund Beecher Wilson. In: *Dictionary of Scientific Biography. Vol XIV. Charles Coulston Gillispie (Ed)*. New York, Charles Scribner's Sons, 1976, pp. 423–436.

Correns, C. G. Mendel's Regel über das Verhalten der Nachkommenschaft der Rassenbastarde. *Ber Dtsch Botanisch Ges* 18:158–168, 1900. (GM 239.1).

Hertwig, W.A.O. Beiträge zur Kenntniss der Bildung, Befruchtung und Theilung des thierischen Eies. *Morph Jb* 1:347–434, 1876. (GM 495).

Jewett, F.B. Presentation of the John J. Carty medal and award to Dr. Edmund Beecher Wilson. *Science* 84:564–565, 1936.

Lyon, M.F. Gene activation in the X-chromosome of the mouse (*Mus musculus I*). *Nature (Lond.)* 190:372–373, 1961. (GM 256.7).

Mendel, G.J. Versuche über Planzen-Hybriden. *Verb natural Vereins Brûnn (1865)* 4:3–47, 1866. (GM 222).

Morgan, T.H. Edmund Beecher Wilson, 1856–1939. *Biographical Memoirs National Academy of Sciences* 21:315–342, 1941 (condensed in *Science* 89:258–259, 1939).

Morgan, T.H. Edmund Beecher Wilson. *Science* 96: 239–242, 1942.

Morgan, T.H., A.H. Sturtevant, H. Muller, and C.B. Bridges. *The mechanism of Mendelian heredity*. New York, H. Holt, 1915. (GM 246).

Muller, H.J. Edmund B. Wilson – an appreciation. *Amer Naturalist* 77:5–37 and 142–172, 1943.

Prévost, J.L. and J.A. Dumas. Deuxième mémoire sur la génération. Rapports de l'oeuf avec la liqueur fécondante. Phénomènes appréciables, résultant de leur action mutuelle. Développement de l'oeuf des batraciens. *Ann Sci Nat (Paris)* 2:100–120, 129–149, 1824. (GM 474.1).

Sedgwick, W.T. and E.B. Wilson. *General biology*. New York, H. Holt and Co., 1886.

Tschermak von Seysenegg, E. Ueber künstliche Kreuzung bei *Pisum sativum. Z landwirtsch Versuchs in Osterreich* 3:465–555, 1900. (GM 239.2).

Van Beneden, E. and A. Neyt. Nouvelles Recherches sur la Fécondation et la Division Mitosique chez l'Ascaride Mégalocephale. Communication Préliminaire. *Bulletins de l'Académie Royale des Sciences, des Lettres et des Beaux-Arts de Belgique* séries 3, 6:238,1887.

Vries, H.M. de. Das Spaltungsqesetz der Bastarde. *Berl Dtsch Bot Ges* 18:83–90, 1900. (GM 239.01).

Wilson, E.B. The cell-lineage of *Neresis*. A contribution to the cytogeny of the annelid body. *J Morphology* 6:361–480, 1892.

Wilson, E.B. Amphioxus and the mosaic theory of development. *Biol Lectures at Marine Biol Lbaoratory, Woods Hole, 1893*. pp. 1–14, 1894.

Wilson, E.B. *The cell in development and inheritance*. New York, Macmillan, 1896. (GM 238).

Wilson, E.B. Mendel's principles of heredity and the maturation of the germ-cells. *Science* 16:991–993, 1902.

Wilson, E.B. Experimental studies on germinal localization. *J Exp Zool* 1:1–72, 1904. (GM 519).

Wilson, E.B. The chromosome in relation to the determination of sex in insects. *Science* 22:500–502, 1905.

Wilson, E.B. Mendelian inheritance and the purity of the gametes. *Science* 23:112–113, 1906.

Wilson, E.B. The chromosomes in relation to the determination of sex. *Science Progress* 16:570–592, 1910.

Wilson, E.B. The bearing of cytological research on heredity. The Croonian Lecture for 1914. *Proc R Soc Lond* 88:333–352, 1914.

Wilson, E.B. *The cell in development and heredity*. New York, Macmillan, 1925.

Friedrich Schatz

Klinische Beiträge zur Physiologie des Fötus. Berlin, Verlag von August Hirschwald, 1900.

Rather than fetal physiology *per se*, the volume *Physiologie des Fötus* by Friedrich Schatz (1841–1920), Professor of Obstetrics at the University of Rostock, chiefly was devoted to placental pathology of the twin-to-twin transfusion syndrome seen in monochorionic "identical" twins. Because with an arterial to venous anastomosis of the placental vasculature between the twins, in the majority of such cases the recipient fetus is significantly larger and edematous, with a greater hemoglobin concentration and blood volume, cardiomegaly, and a fluid-filled bladder. In turn, the donor twin is small and shriveled with an empty bladder. With a relatively high rate of urination into the amniotic cavity by the recipient twin, the mother often presents with polyhydramnios and goes into premature labor (Schatz 1900).

Detailing aspects of comparative size of the twins, their hemoglobin concentrations, hematocrits, and cardiac size, this 712-page volume addresses the varieties of placental vascular anastomosis seen in this condition. It includes 34 full or double page plates, many in color, which depict the placental vascular anatomy seen with this condition. In his analysis, Schatz presented an idealized schema of the placenta with central insertion of the umbilical cord, and in a meridian-like fashion the alternating arteries and veins passing centrifugally to the placental margin. By performing careful injections of wax of different colors, Schatz recognized that most of these cases had an umbilical arteriovenous anastomosis between the placental circulations of the two infants. He demonstrated that in many instances branches of an artery and vein would delve deep into the placenta, into what he termed a "third circulation" through a "villous district" of the cotyledon. Usually these could not be seen directly from the fetal surface. Schatz previously had reported on such a case of placental anastomosis in twins (Schatz 1875), those with differing volumes of amniotic fluid (Schatz 1882), and later explored this concept more fully (Schatz 1886, 1887; see Ludwig 2006).

Many others have described various aspects of the placental and fetal pathology seen in twin to twin transfusion syndrome (Benirschke 1958, 1961; Bergstedt 1957; Falkner et al. 1962; Sacks 1959). In a series of reports, the pathologist Richard L. Naeye of the University of Vermont, detailed many consequences for the fetuses with this condition, including major differences in the relative sizes of several organs of the body (Naeye 1963, 1964a, b). In a study among five twin pairs born from 27 to 30 weeks gestation the body weight difference between recipient and donor was 294 g. For two pairs born near-term, this difference averaged 847 g (Naeye 1963). The recipient twin also demonstrated greater muscle mass about both pulmonary and systemic arteries, as well as significantly larger renal glomeruli (Naeye 1963). In another report of ten such twin pairs from 23 to 30 weeks gestation (which included several cases from previous report), not only was there a discordance of about 200 g in body weight, but significant weight differences were seen in the brain, heart, lungs, liver, and other organs. At the cellular level the number and size of cardiac myocytes were increased significantly in the recipient twin (Naeye 1965). Twin to twin transfusion syndrome has been identified as the most important determinant of growth disparity between identical twins.

In a 233 page four article report, Schatz earlier described the first measurements with a liquid-filled balloon of intrauterine pressures within the uterus of a non-pregnant woman (Schatz 1872a, b, c, d). Other aspects of Schatz's life have been reviewed elsewhere (Ludwig 2006).

References

Benirschke, K. Twin placenta in perinatal mortality. *N Y State Med* J 61:1499–1508, 1961.

Bergstedt, J. Monozygotic twins, one with high erythrocyte values and jaundice, the other with anaemia neonatorum and no jaundice. *Acta Paediatr* 46:201–206, 1957.

Falkner, F., N.D. Banik & R. Westland. Intra-uterine blood transfer between uniovular twins. *Biol Neonat* 4:52–60, 1962.

Naeye, R.L. Human intrauterine parabiotic syndrome and its complications. *N Engl J Med* 268:804–809, 1963.

Naeye, R.L. The fetal and neonatal development of twins. *Pediatrics* 33:546–553, 1964a.

Naeye, R.L. Organ composition in newborn parabiotic twins with speculation regarding neonatal hypoglycemia. *Pediatrics* 34:415–418, 1964b.

Naeye, R.L. Organ abnormalities in a human parabiotic syndrome. *Am J Pathol* 46:829–842, 1965.

Sacks, M.O. Occurrence of anemia and polycythemia in phenotypically dissimilar single-ovum human twins. *Pediatrics* 24:604–608, 1959.

Schatz, F. Beiträge zur physiologischen Geburtskunde. *Arch Gynakol* 3:58–144, 1872a.

Schatz, F. Beiträge zur physiologischen Geburtskunde. *Arch Gynakol* 4:34–111, 1872b.

Schatz, F. Beiträge zur physiologischen Geburtskunde. *Arch Gynakol* 4:193–225, 1872c.

Schatz, F. Beiträge zur physiologischen Geburtskunde. *Arch Gynakol* 4:418–456, 1872d.

Schatz, F. Ueber die während jeder Geburt eintretende relative Verkürzung oder Verlängerung der Nabelschnur und die dadurch unter bestimmten Umständen bedingten Störungen und Gefahren der Geburt. *Arch Gynakol* 8:1–47, 1875.

Schatz, F. Eine besondere Art von einseitiger Polyhydramnie mit anderseitiger Oligohydramnie bei eineiigen Zwillingen. *Arch Gynakol* 19:329–369, 1882.

Schatz, F. Die Gefässverbindungen der Placentakreisläufe eineiiger Zwillinge, ihre Entwickelung und ihre Folgen. *Arch Gynakol* 24:337–399, 1884.

Schatz, F. Die Gefässverbindungen der Placentakreisläufe eineiiger Zwillinge, ihre Entwickelung und ihre Folgen. *Arch Gynakol* 27:1–72, 1886.

Schatz, F. Die Gefässverbindungen der Placentakreisläufe eineiiger Zwillinge, ihre Entwickelung und ihre Folgen. *Arch Gynakol* 30:169–240 and 335–381, 1887.

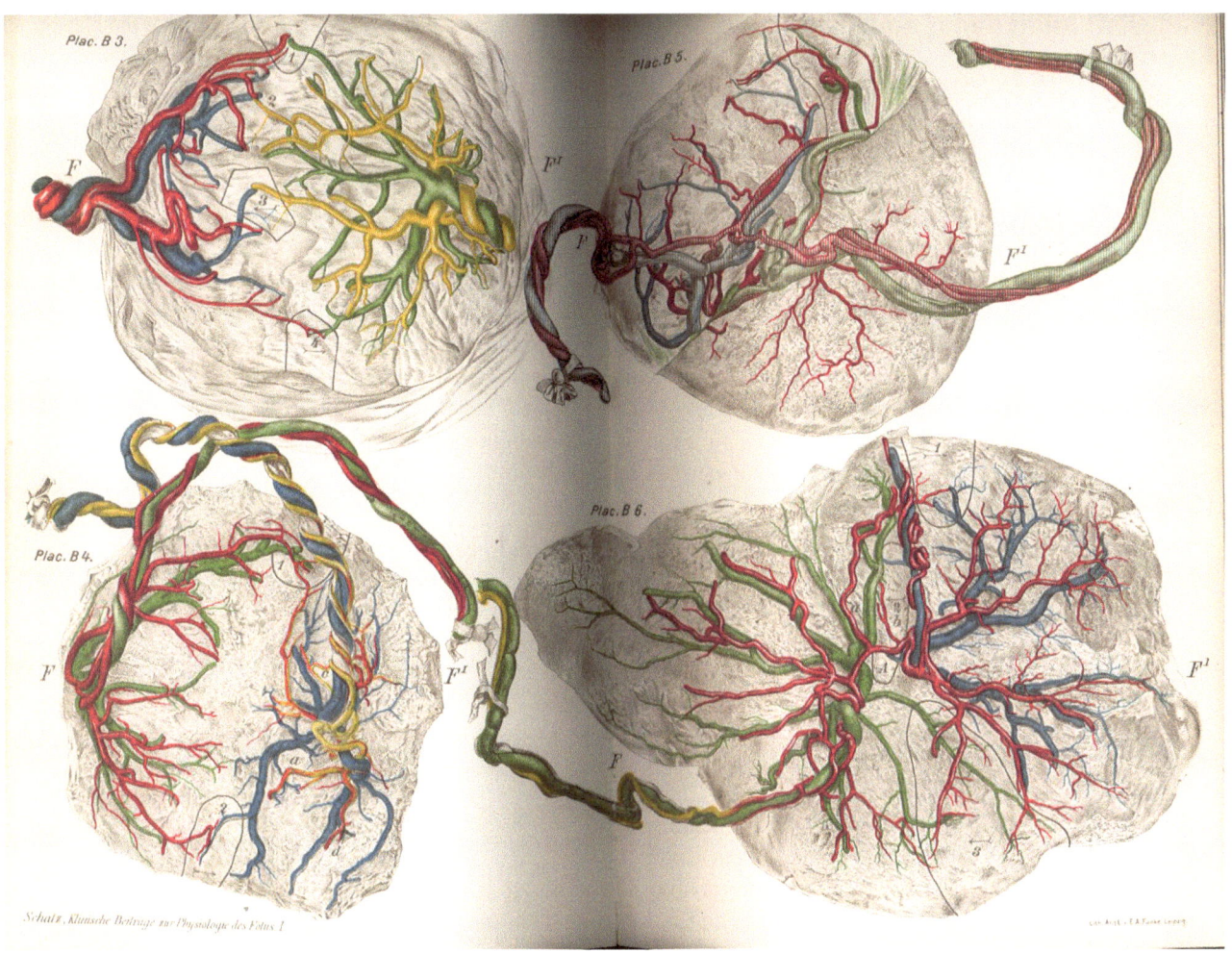

PLATES B 3 TO B 6 FROM SCHATZ, 1900

Franz Karl Julius Keibel and Franklin Paine Mall

Manual of human embryology, 2 vol. Philadelphia, J.B. Lippincott, 1910–1912.

This work by Keibel (1861–1929) and Mall (1862–1917), both students of the embryologist Wilhelm His (1831–1904), expanded upon His' monumental *Anatomie…* of human embryology (His 1880–1885). An exhaustive, encyclopedic work, it was said to "… mark an epoch of accomplishment in the study of human embryology," and at the same time "… it furnishes exceptionally numerous suggestions of many problems yet to be solved, with the most promising lines of attack" (Knower 1911, p. 493). Published simultaneously in German and English, the work illustrates the close relationships of investigators on both sides of the Atlantic that trained under, or were otherwise influenced by, His.

Keibel, who became lecturer at the University of Freiburg, and later at the Universities of Strassburg, Köningsberg, and Berlin, contributed numerous morphologic studies in vertebrate embryology. He is particularly noted for his *Normentafeln zur Entwicklungsgeschichte …* (1897–1938), an encyclopedic survey of the embryonic anatomy of various vertebrates. In Taf. 1 he presents a "normentafel" (normal plate) showing various stages of normal development in the pig, just as His (1880–1885) had done for human embryos (Hopwood 2007). He also contributed to an understanding of many specific aspects of development, such as that of the head and various germ layers (Keibel and Elze 1908).

Mall, a graduate of the University of Michigan School of Medicine (1883), obtained postdoctoral training in Leipzig under Wilhelm His (1831–1904) and Carl Ludwig (1816–1895), with whom he studied the development of the thymus gland and the small intestine. Later, he made important contributions to understanding the anatomy of the liver and the spleen. With the opening of the Johns Hopkins University School of Medicine in 1893, Mall became the first professor of anatomy. There he helped to make the study of anatomy an independent science of its own, and a vital part of medical edu-

cation. In 1914, he became the first director of the Department of Embryology of the Carnegie Institution of Washington, to which he contributed his large collection of human embryos. In addition to his many contributions to anatomy and embryology, Mall became an important teacher and mentor to many of the outstanding anatomists in the United States. He was a cofounder of the *American Journal of Anatomy*, and served as president of the American Association of Anatomists (1906–1908), which he did much to develop as an important scientific society. Mall also played a critical role in development of the full-time system of educators in medicine.

Both Keibel and Mall played important roles in organizing and consolidating the embryological criteria of His throughout the world.

References

Garrison Morton 526.

Corner, G.W. Franklin Paine Mall. In: *Dictionary of Scientific Biography*. Vol IX Charles Coulston Gillipsie (Ed.). New York, Charles Scribner's Sons, 1974, pp. 55–58.

Hill, M.A. (2014) Embryology *Keibel1897 plate01.jpg*. Retrieved December 11, 2014, from https://php.med.unsw.edu.au/embryology/index.php?title=File:Keibel1897_plate01.jpg.

His, W. *Anatomie menschlicher Embryonen* 3 vol. Leipzig, F.C.W. Vogel, 1880–1885.

Hopwood, N. Producing development: the anatomy of human embryos and the norms of Wilhelm His. *Bull Hist Med* 74: 29–79, 2000.

Hopwood, N. A history of normal plates, tables and stages in vertebrate embryology. *Int J Develop Biol* 51: 1–26, 2007.

Keibel, F.K.J. and C. Elze. *Normentafel zur Entwicklungsgeschichte des Menschen*. Jena, Fischer, 1908.

Keibel, F.K.J. *Normentafeln zur Entwicklungsgeschichte der Wirbelthiere*, *Hrsg. Von Franz Keibel*; *16 pts*. Jena, G. Fischer, 1897–1938.

Knower, H.McE. Book review, Human Embryology—Keibel and Mall. *Science* 31: 493–496, 1911.

Mall, F.P. A *contribution to the study of the pathology of early human embryos*; *3 pts*. Baltimore, 1899–1908. Reprinted from: Welch Festschrift, Johns Hopkins Reports, 1900, v. 9, p. 1–68; Vaughan Festschrift, Contributions to Medical Research, Ann Arbor, 1903, p. 12–28; Journal of Morphology, 1908, v. 19, no. 1, p. 1–361. Third contribution has main title: "A study of the causes underlying the origin of human monsters."

Peter, K. Franz Keibel: Ein Nachruf. *Anat Anz* 68: 201–220, 1929.

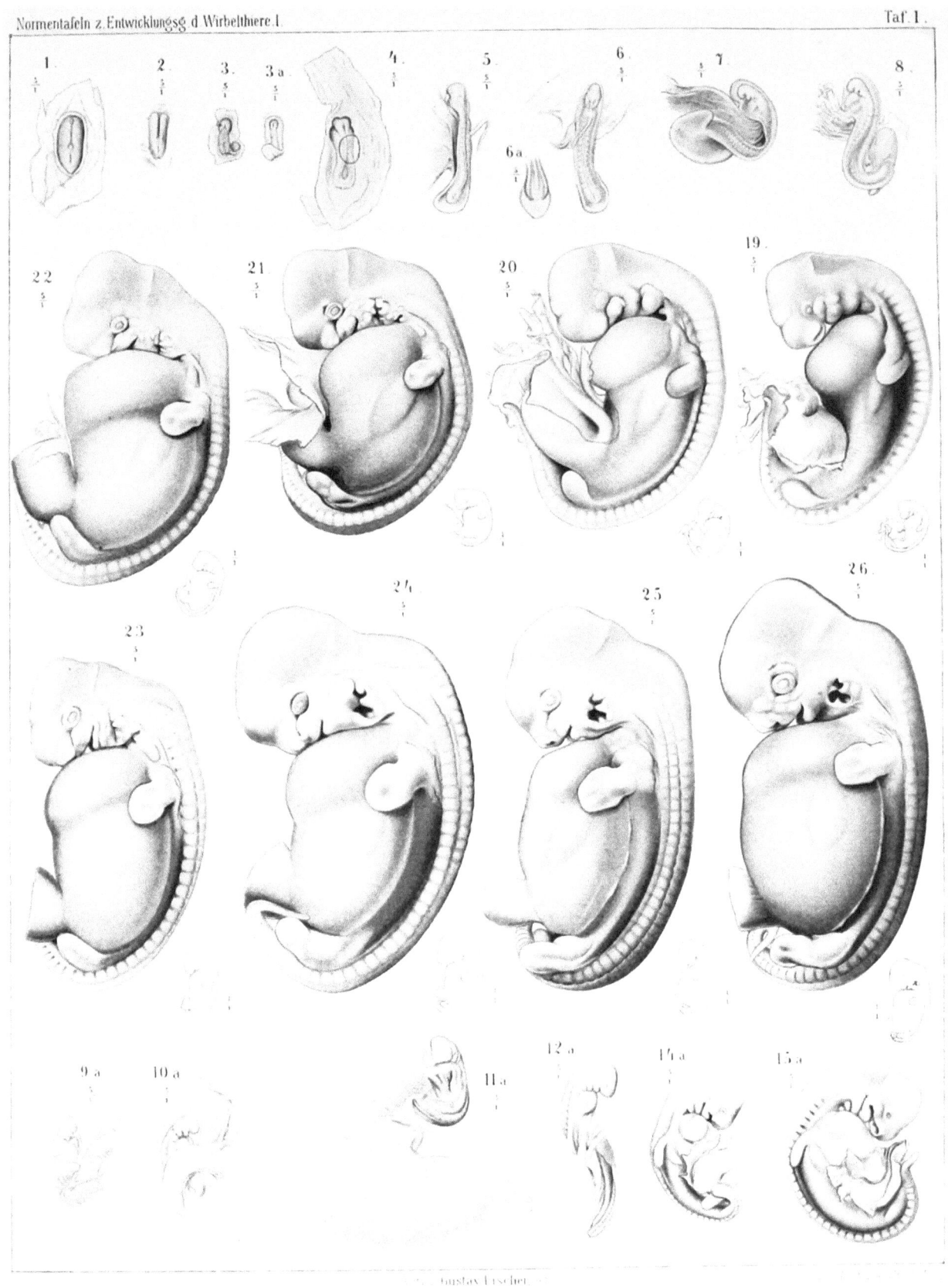

PLATE I FROM KEIBEL 1897 (COURTESY OF HILL, 2014)

Teratology: Monsters and Prodigies

In considering teratology and congenital malformations, the problems of birth defects and monstrosities in humans and animals have attracted attention since earliest times. The question arises, how these anomalies were viewed during the period encompassed by this work, and the manner in which ideas evolved as to their genesis. In earlier times, congenital anomalies were regarded as of supernatural origin. Pictorial records of malformations were made long before man could write, and are to be found on ancient rock carvings. As a naturalist, Aristotle appears to have been the first to examine abnormalities and to view them as a natural biological phenomenon. The Roman scholar Pliny "the elder" (23–79) described many well-known examples, but did not distinguish between those monstrosities that were actual and those legendary. During the middle ages these were treated in the fullest spirit of superstition, and numerous relics of this genre survive. For the most part, human anomalies were believed to represent punishment from God, Divine wrath, or to have been conceived in the wombs of women who commerced with the devil. The belief that such monsters is a consequence of unnatural union between women and male animals, or between men and female animals, is a continued form of the satanic legend. As originally used, the word monster was not pejorative, but, derived from the Latin *monere*, to warn, and *monstrum*, prodigy or portent, indicated a demonstration of something worth viewing. Many held that monsters were a consequence of excess or insufficiency of semen within the womb. Alternatively, it was believed that they were a result of admixture of semen from different fathers. Another popular view had an epigenetic basis, that it was the mother's imagination, from experiences or other influences during pregnancy, that eventuated in this outcome.

Beginning at the end of the sixteenth and in the early seventeenth centuries, those infants with congenital anomalies became interpreted less frequently as a sign of evil inspiring horror, and more as Prodigies [Latin, *prodigium*, omen or monster]. Intellectual curiosity about such abnormalities developed into a more scholarly approach to understanding birth defects and related anomalies. The earliest writers to consider "prodigies" *in extensio* in printed volumes were Conrad Wolffhart [Lycosthenes] (1518–1561; 1557), Schenck von Grafenberg (?–1620; 1609), Fortunio Liceti (1577–1657; 1616), and Ulisse Aldrovandi (1522–1605; 1642).

To a great extent it was the study of embryology that has contributed to a true understanding of such teratological anomalies. In his *Exercitationes de generatione…* (1651), William Harvey first considered monstrosities as legitimate abnormalities of embryonic development. Details of what is known currently about genetic and epigenetic influences on embryonic and fetal development are beyond the scope of this introduction. The embryonic period, from the second through the seventh week of gestation is the most critical and vulnerable for development of malformations, as it is during this time that the major organs are forming. In addition to maternal infections such as syphilis, cytomegalovirus, rubella (German measles), and maternal disorders such as alcoholism, diabetes mellitus with hyperglycemia, hypothyroidism, exposure to

L.D. Longo, L.P. Reynolds, *Wombs with a View*, DOI 10.1007/978-3-319-23567-7_5

excessive ionizing radiation, a number of drugs and chemicals now are known to be teratogens that result in congenital malformations. Also as is widely appreciated, the origin of many malformations is unknown. It was only at the end of the period covered in this volume that beginnings were being made in teratology as an experimental science. Current investigation has established that susceptibility to teratogens is a function of a number of factors, including the specific agent, species and genotype of the individual, stage of development, time of exposure, and others. With the multitudes of children born into the world with congenital heart disease, abnormal eyes and ears, cleft lip and palate, spina bifida, deformed limbs, and other anomalies, our challenge is to understand their biologic basis so that we can work towards their elimination.

References

Gruber, G.B. Studien zur Historik der Teratologie. *Zent allg Path* 105:219–237, 1963.

Smithells, R.W. The challenges of teratology. *Teratology* 22:77–85, 1980.

Conrad Wolffhart [Lycosthenes]

Prodigiorum ac ostentorum chronicon. Basileae: per Henricum Petri, 1557.

In his "Chronicle of Omens and Portents… from the beginning of the world up to these our present times," Conrad Wolffhart (1518–1561) presented a compendium with more than 1500 woodcuts in the text (two of them full-page, one double-page) of freaks, monsters, miracles, disasters, and the wonders of the world. Many of the woodcuts were by Hans Rudolf Manuel Deutsch (1525–1571) and David Kandel (1520–1592), among others. This anthology includes hundreds of accounts of omens and prodigies, and many of which are mythological, listed in chronological order, from the beginning of recorded history to that time. A remarkable book of the Renaissance, this volume contains a most complete collection of illustrations of medieval superstition. Its images show a great variety of human monsters of all types as well as of beasts and other curious animals both actual and mythical, floods, conflagrations, astronomical happenings including comets, earthquakes, and quirky meteorological events. Of particular visual interest is a double-page woodcut of sea monsters (one belching smoke, another crushing a human in a giant claw). Perhaps the most startling woodcut is that of a spaceship (complete with rows of portholes and looking like that of Flash Gordon), supposedly sighted over Arabia in 1479. The overall theme of the volume is that, rather than to be regarded as "freaks," these accounts in the apocalyptic tradition of that age were intended to show miraculous manifestations of divine power, and to serve as solemn warnings of coming events to transpire before the end of the world.

In the plate shown (p. 496), the figure on the left presents a pair of Siamese twins connected at the back, who were born in 1486 in Rohrbach, a village near Heidelberg. The figure on the right shows a more fanciful monster. Other illustrations are of various monsters and "prodigies," including children/twins born with two heads in 1487, one set at Padua and the other in Venice. Lycosthenes associated their births with a major earthquake in the first case, and a hen that died with a large abnormal egg inside in the second.

A Swiss humanist, philosopher, theologian, and professor of grammar, Wolffhart, who took the Greek name of Lycosthenes, borrowed a substantial portion of this chronicle of prodigies and portents from other works, including the *Liber de prodigiis* [Book of Prodigies] of Julius Obsequens (fl. @ 350 AD). Obsequens had been fascinated by many wonders of the world, including unidentified flying objects, and presented several accounts of such phenomena as, "something like a weapon or missile…" that rose from the earth with "a great noise… and soared into the sky." Lycosthenes also added material from a number of other writers.

Shortly before his death in 1559, and only 2 years after the publication of this work, Pope Paul IV (1476–1559) an ultraconservative, rigid supporter of the Inquisition, placed Lycosthenes' entire literary output on the *Index librorum prohibitorum* [Index of prohibited books], a fact that helps to explain the scarcity of this work.

In 1581, Stephen Batman, a cleric of the Church of England, prepared an English translation, *The Doome, warning to all men to the Judgement…*, in which he extended Lycosthenes' chronicle of wondrous accounts. In his work, Batman presented only about 90 woodcuts, some of which were copied from Lycosthenes. As an ardent supporter of the reformation in Britain, Batman presentation of prodigies and portents was motivated, as an argument, using the visible language of God, a portion of his great book of nature, against an alliance of the English monarchy with France, and the reassertion of Spain under Phillip II (1527–1598) (who had launched the Armada against England) with dominance of the Papacy to extinguish Protestantism. Batman intended his re-editing of Lycosthenes as a compendium or reference to the warnings sent to mankind of the coming danger they might avoid if proper action was taken to be obedient to the will of God.

References

Garrison Morton 534.50; see Osler 623, 627, 628.

[Lykosthenes]. *The Doome, warning all men to the iudgemente …* . Translated by Stephen Batman; London, Ralphe Nubery assigned by Henry Bynneman, 1581.

Batman, S. *The doome warning all men to the judgement (1581)…: a facsimile reproduction* — with an introduction by John R. McNair. Delmar, New York; Scholars' Facsimiles & Reprints, 1984.

Obsequens, J. *Liber der prodigiis*. Venetia; Aldus Manutius, 1508.

PLATE I FROM WOLFFHART [LYCOSTHENES] 1557

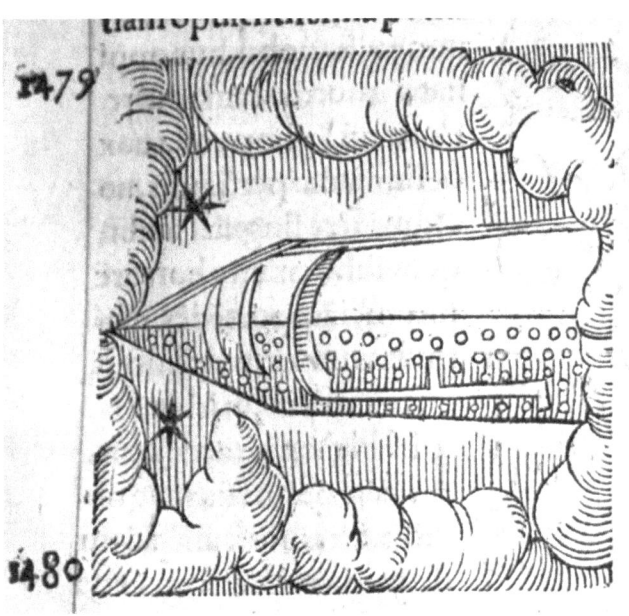

IN Arabia cometa in modum trabis acutiſsimæ, ac uarijs quaſi punctis diſtinctæ, cum falce phœnaria uiſus eſt. Eodem anno totam Carinthiã uaſtarũt Turcę. Cruciferi cõtra Polonos bellum pararunt. In Vngaria fœdus inter Matthiam ac Vladislaum renouatum.

IN Creta inſula (ut à Coccio Sabellico lib. 1.

PLATE FROM WOLFFHART [LYCOSTHENES] 1557

Johann Georg Schenck von Grafenberg

Monstrorum historia memorabilis, monstrosa humanorum partuum miracula, stupendis confirmat onum formulis ab utero materno enata vivis exemplis, observationibus, & picturis, referens. Francofurti, ex officina typogprahica, Matthiae Beckeri, impeus duae Theodoride de Bry, 1609.

Schenck von Grafenberg (birth date unknown–1620), the son of Johann Schenck (von Grafenberg; 1530–1598), was physician to the Count of Hanau-Lichtenberg, and practiced in Hagenau (Alsatia). He published one of the earliest works on teratology. The work is divided into two sections, the first on humans and the second on animals. Most of the abnormalities shown are from the sixteenth century, and for each, the place and date of birth with relevant medical details are given. The engravings are by Theodor de Bry (1528–1598).

Shown on the Plate are several varieties of Siamese twins. That on the left (Number 45) has the trunk, arms, and legs of the twin attached to its lower pectoral region and upper abdomen. The other two sets (47 and 48) are joined at the chest and/or upper abdomen, Set 47 is taken from Lycosthenes (1557), as were others in this collection.

Schenck von Grafenberg also prepared a work on petrification in parts of the human body, such as gallstones, kidney stones, including a petrified fetus (*Lithogenesia...,* [Stone formation...] 1608). He also edited several works of his father, including the monumental *Observationum medicarum...* [Medical observations] (1609). In addition, with Melchiore Guilandini [Wieland] (1519–1589), a German physician and botanist, he produced a catalogue of the botanical gardens at Padua (1608), of which Guilandini was director.

References

Krivatsy 10405; Waller 8595; Wellcome I, 5832.

Guilandini, M. and Schenck von Grafenberg, J.G. *Hortus Patavinus. Cui accessere... conjectanea synonimica plantarum erusditissima....* Frankfurt, Mathaeus Becker, Johann Theodor, and Johann Israel de Bry, [1608].

Schenck von Grafenberg, J.G. *Lithogenesia; sive, De microcosmi membris petrefactis, et de calculis eidem microcosmo per varias matrices innatis....* Francofurti, Ex officina typographica Matthiae Beckeri, simptibus viduae Theodori de Bry, & duorum ejus filiorum, 1608.

Schenck von Grafenberg, J.G. *Biblia iatrica; sive, Bibliotheca medica macta, continuata, consummata, qua velut favissa....* Francofurti, Typis Joannis Spiessii, sumptibus Antonii Hummii, 1609.

Schenck von Grafenberg, J.G. *Monstrorum historia memorabilis, monstrosa humanorum partum miracula....* Francofurti, Ex officina typographica Matthiae Beckeri, impensis viduae Theodori de Bry, & duorum ejus filiorum, 1609.

Schenck von Grafenberg, J. *Observationum medicarum, rararum, novarum, admirabilium, & monstrosarum...* Francofurti, N. Hoffmanni..., 1609.

Wolffhart, C (Lycosthenes). *Prodigiorum ac ostentorum chronicon....* Basileae, per Henricum Petri, 1557.

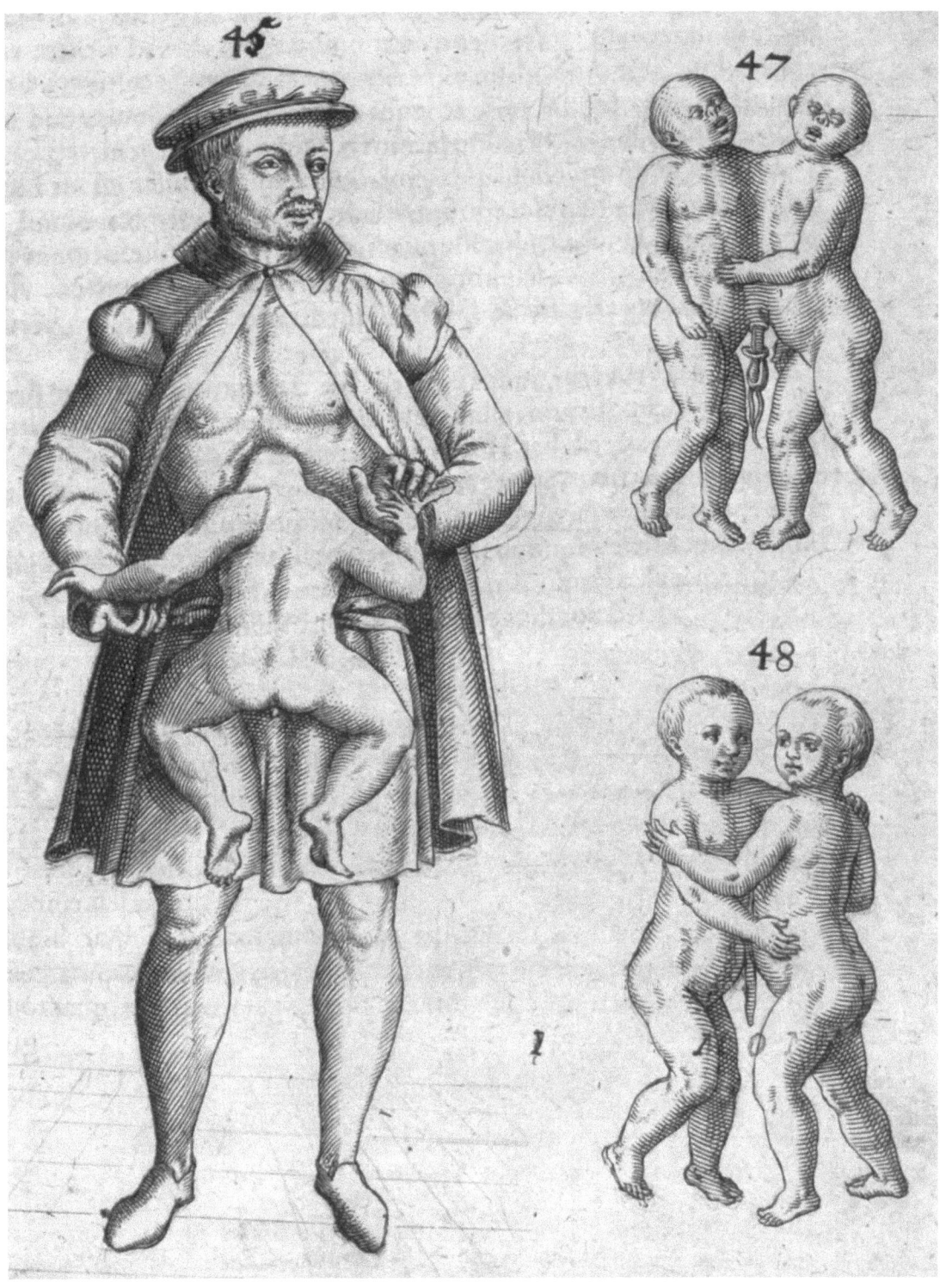

FIGURES 45, 47 AND 48 FROM VON GRAFENBERG, 1609

Fortunio Liceti [Licetus]

De monstrorum caussis, natura et differentiis libri duo .... Patavii, Apud. Paulum Frambottum 1634. [*Colophon*, 1633]

Liceti (1577–1657), Professor of Medicine and of Philosophy at the University of Padua, was one of the earliest to attempt to classify congenital malformations and deformities. His original edition appeared in 1616; however, it included no illustrations. This first illustrated edition, which appeared two decades after the original, includes over seventy copper-plates that depict both actual and imaginary cases. A so called "masterpiece in credulity," this work was reprinted repeatedly. The illustrations depict a host of birth defects and "monsters" or "prodigies," as they were called, both human and animal, including combinations of the two, such as satyrs, bird-men, elephant-men, and so forth. Also included are instances of Siamese twins, hermaphrodites, and cyclops. Liceti emphasized that monstrosities should inspire wonder, not revulsion. He opposed the idea that monsters were a consequence of divine wrath, noting that the Latin verb *monstare* means, "to show;" hence, he argued, they were creatures to be displayed because of their rarity. The frontispiece displays a figure, perhaps that of the author, atop a group of "monsters", both real and imagined, standing in acrobatic manner on one another's shoulders, and displaying a banner with the title of a book. It is signed by the artist Giovanni Battista [Giambattista] Bissoni (1576–1634) and MD Sculps[it].

Liceti's volume is a collection of everything which the imagination of both the ancients and the moderns had been able to relate to human and animal monstrosities. As with the original edition, the cases are presented in chronological order. The illustrations depict curious and fantastic creatures, and most probably was a means of marketing the book, which became very popular. Many of the engravings (many repeated) are copied from broadsheets and sixteenth century books by others. In the plate shown (p 80), with two sets of Siamese twins, the report is copied from that of Lycosthenes (1557), although the drawings are newly made. On the left are twins, born in 1486 in a village near Heidelberg, that are attached at the lower thorax and upper abdomen. On the right, the twins joined at the back were born in Rome in 1493, they lived but a few days, rather than surviving into early childhood as illustrated. Another example of the fascination with monstrous births is depicted in the plate shown (p 182). In 854, in the Empire of Lothary, Duke of Saxony, a woman was reported to have delivered the human-dog pair shown on the right. In the

original version of Lycosthenes (1557), he records that "Immediately after this," the Emperor's death ensued.

A third edition of this work (1665) was edited by Gerhard Blaes [Blasius] (1626–1682), professor of medicine in Amsterdam, who amplified the text to include new varieties of "monsters." Blaes included a new preface, commentaries, and a valuable appendix listing famous freaks discovered after the publication of the original edition of Licetis' work, many of which derive from Thomas Bartholin's (1616–1680) writings. In Blaes' work, the frontispiece displays what is again perhaps Liceti, drawing aside a curtain to reveal a collection of monstrosities around a five-breasted Venus on a pedestal.

Liceti, born near Geneva, is alleged to have received his name Fortunio from having survived a very premature birth. He is said to have measured 5 zoll (inches) at birth. In his comic novel Tristram Shandy, Laurence Sterne (1713–1768) described Liceti's premature birth, and his care in an early incubator:

> And for Licetus Fortunio ... all the world knows he was born a foetus. [He] was no larger than the palm of the hand, but the father, having examined it in his medical capacity, and having found that it was something more than a mere embryo, brought it living to Rapallo, where it was seen by Jerome Bardi and other doctors of the place. They found it was not deficient in anything essential to life, and the father, in order to show his skill, undertook to finish the work of nature and to perfect the formation of the infant by the same artifice as is used in Egypt for the hatching of chickens. He instructed a wet-nurse in all she had to do, and having put his son in an oven, suitably arranged, he succeeded in rearing him, and in making him take on the necessary increase of growth, by the uniformity of the external heat, measured accurately in the degrees of the thermometer, or other equivalent instrument.
>
> (Sterne 1759, 1967, pp. 203–204)

Following his graduation from the University of Bologna, Liceti became professor in several disciplines and universities (logic and Aristotelian physics at Pisa, philosophy at Padua and Bologna, and later the theory of medicine at Padua). He wrote a number of works on embryology (Liceti 1602, 1618, 1630). A proponent of spontaneous generation, Liceti assembled a formidable array of scriptural and classical authority in support of this concept with an exhaustive Aristotelian explanation of its mechanism (Liceti 1602, 1616, 1618). He also was one of the opponents of William Harvey's (1578–1657) concept of the circulation of the blood (1628), and maintained a theory of two circulations, venous and arterial, blood being transmitted through the coronary veins. In 1937, when surrealistic art was popular, a French translation of Liceti's work on monsters was published by François Houssay.

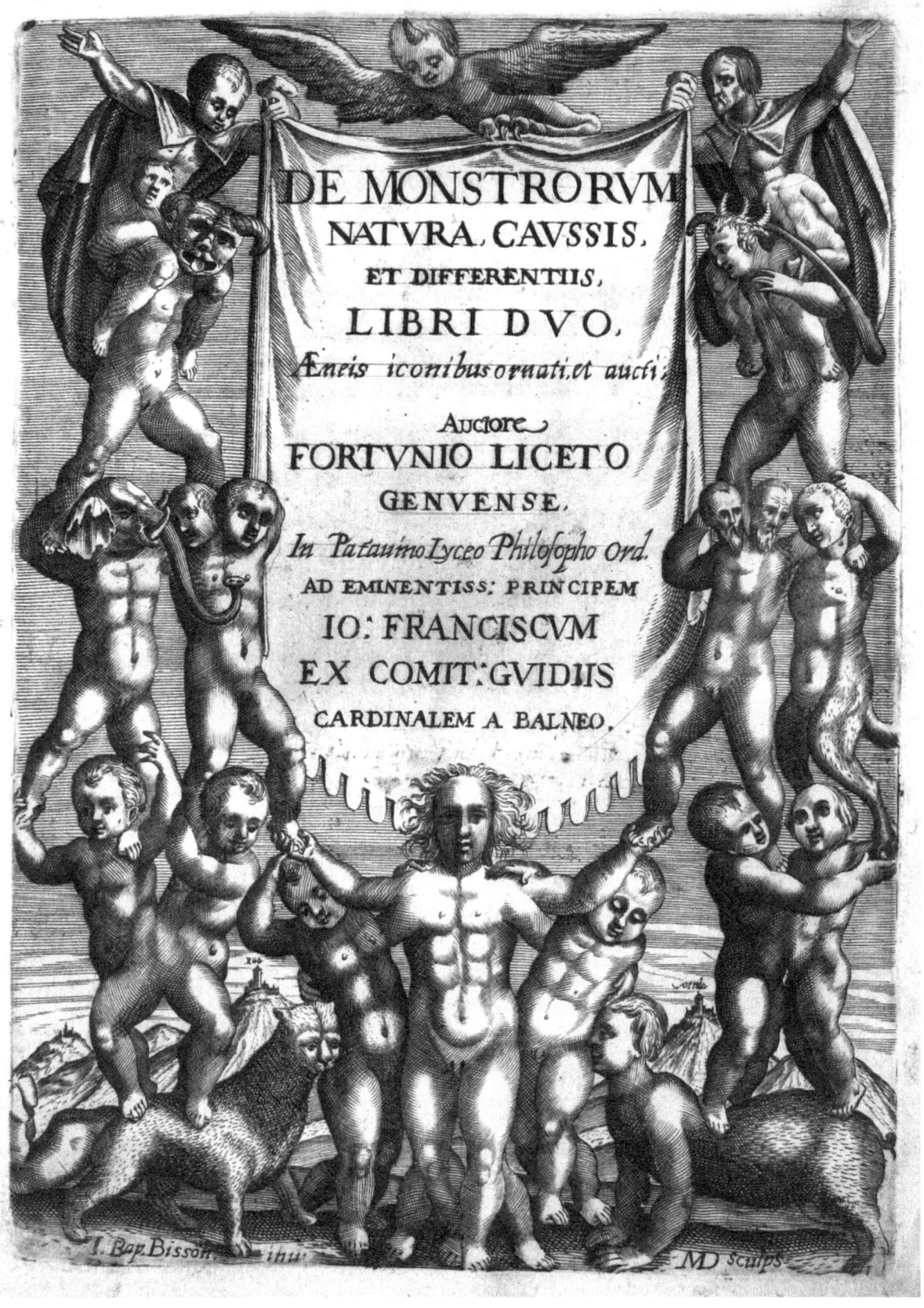

FRONTISPIECE FROM LICETI [LICETUS], 1634

References

See Garrison Morton 534.52; Krivatsy 6958.

Bartholin, T. *De armillis Veterum scheidion. Accessit Olai Wormii De aureo cornu Danico ad Licetum responsio....* Amstelodami, Sumpt. J. Henrici Wetstenii, 1676.

Blaes [Blasius], G. *Anatome animalium*, Amstlodami, J.A. Someren, 1681. (GM 296).

Harvey, W. *Exercitatio anatomica de motu cordis et sanguinis in animalibus.* Francofurti, Sumpt.Guilielmi Fitzeri, 1628.

Houssay, F. *De la nature, des causes, des différences de monstres d'après Fortunio Liceti.* Paris, Collection Hippocrate, 1937.

Liceti F. *De ortu animae humanae libri tres.* Genoa, In aedibus Giuseppe Pavoni, 1602.

Liceti, F. *De monstrorum caussis, natura, et differentiis libri duo.* Patavii, Apud. Casparem Crivellarium, 1616. (Krivatsy 6757).

Liceti, F. *De perfecta constitutione hominis in utero liber unus....* Patavii, Apud. Petrum Bertellium, 1616.

Liceti, F. *De animarum coextensione corpori libri duo, In quibus ex rei natura, consulto semper Aristotele....* Patavii, Apud. Petrum Bertellium, 1616.

Liceti, F. *De Spontaneo Viventium ortu libb: quatuor, in quibus generatione animantium....* Udine, F. Bolzetti, 1618.

Liceti, F. *Allegoria peripatetica de generatione, Amicitia, et Privatione in Aristotelieum aenigma Elia Lelia Crispis.* Patavii, Apud. Gasparem Criuellarium, 1630. (Osler 5050).

Liceti, F. *Encyclopaedia ad Aram Pythiam Pvblili Optatiani Porphyrii.* Patavii, Apud. Casparem Cruellium, 1630.

Liceti, F. *Qui monstra quadam nova & rariora ex recentiorum scriptis addidit.* Editio Novissima [by Blaes, G.]. Amerstdam, Andreas Fisius, 1665.

Liceti, F. *De monstris. Ex recensione Gerardi Blasii, qui monstra....* Amstelodami, Sumpt. Andreae Frisii, 1665. (Krivatsy 6959).

Liceti, F. *De la nature, des causes, des differences, des monstres/*French translation by F. Houssay. Paris, Hippocrate, 1937.

Pilcher Catalogue, p. 135.

Nardi, G. *Apologeticon in Fortunii Liceti mulctram, vel de duplici calore* [Florentiae, Typis Amatoris Massae...], 1638. (Krivatsy 826).

Sterne, L. *The life and opinions of Tristram Shandy, gentleman.* York, 1759. (New York, Airmont Publishing Company, Inc., 1967).

Wolffhart, C [Lycosthenes]. *Prodigiorum ac ostentorum chronicon....* Basileae, per Henricum Petri, 1557.

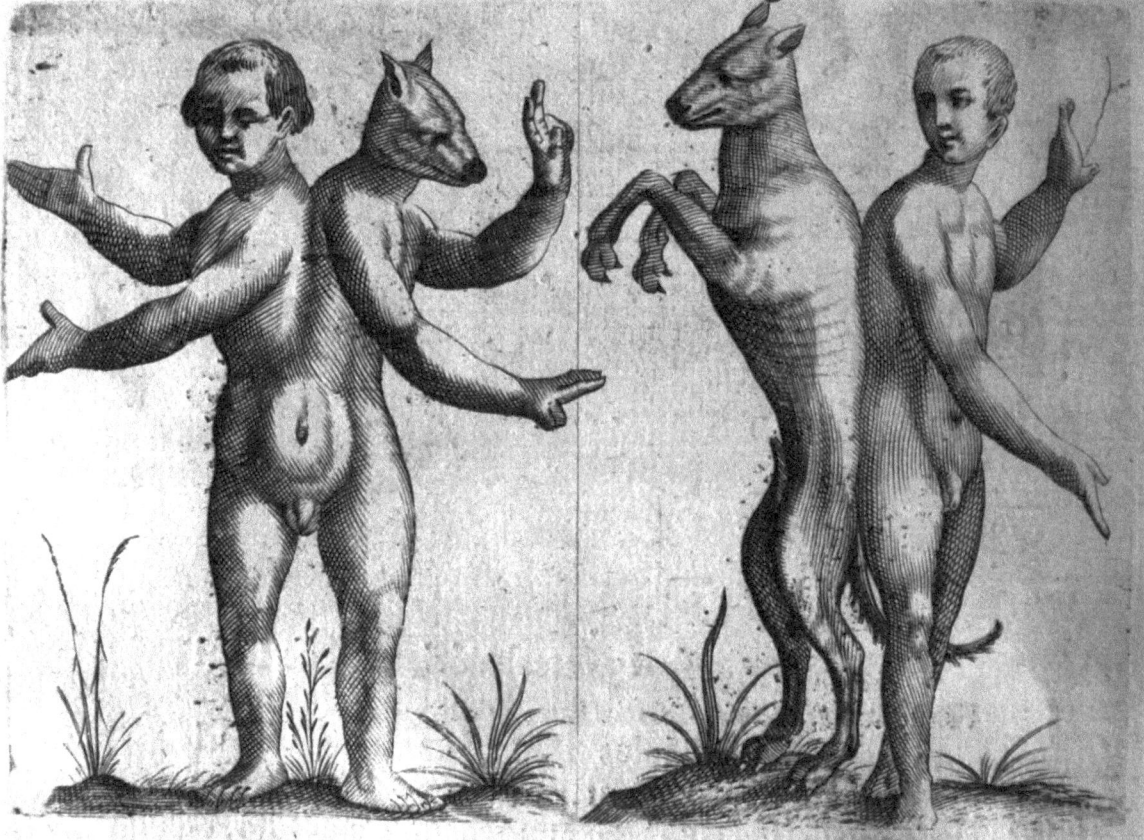

Figures from Liceti [Licetus], 1634

Ulisse Aldrovandi [Aldrovandus]

Monstrorum historia, cum paralipomenis historiae omnium animalium. Bartholomaeus Ambrosinus (ed)... volumen composuit. Marcus Antonius Bernia in lucem edidit .... Bononiae, Typis Nicolai Tebaldini, 1642.

Aldrovandi's (1522–1605) treatise on monsters and prodigies, illustrated with woodcuts of human and animal malformations, as well as anomalous plants, forms a portion of his encyclopedic work on natural history. Although comprising thirteen volumes, only the first three on birds and a fourth on insects were published during his lifetime. In this work, published almost four decades after his death, Aldrovandi presented both observed and mythical descriptions of developmental anomalies, including chimeras and unicorns. The title page presents an illustration of Aldrovandi with curling mustache, and *puttos* that display globes with figures and the captions *sapientiae symbolum* [wise symbols], *fertilitatis indicium* [fruitful information/evidence], *firmitudo bene consuitorum* [strengthen good customs], *aevi perennitas* [lasting forever], and so forth. Among the over 450 woodcuts are 18 that illustrate the fetus *in utero* in various positions, and the placenta. The figures on page 60, illustrating the unnatural position of twins (left) and a fetus with left arm extended (right) are not unlike those of Jacob Rueff (1500–1558) (for instance Rueff, 1554, p. 34 and 37). Page 64 illustrates on aborted fetus, shown as an adult, with placenta and the three-vessel umbilical cord.

Following the lead of Aristotle (384–322 BCE), in his embryologic study of the chick he opened and examined the egg during its incubation period day by day for 20 days, to describe in detail embryonic development. By this technique, Aldrovandi established that the heart is formed in the vitelline sac, rather than in the albumin, as was believed at that time. He also made important observations on the development of teratologic anomalies. Although not profound contributions to the progress of embryology, his studies pointed future investigators to the need for personal direct observation and confirmation, rather than relying on the descriptions of Aristotle, Galen, or others. This laid the groundwork for the subsequent studies of Volcher Coiter (1534–1576; 1572, 1574), Fabricius ab Aquapendente (1533–1619; 1604, 1621), William Harvey (1578–1657; 1651), and Marcello Malpighi (1628–1694; 1673).

Aldrovandi, called the Pliny of his age and the "Bolognese Aristotle," was a classical Renaissance scholar. Born of noble parentage in Bologna, following an adventurous tour of Spain, he returned to Bologna to graduate in medicine and philosophy in 1553. Several years earlier, in 1549, he had been accused of heresy for espousing anti-trinitarian beliefs. He remained in custody in Rome for almost a year, before being absolved. During this time, he became acquainted with the physician Guillaume Rondelet [Rondeletius] (1507–1566) who was in the process of preparing a major work on fishes of the Mediterranean (Rondelet 1554–1555). This experience stimulated Aldrovandi to study natural history. As many other naturalists, to amass a complete collection, he became gripped with a passion to gather and possess specimens of every species known. Holding that "nothing is sweeter than to know all things," he commenced gathering specimens for his own "cabinet," which grew into a major museum with over 18,000 specimens. As a student of many aspects of natural history, he traveled widely to collect the unusual and obscure. Returning to Bologna, in his academic post, Aldrovandi stimulated many students, and in 1561 was appointed full professor of logic and philosophy, as well as of medicine. He wrote on many aspects of natural history, and worked to establish the botanical gardens of Bologna, the *Orto Botanico dell' Università di Bologna*, one of the first in Europe. During this time he prepared an *Antidotario* [Book of Antidotes] an official pharmacopeia (Aldrovandi 1574). This work on monsters was published posthumously by Bartolommeo Ambrosini (1588–1657) who added a number of cases he had observed, and who had succeeded Aldrovandi as director of the Botanical Garden in Bologna. Based upon the specimens and drawings in his collection, many of the illustrations for the copper plate engravings were prepared by the prolific artist to the Medici court, Jacopo Ligozzi (1547–1627) (Aldrovandi 1559–1667).

It was during this period of the Renaissance that modern descriptive and experimental science was being born through the work of investigators such as Aldrovandi. Both the Swedish classifier Carl von Linné [Linnaeus] (1707–1778) and the French naturalist Georges Louis Leclerc *Comte* de Buffon (1707–1788) held him to be the founder of studies in natural history. Aldrovandi assembled a spectacular collection of plants, both living for his botanical garden, and dried for his herbarium. He willed his vast collection of botanical and zoological specimens to the Senate of Bologna. These were conserved in the *Palazzo Pubblica* [Public palace or large building], and then later in the *Palazzo Poggi* [the palace of the Poggi brothers, Alessandro and Cardinal Giovanni]. In the early twentieth century a number of Aldrovandi's specimens were brought together at the *Palazzo Poggi*. Aldrovandi also wrote a survey of the statuary in Rome (1556), and several works in ornithology in which he described their zoological and physiological characteristics. The *Dorsa Aldrovandi* on the moon was named in his honor.

TITLE PAGE FROM ALDROVANDI [ALDROVANDUS], 1642.
COURTESY AMS HISTORICA, UNIVERSITÀ DI BOLOGNA

References

Garrison Morton 534.53; Krivatsy 187.

Aldrovandi, U. *Le antichita de la citta di Roma*. In Venetia, Appresso Giordano Ziletti…, 1556.

Aldrovandi, U. *Opera omnia. 13 vols.* Bononiae, J.G. Bellagamba…, 1559–1667. (GM 290).

Aldrovandi, U. *Antidotarii Bononiensis, siue De vsitata ratione componendorum, miscendorumq[ue] medicamentorum, epitome…*. Bononiae, Apud Ioannem Rossium, 1574.

Aldrovandi, U. *Ornithologiae, hoc est de avibus historiae, libri XII…*. Bononiae, apud Franciscum de Franciscis Senensem, 1599–1603.

Aldrovandi, U. *De animalibus insectis libri septem, cum singulorum inconibus ad viuuum expressis*. Bonon, Apud Ioan. Bapt. Bellagambam, cum consensu superiorum, 1602.

Aldrovandi, U. *Ornithologiae hoc est de avibus historiae libri XII… tomus tertius, ac postremus*. Bononiae, Apud Ioannem Baptistam Bellagambam, MDCIII, 1603.

Coiter, V. *Externarum et internarum principalium humani corporis partium tabulae, atque anatomicae exercitationes observationesque variae, novis, diversis, ac artificiosissimis figuris illustratae, philosophis, medicis, in primis autem anatomico studio addictis summè utiles*. Noribergae, in officina Theodorici Gerlatzeni, 1572. (GM 464.1).

Coiter, V. *Lectiones Gabrielis Fallopii de partibus similaribus humani corporis, ex diversis exemplaribus a Volchero Coiter… collectae; his accessere diversorum animalium sceletorum explicationes iconibus… illustratae… autore eodem Volchero Coiter*. Noribergae, in officina T. Gerlachii, 1575. (GM 284).

Fabricius ab Aquapendente, G. *De formato foetu. Venetiis. per Franciscum Bolzettam 1600*. Colophon, Laurentius Pasquatus, 1604.

Fabricius ab Aquapendente, G. *De formatione ovi, et pulli tractatus accuratissimus*. Patavii [Padua], ex officina Aloysii, Bencii, 1621.

Harvey, W. *Exercitationes de generatione animalium. Quibus accedunt quaedam de partu: de membranis ac humoribus uteri: & de conceptione*. Londoni, Octavian Pulleyn, 1651.

Malpighi, M. *Dissertatio epistolica de formatione pulli in ovo*. London, John Martyn, 1673.

Rondelet, G. *Libri de piscibus marinis, in quibus verae piscium effigies expressae sunt… (Universae aquatilium historiae pars altera, cum veris ipsorum imaginibus…)*. 2 *vols*. Lugduni, apud Matthiam Bonhomme, 1554–1555. (GM 282).

Rueff, J. *Ein schön lustig Trostbüchle von dem Empfengknussen und Geburten der Menschen*.

Tiguri, Apud Frosch [overum], 1554. (GM 6141).

Ruggieri, M. and A. Polizzi. From Aldrovandi's "Homuncio" (1592) to Buffon's girl (1749) and the "Wart Man" of Tilesius (1793): antique illustrations of mosaicism in neurofibromatosis? *J Med Genet* 40:227–232, 2003.

64   Vlyfsis Aldrouandi

Abortus cum membranis, & vafis
vmbilicalibus.

Abor.

PAGE 64 FROM ALDROVANDI [ALDROVANDUS], 1642

[Miscellanea curiosa sive ephemeridum medico-physicarum Germanicarum Academiae Caesareo-Leopoldina Naturae Curiosorum]

Ad Splendiss. S.R.I. Academ. Natur. Cur. Helianthum, Paeonis De Rupicaprarum Internaneis et Aegagropilis, succincta dissertation. Norimbergae, 1683.

The *Miscellanea Curiosa* [a mixture of curiosities] of the German Academy of Sciences Leopoldina, has gone under a number of names which changed over the centuries. Among its 185 "Observations," this first volume of the second decade contains a number of teratology items, with descriptions of various anomalies and pathological conditions. Those observations by D Johannis de Muralto describe complicated obstetrical cases. Plate 26 presents ventral and dorsal views of an infant with incomplete twinning of the head, and a large defect over the thoracic and lumbar spine. Also included are a number of botanical descriptions, with engraved plates and some woodcut text illustrations.

With its long Baroque title, *Collegium Naturae Curiosorum...*, the academy is the oldest for scientific research in Germany, and the first in the world for scientific medicine. The prime mover in the formation of the society in 1652, and its first president, was Johann Laurenz Bausch (1605–1665), *Stadtphyiscus* [city physician] of Schweinfurt, who had studied medicine in Italy and was impressed with the contributions of the *Accademia dei Lincei* in Rome. The publication, which first appeared in 1670, was modeled on the *Philosophical Transactions of the Royal Society of London*. The Preamble of the Academy (1662) presents its purpose, "The glory of God, the enlightenment of the art of healing and the benefit resulting from this for our fellow man be the goal and the only guide ...." This included, "the advancement of medicine and pharmacy through observation; by presenting observations in monographs, and communicating them to the members for correction and further elaboration". Thus, rather than a Scientific Society in the usual sense of holding meetings and conducting business, its chief function was to publish reports of original research by its members, and other discoveries of importance to medicine.

Seeking to achieve the highest possible reputation, the journal was dedicated to the Habsburg Emperor Leopold I (1640–1705) of Vienna, who guided the Holy Roman Empire for more than four decades (1658–1705). In addition to music and the arts, Leopold had great interest in science. In 1677, he formally recognized the Society, accepting the role of patron, and elevating it to the *Sacri Romani Imperatoris Academia Caesaro-Leopoldina Naturae Curiosorum*. The *Ephemerides* [journal], which has survived for three and a half centuries, strove to improve the understanding of dis-

ease and enhance the medical care of patients. In the early twentieth century, the emphasis of the Society was broadened to include essentially all aspects of natural philosophy. *Nunquam otiosus* [never idle] was its motto, and *Soli Deo Gloria* [give Glory only to God] its insignia. The Society chose as its symbol the ship *Argo* from Greek mythology, in which Jason, leader of the Argonauts, sailed in search of the Golden Fleece, the latter signifying scientific truth. Originally, the Society headquarters with its library and collections was located in the city or town of its president; however, in 1878 Halle became its permanent home.

The work and import of the *Collegium* typifies the rise in science during the seventeenth century. To a great extent, existing apart from the universities, members of these societies contributed to the general enlightenment, developed the scientific laboratory, perfected instruments of mensuration, and originated the experimental methods of scientific study. With the Royal Society of London and Rome's *Accademia...*, societies such as the *Collegium* became the *Kulturträger* [bearers of culture] of the second half of the seventeenth century.

References

Büchner, A.E. *Academiae Sacri Romani Imperii Leopoldino-Carolinae naturae curiosorum historia.* Halae Magdeburgicae, Litteris et impensis Ioannis Iustini Gebaueri, 1755.

Büchner, A.E. *Sacrae Caesareae majestatis Mandato et privilegio legas.* Halle, 1776.

Carutti, C.D. *Breve storia della Accademia dei Lincei.* Roma, R. Accademia coi tipi del Salviucci, 1883.

De Muralto, D.J. Miscellanea Curiosa sive Ephemaridum Medico-Physica Academiae Naturae Curiosorum, Sive Ephemeridum Medico-Physicarum Germanicarum Academiae Naturae Curiosorum, Decuriae II. Annus sextus et septimus, Anni MDCLXXV & MDCLXXV. Continens Celeberrimorum Virorum, Tum Medicorum, Tum Medicorum, tum aliorum Eruditorum, in Germania & extra eam, Observationes Medicas, Physicas, Chymicas, Nee non Mathematicas. Cum Appendice Anatomico-Botanico-Chrurgicas. Norimbergae, Wolfgangi Mauritii Endteri. 1675 & 1676.

Ornstein, M. *The role of the scientific societies in the seventeenth century.* New York, 1913, pp. 198–205.

BM(NH): I, 4, under "Academia Caesarea Leopoldino-Carolina Germanica Naturae Curiosorum" lists Decuriae II, Ann. I-X in 5 volumes, 1683–1692. Also see: BMC(Readex): 10, 436–437 for listing of the Academy's publications.

Sprat, T. *The history of the Royal-Society of London, for the improving of natural knowledge....* London, Printed by T.R. for J. Martyn and J. Allestry printers to the Royal Society, 1667.

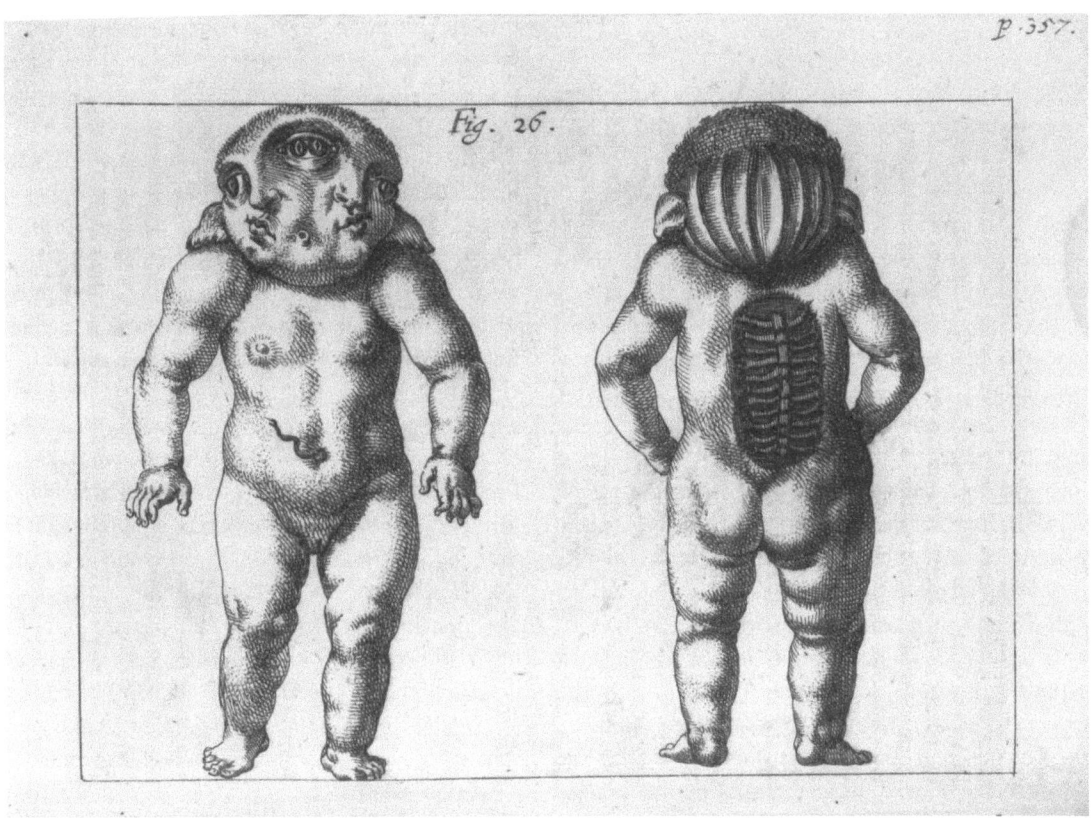

PLATE 26 FROM THE MISCELLANEA CURIOSA [A MIXTURE OF CURIOSITIES] OF THE GERMAN ACADEMY OF SCIENCES LEOPOLDINA 1683

Paul Portal

La Practique des accouchemens soutenue d' un grand nombre d' observations .... Paris, De l'imprimerie de Gabriel Martin, 1685.

Portal (1630–1703), a native of Montpellier, commenced his surgical studies there before moving to Paris, where he studied under François Mauriceau (1637–1709) and Pierre Moreau at the College of France. In 1657, he became *chirurgien accoucheur* of the obstetrical service at the *Hôtel Dieu*, Paris, a post he held for 6 years. During this period about 100 pregnant women were admitted every month. The obstetrical service initially located in the basement, with four or more patients per bed, was under the direction of a nun, *La dame des accouchées*, assisted by several midwives. The consulting male *accoucheurs* were responsible for prescriptions, bleeding, delivery of syphilitic women, and assisting midwives with complicated labours and difficult deliveries. Following his 1663 departure from the *Hôtel Dieu*, his term of service having expired, as *maître chirurgien* [master surgeon] he devoted himself to the field of obstetrics until his death. Based on this experience, he made many contributions to obstetrical science, including demonstrating that podalic version could be done with one foot. He also taught that face presentation usually ran a normal course. Portal, one of the three dominant figures in seventeenth century obstetrics together with Mauriceau and Hendrik van Deventer (1651–1724), taught that dilation of the cervix be left to nature, and that the *accoucheur* adopt a policy of "watchful waiting." Portal also included the first depiction of a placental mole, which he taught could be removed by an intra-uterine finger manipulation.

Portal's work consists of two parts: an initial six chapters on natural and preternatural courses of labor and delivery, followed by 81 "observations" or case reports with astute analysis. For instance, in observation XL, Portal described the delivery in 1671 of an *enfant monstrueux* [monstrous child] (see figure). Illustrated are a kind of monk's cap (A), chorionic and amniotic membranes connected to the infants head (B), inclined forehead (E), the umbilical cord (F), and a distorted hand (G) and foot (H). An additional plate presents another view of this infant. Other plates illustrate imperfect fetuses. A supporter of the doctrine of preformation, the last plate shows a small human figure, a homunculus [little man] representing an embryo, within a large egg. The artist was Reuel while the engraver was Lefebure. The first English edition was published in 1705.

Portal also recognized the importance of low placental implantation, i.e., placenta praevia. In a rather heroic case report (No LXIX), Portal wrote, "January 11th, 1679, I was called out in the morning at 4 O'clock, to deliver a gentlewoman... (she) being about 8 months with child, was seized with a most violent flux of blood, which having continued for 12 or 12 days, she was reduced to a miserable condition. Upon search made with my fingers (well greased) I found the whole vagina or passage filled with clods of coagulated blood, notwithstanding which the flux continued. As soon as I had brought out the clods of blood, I convey'd my fingers further into the inner orifice of the womb, which I found very thin and soft, and so wide, that I could put in three of my fingers foremost; I search'd with one finger first and found the after-burthen foremost, and closely joined round the inner orifice of the womb, which was the occasion of the excessive flux of blood; and as it had reduced the woman to a very low condition, so this join'd to the other circumstances, made me fear the life both of the woman and child... it was resolved to have her deliver'd, notwithstanding the great danger that must needs attend it; but considering her death was infallible, unless she was delivered, I went to work... (and) convey(ed) my hand into the inner orifice of the womb, where I again felt the after-burthen fasten'd to it... I peel'd it off by degrees, and brought it out; and then turning my hand again in the womb, the first thing I met with, was the navel-string, along which I guided my hand first to the child's belly, and then downwards to the thigh, and thence lower to the leg and foot, which I brought out and baptiz'd. Whilst I was pulling this foot the other follow'd, and the whole body after it, as has been observed frequently before. The child being quite alive, the parish priest of the Holy Cross (who had before administered the sacraments unto the mother) had the opportunity of baptizing it, tho' contrary to his and all our expectations. Immediately after the delivery, the woman recovering in some measure her sense... The second day... she complained of a tension and pain in her belly... yet after some time this woman recovered her health, except that 3 weeks after her lying-in, she lost the sight of one of her eyes...."

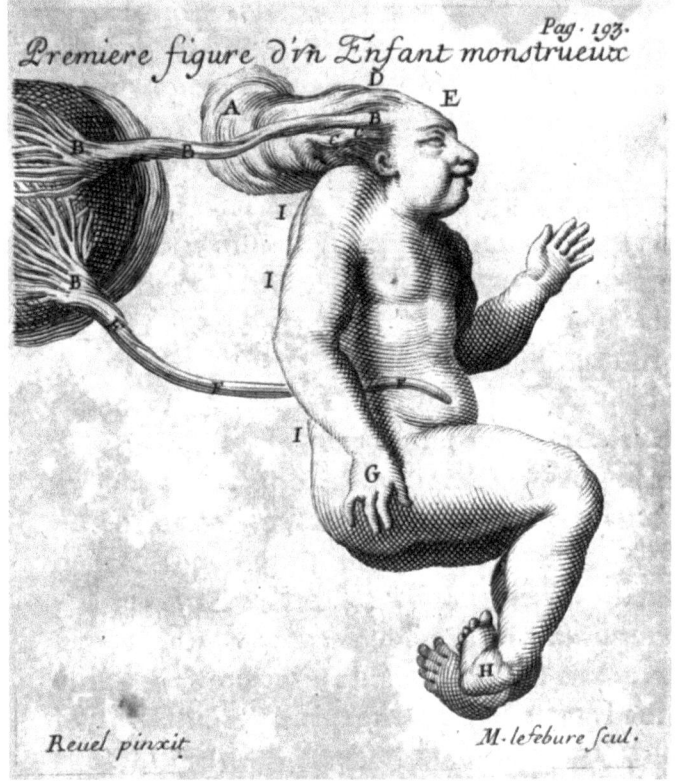

PLATES OF AN **enfant monstrueux** FROM PORTAL **1685**

References

Cutter & Viets, pp. 81–83; Garrison Morton 6148; Krivatsy 9200; Parkinson & Lumb 1948; Ricci, p. 322; Waller 7575.

Cumston, C.G. Paul Portal, his life and treatise on obstetrics, with reflections on the science of the obstetrical art in France from the Renaissance to the 18th century. *Am J Obstet Dis Women Child* 51: 778–804 and 52: 110–124, 1905.

Dunn, P.M. Paul Portal (1630–1703), man-midwife of Paris. *Arch Dis Child Fetal Neonatal Ed* 91: F385–F387, 2006.

Portal, P. *The complete practice of men and women midwives, or the true manner of assisting a woman in childbearing: illustrated with a considerable number of observations.* London, printed for H. Clark, S. Crouch… and J. Taylor, 1705.

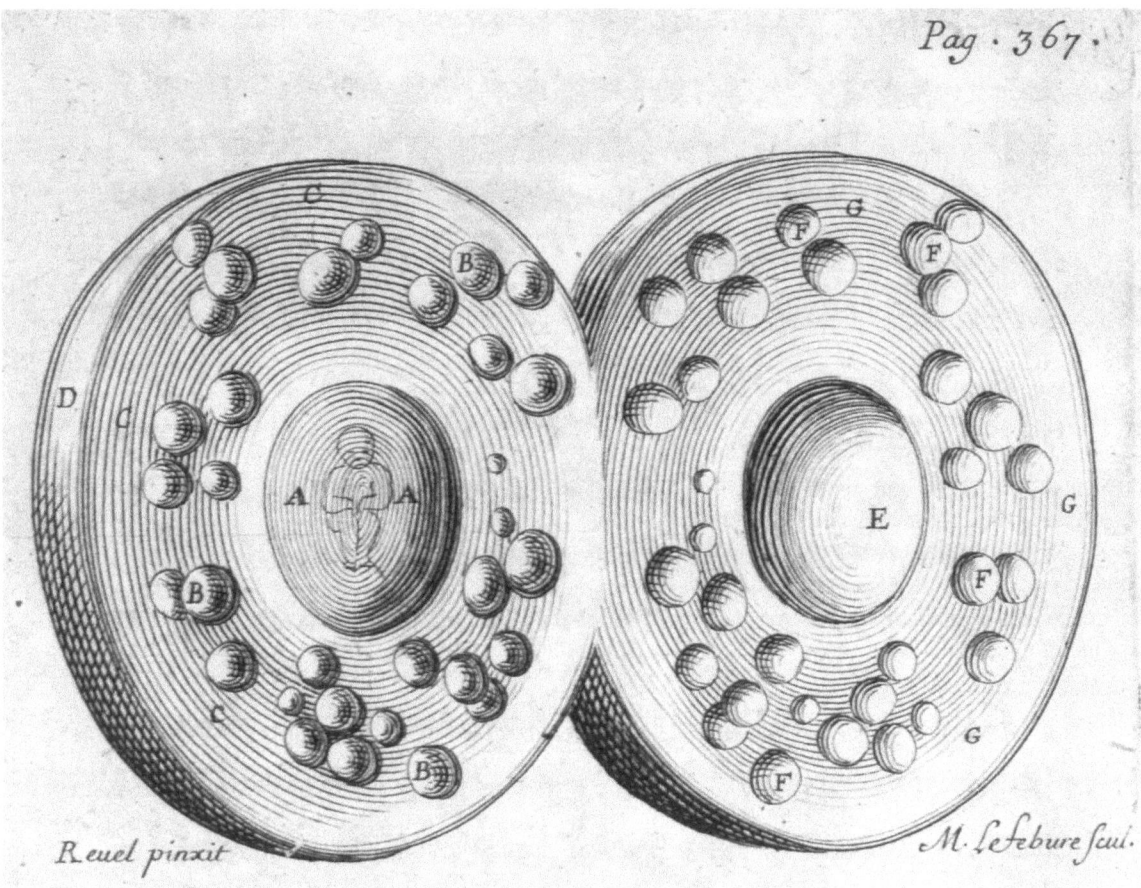

PLATE SHOWING A *ENFANT HOMUNCULUS (LITTLE MAN) INSIDE A HUMAN EGG* FROM PORTAL 1685

Jan Palfijn [Palfyn]

Description Anatomique des Parties de la Femme, qui server Generation; avec un Traite des Monstres, de leur causes, de leur nature, & de leur differences: Et une description anatomique...de deux enfans nes dans la ville de Gand, capital de Flandres le 28. Avril 1703. &c. &c... Lesquels ouvrages on peut considerer comme une suite de l'Accouchement des Femmes. Par Monr. Mauriceau. A Leide, Chez la veve de Bastiaan Schouten, 1708.

Palfijn (1650–1730), a Belgian anatomist-surgeon who was born in Kortrijk, Flanders, and practiced in Ypres, Paris, and Ghent, published an early notable work on the anatomy of the female organs of generation. The volume is in three parts with 11 engraved plates, six of which are folding; including the engraved frontispiece, numerous other engravings in the text of the second part, illustrated with three folding engraved plates. Published here with the first editions in French of his book on the deformed twins of Ghent, showing them from the front (Fig. 1) and the back (Fig. 2) illustrated with four plates, and of Fortunio Liceti's (1577–1657) treatise on monsters (1634), with many remarkable engravings from the original copperplates.

In April, 1703, in the city of Ghent, twins were born united at the lower part of their bodies which created such a stir that Jan Palfyn, because of his great reputation as an anatomist, was selected to dissect them before the magistrate, the directors of the college of medicine and the principal practitioners. A month later another monster was born at Ghent, also dissected by him, in which were found an imperforate anus and vagina together with a double uterus. Palfyn was requested to make a report of his findings, which was published in Flemish in 1703 together with a treatise on the circulation of the blood in the fetus. In 1708 Palfyn published the present French edition. This time the two papers were issued with a work on the anatomy of the female organs of generation and a work on monsters, being Palfyn's rendering into French of both Jan Swammerdam's (1637–1680) *Miraculum naturae* (1672) and Geerhard Blaes' [Blasius] (1626–1682) enlarged edition of Fortunio Licenti's [Licetus] (1577–1657) remarkable work on monsters (1665).... This thick volume was meant to form a supplement to François Mauriceau's (1637–1709) standard textbook on midwifery.

Palfyn invented the *tire-tête* [pull head] also called the *mains de fer* [hands of iron], two curved or spoon-shaped levers held in opposition, but unconnected, to assist in delivering the fetal head. Perhaps because of their difficulty in manipulation, they failed to become used widely by either *accoucheurs* or midwives.

Later workers added an articulation or joint, but within a few years they were replaced by the obstetrical forceps. Although Palfyn exhibited these at a meeting in Paris, they first were illustrated by Lorenz Heister (1683–1758) in his "General System of Surgery" (Heister 1724). Palfyn is remembered in Ghent by the Palfyn Medical Museum and Jan Palfyn Hospital.

References

Blake, p. 336; Cutter & Viets pp. 56–59, 60, 192.

Bibliographie des Geuvres de Jean Palfyn. Publication de l'Université de Gand, C. Vyt, 1888.

Hagelin, O. *The byrth of mankynde otherwyse named the womans booke....* Stockholm, Svenska Lakaresallskapet, 1990, pp. 90–93.

Heister, L. *Chirurgie, in welcher alles, was zur Wund-Artzney gehöret, nach der neuesten und besten Art, grundlich abgehandelt... werden.* Nürnberg, Bey Johann Hoffmanns seel. Erben, 1719. (GM 5576) (Blake p. 203).

Liceti, F. *De monstrorum caussis, natura et differentiis libri duo....* Patavii, Apud. Paulum Frambottum 1634. [*Colophon*, 1633]. (GM 534.52).

Mauriceau, F. *Des maladies des femmes grosses et accouchées. Avec la bonne et veritable méthode de les bien aider en leurs accouchemens naturels, ...* Paris, Chez Jean Henault et al..., Imprimeries de Charles Coignara, 1668.

Palfijn, J. *Nieuwe osteologie, ofte waere, en zeer nauwkeurige beschryving der beenderen van't menschen lichaem.* Ghendt, 1701.

Palfijn, J. *Anatomycke of ontleedkundige beschryving, rakende de wonderbare gesteltenis van eenige uyt....* Ghendt, By d'erfgenamen van Maximiliaen Graet, 1703.

Palfijn, J. *Anatomie du corps humain: avec des remarques utiles aux chirurgiens dans la pratique de leurs opérations....* A Paris, Chez Guillaume Cavelier..., 1726.

Palfijn, J. *Anatomia chirurgica. 3 vols.* Venezia, 1758.

Sondervorst, F.A. Palfijn, chirurgijn van vlaenderen. *Belg Tijdschr Geneesk* 7:337–348, 1951.

Stein, J.B. Jan Palfyn. *Med Record (New York)* 83:47–55, 1913.

Swammerdam, J. *Miraculum naturae; sive, uteri muliebris fabrica....* Lugduni Batavorum, apud Severinum Mathaei, 1672.

PLATE XVI FROM PALFIJN [PALFYN] 1708

Jacobus Denys [Denijs]

Verhandelingen over het ampt der vroed-meesters, en vroed-vrouwen: met aanmerkingen, derzelver Kunst raakende. Getrouwelyk ontdekkende, en leerende zeer noodige Handgreepen, om Baarende Vrouwen kort, en veilig te verlossen, daar de Kinders tegen-natuurlyk, en natuurlyk geleegen zyn ... . Te Leyden, Juriann Wishoff, 1733.

In this extensive Dutch midwifery with 40 chapters, Denys (1681–1741) used numerous case histories to illustrate theory and practice. The plates represent teratological fetuses—five in all, each of whose case history is given. Three of these (each of which is depicted in front and back views) have cephalic misdevelopment, one is hermaphroditic, while the last (the only one depicted in its entirety) has several deformities. Denys was state obstetrician and lecturer in midwifery at Leyden. He also described on benign growth in the vagina.

References

Blake, p. 115.

Denijs, J. *Heelkundige aanmerkingen over den steen der nieren, blaaze, en waterpyp; het snyden der zelven, mitsgaders over de blaas-steek.* Te Leyden, Johannes à Kerkchem, 1730.

Denys, J. *Observationes chirurgicae de calculo renum, vesicae, urethrae, lithotomiâ vesicae puncturâ; in quibus lithotomiae methodum.* Lugduni Batavorum, 1731.

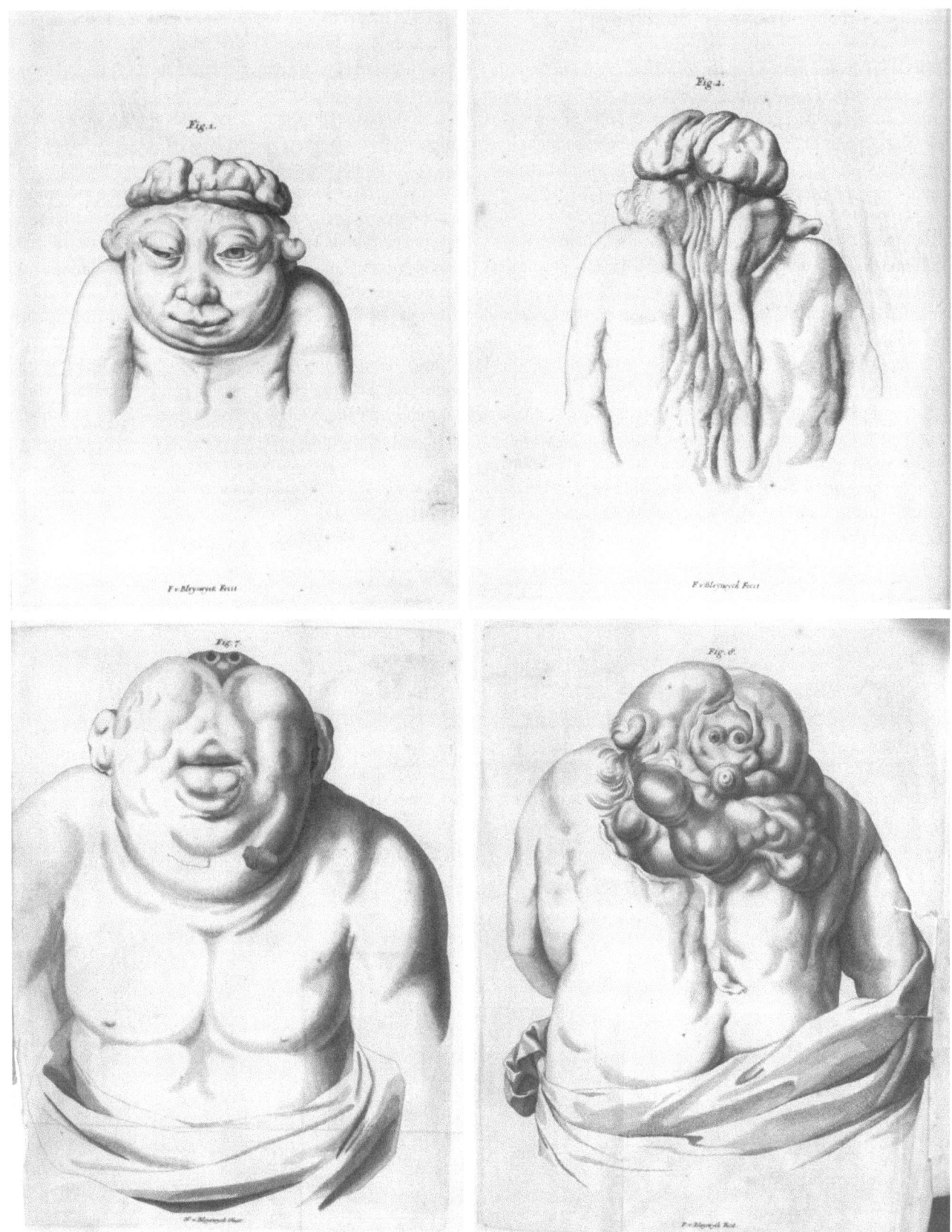

FIG. 1, 2, 7, & 8 SHOWING SHOWING FRONT AND BACK OF TWO PAIRS OF TERATOLOGICAL FETUSES FROM DENYS [DENIJS] 1733

Philipp Adolph Boehmer [Boehmeri]

Observationum anatomicarum rariorum fasciculus notabilia circa uterum humanum continens cum figures ad vivum expessis. with Observationum anatomicarum rariorum fasciculus alter notabilia circa uterum humanum continens cum figures ad vivum expessis. 2 volumes. Halae Magdeburqicac, apud Gebauerum, Joannem Iustinum, 1752–1756.

Boehmer (1717–1789) the son of the well known jurisconsult Justus Henning Boehmer (1664–1749), studied at Halle as a pupil of Friedrich Hoffmann (1660–1742), then practiced obstetrics at Strasbourg. Later, he obtained Johann Friedrich Cassebohm's (1699–1743) chair at the University of Berlin, and became director of the Medical Academy.

The work is divided into seven observations treating various pathological aspects of the female genitalia, including ovarian and tubal conception, several tumors, and deformities of the genital tract. The splendid title-page in red and black with a symbolic cartouche is by Gottfried August Gründler (1710–1775), who also executed the plates. Of interest is the description of a bi-corporal human monster depicted in a full-page engraving, while another is depicted on a composite plate.

Extrauterine pregnancy, whether in the Fallopian tube, ovary, or abdominal cavity, can be a disaster with death of the mother as well as that of the fetus. Plate II shows such a pregnancy in an impregnated ovary (i.e., an ovarian pregnancy), with the fetus' arm extended through a rupture in the ovarian surface. Plate III, Fig I illustrates the same impregnated left ovary (L) after dissection, exposing the fetus and placenta at about 16 weeks gestation, and the sac of which had ruptured. A stylus (C, C) passes from the uterine cervix to the opened uterine apex (A). The thickened uterine decidua (B) in the non-pregnant uterus is particularly striking. The enlarged left Fallopian tube (I) and normal sized right Fallopian tube (E) also are shown. Fig II above illustrates the posterior aspect of the non-pregnant uterus (V) with the ovary and tubes removed, and incised to show the delicate hypertrophic decidua (Z).

References

Blake 52.

D. PHILIPPI ADOLPHI BOEHMERI

MEDICINAE ET ANATOMIAE PROFESSORIS IN REGIA

FRIDERICIANA PVBLICI ORDINARII

IMPERIAL. ACADEM. NATVR. CVRIOS. COLLEGAE

OBSERVATIONVM

ANATOMICARVM

RARIORVM

FASCICVLVS

NOTABILIA

CIRCA VTERVM HVMANVM

CONTINENS

CVM

FIGVRIS AD VIVVM EXPRESSIS.

HALAE MAGDEBVRGICAE

APVD IOANNEM IVSTINVM GEBAVERVM.

CIƆ IƆCC LII.

TITLE PAGE FROM BOEHMER [BOEHMERI] 1752–56

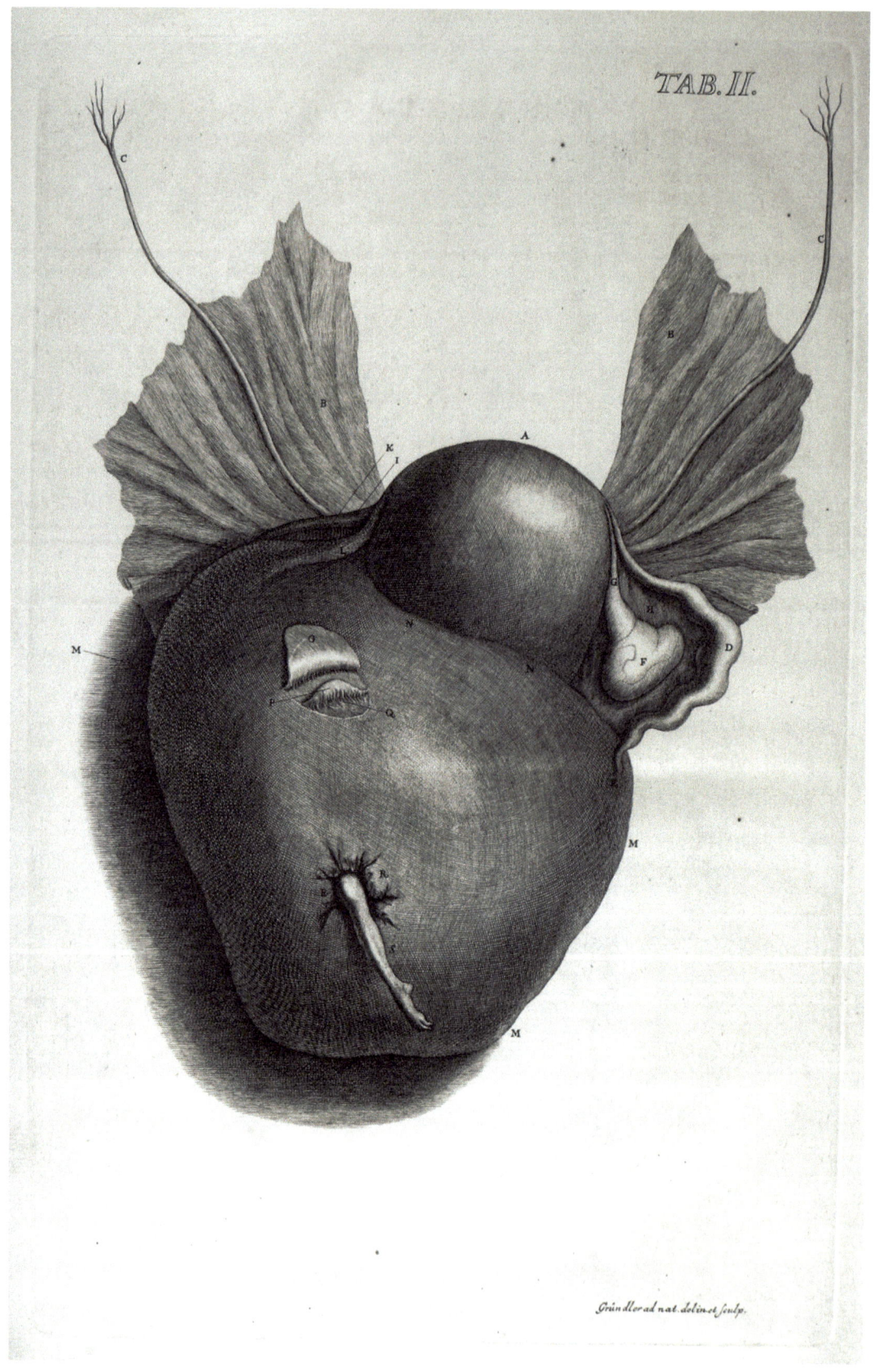

Tab. II from Boehmer [Boehmeri] 1752–56

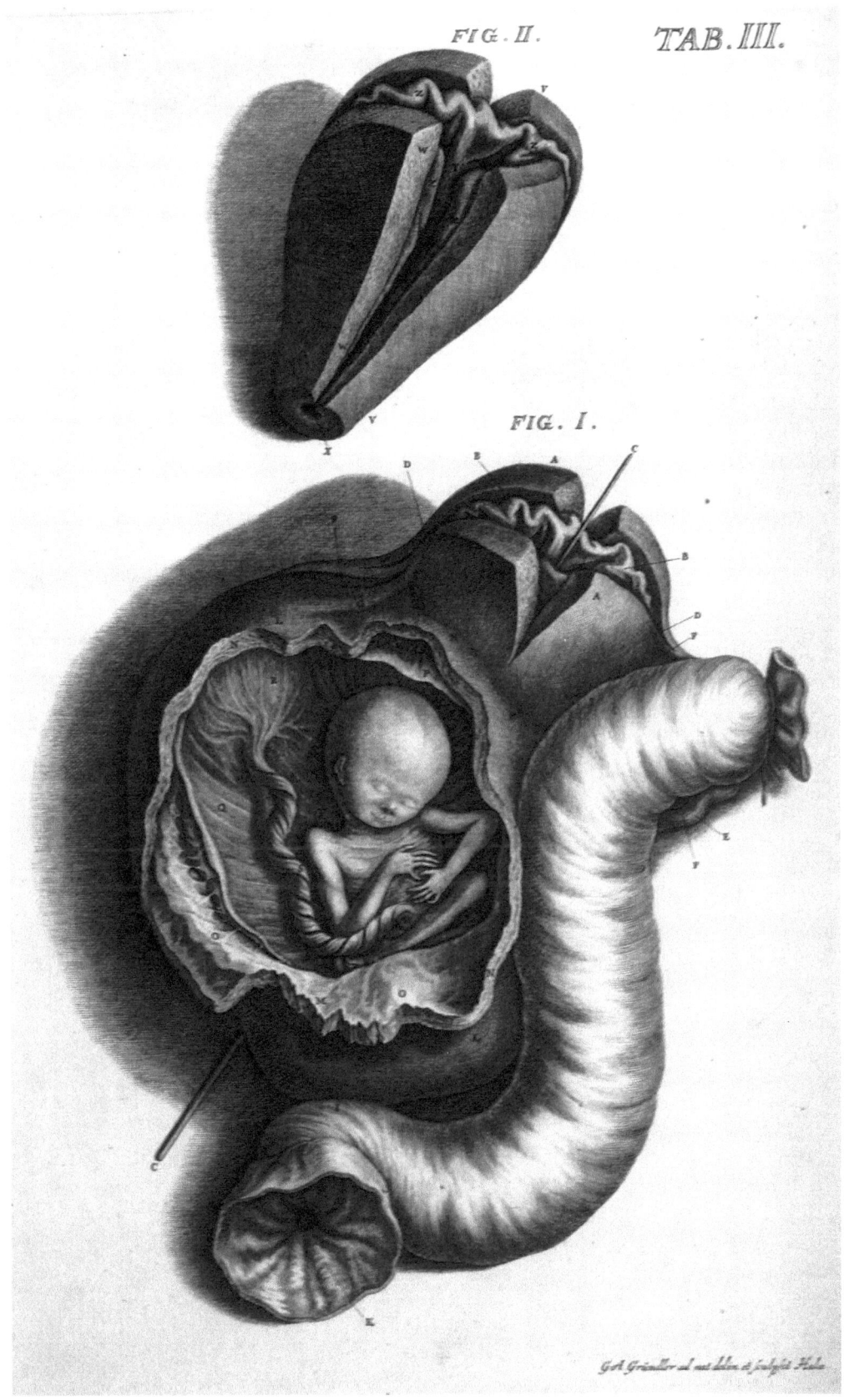

TAB. III, FIG. I AND II FROM BOEHMER [BOEHMERI] 1752–56

Michaele Vincenzo Giacinto Malacarne

De' mostri umani de' caratteri fondamentali su cui se ne potrebbe stabilire la classificazione e delle indicazioni che presentano nel parto. Modena, *Memorie di Matematica e di Fisica della Societa Italiana delle scienze* 9:49–84, 1802.

Malacarne (1744–1816), a leading anatomist and surgeon-obstetrician of the Universities of Torino and Padua, here describes a '*mostri umani*' (human monster) he delivered near-term in Pavia in 1791 and preserved. Plate III illustrates the fetus with opened abdominal cavity (C, umbilical cord; D, liver; H, intestine; L, spleen…). A detailed description of the internal organs is given on pp. 58–59, and an explanation of the three plates (pp. 60–62) is presented in the text.

Malacarne, originally from Saluzzo in the Piedmont region of northwestern Italy, wrote at least one other work on fetal monsters with emphasis on their brains (Malacarne 1807). He is particularly remembered for his contributions to the topographical anatomy of the brain and its comparative study. Malacarne described in detail the structure of the cerebellum, introducing the terms *tonsil* [goiter-like mass], *pyramid* [pointed or cone-shaped structure], *lingula* [tongue], and *uvula* [little grape] (1776). Details from his papers support his acceptance of the discoveries of Luigi Galvani (1737–1798) of "animal electricity" (Galvani 1791), and his experimental contributions to the emerging science of electrophysiology in the understanding the role of electrical currents within the brain, and the nerves in muscle function. Malacarne considered topographical anatomy of the various parts of the brain, and the association and linkage of the nerve tracks to associated regions and nuclei. He held that cerebral activity was directed by harmonious integration with the several brain centers. By studies of cerebral function in cases of brain injury, hydrocephalus, and cretinism, he derived a "hypoplastic" theory of cerebellar malformation as an origin of intellectual deficits (Cherici

2006). With his interest in mental functions and its limitations, Malacarne also wrote "On goiters and the stupidity which in some countries accompanies them" (1789). He concluded that a fatty swelling in the neck, identified by others as the thyroid gland, obstructed blood flow to the brain. In his pursuit of this concept, he invited surgeons in regions affected with a high incidence of goiter to send him the neck and head from corpses being autopsied, that showed evidence of cretinism (Costa 1989).

References

Cherici, C. Vincenzo Malacarne (1744–1816): a researcher in neurophysiology between anatomop-hysiology and electrical physiology of the human brain. *C R Biol* 329:319–329, 2006.

Costa, A. On goiters and the stupidity which in some countries accompanies them. Endeavours of Vincenzo Malacarne, from Saluzzo. *Panminerva Med* 31:97–106, 1989.

Galvani, L. De viribus electricitatis in motu musculari commentarius. *Bonn Sci Art Inst Acad Comment Bologna* 7:363–418, 1791. (GM 593).

Malacarne, M.V.G. *Nuova esposizione della vera struttura del cervelletto umano.* Torino, G. Briolo, 1776. (GM 1382.1).

Malacarne, M.V.G. *Encefalotomia nuova universale. 3 vol.* Torino, Giammichele Briolo, 1780. (Blake p. 284).

Malacarne, M.V.G. *Delle osservazioni in chirurgia.* Torino, Giammichele Briolo, 1784. (Blake p. 284).

Malacarne, M.V.G. *Sui gozzi e sulla stupidità ec dei cretini….* Torino, Stamperia Reale, 1789. (Blake, p. 284, GM 3809).

Malacarne, M.V.G. *Prime linee della chirurgia.* Venezia, [G.A. Pezzana], 1794.

Malacarne, M.V.G. *Oggetti piu interessanti di ostetricia e di storia naturale esistenti nel Museo Ostetricio dell'Università di Padova….* Padova, Stamperia del Seminario, 1807.

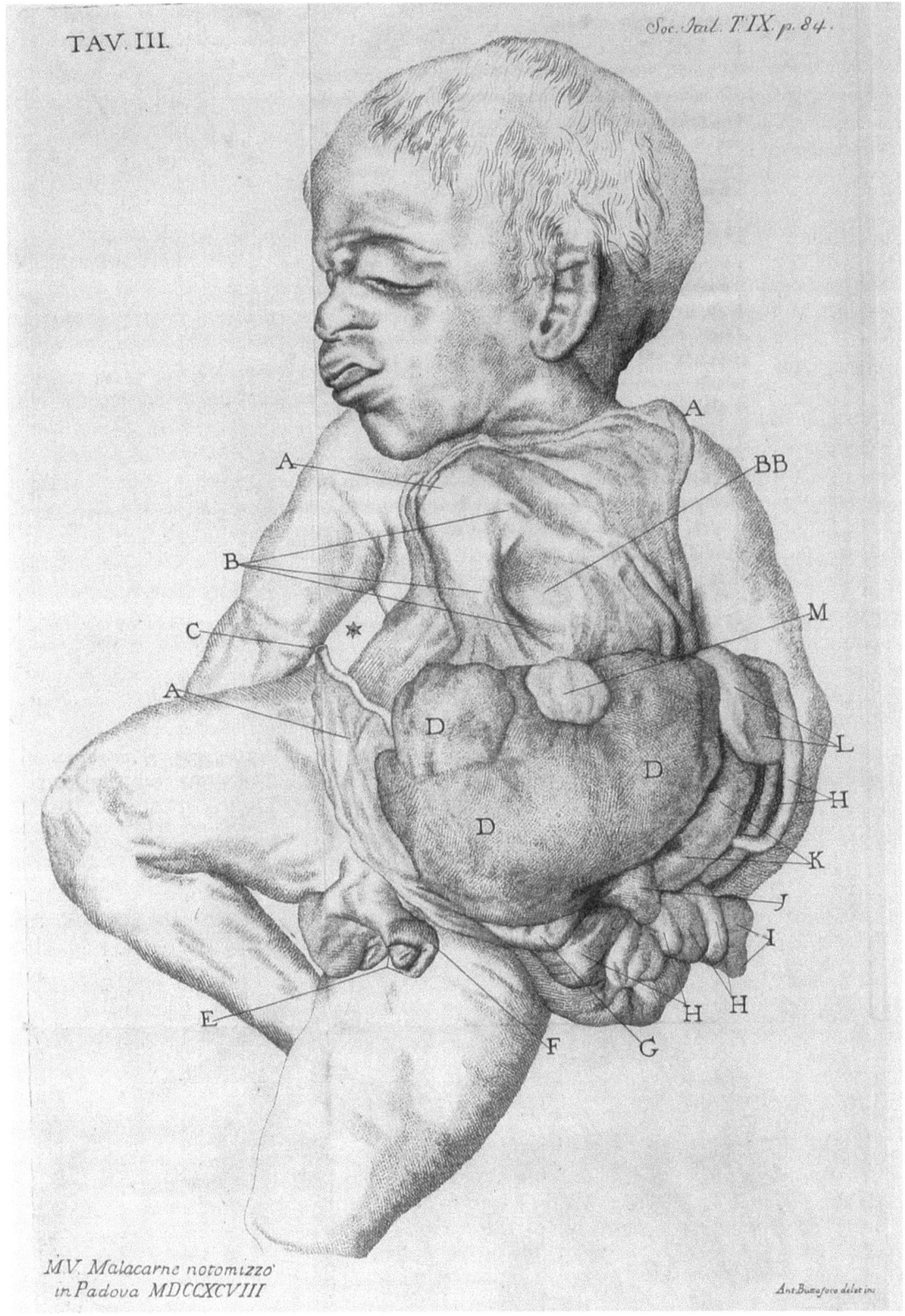

Tav. III from Malacarne 1802

Robert Hooper

The morbid anatomy of the human uterus and its append-ages; with illustrations of the most frequent and important organic diseases to which those viscera are subject. London, Printed for the author…, 1832.

Hooper (1773–1835), a graduate of St. Andrews Medical School in Scotland, and who practiced in London, had a special interest in pathological anatomy, performing more than 4000 autopsies over a 30 year period. In this volume, Hooper included 21 hand colored plates of the normal uterus, and that afflicted with abscess, chronic hematoma, polyps, the neoplasms carcinoma and sarcoma, and other conditions.

Plate XXI illustrates hydatid vesicles of the placenta associated with a hydatiform mole. Hydatid or hydatiform mole is an abnormal pregnancy resulting from a pathologic ovum, with proliferation of the epithelial covering of the chorionic villi and dissolution and cystic cavitation of the avascular stroma of the villi. This results in a mass of grape-like cysts. Hooper states that the drawing was made "…from a portion only of a large mass that filled an ordinary wash-hand basin, and which weighed nearly four pounds." He continued, "It is seen to consist of an immense number of beautifully transparent bladders, filled with a water-like fluid of a yellowish or straw colour…it had very much the appearance of small grapes…. These vesicles are of various sizes; and very small ones are seen hanging from the sides of some of the larger…. They all hang from and are connected by, whitish opaque filaments formed by a firm albumin, apparently unorganized" (p. 67).

In his Case CLXXXVI, William Giffard is believed to have first described hydatidiform mole in Great Britain (Giffard 1734). Hooper also published volumes on the pathology of the brain (Hooper 1826), and other organs, as well as a lexicon of medical terms (Hooper 1811).

References

Giffard, W. *Cases of midwifery. Written by the late Mr William Giffard, surgeon and man-midwife. Revis'd and publish'd by Edward Hody, M.D. and Fellow of the Royal Society.* London, Printed for B. Motte, T. Wotton, and L. Gilliver… and J. Nourse…, Fleet Street, London, 1734.

Hooper, R. *The anatomist's vade mecum: containing the anatomy, physiology, and morbid appearances of the human body.* London, 1798.

Hooper, R. *The anatomist's vade-mecum: containing the anatomy, physiology, morbid appearances etc. of the human body, the art of making preparations etc…. 4th Ed., rev. and enl….* London, Murray and Highley, 1802.

Hooper, R. *The anatomist's vade-mecum, containing the anatomy, physiology, morbid appearances, etc. of the human body: the art of making anatomical preparations, etc…. 5th Ed., to which are now added anatomical, physiological, medical and surgical questions for students.* London, John Murray, 1804.

Hooper, R. *Lexicon medicum; a new medical dictionary; containing an explanation of the terms in anatomy, physiology….* London, Longman, 1811.

Hooper, R. *The morbid anatomy of the human brain.* London, For the Author, 1826. (GM 2284.1).

Hooper, R. *The physician's vade-mecum: containing…* London, 1847.

Munk, W. *The Role of the Royal College of Physicians of London; comprising biographical sketches… Second edition, Revised and Enlarged.* VOL III 1801 to 1825. London, Published by the College, 1878, pp. 29–30.

PLATE **XXI** FROM **H**OOPER **1832**

Willem Vrolik

Tabulae ad illustrandam embryogenesin hominis et mamma-lium tam naturalem quam abnormen. Amsterdam, G.M.P. Londonck, 1844–1849.

Vrolik (1801–1863), in his "Record of embryonic illustrations of man and mammals as much natural and abnormal," presented 100 plates which depict human and animal malformations. Illustrated in Plate 1 are Siamese twins joined at the abdomen.

Illustrated in Tabula LIII are a series of instances in human newborns of *cyclopia* [Greek, circle eye]. This developmental anomaly is characterized by a single orbit, with the globe of the eye rudimentary, apparently normal, enlarged, or duplicated. The nose is either absent or present as a tubular appendage, a *proboscis* [a tubular structure; pro+Greek, to feed or graze], located above the orbit. Based upon their morphology, Vrolik classified Cyclopes into five main types, and concluded that such anomalies were a consequence of defective development of the organs of vision and smell.

Vrolik's father Gerardus (1775–1859) was professor of anatomy and physiology and also of theoretical and clinical obstetrics at the *Athenaeum Illustre*, the predecessor of the University of Amsterdam. There he collected anatomical specimens for what became the *Museum Vrolikianum*. The younger Vrolik studied medicine at the *Athenaeum Illustre*, completing his medical studies at the University of Utrecht (1823). After a decade of medical practice in Amsterdam, he was appointed Professor Extraordinary of Anatomy and Embryology at the University of Gröningen. Here, he had the opportunity to study the anatomical "Cabinet" of the illustrious Peter [Petrus] Camper (1722–1789), which sparked his interest in teratology. Two years later (1831) he returned to Amsterdam as Professor of Anatomy, Physiology, Natural History, and Theoretical Surgery. Preceding his monumental *Tabulae ad illustrandum embryogenesin...*, he published several other important papers on cyclops and teratology (Vrolik 1834, 1840, 1842–1844). Following Vrolik, others have expanded or altered the classification system for cyclops (Baljet et al. 1991, Bock 1989, Duke-Elder 1963, Schwalbe and Joseph 1913). During his years in Amsterdam, he continued to add numerous specimens to the *Museum Vrolikianum*, which still exists. In 1850, largely on the basis of the present work, he was awarded the prize of the Legacy of the Baron de Montyon of the French Academy of Sciences (Baljet 1984, Baljet and Oostra 1998; Baljet et al. 1991).

References

Garrison Morton 534.61.

Baljet, B. Uit de geschiedenis van het Museum Vrolik, de Snijkamer en het Theatrum Anatomicum te Amsterdam. In: *Gids voor het Museum Vrolik*. Universiteit van Amsterdam, 1990, pp. 7–24.

Baljet, B., F. Van Der Werf, & A.J. Otto. Willem Vrolik on cyclopia. *Doc Ophthalmol* 77: 335–368, 1991.

Baljet B. and R-J. Oostra. Historical aspects of the study of malformations in The Netherlands. *Am J Med Genetics* 77: 91–99, 1998.

Bock, C.E. Beschreibung eines atypischen Cyclops. *Klin Mbl Augenheilk* 27: 508–513, 1989.

Duke-Elder, Sir W.S. *System of ophthalmology, Vol. III: Normal and abnormal development, Part 2: Congenital deformities*. London, Kimpton, 1963.

Geoffroy Saint-Hilaire, I. *Histoire générale et particuliére des anomalies de l'organisation chez l'homme et des animaux, 3 Vols and Atlas*. Paris, Ballière, 1832–1837. Vol. 2, pp. 423–424. (GM 534.58).

Nevin, N.C. & A.C. Josephine. *Illustrated guide to malformations of the central nervous system at birth*. London, Churchill Livingstone, 1983.

Oostra, R-J, Baljet, B., Dijkstra, P.F., & R.C.M. Hennekam. Congenital anomalies in the teratological collection of Museum Vrolik in Amsterdam, The Netherlands. I: Syndromes with multiple congenital anomalies. *Am J Med Genetics* 77: 100–115, 1998a.

Oostra, R-J, Baljet, B., Dijkstra, P.F., & R.C.M. Hennekam. Congenital anomalies in the teratological collection of Museum Vrolik in Amsterdam, The Netherlands. II: Skeletal dysplasias. *Am J Med Genetics* 77: 116–134, 1998b.

Schumacher, G.H., H. Gill, & H. Gill. Zur Geschichte angeborener Fehlbildungen unter besonderer Berücksichtigung der Doppelbildungen, 2: Vom 18 bis 20 Jahrhundert. *Anat Anz* 164: 291–303, 1987.

Schwalbe, E. & H. Josephy. Die Cyclopie. In: S*chwalbe E. (Hrsg) Die Morphologie der Missbildungen des Menschen und der Tiere. Teil III: Die Einzelmissbildungen*. Jena: Gustav Fischer, 1913, pp. 204–270.

Vrolik, W. Over den aard en oorsprong der cylopie. *Nieuwe Verhandelingen der Eerste Klasse van het Koninklijk Nederlands Instituu*t 5: 25–112, 1834.

Vrolik, W. *Ontleedkundig onderzoek beschrijving en rangschikking der dubbelde misgeboorten....* Te Amsterdam, bij C.G. Sulpke, 1840.

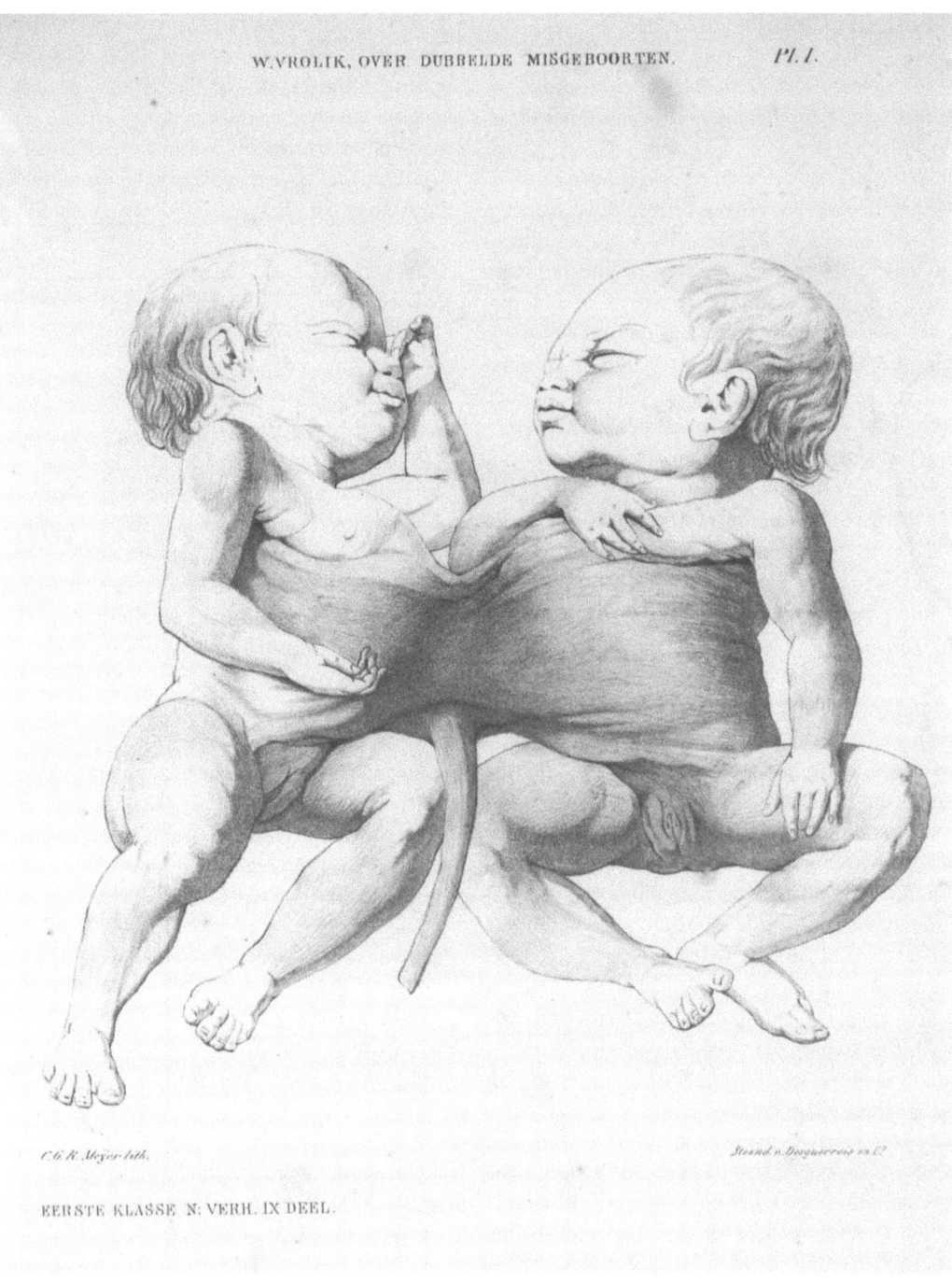

PLATE I FROM VROLIK 1844

Gabriel Madeleine Camille Dareste

Récherches sur la production artificielle des monstruosités, ou essais de tératologénie expérimentale. Paris, Reinwald, 1877.

Congenital anomalies, or "monstrous births," have attracted interest since earliest times. Ascribed by various cultures and times to the "will of god," a sign of the mother's unfilled desires during conception, maternal or paternal transgressions, or other causes, such anomalies have always captured people's imagination. In the early nineteenth century, Mary Wollstonecraft Shelley's (neé Godwin) (1797–1851) *Frankenstein* (Shelley 1818) presented a model of Promethean creative power. It was during this time that the father and son Étienne (1772–1844) and Isidor Geoffroy Saint Hilaire (1805–1861) (the latter who coined the term teratology) attempted to classify such deviations from normal on the basis of objective changes, rather than from possible etiologic causes (Geoffroy Saint Hilaire 1832–1837).

The embryologist Dareste (1822–1899) was the first who attempted to make teratology an experimental science. This became known as teratogenesis, the systematic production of anomalies in the laboratory. Acknowledging his debt to the father and son Étienne and Isidore Geoffroy Saint-Hilaire, as the first embryologists to study in a systematic manner the origin of anomalous births, Dareste attempted to discover in the study of such malformations the possibility of evolutionary variability and innovation. He wrote that although observation allows knowledge of types, "... experimentation, thanks to its creative power, realizes all that is possible; it thus opens up unlimited prospects" (Dareste 1877, p. 40). In his experiments, conducted chiefly in the eggs of the hen, Dareste used a variety of methods to produce anomalies. These included shaking them on a machine (*table à secousses*), covering portions of the shell with varnish and other impermeable substances, and varying the incubation temperature.

In Plate XV Dareste depicts several examples in the chick of double embryos ('monstruosité double') or "Siamese" twins. Figures 1 to 4 show embryos joined at the head, while Figs. 5, 6, and 7 illustrate those joined at the trunk. Plate III shows various abnormalities of vascular development of the chorioallantoic membrane (the chick's equivalent of the mammalian placenta). The artist and the engraver for the sixteen plates were Jacquemin and Karmanski, respectively.

After receiving doctorates in both medicine (1847) and science (1851), Dareste worked as professor of zoology at the University of Lille, and devoted his career to the study of embryology and abnormal development. Beginning in 1872,

he served as professor of ichthyology and herpetology at the *Muséum d'Histoire Naturelle*, Paris. Several years later he became director of the laboratory of teratology, which became attached to the *Ecole des Hautes-Etudes*. In 1877, the year in which this volume was published, the French *Académie des Sciences* awarded Dareste the Grand Prize in physiology for his research in experimental teratology.

References

Garrison Morton 534.65.

Appel, T.A. *The Cuvier-Geoffroy debate. French biology in the decades before Darwin.* Oxford, Oxford University Press, 1987, pp. 125–130.

Canguilhem, G. *The normal and the pathological.* New York, Zone Books, 1989.

Canguilhem, G. La monstruosité et le monstrueux. In: *La connaissance de la vie.* Paris, Vrin, 1992, pp. 171–184.

Cooper, M. Regenerative medicine: stem cells and the science of monstrosity. *Med Humanit* 30: 12–22, 2004.

Dareste, C. *Récherches sur la production artificielle des monstruosités, ou essais de teratologénie expérimentale.* Deuxiéme édition revue of augmentée. Paris, C. Reinwald, 1891.

Geoffroy Saint-Hilaire, É. *Philosophie anatomique. Des monstruosités humaines, ouvrage contenant une classification de monstres.* Paris, Méquignon-Mervis, 1822.

Geoffroy Saint-Hilaire, É. *Philosophie anatomique. Tome II, des Monstruosités.* Paris, Chez l'Auteur, 1818–1822. (GM 534.57).

Geoffroy Saint-Hilaire, I. *Histoire générale et particulière des anomalies de l'organisation chez l'homme et les animaux.... Des monstruosités, des varietés et vices de confirmation, ou traité de teratologie. 3 vols and atlas.* Paris, J.B. Bailliére, 1832–1837. (GM 534.58).

Huet, M.H. *Monstrous Imagination.* Cambridge, MA, Harvard Univ Press, 1993.

O'Neill, E. *Raw material: producing pathology in Victorian culture.* Durham and London, Duke University Press, 2000.

Oppenheimer, J.M. Some historical relationships between teratology and experimental embryology. *Bull Hist Med* 42: 145–159, 1968.

Shelley, M.F. *Frankenstein or the modern Prometheus.* London, Lackington, Hughes, Harding, Mavor & Jones, 1818.

Wolff, E. Les bases de la tératogénèse expérimentale des vertébrés amniotes, de après les résultats de méthods directes. *Arch d'anat., d'histol, et d'emb* 22: 1–375, 1936.

Wolff, E. *La science des monstres.* Paris, Gallimard, 1948.

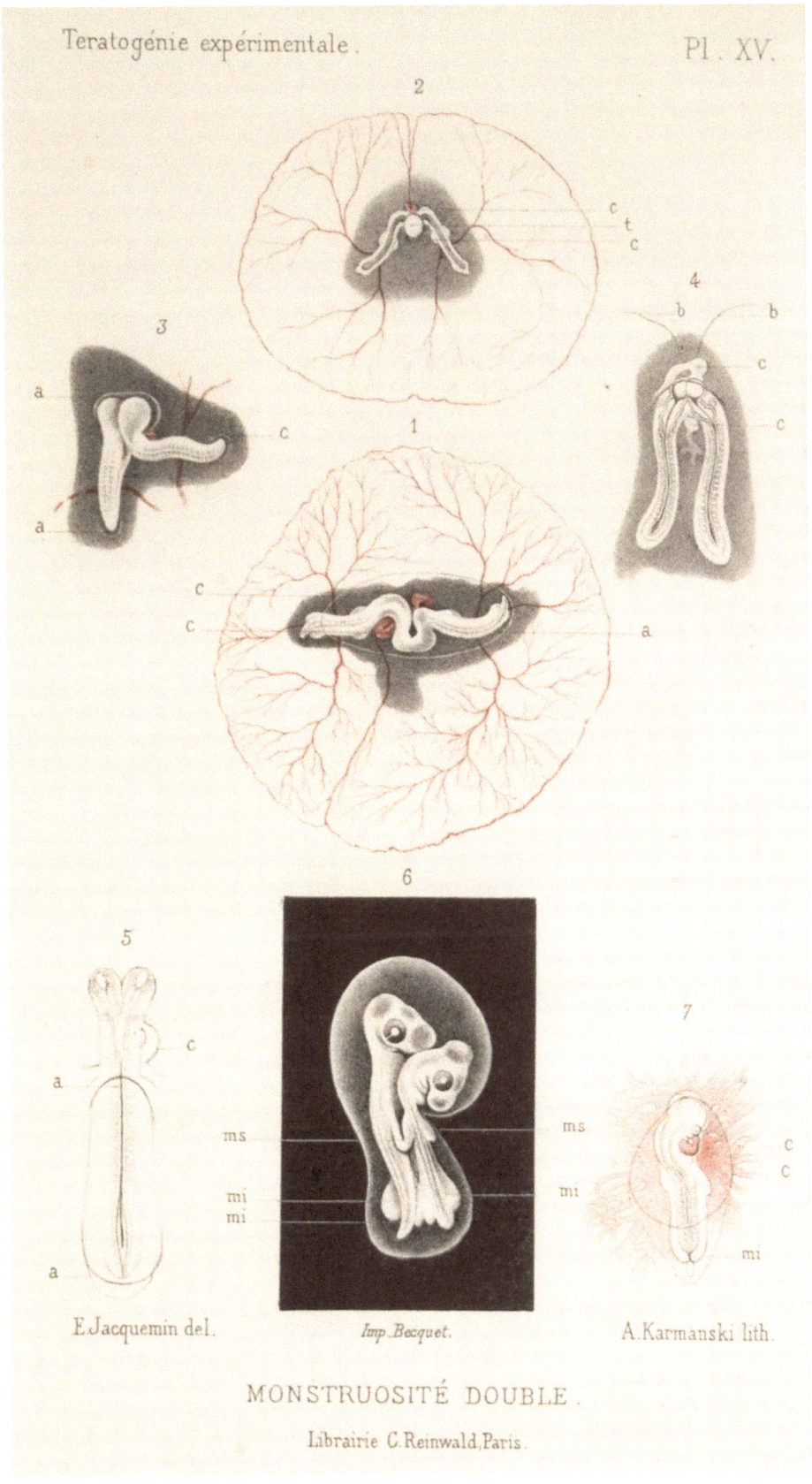

PLATE XV FROM DARESTE 1877

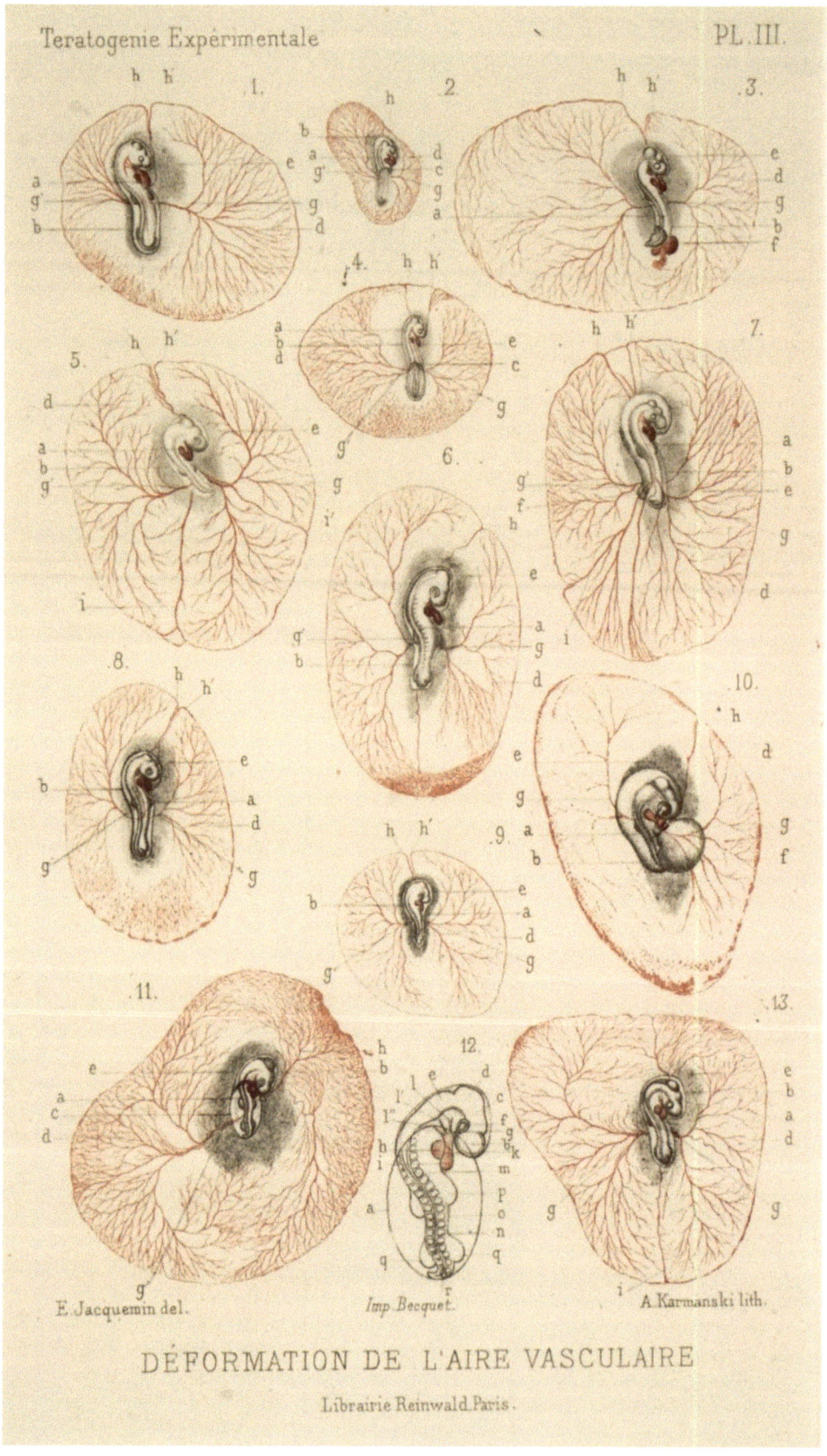

PLATE III FROM DARESTE 1877

Additional Authors of Significance to Teratology

Cesare Taruffi

Storia della teratologia 8 vols. Bologna, Regia Tipographia, 1881–1894.

In this extensive history and bibliography of teratology, Taruffi (1821–1902) described and referred to numerous rare specimens from otherwise unobtainable sources. The section on malformations of individual organs was never completed. In this multi-volume set, Taruffi dedicated himself to describing the history of teratology, to developing a logical classification of congenital malformations, and to explaining the mechanisms of their development. Because of this focus, the volumes thus contained extensive tables and lists of cases, classifications, descriptions, and measurements. Volume 1 for example contains chapters on *Etiologia* (etiology, Chap. V), *Frequenza* (frequency, Chap. 6), and *Classificazione* (classification, Chap. 7), but no illustrations. Although the volumes contained some illustrations, most were taken from other authors, and in only a few cases were taken from Taruffi's own work.

Taruffi graduated from the University of Bologna, as a Bachelor in Surgery (1842) and Medicine (1844). For several years he served as an assistant to the surgeon Francesco Rizzoli (1809–1880). Taruffi took great interest in the *Museo di Patologia* [pathology museum], which had been founded in 1804 by Napoleon Bonaparte (1769–1821), whose troops earlier had destroyed the school of anatomy. Upon assuming the professorship of pathological anatomy and Director of the museum, Taruffi worked to reopen the "Cabinet" of anatomical, embryological, and teratological preparations in 1863. The Institute of Pathology at Bologna, which includes the anatomical museum "Taruffi," preserves the collections of Taruffi and others.

References

Garrison Morton 534.67.

Beckwith, J.B. Museums, antiquarian books, and modern teratology. *Am J Med Genetics* 77: 89–90, 1998.

Ruggeri, F. Il Museo dell'Istituto di Anatomia Umana Normale. In: *I luoghi del conoscere. I laboratory storici e i musei dell'Università di Bologna*. W. Tega (Ed). Bologna, Banca del Monte, 1988, pp. 98–105.

Scarani, P. & G. Lacchini. L'autopsia clinica dell'ottocento a Bologna. Nuove prospettive. *Pathologica* 91: 128, 1999.

Scarani, P., R. de Caro, V. Ottani, M. Raspanti, F. Ruggeri, & A. Ruggeri. Contemporaneous anatomic collections and scientific papers from the 19th century school of anatomy of Bologna: preliminary report. Clin Anat 14: 19–24, 2001.

Epilogue

For the historian of medicine or science, reproductive biologist, obstetrician, pediatrician, midwife, or general reader, the question arises, what, if any, lessons may one learn from this display? To what extent can this survey of our extensive heritage, and especially this focus on illustrations, be of relevance to our twenty-first century world? Quite obviously, points of view will vary widely as to the value of the study of history. One could recite numerous anecdotes along this line. One that appeals is that of Philips the Good (1396–1467), Duke of Burgundy from 1419 until his death. It is said that he had such a reverence for history and its lessons, that every evening before he retired, he had one of his courtiers read to him for two hours from the great writers of the past (Calmette 1962). Others have noted that "… academic historians study the history of medicine to understand *history* better; [physicians and] practical historians study it to understand *medicine* better" (Ziporyn 1985, p. 2713).

It is our hope that this compilation will assist the reader in considering one small fragment of the wealth of scholarship from ages past. We also trust this survey will help to stimulate wonder in the miracle, or interlocking series of miracles, of reproduction. From a purely statistical standpoint, the idea that following conception a healthy, vigorous, sentient human can be formed from a single egg and sperm, and can develop, be born, and go on to a life full of joy and riches, must be highly improbable. As is evident in viewing the images of this volume, we enjoy a rich heritage, not just in seeing works of the "great" men and women, but in addition, gaining appreciation of the intellect and insights of these workers in prior centuries. Clearly their minds were as good as, or better than, ours. Our wish is that, by contemplating the concepts and ideas the individual authors sought to convey, one will gain perspective on the continuity of medical thought. Also, as with fine art, music, and great literature, history is its own reward, enriching one's life, giving insights into achievements of the past, and perhaps, without being too idealistic, giving stimulus to excel as those of old. Engaging in such an epistemological dialogue may help one to understand better the past on its own terms. Some have even made the analogy of history being a Rosetta Stone, to help illuminate the present in a unique manner.

Perhaps another 'lesson' from this history is appreciating the importance of change, and the idea of progress. Hopefully, appreciation of the contributions of the past is of value in our keeping an open and receptive mind. History helps us to better recognize and place in context chance or serendipitous observations that otherwise might pass unnoticed. It was during the late-nineteenth century that Charles Sanders Peirce (1839–1914) and William James (1842–1910) developed the philosophy of "Pragmatism," the theory that the essence of a proposition or course of action lies in its observable consequences, and that its meaning derives from the sum of these. Also during this era, many rather startling advances were being made in microbiology and other areas of medical science that contributed to a greater understanding of the etiology of infections and other diseases. It was as a consequence of these advances that physicians such as William Osler (1849–1919) viewed the course and future of medicine to be onward and upward, with continued progress and enlightenment (Osler 1921)

In addition, an overview such as here presented, may help us to understand better the interrelations of society, culture, and medicine and the healing arts. Rather than events or personalities *per se*, consideration of these works can stimulate us to seek to understand their motivations, and their place in society. It can enlarge one's experience, and enrich one's life. Beyond mastery of isolated facts, we may strive for imaginative interpretation. This is our most fervent hope.

References

Calmette J. The golden age of Burgundy. London: Wiedenfeld; 1962.

Osler W. The evolution of modern medicine. New Haven: Yale University Press; 1921.

Ziporyn T. Historians strive to improve perspective, practice of medicine. JAMA. 1985;254:2713–20.

© Springer International Publishing Switzerland 2016
L.D. Longo, L.P. Reynolds, *Wombs with a View*, DOI 10.1007/978-3-319-23567-7

Afterword

Catherine Y. Spong*

It is remarkable that to date, such a book has not been published. We owe Drs. Lawrence D. Longo and Larry Reynolds much appreciation for taking on this herculean task in creating not only a visually interesting and appealing book but also an incredibly informative one, detailing the progression over time of our understanding of the womb. In *Wombs with a View* we not only are treated to the artistic depiction, but also the background behind the illustrations, providing much needed context.

Although science, technology, and discovery document that major progress has been made and we are much further in our understanding of the gravid uterus, in fact I anticipate that twenty years hence we will reflect and recognize how little in 2016 we actually understand the womb, especially in the human. Adapting current and developing new technologies, as well as currently unimagined advances, will allow us to understand the inner workings, structure and function of the gravid uterus in real time. The *Eunice Kennedy Shriver* National Institute of Child Health and Human Development (NICHD) recently launched the Human Placenta Project with the specific goal to understand over the next decade placental structure and function in human gestation in real time. This project will lead us to insights for the fetus as well, including distinguishing normal from abnormal, and to the possibility of interventions to improve outcome.

Given the critical role of the placenta and well-known evidence that placental dysfunction leads to pregnancy complications, it is essential that this concept of real-time, noninvasive placental assessment in the human be actualized. Current techniques may be considered little advanced from the early illustrations depicted in this beautiful work, yet the opportunities for historic advances in viewing the womb and its contents are now imaginable. Imaging technologies, biosensors, nanoparticles, chips, and assessment of circulating markers, among others will be utilized to assess structure and function in the womb. This should include real time assessment of the activation or suppression of genes, oxygenation status, vascular invasion, cellular pathways, and metabolism, including understanding the trajectory across gestation.

Just as was the case when the first authors and illustrators put pen to paper, understanding the "gravid uterus and its contents" has huge implications for health. By understanding the structure and function of the gravid womb we have the opportunity to optimize pregnancy outcome. This will lead to improved health for both the baby and the mother for the short and long term.

I am energized by the work of the past illustrated in these pages and am encouraged by our cadre of scientists who understand the importance of research in the diseases of pregnancy from which so many people suffer and who are diligently working to provide us new technologies and needed data. It is my hope that in a decade's time we will need a second, and expanded, edition of *Wombs with a View*. I look forward to it!

* Acting Director, Eunice Kennedy Shriver National Institute of Child Health and Human Development, National Institutes of Health, Bethesda, MD, USA.

Index

© Springer International Publishing Switzerland 2016
L.D. Longo, L.P. Reynolds, *Wombs with a View*, DOI 10.1007/978-3-319-23567-7